Lecture Notes in Computer Science 2751

Edited by G. Goos, J. Hartmanis, and J. van Leeuwen

T0223552

Springer
Berlin
Heidelberg
New York
Hong Kong
London
Milan
Paris
Tokyo

Andrzej Lingas   Bengt J. Nilsson (Eds.)

Fundamentals of Computation Theory

14th International Symposium, FCT 2003
Malmö, Sweden, August 12-15, 2003
Proceedings

 Springer

Series Editors

Gerhard Goos, Karlsruhe University, Germany
Juris Hartmanis, Cornell University, NY, USA
Jan van Leeuwen, Utrecht University, The Netherlands

Volume Editors

Andrzej Lingas
Lund University
Department of Computer Science
Box 118, 221 00 Lund, Sweden
E-mail: Andrzej.Lingas@cs.lth.se

Bengt J. Nilsson
Malmö University College
School of Technology and Society
205 06 Malmö, Sweden
E-mail: Bengt.Nilsson@ts.mah.se

Cataloging-in-Publication Data applied for

A catalog record for this book is available from the Library of Congress.

Bibliographic information published by Die Deutsche Bibliothek
Die Deutsche Bibliothek lists this publication in the Deutsche Nationalbibliografie;
detailed bibliographic data is available in the Internet at <http://dnb.ddb.de>.

CR Subject Classification (1998): F.1, F.2, F.4, I.3.5, G.2

ISSN 0302-9743
ISBN 3-540-40543-7 Springer-Verlag Berlin Heidelberg New York

This work is subject to copyright. All rights are reserved, whether the whole or part of the material is
concerned, specifically the rights of translation, reprinting, re-use of illustrations, recitation, broadcasting,
reproduction on microfilms or in any other way, and storage in data banks. Duplication of this publication
or parts thereof is permitted only under the provisions of the German Copyright Law of September 9, 1965,
in its current version, and permission for use must always be obtained from Springer-Verlag. Violations are
liable for prosecution under the German Copyright Law.

Springer-Verlag Berlin Heidelberg New York
a member of BertelsmannSpringer Science+Business Media GmbH

http://www.springer.de

© Springer-Verlag Berlin Heidelberg 2003
Printed in Germany

Typesetting: Camera-ready by author, data conversion by PTP-Berlin GmbH
Printed on acid-free paper   SPIN: 10930632   06/3142   5 4 3 2 1 0

Preface

The papers in this volume were presented at the 14th Symposium on Fundamentals of Computation Theory.

The symposium was established in 1977 as a biennial event for researchers interested in all aspects of theoretical computer science, in particular in algorithms, complexity, and formal and logical methods. The previous FCT conferences were held in the following cities: Poznań (Poland, 1977), Wendisch-Rietz (Germany, 1979), Szeged (Hungary, 1981), Borgholm (Sweden, 1983), Cottbus (Germany, 1985), Kazan (Russia, 1987), Szeged (Hungary, 1989), Gosen-Berlin (Germany, 1991), Szeged (Hungary, 1993), Dresden (Germany, 1995), Kraków (Poland, 1997), Iasi (Romania, 1999), and Riga (Latvia, 2001).

The FCT conferences are coordinated by the FCT steering committee, which consists of B. Chlebus (Denver/Warsaw), Z. Esik (Szeged), M. Karpinski (Bonn), A. Lingas (Lund), M. Santha (Paris), E. Upfal (Providence), and I. Wegener (Dortmund).

The call for papers sought contributions on original research in all aspects of theoretical computer science including design and analysis of algorithms, abstract data types, approximation algorithms, automata and formal languages, categorical and topological approaches, circuits, computational and structural complexity, circuit and proof theory, computational biology, computational geometry, computer systems theory, concurrency theory, cryptography, domain theory, distributed algorithms and computation, molecular computation, quantum computation and information, granular computation, probabilistic computation, learning theory, rewriting, semantics, logic in computer science, specification, transformation and verification, and algebraic aspects of computer science.

There were 73 papers submitted, of which the majority were very good. Because of the FCT format, the program committee could select only 36 papers for presentation. In addition, invited lectures were presented by Sanjeev Arora (Princeton), George Păun (Romanian Academy), and Christos Papadimitriou (Berkeley).

FCT 2003 was held on August 13–15, 2003, in Malmö, and Andrzej Lingas (Lund University) and Bengt Nilsson (Malmö University College) were, respectively, the program committee and the conference chairs.

We wish to thank all referees who helped to evaluate the papers. We are grateful to Lund University, Malmö University College, and the Swedish Research Council for their support.

Lund, May 2003

Andrzej Lingas
Bengt J. Nilsson

Organizing Committee

Bengt Nilsson, *Malmö* (Chair)
Oscar Garrido, *Malmö*
Thore Husfeldt, *Lund*
Mirosław Kowaluk, *Warsaw*

Program Committee

Arne Andersson, *Uppsala*
Stefan Arnborg, *KTH Stockholm*
Stephen Alstrup, *ITU Copenhagen*
Zoltan Esik, *Szeged*
Rusins Freivalds, *UL Riga*
Alan Frieze, *CMU Pittsburgh*
Leszek Gąsieniec, *Liverpool*
Magnus Halldórsson, *UI Reykjavik*
Klaus Jansen, *Kiel*
Juhani Karhumäki, *Turku*
Marek Karpinski, *Bonn*
Christos Levcopoulos, *Lund*
Ming Li, *Santa Barbara*

Andrzej Lingas, *Lund* (Chair)
Jan Małuszyński, *Linköping*
Fernando Orejas, *Barcelona*
Jürgen Prömel, *Berlin*
Rüdiger Reischuk, *Lübeck*
Wojciech Rytter, *Warsaw/NJIT*
Miklos Santha, *Paris-Sud*
Andrzej Skowron, *Warsaw*
Paul Spirakis, *Patras*
Esko Ukkonen, *Helsinki*
Ingo Wegener, *Dortmund*
Pawel Winter, *Copenhagen*
Vladimiro Sassone, *Sussex*

Referees

M. Albert
A. Aldini
J. Arpe
A. Barvinok
C. Bazgan
S.L. Bloom
M. Bläser
M. Bodirsky
B. Bollig
C. Braghin
R. Bruni
A. Bucalo
G. Buntrock
M. Buscemi
B. Chandra
J. Chlebikova
A. Coja-Oghlan
L.A. Cortes
W.F. de la Vega

M. de Rougemont
W. Drabent
S. Droste
C. Durr
M. Dyer
L. Engebretsen
H. Eriksson
L.M. Favrholdt
H. Fernau
A. Ferreira
A. Fishkin
A. Flaxman
D. Fotakis
O. Gerber
G. Ghelli
O. Giel
M. Grantson
J. Gudmundsson
V. Halava

B.V. Halldórsson
L. Hemaspaandra
M. Hermo
M. Hirvensalo
F. Hoffmann
T. Hofmeister
J. Holmerin
J. Hromkovic
L. Ilie
A. Jakoby
T. Jansen
J. Jansson
A. Jarry
M. Jerrum
P. Kanarek
J. Kari
R. Karlsson
J. Katajainen
A. Kiehn

H. Klaudel
B. Klin
B. Konev
S. Kontogiannis
J. Kortelainen
G. Kortsarz
M. Koutny
D. Kowalski
M. Krivelevich
K.N. Kumar
M. Kääriäinen
G. Lancia
R. Lassaigne
M. Latteux
M. Libura
M. Liśkiewicz
K. Loryś
E.M. Lundell
F. Magniez
B. Manthey
M. Margraf
N. Marti-Oliet
M. Mavronicolas
E. Mayordomo

C. McDiarmid
T. Mielikäinen
M. Mitzenmacher
S. Nikoletseas
B.J. Nilsson
U. Nilsson
J. Nordström
H. Ohsaki
D. Osthus
A. Palbom
G. Persiano
T. Petkovic
I. Potapov
C. Priami
E. Prouff
K. Reinert
J. Rousu
M. Sauerhoff
H. Shachnai
J. Shallit
D. Slezak
J. Srba
F. Stephan
O. Sykora

P. Tadepalli
M. Takeyama
A. Taraz
P. Thiemann
M. Thimm
P. Valtr
S.P.M. van Hoesel
J. van Leeuwen
S. Vempala
Y. Verhoeven
E. Venigoda
H. Vogler
B. Vöcking
H. Völzer
A.P.M. Wagelmans
R. Wanka
M. Westermann
A. Wojna
J. Wroblewski
Q. Xin
M. Zachariasen
G. Zhang
G.Q. Zhang
H. Zhang

Table of Contents

Algorithms 2

Networks and Complexity

Computational Biology

Computational Geometry

Computational Models and Complexity

Structural Complexity

Formal Languages

Logic

Proving Integrality Gaps without Knowing the Linear Program

Sanjeev Arora

Princeton University

During the past decade we have had much success in proving (using probabilistically checkable proofs or PCPs) that computing approximate solutions to NP-hard optimization problems such as CLIQUE, COLORING, SET-COVER etc. is no easier than computing optimal solutions.

After the above notable successes, this effort is now stuck for many other problems, such as METRIC TSP, VERTEX COVER, GRAPH EXPANSION, etc.

In a recent paper with Béla Bollobás and László Lovász we argue that NP-hardness of approximation may be too ambitious a goal in these cases, since NP-hardness implies a lowerbound – assuming P $\neq$ NP – on *all* polynomial time algorithms. A less ambitious goal might be to prove a lowerbound on restricted families of algorithms. Linear and semidefinite programs constitute a natural family, since they are used to design most approximation algorithms in practice. A lowerbound result for a large subfamily of linear programs may then be viewed as a lowerbound for a restricted computational model, analogous say to lowerbounds for monotone circuits

The above paper showed that three fairly general families of linear relaxations for vertex cover cannot be used to design a 2-approximation for Vertex Cover. Our methods seem relevant to other problems as well.

This talk surveys this work, as well as other open problems in the field. The most interesting families of relaxations involve those obtained by the so-called *lift and project* methods of Lovász-Schrijver and Sherali-Adams.

Proving lowerbounds for such linear relaxations involve elements of combinatorics (i.e., strong forms of classical Erdős theorems), proof complexity, and the theory of convex sets.

References

1. S. Arora, B. Bollobás, and L. Lovász. Proving integrality gaps without knowing the linear program. *Proc. IEEE FOCS 2002.*
2. S. Arora and C. Lund. Hardness of approximations. In [3].
3. D. Hochbaum, ed. *Approximation Algorithms for NP-hard problems.* PWS Publishing, Boston, 1996.
4. L. Lovász and A. Schrijver. Cones of matrices and setfunctions, and 0-1 optimization. *SIAM Journal on Optimization,* 1:166–190, 1990.
5. H. D. Sherali and W. P. Adams. A hierarchy of relaxations between the continuous and convex hull representations for zeroone programming problems. *SIAM J. Optimization,* 3:411–430, 1990.

A. Lingas and B.J. Nilsson (Eds.): FCT 2003, LNCS 2751, p. 1, 2003.
© Springer-Verlag Berlin Heidelberg 2003

An Improved Analysis of Goemans and Williamson's LP-Relaxation for MAX SAT

Takao Asano

Department of Information and System Engineering
Chuo University, Bunkyo-ku, Tokyo 112-8551, Japan
asano@ise.chuo-u.ac.jp

Abstract. For MAX SAT, which is a well-known NP-hard problem, many approximation algorithms have been proposed. Two types of best approximation algorithms for MAX SAT were proposed by Asano and Williamson: one with best *proven* performance guarantee 0.7846 and the other with performance guarantee 0.8331 if a conjectured performance guarantee of 0.7977 is true in the Zwick's algorithm. Both algorithms are based on their sharpened analysis of Goemans and Williamson's LP-relaxation for MAX SAT. In this paper, we present an improved analysis which is simpler than the previous analysis. Furthermore, algorithms based on this analysis will play a role as a better building block in designing an improved approximation algorithm for MAX SAT. Actually we show an example that algorithms based on this analysis lead to approximation algorithms with performance guarantee 0.7877 and conjectured performance guarantee 0.8353 which are slightly better than the best known corresponding performance guarantees 0.7846 and 0.8331 respectively.

Keywords: Approximation algorithm, MAX SAT, LP-relaxation.

1 Introduction

MAX SAT, one of the most well-studied NP-hard problems, is stated as follows: given a set of clauses with weights, find a truth assignment that maximizes the sum of the weights of the satisfied clauses. More precisely, an instance of MAX SAT is defined by $(\mathcal{C}, w)$, where $\mathcal{C}$ is a set of boolean clauses, each clause $C \in \mathcal{C}$ being a disjunction of literals and having a positive weight $w(C)$. Let $X = \{x_1, \ldots, x_n\}$ be the set of boolean variables in the clauses of $\mathcal{C}$. A *literal* is a variable $x \in X$ or its negation $\bar{x}$. For simplicity we assume $x_{n+i} = \bar{x}_i$ ($x_i = \bar{x}_{n+i}$). Thus, $\bar{X} = \{\bar{x} \mid x \in X\} = \{x_{n+1}, x_{n+2}, \ldots, x_{2n}\}$ and $X \cup \bar{X} = \{x_1, \ldots, x_{2n}\}$. We assume that no literals with the same variable appear more than once in a clause in $\mathcal{C}$. For each $x_i \in X$, let $x_i = 1$ ($x_i = 0$, resp.) if x_i is true (false, resp.). Then, $x_{n+i} = \bar{x}_i = 1 - x_i$ and a clause $C_j = x_{j_1} \vee x_{j_2} \vee \cdots \vee x_{j_{k_j}} \in \mathcal{C}$ can be considered to be a function $C_j = C_j(\boldsymbol{x}) = 1 - \prod_{i=1}^{k_j}(1 - x_{j_i})$ on $\boldsymbol{x} = (x_1, \ldots, x_{2n}) \in \{0, 1\}^{2n}$. Thus, $C_j = C_j(\boldsymbol{x}) = 0$ or 1 for any *truth assignment* $\boldsymbol{x} \in \{0, 1\}^{2n}$ with $x_i + x_{n+i} = 1$ ($i = 1, 2, \ldots, n$) and C_j is *satisfied* if $C_j(\boldsymbol{x}) = 1$.

A. Lingas and B.J. Nilsson (Eds.): FCT 2003, LNCS 2751, pp. 2–14, 2003.
© Springer-Verlag Berlin Heidelberg 2003

The *value* of a truth assignment x is defined to be $F_{\mathcal{C}}(x) = \sum_{C_j \in \mathcal{C}} w(C_j) C_j(x)$. That is, the value of x is the sum of the weights of the clauses in $\mathcal{C}$ satisfied by x. Thus, the goal of MAX SAT is to find an optimal truth assignment (i.e., a truth assignment of maximum value). We will also use MAX kSAT, a restricted version of the problem in which each clause has at most k literals.

MAX SAT is known to be NP-hard and many approximation algorithms for it have been proposed. Håstad [5] has shown that no approximation algorithm for MAX SAT can achieve performance guarantee better than 7/8 unless $P = NP$. On the other hand, Asano and Williamson [1] have presented a 0.7846-approximation algorithm and an approximation algorithm whose performance guarantee is 0.8331 if a conjectured performance guarantee of 0.7977 is true in the Zwick's algorithm [9]. Both algorithms are based on their sharpened analysis of Goemans and Williamson's LP-relaxation for MAX SAT [3].

In this paper, we present an improved analysis which is simpler than the previous analysis by Asano and Williamson [1]. Furthermore, this analysis will lead to approximation algorithms with better performance guarantees if combined with other approximation algorithms which were (and will be) presented. Algorithms based on this analysis will be used as a building block in designing an improved approximation algorithm for MAX SAT. Actually, algorithms based on this analysis lead to approximation algorithms with performance guarantee 0.7877 and conjectured performance guarantee 0.8353 which are slightly better than the best known corresponding performance guarantees 0.7846 and 0.8331 respectively, if combined with the MAX 2SAT and MAX 3SAT algorithms by Halperin and Zwick [6] and the Zwick's algorithm [9], respectively.

To explain our result in more detail, we briefly review the 0.75-approximation algorithm of Goemans and Williamson based on the probabilistic method [3]. Let $x^p = (x_1^p, \ldots, x_{2n}^p)$ be a *random* truth assignment with $0 \leq x_i^p = p_i \leq 1$ $(x_{n+i}^p = 1 - x_i^p = 1 - p_i = p_{n+i})$. That is, x^p is obtained by setting independently each variable $x_i \in X$ to be true with probability p_i (and $x_{n+i} = \bar{x}_i$ to be true with probability $p_{n+i} = 1 - p_i$). Then the probability of a clause $C_j = x_{j_1} \vee x_{j_2} \vee \cdots \vee x_{j_{k_j}} \in \mathcal{C}$ satisfied by the random truth assignment $x^p = (x_1^p, \ldots, x_{2n}^p)$ is $C_j(x^p) = 1 - \prod_{i=1}^{k_j}(1 - x_{j_i}^p)$. Thus, the expected value of the random truth assignment x^p is $F_{\mathcal{C}}(x^p) = \sum_{C_j \in \mathcal{C}} w(C_j) C_j(x^p)$. The probabilistic method assures that there is a truth assignment $x^q \in \{0,1\}^{2n}$ of value at least $F_{\mathcal{C}}(x^p)$. Such a truth assignment x^q can be obtained by the method of conditional probabilities [3].

Using an IP (integer programming) formulation of MAX SAT and its LP (linear programming) relaxation, Goemans and Williamson [3] obtained an algorithm for finding a random truth assignment x^p of value $F_{\mathcal{C}}(x^p)$ at least $\sum_{k \geq 1}(1 - (1 - \frac{1}{k})^k)\hat{W}_k \geq (1 - \frac{1}{e})\hat{W} \approx 0.632\hat{W}$, where e is the base of natural logarithm, $\hat{W}_k = \sum_{C \in \mathcal{C}_k} w(C) C(\hat{x})$, and $F_{\mathcal{C}}(\hat{x}) = \sum_{k \geq 1} \hat{W}_k$ for an optimal truth assignment $\hat{x}$ of $(\mathcal{C}, w)$ ($\mathcal{C}_k$ denotes the set of clauses in $\mathcal{C}$ with k literals). Goemans and Williamson also obtained a 0.75-approximation algorithm by using a hybrid approach of combining the above algorithm with Johnson's algorithm [7]. It finds a random truth assignment of value at least

$$0.750\hat{W}_1 + 0.750\hat{W}_2 + 0.789\hat{W}_3 + 0.810\hat{W}_4 + 0.820\hat{W}_5 + 0.824\hat{W}_6 + \sum_{k \geq 7} \beta_k \hat{W}_k,$$

where $\beta_k = \frac{1}{2}(2 - \frac{1}{2^k} - (1 - \frac{1}{k})^k)$. Asano and Williamson [1] showed that one of the non-hybrid algorithms of Goemans and Williamson finds a random truth assignment x^p with value $F_{\mathcal{C}}(x^p)$ at least

$$0.750\hat{W}_1 + 0.750\hat{W}_2 + 0.804\hat{W}_3 + 0.851\hat{W}_4 + 0.888\hat{W}_5 + 0.915\hat{W}_6 + \sum_{k \geq 7} \gamma_k \hat{W}_k,$$

where $\gamma_k = 1 - \frac{1}{2}(\frac{3}{4})^{k-1}(1 - \frac{1}{3(k-1)})^{k-1}$ for $k \geq 3$ ($\gamma_k > \beta_k$ for $k \geq 3$). Actually, they obtained a 0.7846-approximation algorithm by combining this algorithm with known MAX kSAT algorithms. They also proposed a generalization of this algorithm which finds a random truth assignment x^p with value $F_{\mathcal{C}}(x^p)$ at least

$$0.914\hat{W}_1 + 0.750\hat{W}_2 + 0.750\hat{W}_3 + 0.766\hat{W}_4 + 0.784\hat{W}_5 + 0.801\hat{W}_6 + 0.817\hat{W}_7 + \sum_{k \geq 8} \gamma'_k \hat{W}_k,$$

where $\gamma'_k = 1 - 0.914^k(1 - \frac{1}{k})^k$ for $k \geq 8$. They showed that if this is combined with Zwick's MAX SAT algorithm with conjectured 0.7977 performance guarantee then it leads to an approximation algorithm with performance guarantee 0.8331.

In this paper, we show that another generalization of the non-hybrid algorithms of Goemans and Williamson finds a random truth assignment x^p with value $F_{\mathcal{C}}(x^p)$ at least

$$0.750\hat{W}_1 + 0.750\hat{W}_2 + 0.815\hat{W}_3 + 0.859\hat{W}_4 + 0.894\hat{W}_5 + 0.920\hat{W}_6 + \sum_{k \geq 7} \zeta_k \hat{W}_k,$$

where $\zeta_k = 1 - \frac{1}{4}(\frac{3}{4})^{k-2}$ for $k \geq 3$ and $\zeta_k > \gamma_k$. We also present another algorithm which finds a random truth assignment x^p with value $F_{\mathcal{C}}(x^p)$ at least

$$0.914\hat{W}_1 + 0.750\hat{W}_2 + 0.757\hat{W}_3 + 0.774\hat{W}_4 + 0.790\hat{W}_5 + 0.804\hat{W}_6 + 0.818\hat{W}_7 + \sum_{k \geq 8} \gamma'_k \hat{W}_k.$$

This will be used to obtain a 0.8353-approximation algorithm.

The remainder of the paper is structured as follows. In Section 2 we review the algorithms of Goemans and Williamson [3] and Asano and Williamson [1]. In Section 3 we give our main results and their proofs. In Section 4 we briefly outline improved approximation algorithms for MAX SAT obtained by our main results.

2   MAX SAT Algorithms of Goemans and Williamson

Goemans and Williamson considered the following LP relaxation (GW) of MAX SAT [3]:

$$(GW) \max \sum_{C_j \in \mathcal{C}} w(C_j) z_j$$

$$\text{s.t.} \sum_{i=1}^{k_j} y_{j_i} \geq z_j \qquad \forall C_j = x_{j_1} \vee x_{j_2} \vee \cdots \vee x_{j_{k_j}} \in \mathcal{C}$$

$$y_i + y_{n+i} = 1 \qquad \forall i \in \{1, 2, ..., n\}$$
$$0 \le y_i \le 1 \qquad \forall i \in \{1, 2, ..., 2n\}$$
$$0 \le z_j \le 1 \qquad \forall C_j \in \mathcal{C}.$$

In this formulation, variables $\boldsymbol{y} = (y_i)$ correspond to the literals $\{x_1, \ldots, x_{2n}\}$ and variables $\boldsymbol{z} = (z_j)$ correspond to the clauses $\mathcal{C}$. Thus, variable $y_i = 1$ if and only if $x_i = 1$. Similarly, $z_j = 1$ if and only if C_j is satisfied. The first set of constraints implies that one of the literals in a clause is true if the clause is satisfied and thus IP formulation of this (GW) with $y_i \in \{0, 1\}$ ($\forall i \in \{1, 2, ..., 2n\}$) and $z_j \in \{0, 1\}$ ($\forall C_j \in \mathcal{C}$) exactly corresponds to MAX SAT.

Throughout this paper, let $(\boldsymbol{y}^*, \boldsymbol{z}^*)$ be an optimal solution to this LP relaxation of MAX SAT. Goemans and Williamson set each variable x_i to be true with probability y_i^*. Then a clause $C_j = x_{j_1} \vee x_{j_2} \vee \cdots \vee x_{j_{k_j}}$ is satisfied by this random truth assignment $\boldsymbol{x}^p = \boldsymbol{y}^*$ with probability $C_j(\boldsymbol{y}^*) \ge \left(1 - \left(1 - \frac{1}{k}\right)^k\right) z_j^*$. Thus, the expected value $F(\boldsymbol{y}^*)$ of $\boldsymbol{y}^*$ obtained in this way satisfies

$$F(\boldsymbol{y}^*) = \sum_{C_j \in \mathcal{C}} w(C_j) C_j(\boldsymbol{y}^*) \ge \sum_{k \ge 1} \left(1 - \left(1 - \frac{1}{k}\right)^k\right) W_k^* \ge \left(1 - \frac{1}{e}\right) W^*,$$

where $W^* = \sum_{C_j \in \mathcal{C}} w(C_j) z_j^*$ and $W_k^* = \sum_{C_j \in \mathcal{C}_k} w(C_j) z_j^*$ (note that $W^* = \sum_{C_j \in \mathcal{C}} w(C_j) z_j^* \ge \hat{W} = \sum_{C_j \in \mathcal{C}} w(C_j) \hat{z}_j$ for an optimal solution $(\hat{\boldsymbol{y}}, \hat{\boldsymbol{z}})$ to the IP formulation of MAX SAT). Since $(1 - \frac{1}{e}) \approx 0.632$, this is a 0.632-approximation algorithm for MAX SAT.

Goemans and Williamson [3] also considered three other non-linear randomized rounding algorithms. In these algorithms, each variable x_i is set to be true with probability $f_\ell(y_i^*)$ defined as follows ($\ell = 1, 2, 3$).

$$f_1(y) = \begin{cases} \frac{3}{4}y + \frac{1}{4} & \text{if } 0 \le y \le \frac{1}{3} \\ \frac{1}{2} & \text{if } \frac{1}{3} \le y \le \frac{2}{3} \\ \frac{3}{4}y & \text{if } \frac{2}{3} \le y \le 1, \end{cases}$$

$$f_2(y) = (2a - 1)y + 1 - a \qquad \left(\frac{3}{4} \le a \le \frac{3}{\sqrt[3]{4}} - 1\right),$$

$$1 - 4^{-y} \le f_3(y) \le 4^{y-1}.$$

Note that $f_\ell(y_i^*) + f_\ell(y_{n+i}^*) = 1$ hold for $\ell = 1, 2$ and that $f_3(y_i^*)$ has to be chosen to satisfy $f_3(y_i^*) + f_3(y_{n+i}^*) = 1$. They then proved that all the random truth assignments $\boldsymbol{x}^p = f_\ell(\boldsymbol{y}^*) = (f_\ell(y_1^*), \ldots, f_\ell(y_{2n}^*))$ obtained in this way have the expected values at least $\frac{3}{4}W^*$ and lead to $\frac{3}{4}$-approximation algorithms. Asano and Williamson [1] sharpened the analysis of Goemans and Williamson to provide more precise bounds on the probability of a clause $C_j = x_{j_1} \vee x_{j_2} \vee \cdots \vee x_{j_k}$ with k literals being satisfied (and thus on the expected weight of satisfied clauses in C_k) by the random truth assignment $x^p = f_\ell(y^*)$ for each k (and $\ell = 1, 2$). From now on, we assume by symmetry, $x_{j_i} = x_i$ for each $i = 1, 2, ..., k$ since $f_\ell(x) = 1 - f_\ell(\bar{x})$ and we can set $x := \bar{x}$ if necessary. They considered clause $C_j = x_1 \vee x_2 \vee \cdots \vee x_k$ corresponding to the constraint $y_1 + y_2 + \cdots + y_k \geq z_j$ in the LP relaxation (GW) of MAX SAT, and gave a bound on the ratio of $C_j(f_\ell(y^*))$ to z_j^*, where $C_j(f_\ell(y^*))$ is the probability of clause C_j being satisfied by the random truth assignment $x^p = f_\ell(y^*)$ $(\ell = 1, 2)$. Actually, they analyzed parametrized functions f_1^a and f_2^a with $\frac{1}{2} \leq a \leq 1$ defined as follows:

$$f_1^a(y) = \begin{cases} ay + 1 - a & \text{if } 0 \leq y \leq 1 - \frac{1}{2a} \\ \frac{1}{2} & \text{if } 1 - \frac{1}{2a} \leq y \leq \frac{1}{2a} \\ ay & \text{if } \frac{1}{2a} \leq y \leq 1, \end{cases} \tag{1}$$

$$f_2^a(y) = (2a - 1)y + 1 - a. \tag{2}$$

Note that $f_1 = f_1^{3/4}$ and $f_2 = f_2^a$. Let

$$\gamma_{k,1}^a = 1 - \frac{1}{2} a^{k-1} \left(1 - \frac{1 - \frac{1}{2a}}{k-1} \right)^{k-1}, \qquad \gamma_{k,2}^a = 1 - a^k \left(1 - \frac{1}{k} \right)^k, \tag{3}$$

$$\gamma_k^a = \begin{cases} a & \text{if } k = 1 \\ \min\{\gamma_{k,1}^a, \gamma_{k,2}^a\} & \text{if } k \geq 2 \end{cases} \tag{4}$$

and

$$\delta_k^a = 1 - a^k \left(1 - \frac{2 - \frac{1}{a}}{k} \right)^k. \tag{5}$$

Then their results are summarized as follows.

Proposition 1. [1] For $\frac{1}{2} \leq a \leq 1$, let $C_j(f_\ell^a(y^*)) = 1 - \prod_{i=1}^k (1 - f_\ell^a(y_i^*))$ be the probability of clause $C_j = x_1 \vee x_2 \vee \cdots \vee x_k \in \mathcal{C}$ being satisfied by the random truth assignment $x^p = f_\ell^a(y^*) = (f_\ell^a(y_1^*), \ldots, f_\ell^a(y_{2n}^*))$ $(\ell = 1, 2)$. Then the following statements hold.

1. $C_j(f_1^a(y^*)) = 1 - \prod_{i=1}^k (1 - f_1^a(y_i^*)) \geq \gamma_k^a z_j^*$ and the expected value $F(f_1^a(y^*))$ of $x^p = f_1^a(y^*)$ satisfies $F(f_1^a(y^*)) \geq \sum_{k \geq 1} \gamma_k^a W_k^*$.
2. $C_j(f_2^a(y^*)) = 1 - \prod_{i=1}^k (1 - f_2^a(y_i^*)) \geq \delta_k^a z_j^*$ and the expected value $F(f_2^a(y^*))$ of $x^p = f_2^a(y^*)$ satisfies $F(f_2^a(y^*)) \geq \sum_{k \geq 1} \delta_k^a W_k^*$.
3. $\gamma_k^a > \delta_k^a$ hold for all $k \geq 3$ and for all a with $\frac{1}{2} < a < 1$. For $k = 1, 2$, $\gamma_k^a = \delta_k^a$ $(\gamma_1^a = \delta_1^a = a, \gamma_2^a = \delta_2^a = \frac{3}{4})$ hold.

3 Main Results and Their Proofs

Asano and Williamson did not consider a parametrized function of f_3. In this section we consider a parametrized function f_3^a of f_3 and show that it has better performance than f_1^a and f_2^a. Furthermore, its analysis (proof) is simpler. We also consider a generalization of both f_1^a and f_2^a.

For $\frac{1}{2} \leq a \leq 1$, let f_3^a be defined as follows:

$$f_3^a(y) = \begin{cases} 1 - \frac{a}{(4a^2)^y} & \text{if } 0 \leq y \leq \frac{1}{2} \\[2mm] \frac{(4a^2)^y}{4a} & \text{if } \frac{1}{2} \leq y \leq 1. \end{cases} \tag{6}$$

For $\frac{3}{4} \leq a \leq 1$, let

$$y_a = \frac{1}{a} - \frac{1}{2}. \tag{7}$$

Then the other parametrized function f_4^a is defined as follows:

$$f_4^a(y) = \begin{cases} ay + 1 - a & \text{if } 0 \leq y \leq 1 - y_a \\[2mm] \frac{a}{2}y + \frac{1}{2} - \frac{a}{4} & \text{if } 1 - y_a \leq y \leq y_a \\[2mm] ay & \text{if } y_a \leq y \leq 1. \end{cases} \tag{8}$$

Thus, $f_3^a(y) + f_3^a(1 - y) = 1$ and $f_4^a(y) + f_4^a(1 - y) = 1$ hold for $0 \leq y \leq 1$. Furthermore, f_3^a and f_4^a are both continuous functions which are increasing with y. Thus, $f_3^a(\frac{1}{2}) = f_4^a(\frac{1}{2}) = \frac{1}{2}$. Let ζ_k^a and η_k^a be the numbers defined as follows.

$$\zeta_k^a = \begin{cases} a & \text{if } k = 1 \\ 1 - \frac{1}{4}a^{k-2} & \text{if } k \geq 2, \end{cases} \tag{9}$$

$$\eta_{k,1}^a = \gamma_{k,2}^a = 1 - a^k\left(1 - \frac{1}{k}\right)^k, \qquad \eta_{k,2}^a = \zeta_k^a = 1 - \frac{a^{k-2}}{4}, \tag{10}$$

$$\eta_{k,3}^a = 1 - \frac{a^k}{2}\left(1 - \frac{1 - y_a}{k - 1}\right)^{k-1}, \qquad \eta_{k,4}^a = 1 - \frac{1}{2^k}\left(1 + \frac{a}{2} - \frac{a}{k}\right)^k, \tag{11}$$

$$\eta_k^a = \begin{cases} a & \text{if } k = 1 \\ \min\{\eta_{k,1}^a, \eta_{k,2}^a, \eta_{k,3}^a, \eta_{k,4}^a\} & \text{if } k > 2. \end{cases} \tag{12}$$

Then we have the following theorems for the two parameterized functions f_3^a and f_4^a.

Theorem 1. *For $\frac{1}{2} \leq a \leq \frac{\sqrt{e}}{2} = 0.82436$, the probability of $C_j = x_1 \vee x_2 \vee \cdots \vee x_k \in C$ being satisfied by the random truth assignment $\boldsymbol{x}^p = f_3^a(\boldsymbol{y}^*) = (f_3^a(y_1^*), \ldots, f_3^a(y_{2n}^*))$ is $C_j(f_3^a(\boldsymbol{y}^*)) = 1 - \prod_{i=1}^k (1 - f_3^a(y_i^*)) \geq \zeta_k^a z_j^*$. Thus, the expected value $F(f_3^a(\boldsymbol{y}^*))$ of $\boldsymbol{x}^p = f_3^a(\boldsymbol{y}^*)$ satisfies $F(f_3^a(\boldsymbol{y}^*)) \geq \sum_{k \geq 1} \zeta_k^a W_k^*$.*

Theorem 2. *For $\frac{\sqrt{e}}{2} = 0.82436 \leq a \leq 1$, the probability of $C_j = x_1 \vee x_2 \vee \cdots \vee x_k \in C$ being satisfied by the random truth assignment $\boldsymbol{x}^p = f_4^a(\boldsymbol{y}^*) = (f_4^a(y_1^*), \ldots, f_4^a(y_{2n}^*))$ is $C_j(f_4^a(\boldsymbol{y}^*)) = 1 - \prod_{i=1}^k (1 - f_4^a(y_i^*)) \geq \eta_k^a z_j^*$. Thus, the expected value $F(f_4^a(\boldsymbol{y}^*))$ of $\boldsymbol{x}^p = f_4^a(\boldsymbol{y}^*)$ satisfies $F(f_4^a(\boldsymbol{y}^*)) \geq \sum_{k \geq 1} \eta_k^a W_k^*$.*

Theorem 3. *The following statements hold for γ_k^a, δ_k^a, ζ_k^a, and η_k^a.*

1. *If $\frac{1}{2} \leq a \leq \frac{\sqrt{e}}{2} = 0.82436$, then $\zeta_k^a > \gamma_k^a > \delta_k^a$ hold for all $k \geq 3$.*
2. *If $\frac{\sqrt{e}}{2} = 0.82436 \leq a < 1$, then $\eta_k^a \geq \gamma_k^a > \delta_k^a$ hold for all $k \geq 3$. In particular, if $\frac{\sqrt{e}}{2} = 0.82436 \leq a \leq 0.881611$, then $\eta_k^a > \gamma_k^a > \delta_k^a$ hold for all $k \geq 3$.*
3. *For $k = 1, 2$, $\gamma_k^a = \delta_k^a = \zeta_k^a$ hold if $\frac{1}{2} \leq a \leq \frac{\sqrt{e}}{2} = 0.82436$, and $\gamma_k^a = \delta_k^a = \eta_k^a$ hold if $\frac{\sqrt{e}}{2} = 0.82436 \leq a \leq 1$.*

In this paper, we first give a proof of Theorem 1. It is very simple and we use only the following lemma.

Lemma 1. *If $\frac{1}{2} \leq a \leq \frac{\sqrt{e}}{2} = 0.82436$, then $f_3^a(y) \geq ay$.*

Proof. Let $g(y) \equiv \frac{(4a^2)^y}{4a} - ay$. Then its derivative is $g'(y) = \ln(4a^2)\frac{(4a^2)^y}{4a} - a$. Thus, $g'(y)$ is increasing with y and $g'(1) = a(\ln(4a^2) - 1) \leq 0$, since $\ln(4a^2) \leq \ln(4(\frac{\sqrt{e}}{2})^2) = 1$. This implies that $g'(y) \leq 0$ for all $0 \leq y \leq 1$ and that $g(y)$ is decreasing with $0 \leq y \leq 1$. Thus, $g(y)$ takes a minimum value at $y = 1$, i.e., $g(y) = \frac{(4a^2)^y}{4a} - ay \geq g(1) = \frac{4a^2}{4a} - a = 0$.

Now we are ready to prove the lemma. For $\frac{1}{2} \leq y \leq 1$, we have $f_3(y) - ay = g(y) = \frac{(4a^2)^y}{4a} - ay \geq 0$. For $0 \leq y \leq \frac{1}{2}$, we have

$$f_3(y) - ay = 1 - \frac{a}{(4a^2)^y} - ay = -\frac{(4a^2)^{1-y}}{4a} + a(1-y) + 1 - a$$

$$= -g(1-y) + 1 - a \geq -g(\tfrac{1}{2}) + 1 - a = \frac{1-a}{2} \geq 0$$

since $g(y)$ is decreasing and $g(1-y) \leq g(\frac{1}{2}) = \frac{1-a}{2}$ for $\frac{1}{2} \leq 1 - y \leq 1$. ∎

Proof of Theorem 1. Noting that clause $C_j = x_1 \vee x_2 \vee \cdots \vee x_k$ corresponds to the constraint

$$y_1^* + y_2^* + \cdots + y_k^* \geq z_j^* \tag{13}$$

in the LP relaxation (GW) of MAX SAT, we will show that

$$C_j(f_3^a(\boldsymbol{y}^*)) = 1 - \prod_{i=1}^k (1 - f_3^a(y_i^*)) \geq \zeta_k^a z_j^*$$

for $\frac{1}{2} \leq a \leq \frac{\sqrt{e}}{2} = 0.82436$. By symmetry, we assume $y_1^* \leq y_2^* \leq \cdots \leq y_k^*$. Note that $y_k^* \leq z_j^*$, since otherwise $(\boldsymbol{y}^*, \boldsymbol{z}^*)$ would not be an optimal solution to the LP relaxation (GW) of MAX SAT (if $y_k^* > z_j^*$ then $(\boldsymbol{y}^*, \boldsymbol{z}')$ with $z_j' = y_k^*$ and $z_{j'}' = z_{j'}^*$ ($j' \neq j$) would also be a feasible solution to (GW) and $\sum_{C_{j'} \in C} w(C_{j'}) z_{j'}' > \sum_{C_{j'} \in C} w(C_{j'}) z_{j'}^*$), a contradiction.

If $k = 1$, then we have $C_j(f_3^a(\boldsymbol{y}^*)) = f_3^a(y_1^*) \geq ay_1^* \geq az_j^* = \zeta_1^a z_j^*$ by Lemma 1 and inequality (13).

Next suppose $k \geq 2$. We consider two cases as follows: Case 1: $0 \leq y_k^* \leq \frac{1}{2}$; and Case 2: $\frac{1}{2} < y_k^* \leq 1$.

Case 1: $0 \leq y_k^* \leq \frac{1}{2}$. Since all $y_i^* \leq \frac{1}{2}$ $(i = 1, 2, ..., k)$, we have $f_3^a(y_i^*) = 1 - \frac{a}{(4a^2)^{y_i^*}}$ and $1 - f_3^a(y_i^*) = \frac{a}{(4a^2)^{y_i^*}}$. Thus, we have

$$C_j(f_3^a(\boldsymbol{y}^*)) = 1 - \prod_{i=1}^{k}(1 - f_3^a(y_i^*)) = 1 - \prod_{i=1}^{k} \frac{a}{(4a^2)^{y_i^*}} = 1 - \frac{a^k}{(4a^2)^{\sum_{i=1}^{k} y_i^*}}$$

$$\geq 1 - \frac{a^k}{(4a^2)^{z_j^*}} \geq \left(1 - \frac{a^k}{4a^2}\right)z_j^* = \left(1 - \frac{a^{k-2}}{4}\right)z_j^* = \zeta_k^a z_j^*,$$

where the first inequality follows by inequality (13), and the second inequality follows from the fact that $1 - \frac{a^k}{(4a^2)^{z_j^*}}$ is a concave function in $0 \leq z_j^* \leq 1$.

Case 2: $\frac{1}{2} < y_k^* \leq 1$. Let $y_{k-1}^* > \frac{1}{2}$. Then, since $f_3^a(y_i^*) \geq 1 - a$ $(i = 1, 2, ..., k)$, we have $1 - f_3^a(y_i^*) \leq a$ $(i = 1, 2, ..., k - 2)$, $1 - f_3^a(y_i^*) = 1 - \frac{(4a^2)^{y_i^*}}{4a} \leq \frac{1}{2}$ $(i = k - 1, k)$, and $z_j^* \leq 1$, and $C_j(f_3^a(\boldsymbol{y}^*)) = 1 - \prod_{i=1}^{k}(1 - f_3^a(y_i^*))$ satisfies

$$C_j(f_3^a(\boldsymbol{y}^*)) \geq 1 - a^{k-2}\left(\frac{1}{2}\right)^2 = 1 - \frac{a^{k-2}}{4} \geq \left(1 - \frac{a^{k-2}}{4}\right)z_j^* = \zeta_k^a z_j^*.$$

Thus, we can assume $y_{k-1}^* \leq \frac{1}{2}$. Since $1 - f_3^a(y_i^*) = \frac{a}{(4a^2)^{y_i^*}}$ $(i = 1, 2, ..., k - 1)$, we have

$$C_j(f_3^a(\boldsymbol{y}^*)) = 1 - \prod_{i=1}^{k}(1 - f_3^a(y_i^*)) = 1 - \frac{a^{k-1}}{(4a^2)^{\sum_{i=1}^{k-1} y_i^*}}\left(1 - \frac{(4a^2)^{y_k^*}}{4a}\right)$$

$$\geq 1 - \frac{a^{k-1}}{(4a^2)^{z_j^* - y_k^*}}\left(1 - \frac{(4a^2)^{y_k^*}}{4a}\right) = 1 - \frac{a^{k-1}}{(4a^2)^{z_j^*}}(4a^2)^{y_k^*}\left(1 - \frac{(4a^2)^{y_k^*}}{4a}\right)$$

$$\geq 1 - \frac{a^{k-1}}{(4a^2)^{z_j^*}}a = 1 - \frac{a^k}{(4a^2)^{z_j^*}} \geq \left(1 - \frac{a^{k-2}}{4}\right)z_j^* = \zeta_k^a z_j^*$$

by inequality (13), $y_k^* \leq z_j^*$, $(4a^2)^{y_k^*}(1 - \frac{(4a^2)^{y_k^*}}{4a}) = u(1 - \frac{u}{4a}) \leq a$ with $u = (4a^2)^{y_k^*}$, and the fact that $1 - \frac{a^k}{(4a^2)^{z_j^*}}$ is a concave function in $0 \leq z_j^* \leq 1$. ∎

Proofs of Theorems 2 and 3. Proofs of Theorems 2 and 3 are almost similar to ones in Asano and Williamson [1]. In this sense, proofs may be a little complicated, however, they can be done in a systematic way. Here, we will give only an outline of Proof of Theorem 2. Proof of Theorem 3 is almost similar.

Outline of Proof of Theorem 2. For a clause $C_j = x_1 \vee x_2 \vee \cdots \vee x_k$ corresponding to the constraint $y_1^* + y_2^* + \cdots + y_k^* \geq z_j^*$ as described in Proof of Theorem 1, we will show that $C_j(f_4^a(\boldsymbol{y}^*)) = 1 - \prod_{i=1}^{k}(1 - f_4^a(y_i^*)) \geq \eta_k^a z_j^*$ for $\frac{3}{4} \leq a \leq 1$. We assume $y_1^* \leq y_2^* \leq \cdots \leq y_k^*$ and $y_k^* \leq z_j^*$ holds as described before.

Suppose $k = 1$. Since $f_4^a(y) - ay = 1 - a \geq 0$ for $0 \leq y \leq 1 - y_a$ and $f_4^a(y) - ay = 0$ for $y_a \leq y \leq 1$, we consider the case when $1 - y_a \leq y \leq y_a$.

In this case, $f_4^a(y) - ay = \frac{2-a-2ay}{4}$ is decreasing with $1 - y_a \le y \le y_a$ and we have $f_4^a(y) - ay = \frac{2-a-2ay}{4} \ge f_4^a(y_a) - ay_a = \frac{2-a-2ay_a}{4} = 0$ by Eq.(7). Thus, $C_j(f_4^a(\boldsymbol{y}^*)) = f_4^a(y_1^*) \ge ay_1^* \ge az_j^* = \eta_1^a z_j^*$ by inequality (13).

Next suppose $k \ge 2$. We consider three cases as follows. Case 1: $y_k^* \le 1 - y_a$; Case 2: $1 - y_a < y_k^* \le y_a$; and Case 3: $y_a \le y_k^* \le 1$.

Case 1: $y_k^* \le 1 - y_a$. Since all $y_i^* \le 1 - y_a$ $(i = 1, 2, ..., k)$, $f_4^a(y_i^*) = 1 - a + ay_i^*$ and $1 - f_4^a(y_i^*) = a(1 - y_i^*)$. Thus, $C_j(f_4^a(\boldsymbol{y}^*)) = 1 - \prod_{i=1}^{k}(1 - f_4^a(y_i^*))$ satisfies

$$C_j(f_4^a(\boldsymbol{y}^*)) = 1 - a^k \prod_{i=1}^{k}(1 - y_i^*) \ge 1 - a^k \left(1 - \frac{\sum_{i=1}^{k} y_i^*}{k}\right)^k$$

$$\ge 1 - a^k \left(1 - \frac{z_j^*}{k}\right)^k \ge \left(1 - a^k\left(1 - \frac{1}{k}\right)^k\right) z_j^* = \eta_{k,1}^a z_j^*,$$

where the first inequality follows by the arithmetic/geometric mean inequality, the second by inequality (13), and third by the fact that $1 - a^k(1 - \frac{z_j^*}{k})^k$ is a concave function in $0 \le z_j^* \le 1$.

Case 2: $1 - y_a \le y_k^* \le y_a$. Let ℓ be the number such that $y_\ell^* < 1 - y_a \le y_{\ell+1}^*$ and let $y_A = \sum_{i=1}^{\ell} y_i^*$ and $y_B = \sum_{i=\ell+1}^{k} y_i^*$. Then $k - \ell \ge 1$ and $\ell \ge 0$. If $\ell = 0$ then, $f_4^a(y_i^*) = \frac{1}{2}\left(ay_i^* + 1 - \frac{a}{2}\right)$ $(i = 1, 2, ..., k)$ and for the same reason as in Case 1 above, we have

$$C_j(f_4^a(\boldsymbol{y}^*)) = 1 - \prod_{i=1}^{k}(1 - f_4^a(y_i^*)) = 1 - \left(\frac{1}{2}\right)^k \prod_{i=1}^{k}\left(1 + \frac{a}{2} - ay_i^*\right)$$

$$\ge 1 - \left(\frac{1}{2}\right)^k \left(1 + \frac{a}{2} - \frac{ay_B}{k}\right)^k \ge 1 - \left(\frac{1}{2}\right)^k \left(1 + \frac{a}{2} - \frac{az_j^*}{k}\right)^k$$

$$= \left(1 - \left(\frac{1}{2}\right)^k \left(1 + \frac{a}{2} - \frac{a}{k}\right)^k\right) z_j^* = \eta_{k,4}^a z_j^*.$$

Now suppose $\ell > 0$ and that $y_B \le z_j^*$ (we omit the case when $y_B > z_j^*$, since it can be argued similarly). Then $C_j(f_4^a(\boldsymbol{y}^*)) = 1 - \prod_{i=1}^{k}(1 - f_4^a(y_i^*))$ satisfies

$$C_j(f_4^a(\boldsymbol{y}^*)) = 1 - a^\ell \left(\frac{1}{2}\right)^{k-\ell} \prod_{i=1}^{\ell}(1 - y_i^*) \prod_{i=\ell+1}^{k}\left(1 + \frac{a}{2} - ay_i^*\right)$$

$$\ge 1 - a^\ell \left(\frac{1}{2}\right)^{k-\ell} \left(1 - \frac{y_A}{\ell}\right)^\ell \left(1 + \frac{a}{2} - \frac{ay_B}{k-\ell}\right)^{k-\ell}$$

$$\ge 1 - a^\ell \left(\frac{1}{2}\right)^{k-\ell} \left(1 - \frac{z_j^* - y_B}{\ell}\right)^\ell \left(1 + \frac{a}{2} - \frac{ay_B}{k-\ell}\right)^{k-\ell}$$

$$= 1 - a^\ell \left(\frac{1}{2}\right)^{k-\ell} g(y_B),$$

where $g(y_B) \equiv \left(1 - \frac{z_j^* - y_B}{\ell}\right)^\ell \left(1 + \frac{a}{2} - \frac{ay_B}{k-\ell}\right)^{k-\ell}$. Note that $g(y_B)$ is increasing with y_B. Thus, if $k - \ell \geq 2$ then, by $y_B \leq z_j^*$ and $g(y_B) \leq g(z_j^*)$, we have

$$C_j(f_4^a(\mathbf{y}^*)) \geq 1 - a^\ell \left(\frac{1}{2}\right)^{k-\ell} g(z_j^*) = 1 - a^\ell \left(\frac{1}{2}\right)^{k-\ell} \left(1 + \frac{a}{2} - \frac{az_j^*}{k-\ell}\right)^{k-\ell}$$

$$\geq \left(1 - a^\ell \left(\frac{1}{2}\right)^{k-\ell} \left(1 + \frac{a}{2} - \frac{a}{k-\ell}\right)^{k-\ell}\right) z_j^*$$

$$\geq \left(1 - a^{k-2} \left(\frac{1}{2}\right)^2 \left(1 + \frac{a}{2} - \frac{a}{2}\right)^2\right) z_j^* = \left(1 - \frac{a^{k-2}}{4}\right) z_j^* = \eta_{k,2}^a z_j^*,$$

since $1 - a^\ell \left(\frac{1}{2}\right)^{k-\ell} \left(1 + \frac{a}{2} - \frac{az_j^*}{k-\ell}\right)^{k-\ell}$ is a concave function in $0 \leq z_j^* \leq 1$ and $1 - a^\ell \left(\frac{1}{2}\right)^{k-\ell} \left(1 + \frac{a}{2} - \frac{a}{k-\ell}\right)^{k-\ell}$ is increasing with $k - \ell$ for $\frac{3}{4} \leq a \leq 1$ (which can be shown by Lemma 2.5 in [1]). Similarly, if $k - \ell = 1$, then $y_B = y_k^* \leq y_a$ and

$$C_j(f_4^a(\mathbf{y}^*)) \geq 1 - a^{k-1} \left(\frac{1}{2}\right) g(y_a) = 1 - \frac{a^k}{2} \left(1 - \frac{z_j^* - y_a}{k-1}\right)^{k-1}$$

$$\geq \left(1 - \frac{a^k}{2} \left(1 - \frac{1 - y_a}{k-1}\right)^{k-1}\right) z_j^* = \eta_{k,3}^a z_j^*$$

since $g(y_B) \leq g(y_a) = a \left(1 - \frac{z_j^* - y_a}{k-1}\right)^{k-1}$ by Eq.(7) and $1 - \frac{a^k}{2} \left(1 - \frac{z_j^* - y_a}{k-1}\right)^{k-1}$ is a concave function in $y_a \leq z_j^* \leq 1$ (see Lemma 2.4 in [1]).

Case 3: $y_a \leq y_k^* \leq 1$. If $y_{k-1}^* + y_k^* > 1$ then $(1 - f_4^a(y_{k-1}^*))(1 - f_4^a(y_k^*)) \leq \frac{1}{4}$ and $1 - f_4^a(y_i^*) \leq a$ $(i = 1, 2, ..., k)$ and $C_j(f_4^a(\mathbf{y}^*)) = 1 - \prod_{i=1}^k (1 - f_4^a(y_i^*))$ satisfies

$$C_j(f_4^a(\mathbf{y}^*)) \geq 1 - a^{k-2}(1 - f_4^a(y_{k-1}^*))(1 - f_4^a(y_k^*)) \geq 1 - \frac{a^{k-2}}{4} = \eta_{k,2}^a \geq \eta_{k,2}^a z_j^*.$$

Thus, we can assume $y_{k-1}^* \leq 1 - y_a$. Let $y_A = \sum_{i=1}^{k-1} y_i^*$. Then we have

$$C_j(f_4^a(\mathbf{y}^*)) \geq 1 - a^{k-1}(1 - ay_k^*) \prod_{i=1}^{k-1} (1 - y_i^*) \geq 1 - a^{k-1}(1 - ay_k^*) \left(1 - \frac{y_A}{k-1}\right)^{k-1}$$

$$\geq 1 - a^{k-1}(1 - ay_k^*) \left(1 - \frac{z_j^* - y_k^*}{k-1}\right)^{k-1}$$

$$\geq 1 - a^{k-1}(1 - ay_a) \left(1 - \frac{z_j^* - y_a}{k-1}\right)^{k-1} = 1 - \frac{a^k}{2} \left(1 - \frac{z_j^* - y_a}{k-1}\right)^{k-1}$$

$$\geq \left(1 - \frac{a^k}{2} \left(1 - \frac{1 - y_a}{k-1}\right)^{k-1}\right) z_j^* = \eta_{k,3}^a z_j^*,$$

since $(1 - ay_k^*) \left(1 - \frac{z_j^* - y_k^*}{k-1}\right)^{k-1}$ is decreasing with y_k^* $(y_a \leq y_k^* \leq 1)$.   ∎

4 Improved Approximation Algorithms

In this section, we briefly outline our improved appproximation algorithms for MAX SAT based on a hybrid approach which is described in detail in Asano and Williamson [1]. We use a semidefinite programming relaxation of MAX SAT which is a combination of ones given by Goemans and Williamson [4], Feige and Goemans [2], Karloff and Zwick [8], Halperin and Zwick [6], and Zwick [9]. Our algorithms pick the best solution returned by the four algorithms corresponding to (1) f_3^a in Goemans and Williamson [3], (2) MAX 2SAT algorithm of Feige and Goemans [2] or of Halperin and Zwick [6], (3) MAX 3SAT algorithm of Karloff and Zwick [8] or of Halperin and Zwick [6], and (4) Zwick's MAX SAT algorithm with a conjectured performance guarantee 0.7977 [9]. The expected value of the solution is at least as good as the expected value of an algorithm that uses Algorithm (i) with probability p_i, where $p_1 + p_2 + p_3 + p_4 = 1$.

Our first algorithm picks the best solution returned by the three algorithms corresponding to (1) f_3^a in Goemans and Williamson [3], (2) Feige and Goemans's MAX 2SAT algorithm [2], and (3) Karloff and Zwick's MAX 3SAT algorithm [8] (this implies that $p_4 = 0$). From the arguments in Section 3, the probability that a clause $C_j \in \mathcal{C}_k$ is satisfied by Algorithm (1) is at least $\zeta_k^a z_j^*$, where ζ_k^a is defined in Eq.(9). Similarly, from the arguments in [4,2], the probability that a clause $C_j \in \mathcal{C}_k$ is satisfied by Algorithm (2) is

$$\text{at least} \quad 0.93109 \cdot \frac{2}{k} z_j^* \quad \text{for } k \geq 2, \quad \text{and at least} \quad 0.97653 z_j^* \quad \text{for } k = 1.$$

By an analysis obtained by Karloff and Zwick [8] and an argument similar to one in [4], the probability that a clause $C_j \in \mathcal{C}_k$ is satisfied by Algorithm (3) is

$$\text{at least} \quad \frac{3}{k}\frac{7}{8} z_j^* \quad \text{for } k \geq 3, \quad \text{and at least} \quad 0.87856 z_j^* \quad \text{for } k = 1, 2.$$

Suppose that we set $a = 0.74054$, $p_1 = 0.7861$, $p_2 = 0.1637$, and $p_3 = 0.0502$ ($p_4 = 0$). Then

$$ap_1 + 0.97653p_2 + 0.87856p_3 \geq 0.7860 \quad \text{for } k = 1,$$
$$\frac{3}{4}p_1 + 0.93109p_2 + 0.87856p_3 \geq 0.7860 \quad \text{for } k = 2,$$
$$\zeta_k^a p_1 + \frac{2 \times 0.93109}{k}p_2 + \frac{3}{k}\frac{7}{8}p_3 \geq 0.7860 \quad \text{for } k \geq 3.$$

Thus this is a 0.7860-approximation algorithm. Note that the algorithm in Asano and Williamson [1] picking the best solution returned by the three algorithms corresponding to (1) f_1^a with $a = \frac{3}{4}$ in Goemans and Williamson [3], (2) Feige and Goemans [2], and (3) Karloff and Zwick [8] only achieves the performance guarantee 0.7846.

Suppose next that we use three algorithms (1) f_3^a in Goemans and Williamson [3], (2) Halperin and Zwick's MAX 2SAT algorithm [6], and (3) Halperin and Zwick's MAX 3SAT algorithm [6] instead of Feige and Goemans [2] and Karloff and Zwick [8]. If we set $a = 0.739634$, $p_1 = 0.787777$, $p_2 = 0.157346$, and $p_3 = 0.054877$, then we have

$$ap_1 + 0.9828p_2 + 0.9197p_3 \geq 0.7877 \quad \text{for } k = 1,$$

$$\frac{3}{4}p_1 + 0.9309p_2 + 0.9197p_3 \geq 0.7877 \quad \text{for } k = 2,$$

$$\zeta_k^a p_1 + \frac{2 \times 0.9309}{k}p_2 + \frac{3}{k}\frac{7}{8}p_3 \geq 0.7877 \quad \text{for } k \geq 3.$$

Thus we have a 0.7877-approximation algorithm for MAX SAT (note that the performance guarantees of Halperin and Zwick's MAX 2SAT and MAX 3SAT algorithms are based on the numerical evidence [6]).

Suppose finally that we use two algorithms (1) f_4^a in Goemans and Williamson [3] and (4) Zwick's MAX SAT algorithm with a conjectured performance guarantee 0.7977 [9]. If we set $a = 0.907180$, $p_1 = 0.343137$ and $p_4 = 0.656863$ ($p_2 = p_3 = 0$), then the probability of clause C_j with k literals being satisfied can be shown to be at least $0.8353z_j^*$ for each $k \geq 1$. Thus, we can obtain a 0.8353-approximation algorithm for MAX SAT if a conjectured performance guarantee 0.7977 is true in Zwick's MAX SAT algorithm [9,1].

Remarks. As described above, algorithms based on f_3^a and f_4^a can be used as a building block for designing an improved approximation algorithm for MAX SAT. We have examined several other parameterized functions including ones in Asano and Williamson [1] and we are sure that algorithms based on f_3^a and f_4^a are almost the best as such a building block among functions of using an optimal solution (y^*, z^*) to Goemans and Williamson's LP relaxation for MAX SAT.

Acknowledgments. I would like to thank Prof. B. Korte of Bonn University for having invited me to have stayed in his institute and done this work. I also thank Dr. D.P. Williamson for useful comments. This work was supported in part by 21st Century COE Program: Research on Security and Reliability in Electronic Society, Grant in Aid for Scientific Research of the Ministry of Education, Science, Sports and Culture of Japan, The Institute of Science and Engineering of Chuo University, and The Telecommunications Advancement Foundation.

References

1. T. Asano and D.P. Williamson, Improved approximation algorithms for MAX SAT, *Journal of Algorithms* 42, pp.173–202, 2002.
2. U. Feige and M.X. Goemans, Approximating the value of two prover proof systems, with applications to MAX 2SAT and MAX DICUT, In *Proc. 3rd Israel Symposium on Theory of Computing and Systems*, pp. 182–189, 1995.
3. M.X. Goemans and D.P. Williamson, New 3/4-approximation algorithms for the maximum satisfiability problem, *SIAM Journal on Discrete Mathematics* 7, pp. 656–666, 1994.

4. M.X. Goemans and D.P. Williamson, Improved approximation algorithms for maximum cut and satisfiability problems using semidefinite programming, *Journal of the ACM* 42, pp. 1115–1145, 1995.
5. J. Håstad, Some optimal inapproximability results, In *Proc. 28th ACM Symposium on the Theory of Computing*, pp. 1–10, 1997.
6. E. Halperin and U. Zwick, Approximation algorithms for MAX 4-SAT and rounding procedures for semidefinite programs, *Journal of Algorithms* 40, pp. 184–211, 2001.
7. D.S. Johnson, Approximation algorithms for combinatorial problems, *Journal of Computer and Systems Science* 9, pp. 256–278, 1974.
8. H. Karloff and U. Zwick, A 7/8-approximation algorithm for MAX 3SAT?, In *Proc. 38th IEEE Symposium on the Foundations of Computer Science*, pp. 406–415, 1997.
9. U. Zwick, Outward rotations: a tool for rounding solutions of semidefinite programming relaxations, with applications to MAX CUT and other problems, In *Proc. 31st ACM Symposium on the Theory of Computing*, pp. 679–687, 1999.

Certifying Unsatisfiability of Random $2k$-SAT Formulas Using Approximation Techniques

Amin Coja-Oghlan[1], Andreas Goerdt[2], André Lanka[2], and Frank Schädlich[2]

[1] Humboldt-Universität zu Berlin, Institut für Informatik
Unter den Linden 6, 10099 Berlin, Germany
coja@informatik.hu-berlin.de
[2] Technische Universität Chemnitz, Fakultät für Informatik
Straße der Nationen 62, 09107 Chemnitz, Germany
{goerdt,lanka,frs}@informatik.tu-chemnitz.de

Abstract. Let k be an even integer. We investigate the applicability of approximation techniques to the problem of deciding whether a random k-SAT formula is satisfiable. Let n be the number of propositional variables under consideration. First we show that if the number m of clauses satisfies $m \geq Cn^{k/2}$ for a certain constant C, then unsatisfiability can be certified efficiently using (known) approximation algorithms for MAX CUT or MIN BISECTION. In addition, we present an algorithm based on the Lovász ϑ function that within polynomial expected time decides whether the input formula is satisfiable, provided $m \geq Cn^{k/2}$. These results improve previous work by Goerdt and Krivelevich [14]. Finally, we present an algorithm that approximates random MAX 2-SAT within expected polynomial time.

1 Introduction

The *k-SAT problem* is to decide whether a given k-SAT formula is satisfiable or not. Since it is well-known that the k-SAT problem is $\mathcal{NP}$-complete for $k \geq 3$, it is natural to ask for algorithms that can handle *random* formulas efficiently. Given a set of n propositional variables and a function $c = c(n)$, a random k-SAT instance is obtained by picking c k-clauses over the set of n variables uniformly at random and independently of each other. Part of the recent interest in random k-SAT is due to the interesting threshold behavior, in that there exist values $c_k = c_k(n)$ such that random k-SAT instances with at most $(1 - \varepsilon) \cdot c_k \cdot n$ random clauses are satisfiable with high probability, whereas for at least $(1 + \varepsilon) \cdot c_k \cdot n$ random clauses we have unsatisfiability with high probability. (Here, "with high probability" or "whp." means "with probability tending to 1 as n, the number of variables, tends to infinity"). In particular, according to current knowledge $c_k = c_k(n)$ lies in a bounded interval depending on k only. However, it is not known whether the threshold really is a constant independent of n, cf. [10]. In this paper, we are concerned with values of $c(n)$ well above the threshold, and the problem is to certify *efficiently* that a random formula is unsatisfiable.

A. Lingas and B.J. Nilsson (Eds.): FCT 2003, LNCS 2751, pp. 15–26, 2003.
© Springer-Verlag Berlin Heidelberg 2003

There are two different types of algorithms for deciding whether a random k-SAT formula is satisfiable or not. First, there are algorithms that on *any* input formula have a polynomial running time, and that whp. give the correct answer, "satisfiable" or "unsatisfiable". However, with probability $o(1)$, the algorithm may give an inconclusive answer. Hence, the algorithm never makes an incorrect decision. We shall refer to algorithms of this type as *efficient certification algorithms*. Note that the trivial constant time algorithm always returning "unsatisfiable" is not an efficient certification algorithm in our sense because it gives an incorrect answer in some (rare) cases. Secondly, there are algorithms that *always* answer correctly (either "satisfiable" or "unsatisfiable"), and that applied to a random formula have a polynomial *expected* running time.

Let us emphasize that although an efficient certification algorithm may give an inconclusive answer in some (rare) cases, such an algorithm is still *complete* in the following sense. Given a random k-SAT instance such that the number of clauses is above the satisfiability threshold, whp. the algorithm will indeed give the correct answer ("unsatisfiable" in the present case). Note that no polynomial time algorithm can answer "unsatisfiable" on all unsatisfiable inputs; completeness only refers to a subset whose probability tends to 1.

Any certification algorithm can be turned into a satisfiability algorithm that answers correctly on any input, simply by invoking an enumeration procedure in case that the efficient certification procedure gives an inconclusive answer. However, an algorithm obtained in this manner will not run in polynomial expected time in general. For the probability of an inconclusive answer may be too large (even though it is $o(1)$). Thus, asking for polynomial expected running time is a rather strong requirement.

From [11] and [14] it is essentially known that for random k-SAT instances with $\text{Poly}(\log n) \cdot n^{k/2}$ clauses we can efficiently certify unsatisfiability, in case of even k. For odd k we need $n^{(k/2)+\varepsilon}$ random clauses. Hence, it is an obvious problem to design algorithms that can certify unsatisfiability of random formulas efficiently for smaller numbers of clauses than given in [11,14]. To make further progress on this question, new techniques seem to be necessary. Therefore, in this paper we investigate what various algorithmic techniques contribute to the random k-SAT problem. We achieve some improvements for the case of even k, removing the polylogarithmic factor and achieving an algorithm with a polynomial expecteded running time.

Based on reductions from 4-SAT instances to instances of graph theoretic optimization problems we obtain efficient certification algorithms applying known approximation algorithms for the case of at least $C \cdot n^2$ 4-clauses. Similar constructions involving approximation algorithms can be found in [6] or [13]. We present two different certification algorithms. One applies the MAX CUT approximation algorithm of Goemans and Williamson [12]. The other one employs the MIN BISECTION approximation algorithm of Feige and Krauthgamer [8]. Since the MAX CUT approximation algorithm is based on semidefinite programming, our first algorithm is not purely combinatorial. In contrast, the application of the MIN BISECTION algorithm yields a combinatorial algorithm. We state our result only for $k = 4$, but it seems to be only a technical matter to extend it to arbitrary even numbers k and $C \cdot n^{k/2}$ clauses.

Moreover, we obtain the first algorithm for deciding satisfiability of random k-SAT formulas with at least $C \cdot n^{k/2}$ random clauses in expected polynomial time (k even). Indeed, the algorithm can even handle *semirandom* formulas, cf. Sec. 4 for details. Since the algorithm is based on computing the Lovász number ϑ, it is not purely combinatorial. The analysis is based on a recent estimate on the probable value of the ϑ-function of sparse random graphs [4].

The paper [2] is also motivated by improving the $n^{k/2}$ barrier. Further, in [9] another algorithm is given that certifies unsatisfiability of random 2k-SAT formulas consisting of at least $Cn^{k/2}$ clauses with probability tending to 1 as $C \rightarrow \infty$.

Though the decision version of the 2-SAT problem ("given a 2-SAT formula, is there a satisfying assignment?") can be solved in polynomial time, the optimization version MAX 2-SAT ("given a 2-SAT formula, find an assignment that satisfies the maximum number of clauses") is NP-hard. Therefore, we present an algorithm that approximates MAX 2-SAT in expected polynomial time. The algorithm is based on a probabilistic analysis of Goemans' and Williamson's semidefinite relaxation of MAX 2-SAT [12]. Concerning algorithms for worst case instances cf. [7].

In Section 2 we give our certification algorithms and in Section 3 we state the theorem crucial for their correctness. Section 4, which is independent of Sections 2 and 3, deals with the expected polynomial time algorithm. Finally, in Section 5 we consider the MAX 2-SAT problem.

2 Efficient Certification of Unsatisfiability

Given a set of n propositional variables, $\text{Var} = \text{Var}_n - \{v_1, \ldots, v_n\}$, a literal over Var is a variable v_i or a negated variable $\neg v_i$. A k-clause is an ordered k-tuple $l_1 \vee l_2 \vee \ldots \vee l_k$ of literals such that the variables underlying the literals are distinct. A k-SAT instance is a set of k-clauses. We think of a k-SAT instance as $C_1 \wedge C_2 \wedge \ldots \wedge C_m$ where each C_i is a k-clause. Given a truth value assignment a of Var, we can assign true or false to a k-SAT instance as usual. We let T_a be the set of variables x with $a(x) = \text{true}$ and F_a the set of variables x with $a(x) = \text{false}$. The probability space $\text{Form}_{n,k,p}$ is the probability space of k-SAT instances obtained by picking each k-clause with probability p independently.

A k-uniform hyperedge or simply k-tuple over the vertex set V is a vector $(x_1, x_2, \ldots, x_k)$ where the $x_i \in V$ are *all distinct*. $H = (V, E)$ is a k-uniform hypergraph if E is a set of k-tuples over the vertex set V. In the context of k-uniform hypergraphs we use the notion of *type* in the following sense: Let $X_1, X_2, \ldots, X_k \subseteq V$, a k-tuple $(x_1, x_2, \ldots, x_k)$ is of type $(X_1, X_2, \ldots, X_k)$ if we have for all i that $x_i \in X_i$. A random hypergraph $H \in HG_{n,k,p}$ is obtained by picking each of the possible $(n)_k$ k-tuples with probability p, independently.

Let S be a set of k-clauses over the set of variables Var, as defined above. The hypergraph $H = (V, E)$ associated to S is defined by $V = \text{Var}$ and

$(x_1, x_2, x_3, \ldots, x_k) \in E$ if and only if there is a k-clause $l_1 \vee l_2 \vee \ldots \vee l_k \in S$ such that for all i $l_i = x_i$ or $l_i = \neg x_i$. In case of even k, the graph $G = (V, E)$ associated to S is defined by $V = \{(x_1, \ldots, x_{k/2}) \mid x_i \in \text{Var and } x_i \neq x_j \text{ for } i \neq j\}$ and $\{(x_1, x_2, \ldots, x_{k/2}), (x_{(k/2)+1}, \ldots, x_k)\} \in E$ if and only if there is a k-clause $l_1 \vee l_2 \vee \ldots \vee l_k \in S$ such that the variable underlying l_i is x_i.

The following asymptotic abbreviations are used: $f(n) \sim_s g(n)$ iff there is an $\varepsilon > 0$ such that $f(n) = g(n) \cdot (1 + O(1/n^\varepsilon))$. Here $\sim_s$ stands for strong asymptotic equality. Similarly we use $f(n) = so(g(n))$ iff $f(n) = O(1/n^\varepsilon) \cdot g(n)$. We say $f(n)$ is negligible iff $f(n) = so(1)$.

Parity properties analogous to the next theorem have been proved in [6] for 3-SAT instances with a linear number of clauses and in [13] for 4-SAT instances. But in the proof of [13] it is important that the probability of each clause is $p \leq 1/n^{2+\varepsilon}$ where $\varepsilon > 0$ is a constant. This implies that the number of occurrences of two given literals in several clauses of a random formula is small. This is not any more the case for $p = C/n^2$ and some complications arise.

Theorem 1 (Parity Theorem). *For a random $F \in \text{Form}_{n,4,p}$ where $p = C/n^2$ and C is a sufficiently large constant, we can efficiently certify the following properties.*

(a) Let $S \subseteq F$ be the subset of all clauses of F corresponding to one of the 16 possibilities of placing negated and non-negated variables into the four slots of clauses available. Let $G = (V, E)$ be the graph associated to S. Then $|S| = C \cdot n^2 \cdot (1 + so(1))$ and $|E| = C \cdot n^2 \cdot (1 + so(1))$.

(b) For all satisfying assignments a of F we have that $|T_a| \sim_s (1/2) \cdot n$ and $|F_a| \sim_s (1/2) \cdot n$.

(c) Let S be the set of clauses of F consisting only of non-negated variables. Let H be the hypergraph associated to S. For all satisfying assignments a of F the number of 4-tuples of H of each of the 8 types (T_a, T_a, T_a, F_a), (T_a, T_a, F_a, T_a), (T_a, F_a, T_a, T_a), (F_a, T_a, T_a, T_a), (F_a, F_a, F_a, T_a), (F_a, F_a, T_a, F_a), (F_a, T_a, F_a, F_a), (T_a, F_a, F_a, F_a) is $(1/8) \cdot C \cdot n^2 \cdot (1 + so(1))$. The same statement applies when S is one of the remaining seven subsets of clauses of F which have a given even number of negated variables in a given subset of the four slots available.

(d) Let H be the hypergraph associated to those clauses of F whose first slot contains a negated variable and whose remaining three slots contain non-negated variables. The number of 4-tuples of H of each of the 8 types (T_a, T_a, T_a, T_a), (T_a, T_a, F_a, F_a), (T_a, F_a, T_a, F_a), (T_a, F_a, F_a, T_a), (F_a, F_a, F_a, F_a), (F_a, F_a, T_a, T_a), (F_a, T_a, F_a, T_a), (F_a, T_a, T_a, F_a) is $(1/8) \cdot C \cdot n^2 \cdot (1 + so(1))$. A statement analogously to (c) applies.

The technical notion type of a 4-tuple of a hypergraph is defined above. Statement (b) means that we have an $\varepsilon > 0$ such that we can certify that all assignments a with $|T_a| \geq (1/2) \cdot n \cdot (1 + 1/n^\varepsilon)$ or $|F_a| \geq (1/2) \cdot n \cdot (1 + 1/n^\varepsilon)$ do not satisfy a random F. Similarly for the remaining statements. Of course probabilistically there should be no satisfying assignment.

Given a graph $G = (V, E)$, a cut is a partition of V into two subsets V_1 and V_2. The MAX CUT problem is the problem to maximize the number of crossing edges, that is the number of edges with one endpoint in V_1 and the other endpoint in V_2. There is a polynomial time approximation algorithm which, given G, finds a cut such that the number of crossing edges is guaranteed to be at least $0.87 \cdot \mathrm{Opt}(G)$, see [12]. Note that the algorithm is deterministic.

Algorithm 2. Certifies unsatisfiability. The input is a 4-SAT instance F.

1. Certify the properties as stated in Theorem 1.

2. Let S be the subset of all clauses of F containing only non-negated variables. We construct the graph $G = (V, E)$ as defined above, associated to S.

3. Apply the MAX CUT approximation algorithm to G.

4. If the cut found in 3. contains at most $0.86 \cdot |E|$ edges the output is "unsatisfiable", otherwise the algorithm gives an inconclusive answer.

Theorem 3. *When applying Algorithm 2 to a $F \in Form_{n,4,p}$ where $p = C/n^2$ and C is sufficiently large the algorithm efficiently certifies the unsatisfiability of F.*

Proof. To show that the algorithm is correct, let F be any satisfiable 4-SAT instance. Let a be a satisfying truth value assignment of F. Then Theorem 1 (c) implies that G has a cut comprising almost all edges and the approximation algorithm finds sufficiently many edges, so that we do not answer "unsatisfiable". Completeness follows from Theorem 1 (c) and the fact that when C is a sufficiently large constant any cut of G has at most slightly more than a fraction of $1/2$ of all edges with high probability.   □

At this point we know that an algorithm efficiently certifying unsatisfiability exists, because there exist suitable $so(1)$-terms as we know from our theorems and considerations.

Given a graph $G = (V, E)$, where $|V|$ is even. A bisection of G is a partition of V into two subsets V_1 and V_2 with $|V_1| = |V_2| = |V|/2$. The MIN BISECTION problem is the problem to minimize the number of crossing edges. There is a polynomial time approximation algorithm which, given G, finds a bisection such that the number of crossing edges is guaranteed to be at most $O((\log n)^2) \cdot \mathrm{Opt}(G)$, $|V| = n$, see [8].

Algorithm 4. Certifies unsatisfiability. The input is a 4-SAT instance F.

1. Certify the properties as stated in Theorem 1.

2. Let S be the subset of all clauses of F whose first literal is a negated variable and whose remaining literals are non-negated variables. We construct the graph $G = (V, E)$ associated to this set S. Check if the maximal degree of G is at most $3 \cdot \ln n$.

3. Apply the MIN BISECTION approximation algorithm to G.

4. If the bisection found contains at least $(1/3) \cdot |E|$ edges, then the output is "unsatisfiable", otherwise inconclusive.

Theorem 3 applies analogously to Algorithm 4. Now, the proof relies on Theorem 1 (d).

3   Proof of the Parity Theorem

We present the algorithms to prove Theorem 1. To deal with the problem of multiple occurrences of pairs of variables in several clauses we need to work with labelled (multi-)graphs and labelled (multi-)hypergraphs. Here the edges between vertices are distinguished by labels.

Let $H = (V, E)$ be a standard 4-uniform hypergraph. When speaking of the projection of H onto coordinates 1 and 2 we think of H as a labelled multi-graph in which the labelled edge $\{x_1, x_2\}_{(x_1, x_2, x_3, x_4)}$ is present if and only if $(x_1, x_2, x_3, x_4) \in E$. We denote this projection by $G = (V, E)$.

Let $e = |E|$, $V = \{1, \ldots, n\}$, $X \subseteq V$, and $Y = V \setminus X$. We denote the number of labelled edges of G with one endpoint in X and the other endpoint in Y by $e(X, Y)$. Similarly $e(X)$ is the number of labelled edges with both endpoints from X. In an asymptotic setting we use our terminology from Section 2 and say that G has negligible discrepancy iff for all $X \subseteq V$ with $|X| = \alpha \cdot n$ where $\beta \leq \alpha \leq 1 - \beta$ and $Y = V \setminus X$ $e(X) \sim_s e\alpha^2$ and $e(X, Y) \sim_s 2e\alpha(1 - \alpha)$. Here $\beta > 0$ is a constant. This extends the discrepancy notion from page 71ff. of [3] to multigraphs. The $n \times n$-matrix $A = A_G$ is the adjacency matrix of G, where $A(x, y)$ is the number of labelled edges between x and y. As A is real valued and symmetric, A has n different eigenvectors and corresponding real eigenvalues which we consider ordered as $\lambda_{1,A} \geq \lambda_{2,A} \geq \cdots \geq \lambda_{n,A}$. We let $\lambda = \lambda_A = \max_{2 \leq i \leq n} |\lambda_{i,A}|$. In an asymptotic context we speak of strong eigenvalue separation with respect to a constant k. By this we mean that $\sum_{i=2}^{n} \lambda_i^k = so(\lambda_1^k)$. When k is even and constant, strong eigenvalue separation implies in particular that $\lambda = so(\lambda_1)$. It is known that for any $k \geq 0$ $\mathrm{Trace}(A^k) = \sum_{x=1}^{n} A^k(x, x) = \sum_{i=1}^{n} \lambda_{i,A}^k$. Moreover, the $\mathrm{Trace}(A^k)$ is equal to the number of closed walks of length k, that is k steps, in G.

The degree of the vertex x in G d_x is the number of labelled edges in which x occurs. The $n \times n$-matrix $L = L_G$ is a normalized adjacency matrix, it is related to the Laplacian matrix. We have $L(x, y) = A(x, y)/\sqrt{d_x d_y}$. As $L = L_G$ is real valued and symmetric, too, we use all the eigenvalue notation introduced for A analogously for L. Here $\lambda_{1,L} = 1$ is known. Let $d = d(n)$ be given. In an asymptotic context we say that G is almost d-regular, if for any vertex x of G $d_{x,G} = d(n) \cdot (1 + so(1))$. Theorem 5.1 and its corollaries on page 72/73 of [3] imply the following fact.

Fact 5. Let $G = (V, E)$ where $V = \{1, \ldots, n\}$ be a projection onto two coordinates of the 4-uniform hypergraph $H = (V, E)$ with $e = |E|$. Let G be almost d-regular, let $\beta \leq \alpha \leq 1 - \beta$ where $\beta > 0$ is a constant, and let $X \subseteq V$ with $|X| = \alpha n$. Then we have,

(a) $|e(X) - e\alpha^2| \leq \lambda_L \cdot e \cdot \alpha \cdot (1 + so(1))$,

(b) $|e(X, Y) - 2e\alpha(1 - \alpha)| \leq \lambda_L \cdot 2 \cdot e \cdot \sqrt{\alpha \cdot (1 - \alpha)} \cdot (1 + so(1))$ for $Y = V \setminus X$.

We need methods to estimate λ_L, they are provided by the next lemma.

Lemma 6. Let G be the projection onto two given coordinates of the 4-uniform hypergraph $H = (V, E)$ where $V = \{1, \ldots, n\}$. If G is almost d-regular and

A_G has strong eigenvalue separation with respect to a given constant k, then L_G has strong eigenvalue separation with respect to k.

Proof. Let W be the number of closed walks of length k in G. Then $W = \text{Trace}(A^k)$ and $\text{Trace}\left(L_G^k\right) = \sum_{x=1}^{n} L_G^k(x,x)$. An inductive argument shows that $\text{Trace}\left(L_G^k\right) \leq W \cdot (1/d)^k \cdot (1 + so(1))$. Then we get, $\sum_{i=1}^{n} \lambda_{i,L_G}^k \leq \left(\sum_{i=1}^{n} \lambda_{i,A_G}^k\right) \cdot \left(\frac{1}{d}\right)^k \cdot (1+so(1))$. As $\lambda_{1,L_G} = 1$, whereas $\lambda_{1,A_G}^k = d^k \cdot (1+so(1))$ we get that $\sum_{i=2}^{n} \lambda_{i,L_G}^k = so(1)$. Note that $\lambda_{1,A}$ is always at most the maximal degree of G and at least the minimal degree. $\qquad\square$

We collect some probabilistic properties of labelled projections when H is a random hypergraph. The proof follows known principles.

Lemma 7. *Let $p = c/n^2$ where c is a sufficiently large constant and let $H = (V,E)$ be a hypergraph from $HG_{n,4,p}$. Let $G = (V,E)$ be a labelled projection of H onto two coordinates. (a) Let $d = d(n) = 2 \cdot c \cdot n$. Then G is almost d-regular with probability at least $1 - e^{-\Omega(n^\varepsilon)}$ for a constant $\varepsilon > 0$. (b) The adjacency matrix $A = A_G$ has strong Eigenvalue separation with respect to $k = 4$ with high probability.*

Algorithm 8. Efficiently certifies negligible discrepancy with respect to a given constant β of projection graphs. Input is a 4-uniform hypergraph $H = (V,E)$. Let $G = (V,E)$ be the projection onto two given coordinates of H. Check almost d-regularity of G and check for the adjacency matrix A of G if $\text{Trace}\left(A^4\right) = d^4 \cdot (1 + so(1))$.

The correctness of the algorithm follows from Fact 5, the completeness when considering $HG_{n,4,p}$, where $p = C/n^2$, C sufficiently large, from Lemma 7 and Fact 5.

We need to certify discrepancy properties of projections onto 3 given coordinates of a random 4-uniform hypergraph from $HG_{n,4,p}$ where $p = c/n^2$. Let $H = (V,E)$ be a standard 4-uniform hypergraph. When speaking of the projection of H onto coordinates 1, 2, and 3, we think of H as a labelled 3-uniform hypergraph $G = (V,E)$ in which the labelled 3-tuple $(x_1,x_2,x_3)_{(x_1,x_2,x_3,x_4)}$ is present if $(x_1,x_2,x_3,x_4) \in E$. We restrict attention to the projection onto coordinates 1, 2 and 3 in the following. For $X,Y,Z \subseteq V$ we define $e_G(X,Y,Z) = |\{(x,y,z,-) \in E \mid (x,y,z) \text{ is of type } (X,Y,Z)\}|$. For the notion of type we refer to the beginning of Section 2. With $n = |V|$ and $e = |E|$ we say that the projection G has negligible discrepancy with respect to β if for all X with $|X| = \alpha n$, $\beta \leq \alpha \leq 1 - \beta$, and $Y = V \backslash X$ we have that $e_G(X,X,X) \sim_s \alpha^3 \cdot e$, $e_G(X,Y,X) \sim_s \alpha^2(1-\alpha) \cdot e$ and analogously for the remaining 6 possibilities of placing X and Y. For $1 \leq i \leq 3$ and $x \in V$ we let $d_{x,i}$ be the number of 4-tuples in E which have x in the i'th slot. Given $d = d(n)$, we say that G is almost d-regular if and only if $d_{x,i} = d \cdot (1 + so(1))$ for all $x \in V$ and all $i = 1,2,3$. We assign labelled product graphs to G.

Definition 9 (Labelled product). *Let $G = (V,E)$ be the projection onto coordinates 1, 2, and 3 of the 4-uniform hypergraph $H = (V,E)$.*

The labelled product of G with respect to the first coordinate is the labelled graph $P = (W, F)$, where $W = V \times V$ and F is defined as: For $x_1, x_2, y_1, y_2 \in V$ with $(x_1, y_1) \neq (x_2, y_2)$ we have $\{(x_1, y_1), (x_2, y_2)\}_{(h,k)} \in F$ iff $h = (z, x_1, x_2, -) \in E$ and $k = (z, y_1, y_2, -) \in E$ and (!) $h \neq k$.

If the projection G is almost d-regular the number of labelled edges of the product is $n \cdot d^2 \cdot (1 + so(1))$ provided $d \geq n^\epsilon$ for constant $\epsilon > 0$. Discrepancy notions for labelled products are totally analogous to those for labelled projection graphs defined above. Theorem 10 is an adaption of Theorem 3.2 in [13].

Theorem 10. *Let $\epsilon > 0$ and $d = d(n) \geq n^\epsilon$. Let $G = (V, E)$ with $|V| = n$ be the labelled projection hypergraph onto coordinates 1, 2 and 3 of the 4-uniform hypergraph $H = (V, E)$. Assume that G and H have the following properties. 1. G is almost d-regular. 2. The labelled projection graphs of H onto any two of the coordinates $1, 2$, and 3 have negligible discrepancy with respect to $\beta > 0$. 3. The labelled products of G have negligible discrepancy with respect to β^2. Then the labelled projection G has negligible discrepancy with respect to β.*

Lemma 11. *Let $H = (V, E)$ be a random hypergraph from $HG_{n,4,p}$ where $p = c/n^2$ and c is sufficiently large. Let G be the labelled projection of H onto the coordinates 1, 2, and 3. Let $P = (W, F)$ be the labelled product with respect to the first coordinate of G. Then we have*

(a) P is almost d-regular, where $d = 2 \cdot c^2 \cdot n$, with probability $1 - n^{-\Omega(\log \log n)}$.

(b) The adjacency matrix A_P has strong eigenvalue separation with respect to $k = 6$.

Proof. (a) We consider the vertex $(x_1, y_1) \in W$. First, assume that $x_1 \neq y_1$. We introduce the random variables,

$$X_z = |\{(z, x_1, -, -) \in E\}|, Y_z = |\{(z, y_1, -, -) \in E\}|$$
$$X'_z = |\{(z, -, x_1, -) \in E\}|, \; Y'_z = |\{(z, -, y_1, -) \in E\}|$$

and finally $D = \sum_z X_z \cdot Y_z + \sum_z X'_z \cdot Y'_z$. Then D is the degree of the vertex (x_1, y_1) in the labelled product. The claim follows with Hoeffding's bound [16], page 104, Theorem 7. For $x_1 = y_1$ we can argue similarly.

(b) Applying standard techniques we get $E[\text{Trace}(A_P^6)] = (2c^2 n)^6 + so(n^6)$. which with (a) implies strong Eigenvalue separation with respect to $k = 6$ with high probability. □

Algorithm 12. Certifies negligible discrepancy of labelled projections onto 3 coordinates of 4-uniform hypergraphs. The input is a 4-uniform hypergraph $H = (V, E)$. Let $G = (V, E)$ be the projection of H onto the coordinates $1, 2$, and 3.

1. Check if there is a suitable d such that G is almost d-regular. That is check if $d_{x,i} = d \cdot (1 + so(1))$ for all vertices x and all $i = 1, 2, 3$.

2. Check if the labelled projections onto any two of the coordinates $1, 2, 3$ of H have negligible discrepancy. Apply Algorithm 8.

3. Check if the products of G are almost d-regular with $d = 2c^2 n$.

4. For each of the 3 labelled products P of G check if Trace $\left(A_P^6\right) = (2c^2n)^6 \cdot (1 + so(1))$ where A_P is the adjacency matrix of P.

5. Successful certification for G iff all checks are positive.

Correctness of the algorithm follows with Theorem 10. Completeness for $HG_{n,k,p}$ with $p = C/n^2$ and C sufficiently large with Theorem 10, Lemma 11 whose proof shows that the property concerning the trace holds with high probability and implies strong eigenvalue separation.

Now we can prove Theorem 1. Theorem 1 (a) is trivial. Concerning Theorem 1 (b) we consider the following algorithm.

Algorithm 13. Certifies Theorem 1 (b). The input is a 4-SAT instance F. Let $H = (V, E)$ be the hypergraph associated to the subset of clauses which consist of unnegated variables only.

1. Check that the labelled projection of H onto coordinates $1, 2, 3$ has negligible discrepancy.

2. Check that the labelled projection of H onto coordinates $2, 3, 4$ has negligible discrepancy.

3. Do the same as 1. and 2. for the hypergraph associated to the clauses consisting only of negated variables.

4. If all checks have been successful, certify Theorem 1 (b).

Let F be any 4-SAT instance such that the algorithm is successful. Let a be an assignment with $|F_a| \geq (1/2) \cdot n \cdot (1 + \delta)$ where $\delta = \delta(n) > 0$ is not negligible in the sense of Section 2 (for example $\delta = 1/\log n$). From Step 1 we know that the fraction of 4-tuples of H of type $(F_a, F_a, F_a, -)$ is $((1/2) \cdot (1 + \delta))^3 \cdot (1 + so(1))$. Under the assumption that a satisfies F, the empty slot is filled with a variable from T_a. From Step 2 we know that the fraction of 4-tuples of H of type $(-, F_a, F_a, T_a)$ is $((1/2) \cdot (1 + \delta))^2 \cdot (1/2) \cdot (1 - \delta)$. As δ is not negligible this contradicts negligible discrepancy of the labelled projection onto coordinates 2, 3, and 4 of H. In the same way we can exclude assignments with more variables set to true than false because Step 3 is successful. Therefore the algorithm is correct. For random F the hypergraphs constructed are random hypergraphs and the completeness of Algorithm 12 implies the completeness of the algorithm.

Concerning Theorem 1 (c) we consider the following algorithm.

Algorithm 14. certifies parity properties. The input is a 4-SAT instance F.

1. Invoke Algorithm 13.

2. Let H be the hypergraph associated to the clauses of F consisting only of non-negated variables.

3. Certify that all 4 labelled projections onto any 3 different coordinates of H have negligible discrepancy (wrt. a suitable $\beta > 0$).

4. Certify that all 6 labelled projections onto any two coordinates of H have negligible discrepancy.

5. Certify Theorem 1 (c) if all preceeding checks are successful.

Correctness and completeness follow similarly as for the preceding algorithm. Those cases of Theorem 1 which are left open by now can be treated analogously and the Parity Theorem is proved.

4   Deciding Satisfiability in Expected Polynomial Time

Let $\mathrm{Var} = \mathrm{Var}_n = \{x_1, \ldots, x_n\}$ be a set of variables, and let $\mathrm{Form}_{n,k,m}$ denote a k-SAT formula chosen uniformly at random among all $(2n)^{k \cdot m}$ possibilities. Further, we consider *semirandom* formulas $\mathrm{Form}_{n,k,m}^+$, which are made up of a random share and a worst case part added by an adversary:

1. Choose $F_0 = C_1 \wedge \cdots \wedge C_m = \mathrm{Form}_{n,k,m}$ at random.
2. An adversary picks any formula $F = \mathrm{Form}_{n,k,m}^+$ over Var in which at least one copy of each C_i, $i = 1, \ldots, m$, occurs.

Note that in general we cannot reconstruct F_0 from F. We say that an algorithm $\mathcal{A}$ *has a polynomial expected running time applied to* $\mathrm{Form}_{n,k,m}^+$ if the expected running time remains bounded by a polynomial in the input length regardless of the decisions of the adversary.

Theorem 15. *Let $k \geq 4$ be an even integer. Suppose that $m \geq C \cdot 2^k \cdot n^{k/2}$, for some sufficiently large constant $C > 0$. There exists an algorithm DecideSAT that satisfies the following conditions.*

1. *Let F be any k-SAT instance over Var. If F is satisfiable, then DecideSAT(F) will find a satisfying assignment. Otherwise DecideSAT(F), will output "unsatisfiable".*
2. *Applied to $\mathrm{Form}_{n,k,m}^+$, DecideSAT runs in polynomial expected time.*

DecideSAT exploits the following connection between the k-SAT problem and the maximum independent set problem. Let $V = \{1, \ldots, n\}^{k/2}$, and $\nu = n^{k/2}$. Given any k-SAT instance F over Var_n we define two graphs $G_F = (V, E_F)$, $G_F' = (V, E_F')$ as follows. We let $\{(v_1, \ldots, v_{k/2}), (w_1, \ldots, w_{k/2})\} \in E_F$ iff the k-clause $x_{v_1} \vee \cdots \vee x_{v_{k/2}} \vee x_{w_1} \vee \cdots \vee x_{w_{k/2}}$ occurs in F. Similarly, $\{(v_1, \ldots, v_{k/2}), (w_1, \ldots, w_{k/2})\} \in E_F'$ iff the k-clause $\neg x_{v_1} \vee \cdots \vee \neg x_{v_{k/2}} \vee \neg x_{w_1} \vee \cdots \vee \neg x_{w_{k/2}}$ occurs in F. Let $\alpha(G)$ denote the independence number of a graph G.

Lemma 16. *[14] If F is satisfiable, then $\max\{\alpha(G_F), \alpha(G_F')\} \geq 2^{-k/2} n^{k/2}$.*

Let $G_{\nu,\mu}$ denote a graph with ν vertices and μ edges, chosen uniformly at random. We need the following slight extension of a lemma from [14].

Lemma 17. *Let $F \in \mathrm{Form}_{n,k,m}$ be a random formula.*

1. *Conditioned on $|E(G_F)| = \mu$, the graph G_F is uniformly distributed; i.e. $G_F = G_{\nu,\mu}$. A similar statement holds for G_F'.*
2. *Let $\varepsilon > 0$. Suppose that $2^k \cdot n^{k/2} \leq m \leq n^{k-1}$. Then with probability at least $1 - \exp(-\Omega(m))$ we have $\min\{|E(G_F)|, |E(G_F')|\} \geq (1 - \varepsilon) \cdot 2^{-k} \cdot m$.*

Thus, our next aim is to bound the independence number of a semirandom graph efficiently. Let $0 \le \mu \le \binom{\nu}{2}$. The semirandom graph $G_{\nu,\mu}^+$ is produced in two steps: First, choose a random graph $G_0 = G_{\nu,\mu}$. Then, an adversary adds to G_0 arbitrary edges, thereby completing $G = G_{\nu,\mu}^+$. We employ the *Lovász number* ϑ, which can be seen as a semidefinite programming relaxation of the independence number. Indeed, $\vartheta(G) \ge \alpha(G)$ for any graph G, and $\vartheta(G)$ can be computed in polynomial time [15]. Our algorithm DecideMIS, which will output "typical", if the independence number of the input graph is "small", and "not typical" otherwise, is based on ideas invented in [4,5].

Algorithm 18. DecideMIS(G, μ)

Input: A graph G of order ν, and a number μ. *Output:* "typical" or "not typical".

1. If $\vartheta(G) \le C'\nu(2\mu)^{-1/2}$, then terminate with output "typical". Here C' denotes some sufficiently large constant.
2. If there is no subset S of V, $|S| = 25\ln(\mu/\nu)\nu/\mu$, such that $|V \setminus (S \cup N(S))| > 12\nu(2\mu)^{-1/2}$, then output "typical" and terminate.
3. Check whether in G there is an independent set of size $12\nu(2\mu)^{-1/2}$. If this is not the case, then output "typical". Otherwise, output "not typical".

Proposition 19. *For any G, if DecideMIS(G, μ) outputs "typical", then we have $\alpha(G) \le C'\nu(2\mu)^{-1/2}$. Moreover, the probability that DecideMIS$(G_{\nu,\mu}^+, \mu)$ outputs "not typical" is $< \exp(-\nu)$. Applied to $G_{\nu,\mu}^+$, DecideMIS has a polynomial expected running time, provided $\mu \ge C''\nu$, for some constant $C'' > 0$.*

Proof. The proof goes along the lines of [5] and is based on the following facts (cf. [4]): Whp. we have $\vartheta(G_{\nu,\mu}) \le c_1\nu(2\mu)^{-1/2}$. Moreover, if M is a median of $\vartheta(G_{\nu,\mu})$, and if $\xi > 0$, then $\mathrm{Prob}[\vartheta(G_{\nu,p}) \ge M + \xi] \le 30\nu\exp(-\xi^2/(5M + 10\xi))$. To handle $G_{\nu,\mu}^+$, we make use of the monotonicity of ϑ (cf. [15]). $\square$

Algorithm 20. DecideSAT(F)

Input: A k-SAT formula F over Var_n.

Output: Either a satisfying assignment of F or "unsatisfiable".

1. Let $\mu = 2^{-k-1}m$. If both DecideMIS(G_F, μ) and DecideMIS(G_F', μ) answer "typical", then terminate with output "unsatisfiable".
2. Enumerate all 2^n assignments and look for a satisfying one.

Thus, Thm. 15 follows from Lemmas 16, 17 and Prop. 19.

5   Approximating Random MAX 2-SAT

Theorem 21. *Suppose that $m = Cx^2n$ for some large constant $C > 0$ and some constant $x > 0$. There is an algorithm ApxM2S that approximates MAX 2-SAT within a factor of $1 - 1/x$ for any formula $C \in \mathrm{Form}_{n,2,m}$ such that the expected running time of ApxM2S$(\mathrm{Form}_{n,2,m})$ is polynomial.*

The analysis of ApxM2S is based on the probabilistic analysis of the SDP relaxation SMS of MAX 2-SAT of Goemans and Williamson [12] (details omitted).

Algorithm 22. ApxM2S($\mathcal{C}$)
Input: An instance $\mathcal{C} \in \text{Form}_{n,2,m}$ of MAX 2-SAT.
Output: An assignment of $x_1, \ldots, x_n$.

1. Check whether the assignment x_i =true for all i satisfies at least $3m/4 - c_1\sqrt{mn}$ clauses of $\mathcal{C}$. If this is not the case, then go to 3. Here c_1 denotes some suitable constant.
2. Compute SMS($\mathcal{C}$). If SMS($\mathcal{C}$) $\leq 3m/4 + c_2\sqrt{mn}$, then output the assignment x_i =true for all i and terminate. Here c_2 denotes some suitable constant.
3. Enumerate all 2^n assignments of $x_1, \ldots, x_n$ and output an optimal solution.

References

1. Alon, N., Spencer J.: The Probabilistic Method. John Wiley and Sons 1992.
2. Ben-Sasson, E., Bilu, Y.: A Gap in Average Proof Complexity. ECCC 003 (2002).
3. Chung, F.R.K.: Spectral Graph Theory. American Mathematical Society 1997.
4. Coja-Oghlan, A.: The Lovasz number of random graphs. Hamburger Beiträge zur Mathematik **169**.
5. Coja-Oghlan, A., Taraz, A.: Colouring random graphs in expected polynomial time. Proc. STACS 2003, Springer LNCS **2607** 487–498.
6. Feige, U.: Relations between average case complexity and approximation complexity. Proc. 34th STOC (2002) 310–332.
7. Feige, U., Goemans, M. X.: Approximating the value of two prover proof systems, with applications to MAX 2SAT and MAX DICUT. Proc. 3rd Israel Symp. on Theory of Computing and Systems (1995) 182–189.
8. Feige, U., Krauthgamer, R.: A polylogarithmic approximation of the minimum bisection. Proc. 41st FOCS (2000) 105–115.
9. Feige, U., Ofek, E.: Spectral techniques applied to sparse random graphs, report MCS03-01, Weizmann Institute of Science (2003).
10. Friedgut., E.: Necessary and Sufficient Conditions for Sharp Thresholds of Graph Properties and the k-SAT problem. J. Amer. Math. Soc. **12** (1999) 1017–1054.
11. Friedman, J., Goerdt, A.: Recognizing more Unsatisfiable Random 3-SAT Instances efficiently. Proc. ICALP 2001, Springer LNCS **2076** 310–321.
12. Goemans, M.X., Williamson, D.P.: Improved approximation algorithms for maximum cut and satisfiability problems using semidefinite programming. J. ACM **42** 1115–1145.
13. Goerdt, A., Jurdzinski, T.: Some Results on Random Unsatisfiable k-SAT Instances and Approximation Algorithms Applied to Random Structures. Proc. MFCS 2002, Springer LNCS **2420** 280–291.
14. Goerdt, A., Krivelevich, M.: Efficient recognition of random unsatisfiable k-SAT instances by spectral methods. Proc. STACS 2001, Springer LNCS **2010** 294–304.
15. Grötschel, M., Lovász, L., Schrijver, A.: Geometric algorithms and combinatorial optimization. Springer 1988.
16. Hofri, M.: Probabilistic Analysis of Algorithms. Springer 1987.

Inapproximability Results for Bounded Variants of Optimization Problems

Miroslav Chlebík[1] and Janka Chlebíková[2][*]

[1] Max Planck Institute for Mathematics in the Sciences
Inselstraße 22-26, D-04103 Leipzig, Germany
[2] Christian-Albrechts-Universität zu Kiel
Institut für Informatik und Praktische Mathematik
Olshausenstraße 40, D-24098 Kiel, Germany
jch@informatik.uni-kiel.de

Abstract. We study small degree graph problems such as MAXIMUM INDEPENDENT SET and MINIMUM NODE COVER and improve approximation lower bounds for them and for a number of related problems, like MAX-B-SET PACKING, MIN-B-SET COVER, MAX-MATCHING in B-uniform 2-regular hypergraphs. For example, we prove NP-hardness factor of $\frac{95}{94}$ for MAX-3DM, and factor of $\frac{48}{47}$ for MAX-4DM; in both cases the hardness result applies even to instances with exactly two occurrences of each element.

1 Introduction

This paper deals with combinatorial optimization problems related to bounded variants of MAXIMUM INDEPENDENT SET (MAX-IS) and MINIMUM NODE COVER (MIN-NC) in graphs. We improve approximation lower bounds for small degree variants of them and apply our results to even highly restricted versions of set covering, packing and matching problems, including MAXIMUM-3-DIMENSIONAL-MATCHING (MAX-3DM).

It has been well known that MAX-3DM is MAX SNP-complete (or APX-complete) even when restricted to instances with the number of occurrences of any element bounded by 3. To the best of our knowledge, the first inapproximability result for bounded MAX-3DM with the bound 2 on the number of occurrences of any elements in triples, appeared in our paper [5], where the first explicit approximation lower bound for MAX-3DM problem is given. (For less restricted matching problem, MAX 3-SET PACKING, the similar inapproximability result for instances with 2 occurrences follows directly from hardness results for MAX-IS problem on 3-regular graphs [2], [3]). For B-DIMENSIONAL MATCHING problem with $B \geq 4$ the lower bounds on approximability were recently proven by Hazan, Safra and Schwartz [12]. A limitation of their method, as their explicitly state, is that it does not provide an inapproximability factor for

---

[*] The author has been supported by EU-Project ARACNE, Approximation and Randomized Algorithms in Communication Networks, HPRN-CT-1999-00112.

A. Lingas and B.J. Nilsson (Eds.): FCT 2003, LNCS 2751, pp. 27–38, 2003.
© Springer-Verlag Berlin Heidelberg 2003

3-DIMENSIONAL MATCHING. But just inapproximability factor for 3-dimensional case is of major interest, as it allows the improvement of hardness of approximation factors for several problems of practical interest, e.g. scheduling problems, some (even highly restricted) cases of Generalized Assignment problem, and other packing problems.

This fact, and an important role of small degree variants of MAX-IS (MIN-NC) problem as intermediate steps in reductions to many other problems of interest, are good reasons for trying to push our technique to its limits. We build our reductions on a restricted version of MAXIMUM LINEAR EQUATIONS over $\mathbb{Z}_2$ with 3 variables per equation and with the (large) constant number of occurrences of each variable. Recall that this method, based on the deep Håstad's version of PCP theorem, was also used to prove ($\frac{117}{116} - \varepsilon$)-approximability lower bound for TRAVELING SALESMAN problem by Papadimitriou and Vempala [14], and for our lower bound of $\frac{96}{95}$ for STEINER TREE problem in graphs [6].

In this paper we optimize our equation gadgets and their coupling via a *consistency amplifier*. The notion of consistency amplifier varies slightly from problem to problem. Generally, they are graphs with suitable expanding (or mixing) properties. Interesting quantities, in which our lower bounds can be expressed, are parameters of consistency amplifiers that provably exist.

Let us explain how our inapproximability results for bounded variants of MAX-IS and MIN-NC, namely B-MAX-IS and B-MIN-NC, imply the same bounds for some set packing, set covering and hypergraph matching problems. MAX SET PACKING (resp. MIN SET COVER) is the following: Given a collection $\mathcal{C}$ of subsets of a finite set S, find a maximum (resp., minimum) cardinality collection $\mathcal{C}' \subseteq \mathcal{C}$ such that each element in S is contained in at most one (resp., in at least one) set in $\mathcal{C}'$. If each set in $\mathcal{C}$ is of size at most B, we speak about B-SET PACKING (res. B-SET COVER).

It may be phrased also in hypergraph notation; the set of nodes is S and elements of $\mathcal{C}$ are hyperedges. In this notation a set packing is just a matching in the corresponding hypergraph. For a graph $G = (V, E)$ we define its dual hypergraph $\widetilde{G} = (E, \widetilde{V})$ whose node set is just E, $\widetilde{V} = \{\widetilde{v} : v \in V\}$, and for each $v \in V$ hyperedge $\widetilde{v}$ consists of all $e \in E$ such that $v \in e$ in G. Hypergraph $\widetilde{G}$ defined by this duality is clearly 2-regular, each node of $\widetilde{G}$ is contained exactly in two hyperedges. G is of maximum degree B iff $\widetilde{G}$ is of dimension B, in particular G is B-regular iff $\widetilde{G}$ is B-uniform. Independent sets in G are in one-to-one correspondence with matchings in $\widetilde{G}$ (hence with set packings, in set-system notation), and node covers in G with set covers for $\widetilde{G}$. Hence any approximation hardness result for B-MAX-IS translates via this duality to the one for MAX-B-SET PACKING (with exact 2 occurrences), or to MAX MATCHING in 2-regular B-dimensional hypergraphs. Similar is the relation of results on B-MIN-NC to MIN-B-SET COVER problem.

If G is B-regular edge B-colored graph, then $\widetilde{G}$ is, moreover, B-partite with balanced B-partition determined by corresponding color classes. Hence independent sets in such graphs correspond to B-dimensional matchings in natural way. Hence any inapproximability result for B-MAX-IS problem restricted to

B-regular edge-B-colored graphs translates directly to inapproximability result for MAX-B-DIMENSIONAL MATCHING (MAX-B-DM), even on instances with exact two occurrences of each element.

Our results for MAX-3DM and MAX-4DM nicely complement recent results of [12] on MAX-B-DM given for $B \geq 4$. To compare our results with their for $B = 4$, we have better lower bound ($\frac{48}{47}$ vs. $\frac{54}{53} - \varepsilon$) and our result applies even to highly restricted version with two occurrences. On the other hand, their hard gap result has *almost perfect completeness*.

The main new explicit NP-hardness factors of this contribution are summarized in the following theorem. In more precise parametric way they are expressed in Theorems 3, 5, 6. Better upper estimates on parameters from these theorems immediately improve lower bounds given bellow.

Theorem. It is NP-hard to approximate:

- MAX-3DM and MAX-4DM to within $\frac{95}{94}$ and $\frac{48}{47}$ respectively, both results apply to instances with exactly two occurrences of each element;
- 3-MAX-IS (even on 3-regular graphs) and MAX TRIANGLE PACKING (even on 4-regular line graphs) to within $\frac{95}{94}$;
- 3-MIN-NC (even on 3-regular graphs) and MIN-3-SET COVER (with exactly two occurrences of each element) to within $\frac{100}{99}$;
- 4-MAX-IS (even on 4-regular graphs) to within $\frac{48}{47}$;
- 4-MIN-NC (even on 4-regular graphs) and MIN-4-SET COVER (with exactly two occurrences) to within $\frac{53}{52}$;
- B-MIN-NC ($B \geq 3$) to within $\frac{7}{6} - 12\frac{\log B}{B}$.

Preliminaries

Definition 1. MAX-E3-LIN-2 *is the following optimization problem: Given a system I of linear equations over $\mathbb{Z}_2$, with exactly 3 (distinct) variables in each equation. The goal is to maximize, over all assignments φ to the variables, the ratio $\frac{sat(\varphi)}{|I|}$, where $sat(\varphi)$ is the number of equations of I satisfied by φ.*

We use the notation Ek-MAX-E3-LIN-2 for the same maximization problem, where each variable occurs exactly k times. The following theorem follows from Håstad's results [11], see [5] for more details

Theorem 1. *For every $\varepsilon \in \left(0, \frac{1}{4}\right)$ there is a constant $k(\varepsilon)$ such that for every $k \geq k(\varepsilon)$ the following problem is NP-hard: given an instance of Ek-MAX-E3-LIN-2, decide whether the fraction of more than $(1 - \varepsilon)$ or less than $(\frac{1}{2} + \varepsilon)$ of all equations is satisfied by the optimal (i.e. maximizing) assignment.*

To use all properties of our equation gadgets, the order of variables in equations will play a role. We denote by E$[k, k, k]$-MAX-E3-LIN-2 those instances of E3k-MAX-E3-LIN-2 for which each variable occurs exactly k times as the first variable, k times as the second variable and k times as the third variable in equations. Given an instance I_0 of Ek-MAX-E3-LIN-2 we can easily transform it into an instance I of $E[k, k, k]$-MAX-E3-LIN-2 with the same optimum, as follows: for any equation $x + y + z = j$ of I_0 we put in I the triple of equations $x + y + z = j$, $y + z + x = j$, and $z + x + y = j$. Hence the same NP-hard gap as in Theorem 1 applies for $E[k, k, k]$-MAX-E3-LIN-2 as well. We describe several reductions from $E[k, k, k]$-MAX-E3-LIN-2 to bounded occurrence instances of NP-hard problems that preserve the hard gap of $E[k, k, k]$-MAX-E3-LIN-2.

2 Consistency Amplifiers

As a parameter of our reduction for B-MAX-IS (or B-MIN-NC) ($B \geq 3$), and MAX-3DM, we will use a graph H, so called *consistency 3k-amplifier*, with the following structure:

(i)   The degree of each node is at most B.
(ii) There are $3k$ pairs of *contact nodes* $\{(c_0^i, c_1^i) : i = 1, 2, \ldots, 3k\}$.
(iii) The degree of any contact node is at most $B - 1$.
(iv) The first $2k$ pairs of contact nodes $\{(c_0^i, c_1^i) : i = 1, 2, \ldots, 2k\}$ are *implicitly linked* in the following sense: whenever J is an independent set in H, there is an independent set J' in H such that $|J'| \geq |J|$, a contact node c can belong to J' only if $c \in J$, and for any $i = 1, 2, \ldots, 2k$ at most one node of the pair (c_0^i, c_1^i) belongs to J'.
(v)   *The consistency property:* Let us denote $C_j := \{c_j^1, c_j^2, \ldots, c_j^{3k}\}$ for $j \in \{0, 1\}$, and $M_j := \max\{|J| : J$ is an independent set in H such that $J \cap C_{1-j} = \emptyset\}$. Then $M_1 = M_2 (:= M(H))$, and for every $\psi : \{1, 2, \ldots, 3k\} \to \{0, 1\}$ and for every independent set J in $H \setminus \{c_{1-\psi(i)}^i : i = 1, 2, \ldots, 3k\}$ we have $|J| \leq M(H) - \min\{|\{i : \psi(i) = 0\}|, |\{i : \psi(i) = 1\}|\}$.

Remark 1. Let $j \in \{0, 1\}$ and J be any independent set in $H \setminus C_{1-j}$ such that $|J| = M(H)$, then $J \supseteq C_j$. To show that, assume that for some $l \in \{1, 2, \ldots, 3k\}$ $c_j^l \notin J$. Define $\psi : \{1, 2, \ldots, 3k\} \to \{0, 1\}$ by $\psi(l) = 1 - j$, and $\psi(i) = j$ for $i \neq l$. Now (v) above says $|J| < M(H)$, a contradiction. Hence, in particular, C_j is an independent set in H.

To obtain better inapproximability results we use equation gadgets that require some further restrictions on degrees of contact nodes of a consistency $3k$-amplifier: (iii-1) For B-MAX-IS, $B \geq 6$, the degree of any contact node is at most $B - 2$. (iii-2) For B-MAX-IS, $B \in \{4, 5\}$, the degree of any contact node c_j^i with $i \in \{1, \ldots, k\}$ is at most $B - 1$, the degree of c_j^i with $i \in \{k + 1, \ldots, 3k\}$ is at most $B - 2$, where $j = 1, 2$.

For integers $B \geq 3$ and $k \geq 1$ let $\mathcal{G}_{B,k}$ stand for the set of corresponding consistency $3k$-amplifiers. Let $\mu_{B,k} := \min\{\frac{M(H)}{k} : H \in \mathcal{G}_{B,k}\}$, $\lambda_{B,k} := \min\{\frac{|V(H)| - M(H)}{k} : H \in \mathcal{G}_{B,k}\}$ (if $\mathcal{G}_{B,k} = \emptyset$, let $\lambda_{B,k} = \mu_{B,k} = \infty$), $\mu_B = \underline{\lim}_{k \to \infty} \mu_{B,k}$, and $\lambda_B = \underline{\lim}_{k \to \infty} \lambda_{B,k}$. The parameters μ_B and λ_B play a role of quantities in which our inapproximability results for B-MAX-IS and B-MIN-NC can be expressed. To obtain explicit lower bounds on approximability requires to find upper bounds on those parameters.

In what follows we describe some methods how consistency $3k$-amplifiers can be constructed. We will confine ourselves to highly regular amplifiers. This ensures that our inapproximability results apply to B-regular graphs for small values of B. We will look for a consistency $3k$-amplifier H as a bipartite graph with bipartition (D_0, D_1), where $C_0 \subseteq D_0$, $C_1 \subseteq D_1$ and $|D_0| = |D_1|$. The idea is that if D_j $(j = 0, 1)$ is significantly larger than $3k$ $(= |C_j|)$ then suitable probabilistic model of constructing bipartite graphs with bipartition (D_0, D_1) and prescribed degrees, will produce with high probability a graph H with good "mixing properties" that ensures the consistency property with $M(H) = |D_j|$. We will not develop probabilistic model here, rather we will rely on what has already been proved (using similar methods) for amplifiers. The starting point to our construction of consistency $3k$-amplifiers will be amplifiers, which were studied by Berman & Karpinski [3], [4] and Chlebík & Chlebíková [5].

Definition 2. *A graph $G = (V, E)$ is a $(2,3)$-graph if G contains only the nodes of degree 2 (contacts) and 3 (checkers). We denote Contacts $= \{v \in V : \deg_G(v) = 2\}$, and Checkers $= \{v \in V : \deg_G(v) = 3\}$. Furthermore, a $(2,3)$-graph G is an* amplifier *if for every $A \subseteq V$: $|\text{Cut } A| \geq |\text{Contacts} \cap A|$, or $|\text{Cut } A| \geq |\text{Contacts} \setminus A|$, where $\text{Cut } A = \{\{u, v\} \in E: \text{exactly one of nodes } u \text{ and } v \text{ is in } A\}$. An amplifier G is called a (k, τ)-amplifier if $|\text{Contacts}| = k$ and $|V| = \tau k$.*

To simplify proofs we will use in our constructions only such (k, τ)-amplifiers which contain no edge between contact nodes. Recall, that the infinite families of amplifiers with $\tau = 7$ [3], and even with $\tau \leq 6.9$ constructed in [5], are of this kind.

The consistency $3k$-amplifier for $B = 3$. Let a $(3k, \tau)$-amplifier $G = (V(G), E(G))$ from Definition 2 be fixed, and $x^1, \ldots, x^{3k}$ be its contact nodes. We assume, moreover, that there is a matching in G consisting of nodes $V(G) \setminus \{x^{2k+1}, \ldots, x^{3k}\}$. Let us point out that both, the wheel-amplifiers with $\tau = 7$ [3], and also their generalization given in [5] with $\tau \leq 6.9$, clearly contain such matchings.

Let one such matching $\mathcal{M} \subseteq E(G)$ be fixed from now on. Each node $x \in V(G)$ is replaced with a small gadget A_x. The gadget of $x \in V(G) \setminus \{x^{2k+1}, \ldots, x^{3k}\}$ is a path of 4 nodes x_0, X_1, X_0, x_1 (in this order). For $x \in \{x^{2k+1}, \ldots, x^{3k}\}$ we take as A_x a pair of nodes x_0, x_1 without an edge. Denote $E_x := \{x_0, x_1\}$ for each $x \in V(G)$, and $F_x := \{X_0, X_1\}$ for $x \in V(G) \setminus \{x^{2k+1}, \ldots, x^{3k}\}$. The union of gadgets A_x (over all $x \in V(G)$) contains already all nodes of our consistency $3k$-amplifier H, and some of its edges. Now we identify the remaining edges of H. For each edge $\{x, y\}$ of G we connect corresponding gadgets A_x, A_y with a pair of edges in H, as follows: if $\{x, y\} \in \mathcal{M}$, we connect X_0 with Y_1 and X_1 with Y_0; if $\{x, y\} \in E(G) \setminus \mathcal{M}$, we connect x_0 with y_1, and x_1 with y_0.

Having this done, one after another for each edge $\{x, y\} \in E(G)$, we obtain the consistency $3k$-amplifier $H = (V(H), E(H))$ with contact nodes x^i_j determined by contact nodes x^i of G, for $j \in \{0, 1\}$, $i \in \{1, 2, \ldots, 3k\}$. The proof of all conditions from the definition of a consistency $3k$-amplifier can be found in [7]. Hence, $\mu_3 \leq 40.4$, $\lambda_3 \leq 40.4$ follows from this construction.

The construction of the consistency amplifier for $B = 4$ is similar and can be also found in [7]. In this case $\mu_4 \leq 21.7$, $\lambda_4 \leq 21.7$ follows from the construction. We do not try to optimize our estimates for $B \geq 5$ in this paper, we are mainly focused on cases $B = 3$ and $B = 4$. For larger B we provide our inapproximability results based on small degree amplifiers constructed above. Of course, one can expect that amplifiers with much better parameters can be found for these cases by suitable constructions. We only slightly change the consistency $3k$-amplifier H constructed for case $B = 4$ to get some (very small) improvement for $B \geq 5$ case. Namely, also for $x \in \{x^{k+1}, x^{k+2}, \ldots, x^{2k}\}$ we take as A_x a pair of nodes connected by an edge. The corresponding c_0^i, c_1^i nodes of H will have degree 3 in H, but we will have now $M(H) = 3\tau k$. The same proof of consistency for H will work. This consistency amplifier H will be clearly simultaneously a consistency $3k$-amplifier for any $B \geq 5$. In this way we get the upper bound $\mu_B \leq 20.7$, $\lambda_B \leq 20.7$ for any $B \geq 5$.

3   The Equation Gadgets

In the reduction to our problems we use the equation gadgets G_j for equations $x + y + z = j$, $j = 0, 1$. To obtain better inapproximability results, we use slightly modified equation gadgets for distinct value of B in B-MAX-IS problem (or B-MIN-NC problem). For $j \in \{0, 1\}$ we define equation gadgets $G_j[3]$ for 3-MAX-IS problem (Fig. 1), $G_j[4]$ for 4(5)-MAX-IS (Fig. 2(i)), $G_j[6]$ for B-MAX-IS $B \geq 6$ (Fig. 2(ii)). In each case the gadget $G_1[*]$ can be obtained from $G_0[*]$ replacing each $i \in \{0, 1\}$ in indices and labels by $1 - i$.

For each $u \in \{x, y, z\}$ we denote by F_u the set of all accented u-nodes from G_j (hence F_u is a subset of $\{u_0', u_1', u_0'', u_1''\}$), and $F_u := \emptyset$ if G_j does not contain any accented u-node; $T_u := F_u \cup \{u_0, u_1\}$. For a subset A of nodes of G_j and any independent set J in G_j we will say that J is pure in A if all nodes of $A \cap J$ have the same lower index (0 or 1). If moreover, $A \cap J$ consists exactly of all nodes of A of one index, we say that J is full in A.

The following theorem describes basic properties of equation gadgets, the proof can be found in [7].

Theorem 2. *Let G_j ($j \in \{0, 1\}$) be one of the following gadgets: $G_j[3]$, $G_j[4]$, or $G_j[6]$, corresponding to an equation $x + y + z = j$. Let J be an independent set in G_j such that for each $u \in \{x, y\}$ at most one of two nodes u_0 and u_1 belongs to J. Then there is an independent set J' in G_j with the following properties:*

(I) $|J'| \geq |J|$,
(II)   for each $u \in \{x, y\}$ it holds $J' \cap \{u_0, u_1\} = J \cap \{u_0, u_1\}$,
(III) $J' \cap \{z_0, z_1\} \subseteq J \cap \{z_0, z_1\}$ and $|J' \cap \{z_0, z_1\}| \leq 1$,
(IV) J' contains (exactly) one special node, say $\psi(x)\psi(y)\psi(z)$. Furthermore, J' is pure in T_u and full in F_u.

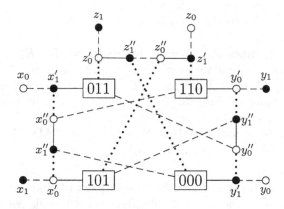

Fig. 1. The equation gadget $G_0 := G_0[3]$ for 3-MAX-IS and MAX-3DM.

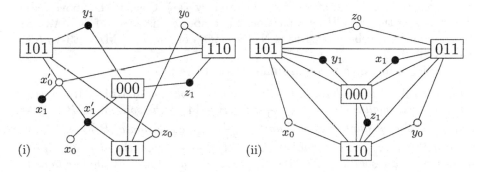

Fig. 2. The equation gadget (i) $G_0 := G_0[4]$ for B-MAX-IS, $B \in \{4,5\}$, (ii) $G_0 := G_0[6]$ for B-MAX-IS ($B \geq 6$).

4   Reduction for B-MAX-IS and B-MIN-NC

For arbitrarily small fixed $\varepsilon > 0$ consider k large enough such that conclusion of Theorem 1 for $E[k, k, k]$-MAX-E3-LIN-2 is satisfied. Further, let a consistency $3k$-amplifier H have $\frac{M(H)}{k}$ (resp. $\frac{|V(H)|-M(H)}{k}$) as close to μ_B (resp. λ_B) as we need. Keeping one consistency $3k$-amplifier H fixed, our reduction $f (= f_H)$ from $E[k, k, k]$-MAX-E3-LIN-2 to B-MAX-IS (resp., B-MIN-NC) is as follows: Let I be an instance of $E[k, k, k]$-MAX-E3-LIN-2, $\mathcal{V}(I)$ be the set of variables of I, $m := |\mathcal{V}(I)|$. Hence I has mk equations, each variable $u \in \mathcal{V}(I)$ occurs exactly in $3k$ of them: k times as the first variable, k times as the second one, and k times as the third variable in the equation. Assume, for convenience, that equations are numbered by $1, 2, \ldots, mk$. Given variable $u \in \mathcal{V}(I)$ and $s \in \{1, 2, 3\}$ let $r_s^1(u) < r_s^2(u) < \cdots < r_s^k(u)$ be the numbers of equations in which variable u occurs as the s-th variable. On the other hand, if for fixed $r \in \{1, 2, \ldots, mk\}$ the r-th equation is $x + y + z = j$ ($j \in \{0, 1\}$), there are uniquely determined numbers $i(x,r)$, $i(y,r)$, $i(z,r) \in \{1,2,\ldots,k\}$ such that $r_1^{i(x,r)}(x) = r_2^{i(y,r)}(y) = r_3^{i(z,r)}(z) = r$.

Take m disjoint copies of H, one for each variable. Let H_u denote a copy of H that correspondents to a variable $u \in \mathcal{V}(I)$. The corresponding contacts are in H_u denoted by $C_j(u) = \{u_j^i : i = 1,2,\ldots,3k\}$, $j = 0,1$. Now we take mk disjoint copies of equation gadgets G^r, $r \in \{1,2,\ldots,mk\}$. More precisely, if the r-th equation reads as $x + y + z = j$ ($j \in \{0,1\}$) we take as G^r a copy of $G_j[3]$ for 3-MAX-IS (or $G_j[4]$ for 4(5)-MAX-IS or $G_j[6]$ for B-MAX-IS, $B \geq 6$). Then the nodes x_0, x_1, y_0, y_1, z_0, z_1 of G^r are identified with nodes $x_0^{i(x,r)}$, $x_1^{i(x,r)}$ (of H_x), $y_0^{k+i(y,r)}$, $y_1^{k+i(y,r)}$ (of H_y), $z_0^{2k+i(z,r)}$, $z_1^{2k+i(z,r)}$ (of H_z), respectively. It means that in each H_u the first k-tuple of pairs of contacts corresponds to the occurrences of u as the first variable, the second k-tuple corresponds to the occurrences as the second variable, and the third one occurrences as the last variable. Making the above identification for all equations, one after another, we get a graph of degree at most B, denoted by $f(I)$. Clearly, the above reduction f (using the fixed H as a parameter) to special instances of B-MAX-IS is polynomial. It can be proved that NP-hard gap of $\mathrm{E}[k,k,k]$-MAX-E3-LIN-2 is preserved ([7]).

The following main theorem summarizes the results

Theorem 3. *It is* NP*-hard to approximate: the solution of* 3-MAX-IS *to within any constant smaller than* $1 + \frac{1}{2\mu_3 + 13}$*; for* $B \in \{4,5\}$ *the solution of* B-MAX-IS *to within any constant smaller than* $1 + \frac{1}{2\mu_B + 3}$*, the solution of* B-MAX-IS, $B \geq 6$*, to within any constant smaller than* $1 + \frac{1}{2\mu_B + 1}$*. Similarly, it is* NP*-hard to approximate the solution of* 3-MIN-NC *to within any constant smaller than* $1 + \frac{1}{2\lambda_3 + 18}$*, for* $B \in \{4,5\}$ *the solution of* B-MIN-NC *to within any constant smaller than* $1 + \frac{1}{2\lambda_B + 8}$*, the solution of* B-MIN-NC, $B \geq 6$*, to within any constant smaller than* $1 + \frac{1}{2\lambda_B + 6}$*.*

Using our upper bounds given for μ_B, λ_B for distinct value of B we obtain

Corollary 1. *It is* NP*-hard to approximate the solution of* 3-MAX-IS *to within* 1.010661 ($> \frac{95}{94}$)*; the solution of* 4-MAX-IS *to within* 1.0215517 ($> \frac{48}{47}$)*, the solution of* 5-MAX-IS *to within* 1.0225225 ($> \frac{46}{45}$) *and the solution of* B-MAX-IS, $B \geq 6$ *to within* 1.0235849 ($> \frac{44}{43}$)*. Similarly, it is* NP*-hard to approximate the solution of* 3-MIN-NC *to within* 1.0101215 ($> \frac{100}{99}$)*; the solution of* 4-MIN-NC *to within* 1.0194553 ($> \frac{53}{52}$)*; the solution of* 5-MIN-NC *to within* 1.0202429 ($> \frac{51}{50}$) *and* B-MIN-NC, $B \geq 6$*, to within* 1.021097 ($> \frac{49}{48}$)*. For each* $B \geq 3$*, the corresponding result applies to* B*-regular graphs as well.*

5   Asymptotic Approximability Bounds

This paper is focused mainly on graphs of very small degree. In this section we discuss also the asymptotic relation between hardness of approximation and degree for INDEPENDENT SET and NODE COVER problem in bounded degree graphs.

For the INDEPENDENT SET problem in the class of graphs of maximum degree B the problem is known to be approximable with performance ratio arbitrarily close to $\frac{B+3}{5}$ (Berman & Fujito, [2]). But asymptotically better ratios can be achieved by polynomial algorithms, currently the best one approximates to within a factor of $O(B \log \log B / \log B)$, as follows from [1], [13]. On the other hand, Trevisan [15] has proved NP-hardness to approximate the solution to within $B/2^{O(\sqrt{\log B})}$.

For the NODE COVER problem the situation is more challenging, even in general graphs. A recent result of Dinur and Safra [10] shows that for any $\delta > 0$ the MINIMUM NODE COVER problem is NP-hard to approximate to within $10\sqrt{5} - 21 - \delta$. One can observe that their proof can give hardness result also for graphs with (very large) bounded degree $B(\delta)$. This follows from the fact that after their use of Raz's parallel repetition (where each variable appears in only a constant number of tests), the degree of produced instances is bounded by a function of δ. But the dependence of $B(\delta)$ on δ in their proof is really very complicated. The earlier $\frac{7}{6} - \delta$ lower bound proved by Håstad [11] was extended by Clementi & Trevisan [9] to graphs with bounded degree $B(\delta)$.

Our next result improve on their; it has better trade-off between non-approximability and the degree bound. There are no hidden constants in our asymptotic formula, and it provides good explicit inapproximability results for degree bound B starting from few hundreds. First we need to introduce some notation.

Notation. Denote $F(x) := -x \log x - (1 - x) \log(1 - x)$, $x \in (0,1)$, where $\log$ means the natural logarithm. Further, $G(c,t) := (F(t) + F(ct))/(F(t) - ctF(\frac{1}{c}))$ for $0 < t < \frac{1}{c} < 1$, $g(t) := G(\frac{1-t}{t}, t)$ for $t \in (0, \frac{1}{2})$. More explicitly, $g(t) = 2[-t \log t - (1 - t) \log(1 - t)]/[-2(1 - t) \log(1 - t) + (1 - 2t) \log(1 - 2t)]$. Using Taylor series of the logarithm near 1 we see that the denominator here is $t^2 \cdot \sum_{k=0}^{\infty} \frac{2^{k+2}-2}{(k+1)(k+2)} t^k > t^2$, and $-(1-t)\log(1-t) = t - t^2 \sum_{k=0}^{\infty} \frac{1}{(k+1)(k+2)} t^k < t$, consequently $g(t) < \frac{2}{t}(1 \mid \log \frac{1}{t})$.

For large enough B we look for $\delta \in (0, \frac{1}{6})$ such that $3\lfloor g(\frac{\delta}{2}) \rfloor + 3 \le B$. As $g(\frac{1}{12}) \approx 75.62$ and g is decreasing in $(0, \frac{1}{12})$, we can see that for $B \ge 228$ any $\delta > \delta_B := 2g^{-1}(\lfloor \frac{B}{3} \rfloor)$ will do. Trivial estimates on δ_B (using $g(t) < \frac{2}{t}(1 + \log \frac{1}{t})$) are $\delta_B < \frac{12}{B-3}(\log(B - 3) + 1 - \log 6) < \frac{12 \log B}{B}$.

We will need the following lemma about regular bipartite expanders to prove the Theorem 4 (see [7] for proofs).

Lemma 1. *Let* $t \in (0, \frac{1}{2})$ *and* d *be an integer for which* $d > g(t)$. *For every sufficiently large positive integer* n *there is a* d-*regular* n *by* n *bipartite graph* H *with bipartition* (V_0, V_1), *such that for each independent set* J *in* H *either* $|J \cap V_0| \le tn$, *or* $|J \cap V_1| \le tn$.

Theorem 4. *For every* $\delta \in (0, \frac{1}{6})$ *it is* NP-*hard to approximate* MINIMUM NODE COVER *to within* $\frac{7}{6} - \delta$ *even in graphs of maximum degree* $\le 3\lfloor g(\frac{\delta}{2}) \rfloor + 3 \le 3\lceil \frac{4}{\delta}(1 + \log \frac{2}{\delta}) \rceil$. *Consequently, for any* $B \ge 228$ *it is* NP-*hard to approximate* B-MIN-NC *to within any constant smaller than* $\frac{7}{6} - \delta_B$, *where* $\delta_B := 2g^{-1}(\lfloor \frac{B}{3} \rfloor) < \frac{12}{B-3}(\log(B - 3) + 1 - \log 6) < 12\frac{\log B}{B}$.

Typically, the methods used for asymptotic results cannot be used for small values of B to achieve interesting lower bounds. Therefore we work on new techniques that improve the results of Berman & Karpinski [3] and Chlebík & Chlebíková [5].

6   MAX-3DM and Other Problems

Clearly, the restriction of B-MAX-IS problem to edge-B-colored B-regular graphs is a subproblem of MAXIMUM B-DIMENSIONAL MATCHING (see [5] for more details). Hence we want to prove that our reduction to B-MAX-IS problem can produce as instances edge-B-colored B-regular graphs. In this contribution we present results for $B = 3, 4$. For the equation $x + y + z = j$ ($j \in \{0, 1\}$) of $E[k, k, k]$-MAX-E3-LIN-2 we will use an equation gadget $G_j[B]$, see Fig. 1 and Fig. 2(i). The basic properties of these gadgets are described in Theorem 2.

Maximum 3-Dimensional Matching

As follows from Fig. 1 a gadget $G_0[3]$ can be edge-3-colored by colors a, b, c in such way that all edges adjacent to nodes of degree one (contacts) are colored by one fixed color, say a (for $G_1[3]$ we take the corresponding analogy). As an amplifier of our reduction $f = f_H$ from $E[k, k, k]$-MAX-E3-LIN-2 to MAX-3DM we use a consistency $3k$-amplifier $H \in \mathcal{G}_{3,k}$ with some additional properties: degree of any contact node is exactly 2, degree of any other node is 3 and moreover, a graph H is an edge-3-colorable by colors a, b, c in such way that all edges adjacent to contact nodes are colored by two colors b and c. Let $\mathcal{G}_{3DM,k} \subseteq \mathcal{G}_{3,k}$ be the class of all such amplifiers. Denote $\mu_{3DM,k} = \min\{\frac{M(H)}{k} : H \in \mathcal{G}_{3DM,k}\}$ and $\mu_{3DM} := \underline{\lim}_{k \to \infty} \mu_{3DM,k}$.

We use the same construction for consistency $3k$-amplifiers as was presented for 3-MAX-IS, but now we have to show that produced graph H fulfills conditions about coloring of edges. For fixed $(3k, \tau)$-amplifier G and the matching $\mathcal{M} \subseteq E(G)$ of nodes $V(G) \setminus \{x^{2k+1}, \ldots, x^{3k}\}$ we define edge coloring in two steps: (i) Take preliminary the following edge coloring: for each $\{x, y\} \in \mathcal{M}$ we color the corresponding edges in H as depicted on Fig. 3(i). The remaining edges of H are easily 2-colored by colors b and c, as the rest of the graph is bipartite and of degree at most 2. So, we have a proper edge-3-coloring but some edges adjacent to contacts are colored by color a. It will happen exactly if $x \in \{x^1, x^2, \ldots, x^{2k}\}$, $\{x, y\} \in \mathcal{M}$. (We assume that no two contacts of G are adjacent, hence y is a checker node of G.) Clearly, one can ensure that in the above extension of coloring of edges by colors c and b both other edges adjacent to x_0 and x_1 have the same color. (ii) Now we modify our edge coloring in all these violating cases as follows. Fix $x \in \{x^1, \ldots, x^{2k}\}$, $\{x, y\} \in \mathcal{M}$, and let both other edges adjacent to x_0 and x_1 have assigned color b. Then change coloring according Fig. 3(ii). The case when both edges have assigned color c, can be solved analogously (see Fig. 3(iii)). From the construction follows $\mu_{3DM} \leq 40.4$.

Keeping one such consistency $3k$-gadget H fixed, our reduction f ($= f_H$) from $E[k, k, k]$-MAX-E3-LIN-2 is exactly the same as for B-MAX-IS described

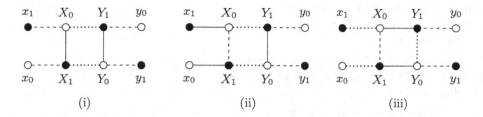

Fig. 3. a color: dashed line, b color: dotted line, c color: solid line

in Section 4. Let us fix an instance I of $E[k, k, k]$-MAX-E3-LIN-2 and consider an instance $f(I)$ of 3-MAX-IS. As $f(I)$ is edge 3-colored 3-regular graph, it is at the same time an instance of 3DM with the same objective function. We can show how the NP-hard gap of $E[k, k, k]$-MAX-E3-LIN-2 is preserved exactly in the same way as for 3-MAX-IS. Consequently it is NP-hard to approximate the solution of MAX-3DM to within $1 + (1 - 4\varepsilon)(\frac{2M(H)}{k} + 13 + 2\varepsilon)$, even on instances with each element occurring in exactly two triples.

Maximum 4-Dimensional Matching

We will use the following edge-4-coloring of our gadget $G_0[4]$ in Fig. 2(i) (analogously for $G_1[4]$): a-colored edges $\{x'_0, \boxed{101}\}$, $\{x'_1, \boxed{011}\}$, $\{y_1, \boxed{000}\}$, $\{y_0, \boxed{110}\}$; b-colored edges $\{x'_0, \boxed{110}\}$, $\{x'_1, \boxed{000}\}$, $\{y_1, \boxed{101}\}$, $\{y_0, \boxed{011}\}$; c-colored edges $\{x_1, x'_0\}$, $\{x_0, x'_1\}$, $\{\boxed{101}, \boxed{110}\}$, $\{z_0, \boxed{011}\}$, $\{z_1, \boxed{000}\}$; d-colored edges $\{x'_0, x'_1\}$, $\{\boxed{000}, \boxed{011}\}$, $\{z_0, \boxed{101}\}$, $\{z_1, \boxed{110}\}$. Now we will show that an edge-4-coloring of a consistency $3k$-amplifier H exists that fit well with the above coloring of equation gadgets. We suppose that the $(3k, \tau)$-amplifier G from which H was constructed has a matching $\mathcal{M}$ of all checkers. (This is true for amplifiers of [3] and [5]). The color d will be used for edges $\{x_0, x_1\}$, $x \in V(G) \setminus \{x^{2k+1}, \ldots, x^{3k}\}$. Also, for any $x \in \{x^{k+1}, \ldots, x^{2k}\}$, the corresponding $\{X_0, X_1\}$ edge will have color d too. The color c will be reserved for coloring edges of H "along the matching $\mathcal{M}$", i.e. if $\{x, y\} \in \mathcal{M}$, edges $\{x_0, y_1\}$ and $\{x_1, y_0\}$ have color c. Furthermore, for $x \in \{x^{k+1}, \ldots, x^{2k}\}$ the corresponding edges $\{x_0, X_1\}$ and $\{x_1, X_0\}$ will be of color c too. The edges that are not colored by c and d form a 2-regular bipartite graph, hence they can be edge 2-colored by colors a and b. The above edge 4-coloring of H and $G_j[4]$ ($j \in \{0, 1\}$) ensures that instances produced in our reduction to 4-MAX-IS are edge-4-colored 4-regular graphs.

The following theorem summarizes both achieved results:

Theorem 5. *It is* NP-*hard to approximate the solution of* MAX-3DM *to within any constant smaller than* $1 + \frac{1}{2\mu_{3DM}+13} > 1.010661 > \frac{95}{94}$, *and the solution of* MAX-4-DM *to within* 1.0215517 $(> \frac{48}{47})$. *The both inapproximability results hold also on instances with each element occurring in exactly two triples, resp. quadruples.*

Lower bound for MIN-B-SET COVER follows from that of B-MIN-NC, as was explained in Introduction. It is also easy to see that instances obtained by our reduction for 3-MAX-IS are 3-regular triangle-free graphs. Hence, we get the same lower bound for MAXIMUM TRIANGLE PACKING by simple reduction (see [5] for more details).

Theorem 6. *It is* NP-*hard to approximate the solution of the problems* MAXIMUM TRIANGLE PACKING *(even on 4-regular line graphs) to within any constant smaller than* $1 + \frac{1}{2\mu_3 + 13} > 1.010661 > \frac{95}{94}$, MIN-3-SET COVER *with exactly two occurrences of each element to within any constant smaller than* $1 + \frac{1}{2\lambda_3 + 13} > 1.0101215 > \frac{100}{99}$; *and* MIN-4-SET COVER *with exactly two occurrences of each element to within any constant smaller than* $1 + \frac{1}{2\lambda_4 + 8} > 1.0194553 > \frac{53}{52}$.

Conclusion Remarks. A plausible direction to improve further our inapproximability results is to give better upper bounds on parameters λ_B, μ_B. We think that there is still a potential for improvement here, using a suitable probabilistic model for the construction of amplifiers.

References

1. N. Alon and N. Kahale: *Approximating the independent number via the θ function*, Mathematical Programming **80**(1998), 253–264.
2. P. Berman and T. Fujito: *Approximating independent sets in degree 3 graphs*, Proc. of the 4th WADS, LNCS **955**, 1995, Springer, 449–460.
3. P. Berman and M. Karpinski: *On Some Tighter Inapproximability Results, Further Improvements*, ECCC Report TR98-065, 1998.
4. P. Berman and M. Karpinski: *Efficient Amplifiers and Bounded Degree Optimization*, ECCC Report TR01-053, 2001.
5. M. Chlebík and J. Chlebíková: *Approximation Hardness for Small Occurrence Instances of NP-Hard Problems*, Proc. of the 5th CIAC, LNCS **2653**, 2003, Springer (also ECCC Report TR02-73, 2002).
6. M. Chlebík and J. Chlebíková: *Approximation Hardness of the Steiner Tree Problem on Graphs*, Proc. of the 8th SWAT, LNCS **2368**, 2002, Springer, 170–179.
7. M. Chlebík and J. Chlebíková: *Inapproximability results for bounded variants of optimization problems*, ECCC Report TR03-26, 2003.
8. F. R. K. Chung: *Spectral Graph Theory*, CBMS Regional Conference Series in Mathematics, AMS, 1997, ISSN 0160-7642, ISBN 0-8218-0315-8.
9. A. Clementi and L. Trevisan: *Improved non-approximability results for vertex cover with density constraints*, Theoretical Computer Science **225**(1999), 113–128.
10. I. Dinur and S. Safra: *The importance of being biased*, ECCC Report TR01-104, 2001.
11. J. Håstad: *Some optimal inapproximability results*, Journal of ACM **48**(2001), 798–859.
12. E. Hazan, S. Safra and O. Schwartz: *On the Hardness of Approximating k-Dimensional Matching*, ECCC Report TR03-20, 2003.
13. D. Karger, R. Motwani and M. Sudan: *Approximate graph coloring by semi-definite programming*, Journal of the ACM **45(2)**(1998), 246–265.
14. C. H. Papadimitriou and S. Vempala: *On the Approximability of the Traveling Salesman Problem*, In Proc. 32nd ACM Symposium on Theory of Computing, Portland, 2000.
15. L. Trevisan: *Non-approximability results for optimization problems on bounded degree instances*, In Proc. 33rd ACM Symposium on Theory of Computing, 2001.

Approximating the Pareto Curve with Local Search for the Bicriteria TSP(1,2) Problem*

(Extended Abstract)

Eric Angel, Evripidis Bampis, and Laurent Gourvès

LaMI, CNRS UMR 8042, Université d'Évry Val d'Essonne, France

Abstract. Local search has been widely used in combinatorial optimization [3], however in the case of multicriteria optimization almost no results are known concerning the ability of local search algorithms to generate "good" solutions with performance guarantee. In this paper, we introduce such an approach for the classical traveling salesman problem (TSP) problem [13]. We show that it is possible to get in linear time, a $\frac{3}{2}$-approximate Pareto curve using an original local search procedure based on the 2-opt neighborhood, for the bicriteria TSP(1,2) problem where every edge is associated to a couple of distances which are either 1 or 2 [12].

1 Introduction

The traveling salesman problem (TSP) is one of the most popular problems in combinatorial optimization. Given a complete graph where the edges are associated with a positive distance, we search for a cycle visiting each vertex of the graph exactly once and minimizing the total distance. It is well known that the TSP problem is NP-hard and it cannot be approximated within a bounded approximation ratio, unless $P=NP$. However, for the metric TSP (i.e. when the distances satisfy the triangle inequality), Christofides proposed an algorithm with performance ratio 3/2 [1]. For more than 25 years, many researchers attempted to improve this bound but with no success. Papadimitriou and Yannakakis [12] studied a more restrictive version of the metric TSP, the case where all distances are either one or two, and they achieved a 7/6 approximation algorithm. This problem, known as the $TSP(1,2)$ problem, remains NP-hard, it is in fact this version of TSP that was shown NP-complete in the original reduction of Karp [2]. The $TSP(1,2)$ problem is a generalization of the hamiltonian cycle problem since we are asking for the tour of the graph that contains the fewest possible non-edges (edges of weight 2). More recently, Monnot et al. obtained results for the $TSP(1,2)$ with respect to the *differential* approximation ratio [8,9].

In this paper, we consider the bicriteria $TSP(1,2)$ problem which is a special case of the multicriteria TSP problem [14] in which every edge is associated to a

* Research partially supported by the thematic network APPOL II (IST 2001-32007) of the European Union, and the France-Berkeley Fund project MULT-APPROX.

A. Lingas and B.J. Nilsson (Eds.): FCT 2003, LNCS 2751, pp. 39–48, 2003.
© Springer-Verlag Berlin Heidelberg 2003

couple of distances which are either 1 or 2, i.e. each edge can take a value from the set $\{(1,1), (1,2), (2,1), (2,2)\}$. As an application consider two undirected graphs G_1 and G_2 on the same set V of n vertices. Does there exists a hamiltonian cycle which is common for both graphs? This problem can be formulated as a special case of the bicriteria traveling salesman problem we consider. Indeed, for $G = G_1$ or G_2 let $\delta_G([i,j]) = 1$ if there is an edge between vertices i and j in graph G and let $\delta_G([i,j]) = 0$ otherwise. We form a bicriteria TSP instance in a complete graph in the following way: consider any couple of vertices $\{i,j\} \in V^2$, we set the cost of edge $[i,j]$ to be $c([i,j]) = (2 - \delta_{G_1}([i,j]), 2 - \delta_{G_2}([i,j]))$. Then there exists a hamiltonian cycle common for both graphs if and only if there exists a solution for the bicriteria TSP achieving a cost (n,n). Here, we study the optimization version of this bicriteria TSP in which we look for a common "hamiltonian cycle" using the fewest possible non-edges in each graph, i.e. we are seeking a hamiltonian cycle in the complete graph of the $TSP(1,2)$ instance minimizing the cost of both coordinates. A solution of our problem is evaluated with respect to two different optimality criteria (see [5] for a recent book on multicriteria optimization). Here, we are interested in the trade-off between the different objective functions which is captured by the set of all possible solutions which are not dominated by other solutions (the so-called *Pareto curve*). Since the monocriterion $TSP(1,2)$ problem is NP-hard, determining whether a point belongs to the Pareto curve is NP-hard. Papadimitriou and Yannakakis [11] considered an approximate version of the Pareto curve, the so-called $(1 + \varepsilon)$-*approximate Pareto curve*. Informally, an $(1+\varepsilon)$-Pareto curve is a set of solutions that dominates all other solutions approximately (within a factor $1+\varepsilon$) in all the objectives. In other words, for every other solution, the considered set contains a solution that is as good approximately (within a factor $1 + \varepsilon$) in all objectives.

We propose a bicriteria *local search* procedure using the *2-opt neighborhood* which finds a 3/2-approximate Pareto curve (notice that a 2-approximate Pareto curve can be trivially constructed, just consider any tour). Interestingly, Khanna et al. [7] have shown that a local search algorithm using the 2-opt neighborhood achieves a 3/2 performance ratio, for the monocriterion $TSP(1,2)$ problem. We furthermore show that the gap between the cost of a local optimum produced by our local search procedure when compared to a solution of the exact Pareto curve is 3/2, and thus our result is tight. Up to the best of our knowledge, no results were known about the ability of local search algorithms to provide *good* (from the approximation –with performance guarantee– point of view) solutions in the area of multicriteria optimization.

1.1   Definitions

Given an instance of a multicriteria minimization problem, with $\gamma \geq 1$ objective functions $G_i, i = 1, \ldots, \gamma$, its Pareto curve P is the set of all γ-vectors (cost vectors) such that for each $v = (v_1, \ldots, v_\gamma) \in P$,

1. there exists a feasible solution s such that $G_i(s) = v_i$ for all i, and
2. there is no other feasible solution s' such that $G_i(s') \leq v_i$ for all i, with a strict inequality for some i.

For the ease of presentation, we will sometimes use P to denote a set of solutions which achieve these values. (If there is more than one solution with the same v_i values, P contains one of them.) Since for the problem we consider computing the (exact) Pareto curve is infeasible in polynomial time (unless $P=NP$), we consider an approximation. Given $\varepsilon > 0$, an $(1+\varepsilon)$-*approximate Pareto curve*, denoted $P_{(1+\varepsilon)}$, is a set of cost vectors of feasible solutions such that for every feasible solution s of the problem there is a solution s' with cost vector from $P_{(1+\varepsilon)}$ such that $G_i(s') \leq (1 + \varepsilon)G_i(s)$ for all $i = 1, ..., \gamma$.

2   Bicriteria Local Search

We consider the bicriteria $TSP(1, 2)$ with n cities. For an edge e, we shall denote by $c(e) \in \{(1,1), (1,2), (2,1), (2,2)\}$ its cost, and $c(e) = (c_1(e), c_2(e))$. The objective is to find a tour T (set of edges) minimizing $G_1(T) = \sum_{e \in T} c_1(e)$ and $G_2(T) = \sum_{e \in T} c_2(e)$. In the following we develop a local search based procedure in order to find a 3/2-approximate Pareto curve for this bicriteria problem.

We shall use the well known 2-opt neighborhood for the traveling salesman problem [4]. Given a tour T, its neighborhood $\mathcal{N}(T)$, is the set of all the tours which can be obtained from T by removing two non adjacent edges from T ($a = [x, y]$ and $b = [u, v]$ in Figure 1) and inserting two new edges ($c = [y, v]$ and $d = [x, u]$ in Figure 1) in order to obtain a new tour.

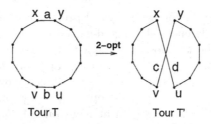

Fig. 1. The 2-opt move.

In the bicriteria setting there is a difficulty to define properly what is a local optimum. The natural preference relation over the set of tours, denoted $\prec_n$, is defined as follows.

Definition 1. *Let T and T' be two tours. One has $T' \prec_n T$ iff*

- $G_1(T') \leq G_1(T)$ *and* $G_2(T') < G_2(T)$, *or*
- $G_1(T') < G_1(T)$ *and* $G_2(T') \leq G_2(T)$.

If we consider this natural preference relation in order to define the notion of local optimum i.e. if we say that a tour T is a local optimum tour with respect to the 2-opt neighborhood whenever there does not exist a tour $T' \in \mathcal{N}(T)$ such that $T' \prec_n T$, then there exist instances for which a local optimum tour gives a performance guarantee strictly worse than 3/2 for one criterion.

Indeed, in Figure 2, the exact Pareto curve of the depicted instance contains only the tour $abcdefghij$ of weight $(10, 10)$. Thus, a 3/2-approximate Pareto curve of the instance should contain a single tour of weight strictly less than 16 for both criteria. Tours $aebicdfghj$ and $adjigfecbh$ are both local optima with respect to $\prec_n$ and their weights are respectively $(16, 10)$ and $(10, 16)$ (see Figure 2). Thus, using local optima with respect to $\prec_n$ is not appropriate to compute a 3/2-approximate Pareto curve of the considered problem (more details are given in the full paper).

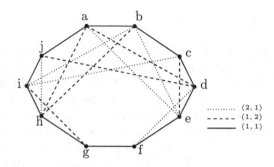

Fig. 2. Non represented egdes have a weight $(2, 2)$.

Hence, we introduce the following partial preference relations among the set of two edges. These preference relations, denoted by $\prec_1$ and $\prec_2$, are defined in Figure 3. The set of the ten possible couples of cost-vectors of the edges has been partitioned into three sets S_1, S_2 and S_3, and for any $s_1 \in S_1$, $s_2 \in S_2$, $s_3 \in S_3$, we have $s_1 \prec_1 s_2$, $s_1 \prec_1 s_3$ and $s_2 \prec_1 s_3$. Intuitively, preference relation $\prec_1$ (resp. $\prec_2$) means: pairs with at least one (1,1)-weighted edge in front of all others, and among the rest, pairs with at least one (1,2)-weighted edge (resp. (2,1)-weighted edge) in front.

Definition 2. *We say that the tour T is a local optimum tour with respect to the 2-opt neighborhood and the preference relation $\prec_1$ if there does not exist a tour $T' \in \mathcal{N}(T)$, obtained from T by removing edges a, b and inserting edges c, d, such that $\{c, d\} \prec_1 \{a, b\}$.*

A similar definition holds for the preference relation $\prec_2$.

We consider the following local search procedure:

BICRITERIA LOCAL SEARCH (BLS):

1. Let s_1 be a 2-opt local optimum tour with the preference relation $\prec_1$.
2. Let s_2 be a 2-opt local optimum tour with the preference relation $\prec_2$.
3. If $s_1 \prec_n s_2$ output $\{s_1\}$, if $s_2 \prec_n s_1$ output $\{s_2\}$, otherwise output $\{s_1, s_2\}$.

In order to find a local optimum tour, we start from an arbitary solution (say s). We look for a solution s' in the 2-opt neighborhood of s such that $s' \prec_1 s$ (resp. $s' \prec_2 s$) and replace s by s'. The procedure stops when such a solution s' does not exist, meaning that the solution s is a local optimum with respect to the preference relation $\prec_1$ (resp. $\prec_2$).

Notice that the proposed 2-opt neighborhood local search algorithm does not collapse to the traditional 2-opt neighborhood local search when applied to the monocriterion special case TSP with $c_1(e) = c_2(e)$ for all edges e. In this case our BLS algorithm does not replace a pair of edges with weights (1,1) and (2,2) by a pair of edges edges with weights (1,1) and (1,1), even if this move improves the quality of the tour. However allowing such moves does not improve the performance guarantee as the example in Figure 7 shows.

In the next section, we prove the next two theorems.

Theorem 1. *The set of solution(s) returned by the Bicriteria Local Search (BLS) procedure is a 3/2-approximate Pareto curve for the multicriteria TSP problem with distances one and two. Moreover, this bound is asymptotically sharp.*

Theorem 2. *The number of 2-opt moves performed by BLS is $O(n)$.*

3   Analysis of BLS

The idea of the proof of Theorem 1 is based (as in [7]) on the comparison of the number of the different types of cost vectors in the obtained local optimum solution(s) with the corresponding numbers with any other feasible solution (including the optimal one). In the following we assume that T is any 2-opt local optimal tour with respect to the preference relation $\prec_1$. The tour O is any fixed tour (in particular, one of the exact Pareto curve). Let us denote by x (resp. y, z and t) the number of (1,1) (resp. (1,2), (2,1) and (2,2)) edges in tour T. We denote with a prime the same quantities for the tour O.

Lemma 1. *With the preference relation $\prec_1$ one has $x \geq x'/2$.*

Proof. Let U_O (resp. U_T) be the set of $(1,1)$ edges in the tour O (resp. local optimum tour T). We define a function $f : U_O \to U_T$ in the following way. Let e be an edge in U_O. If $e \in U_T$ then $f(e) = e$. Otherwise let e' and e'' be the two edges adjacent to e in the tour T as depicted in Figure 4 (we assume an arbitrary orientation of T and consider that the only edges adjacent to e are e' and e'' and not e^4 and e^5). Let e''' be the edge forming a cycle of length 4 with e, e' and e'' (see Figure 4). We claim that there is at least one edge among e' and e'' with a weight $(1,1)$ and define $f(e)$ to be one of those edges (possibly chosen arbitrarily). Otherwise, we have $\{e, e'''\} \in S_1$ and $\{e', e''\} \in S_2 \cup S_3$ (see Figure 3), contradicting the fact that T is a local optimum with respect to the preference relation $\prec_1$. Now observe that for a given edge $e' \in U_T$, there can be at most two edges $e \in U_O$ such that $f(e) = e'$. Such a case occurs in Figures 5 and 6. Therefore we have $|U_T| \geq |U_O|/2$.   □

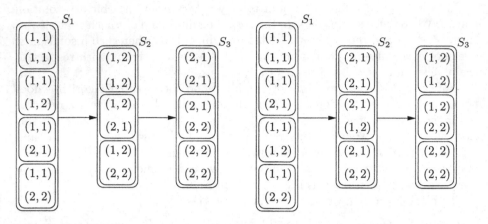

(a) The preference relation $\prec_1$.   (b) The preference relation $\prec_2$.

Fig. 3. The two preference relations $\prec_1$ and $\prec_2$.

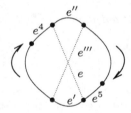

Fig. 4. The local optimal tour T (arbitrarily oriented).

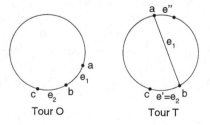

Tour O   Tour T

Fig. 5. $f(e_1) = f(e_2) = e'$ with $e_1, e_2 \in O$ and $e' \in T$

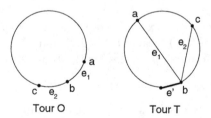

Tour O   Tour T

Fig. 6. $f(e_1) = f(e_2) = e'$ with $e_1, e_2 \in O$ and $e' \in T$

Lemma 2. *With the preference relation $\prec_2$ one has $x \geq x'/2$.*

Proof. The proof of Lemma 2 is symmetric to the one of Lemma 1, just assume that T is any 2-opt local optimal tour with respect to the preference relation $\prec_2$. $\qquad\square$

Lemma 3. *With the preference relation $\prec_1$ one has $x + y \geq (x' + y')/2$.*

Proof. Let U_O (resp. U_T) be the set of $(1,1)$ and $(1,2)$ edges in the tour O (resp. local optimum tour T). We define a function $f : U_O \to U_T$ in the following way. Let e be an edge in U_O. If $e \in U_T$ then $f(e) = e$. Otherwise let e' and e'' be the two edges adjacent to e in the tour T as depicted in Figure 4 (we assume an arbitrary orientation of T as in the proof of Lemma 1). Let e''' be the edge forming a cycle of length 4 with e, e' and e'' (see Figure 4). We claim that there is at least one edge among e' and e'' with a weight $(1,1)$ or $(1,2)$ and define $f(e)$ to be one of those edges (possibly chosen arbitrarily). Otherwise, we have $\{e, e'''\} \in S_1 \cup S_2$ and $\{e', e''\} \in S_3$ (see Figure 3), contradicting the fact that T is a local optimum with respect to the preference relation $\prec_1$. Now observe that for a given edge $e' \in U_T$, there can be at most two edges $e \in U_O$ such that $f(e) = e'$. Therefore we have $|U_T| \geq |U_O|/2$. $\qquad\square$

Proposition 1. *If the tour O has a cost $(X, X + \alpha)$ with X a positive integer $(n \leq X \leq 2n)$ and $n \geq \alpha \geq 0$, then the solution T achieves a performance guarantee of $3/2$ relatively to the solution O for both criteria.*

Proof. Let (C_O^1, C_O^2) be the cost of the tour O and (C_T^1, C_T^2) be the cost of the tour T. We have $C_T^1 = 2n - x - y$, $C_O^1 = 2n - x' - y'$ and $C_T^2 = 2n - x - z$, $C_O^2 = 2n - x' - z'$. Let us consider the first coordinate. We want to show that $\frac{C_T^1}{C_O^1} = \frac{2n - x - y}{2n - x' - y'} \leq \frac{3}{2}$. Using Lemma 3 we get $\frac{2n - x - y}{2n - x' - y'} \leq \frac{2n - \frac{x'}{2} - \frac{y'}{2}}{2n - x' - y'}$.

Now we have to show

$$\frac{2n - \frac{x'}{2} - \frac{y'}{2}}{2n - x' - y'} \leq \frac{3}{2} \iff 4n - x' - y' \leq 6n - 3x' - 3y'$$

$$\iff 2x' + 2y' \leq 2n$$

$$\iff x' + y' \leq n,$$

which is true since $x' + y' + z' + t' = n$ and $z', t' \geq 0$. We consider now the second coordinate. Since the tour O has a cost $(X, X + \alpha)$, it means that $C_O^2 = C_O^1 + \alpha$ and therefore $z' = y' - \alpha$. We have to show

$$\frac{2n - x - z}{2n - x' - z'} \leq \frac{3}{2} \iff 4n - 2x - 2z \leq 6n - 3x' - 3z'$$

$$\iff 3x' - 2x + 3z' - 2z \leq 2n$$

$$\iff 3x' - 2x + 3y' - 3\alpha - 2z \leq 2(x' + y' + z' + t')$$

$$\iff x' - 2x - y' - \alpha - 2z \leq 2t',$$

which is true since $x' - 2x \leq 0$ by Lemma 1. $\qquad\square$

We assume now that T is any 2-opt local optimal tour with respect to the preference relation $\prec_2$. The tour O is any fixed tour. In a similar way as in the case of Lemma 3 we can prove:

Lemma 4. *With the preference relation $\prec_2$ one has $x + z \geq (x' + z')/2$.*

Proof. The proof of Lemma 4 is symmetric to the one of Lemma 3.   □

Proposition 2. *If the tour O has a cost $(X + \alpha, X)$ with X a positive integer $(n \leq X \leq 2n)$ and $\alpha > 0$, then the solution T achieves a performance guarantee of $3/2$ relatively to the solution O for both criteria.*

Proof. The proof of Proposition 2 is symmetric to the one of Proposition 1, using Lemma 4 and Lemma 2 instead of Lemma 3 and Lemma 1.   □

Now, we are ready to prove Theorems 1 and 2.

Proof of Theorem 1.

Proof. Let s be an arbitrary tour. If s has a cost $(X, X + \alpha)$, $\alpha \geq 0$, then using Proposition 1 the solution s_1 $3/2$-approximately dominates the solution s. Otherwise, s has a cost $(X + \alpha, X)$, $\alpha > 0$, and using Proposition 2 the solution s_2 $3/2$-approximately dominates the solution s.

To see that this bound is asymptotically sharp consider the instance depicted in Figure 7. The tour $s_1 s_2 \ldots s_{2n} s_1$ is a local optimum with respect to $\prec_1$ and $\prec_2$, and it has a weight $n \times (1,1) + n \times (2,2) = (3n, 3n)$, whereas the optimal tour

$$s_1 s_3 s_2 s_4 s_{2n-1} \cdots s_{n-1} s_{n+4} s_n s_{n+3} s_{n+1} s_{n+2} s_2 s_1$$

has a weight $(2n - 1) \times (1,1) + (2,2) = (2n + 1, 2n + 1)$.   □

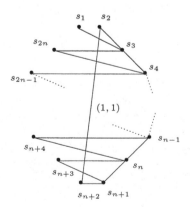

Fig. 7. The edges represented have a weight $(1,1)$, whereas non represented edges have a weight $(2,2)$.

Proof of Theorem 2.

Proof. Let T be a tour. Let $F_1(T) = 3x + y$ with x (resp. y) the number of $(1, 1)$ edges (resp. $(1, 2)$ edges) of T. We assume that one 2-opt move, with respect to $\prec_1$, transforms T to T'. Then it is easy to see that one has $F_1(T') \geq F_1(T) + 1$ for any such 2-opt move. Indeed, each 2-opt move with respect to $\prec_1$ increases either the number of $(1, 2)$ without decreasing the number of $(1, 1)$, or increases the number of $(1, 1)$ edges by decreasing the number of $(1, 2)$ edges by at most two. Since $0 \leq F_1(T) \leq 3(x + y) \leq 3n$ and $F_1(T) \in \mathbf{N}$, a local search which uses $\prec_1$ converges to a local optimum solution in less than $3n$ steps.

One can use the same proof with $\prec_2$, just assume that $F_2(T) = 3x + z$ with x (resp. z) the number of $(1, 1)$ edges (resp. $(2, 1)$ edges) of a tour T. □

4   Concluding Remarks

In this paper we proposed a bicriteria local search procedure based on the standard 2-opt neighborhood which allowed to get a 3/2-approximate Pareto curve for the bicriteria $TSP(1, 2)$. Our results can be extended to the $TSP(a, a + \delta)$ with $a \in \mathbf{R}^{+*}$ and $0 \leq \delta \leq a$. In that case we obtain an $1 + \frac{\delta}{2a}$-approximate Pareto curve. Since Chandra et al. [6] have shown that for the TSP satisfying the triangle inequality, the worst-case performance ratio of 2-opt (resp. k-opt) local search is at most $4\sqrt{n}$ and at least $\frac{1}{4}\sqrt{n}$ (resp. $\frac{1}{4}n^{\frac{1}{2k}}$), our constant approximation result cannot be extended for the metric case. It would be however interesting to establish lower and upper bounds for this more general case.

Our results can also be applied to the bicriteria version of the $MAX\ TSP$ $(1, 2)$ problem. In this problem, the objective is the maximization of the length of the tour. For the monocriterion case the best approximation algorithm known has a performance ratio of 7/8 [8,9] (the previously known approximation algorithm had a performance ratio of 3/4 [10]). We can obtain for the bicriteria case a 2/3-approximate Pareto curve in the following way. The idea is to modify the instance by replacing each edge $(2,2)$ by an edge $(1,1)$, each edge $(1,1)$ by and edge $(2,2)$, and each edge $(1,2)$ by an edge $(2,1)$ and *vice et versa*. It can be shown that obtaining a 3/2-approximate Pareto curve for the bicriteria MIN $TSP(1, 2)$ on this modified instance yields a 2/3-approximate Pareto curve for the bicriteria $MAX\ TSP(1, 2)$ on the original instance. This is equivalent to say that we work on the original instance, but using modified preference relations $\prec'_1$ and $\prec'_2$ obtained from $\prec_1$ and $\prec_2$ by replacing each edge $(2,2)$ by an edge $(1,1)$, each edge $(1,2)$ by an edge $(2,1)$, and *vice et versa*.

An interesting question is whether it is possible to obtain constant approximation ratios for the more general k-criteria $TSP(1, 2)$ problem (for $k > 2$). It seems that our approach cannot be directly applied to this case.

References

1. N. Christofides. Worst-Case analysis of a new heuristic for the traveling salesman problem. Technical Report, GSIA, Carnegie Mellon University, 1976.
2. R.M. Karp. Reducibility among combinatorial problems. Complexity of Computer Computations, R.E. Miller and J.W. Thatcher (Eds.), Pluner, NY, 1972.
3. E. Aarts and J.K. Lenstra, Local search in combinatorial optimization, John Wiley and Sons, 1997.
4. D.S. Johnson and L.A. McGeoch, The traveling salesman problem: a case study in Local Optimization, chapter in Local search in combinatorial optimization, E. Aarts and J.K. Lenstra (eds.), John Wiley and Sons, 1997.
5. M. Ehrgott Multicriteria Optimization, Lecture Notes in Economics and Mathematical Systems, vol. 491, Springer, 2000.
6. B. Chandra, H. Karloff and C. Tovey, New results on the old k-opt algorithm for the TSP, SIAM Journal on Computing, 28(6), 1998–2029, 1999.
7. S. Khanna, R. Motwani, M. Sudan and V. Vazirani, On syntactic versus computational views of approximability, SIAM Journal on Computing, 28(1), 164–191, 1998.
8. J. Monnot, Differential approximation results for the traveling salesman and related problems, Information Processing Letters, 82(5), 229–235, 2002.
9. J. Monnot, V. Th. Paschos and S. Toulouse, Differential approximation results for the traveling salesman problem with distances 1 and 2, European Journal of Operational Research, 145, 557–568, 2003.
10. A.I. Serdyukov, An algorithm with an estimate for the traveling salesman problem of the maximum, Upravlyaemye Sistemy, 25, 80–86, 1984.
11. C.H. Papadimitriou and M. Yannakakis, On the approximability of trade-offs and optimal access of web sources, Proceedings 41th Annual IEEE Symposium on Foundations of Computer Science, 86–92, 2000.
12. C.H. Papadimitriou and M. Yannakakis. The traveling salesman problem with distances one and two. In *Mathematics of Operations Research*, 18(1), 1–11, 1993.
13. C.H. Papadimitriou, S. Vempala. On the approximability of the traveling salesman problem. Proc. STOC'00, 126–133, 2000.
14. A. Gupta, A. Warburton. Approximation methods for multiple criteria traveling salesman problems, Towards Interactive and Intelligent Decision Support Systems, Proc. of the 7th International Conference on Multiple Criteria Decision Making, (Y. Sawaragi Ed.), Springer Verlag, 211–217, 1986.

Scheduling to Minimize Max Flow Time: Offline and Online Algorithms*

Monaldo Mastrolilli

IDSIA, Galleria 2, 6928 Manno, Switzerland
monaldo@idsia.ch

Abstract. We investigate the max flow scheduling problem in the off-line and on-line setting. We prove positive and negative theoretical results. In the off-line setting, we address the unrelated parallel machines model and present the first known fully polynomial time approximation scheme, when the number of machines is fixed. In the on-line setting and when the machines are identical, we analyze the *First In First Out* (FIFO) heuristic when preemption is allowed. We show that FIFO is an on-line algorithm with a $(3 - 2/m)$-competitive ratio. Finally, we present two lower bounds on the competitive ratio of deterministic on-line algorithms.

1 Introduction

The m-machine scheduling problem is one of the most widely-studied problems in computer science, with an almost limitless number of variants ([3,6,12,18] are surveys). The most common objective function is the *makespan*, which is the length of the schedule, or equivalently the time when the last job is completed. This objective function formalizes the viewpoint of the *owner* of the machines. If the makespan is small, the utilization of his machines is high; this captures the situation when the benefits of the owner are proportional to the work done. If we turn our attention to the viewpoint of a *user*, the time it takes to finish individual jobs may be more important; this is especially true in interactive environments. Thus, if many jobs that are released early are postponed at the end of the schedule, it is unacceptable to the user of the system even if the makespan is optimal.

For that reason other objective functions are studied. With this aim, a well-studied objective function is the *total flow time* [1,13,17]. The flow time of a job is the time the job is in the system, i.e., the completion time minus the time when it becomes first available. The above mentioned objective function is the sum of these values over all jobs. The *Shortest Remaining Processing Times* (SRPT) heuristic produces a schedule with optimum total flow time (see [12]) when there is a single processor. Unfortunately, this heuristic has the well-known

---

* Supported by the "Metaheuristics Network", grant HPRN-CT-1999-00106, and by Swiss National Science Foundation project 20-63733.00/1, "Resource Allocation and Scheduling in Flexible Manufacturing Systems".

A. Lingas and B.J. Nilsson (Eds.): FCT 2003, LNCS 2751, pp. 49–60, 2003.
© Springer-Verlag Berlin Heidelberg 2003

drawback that it leads to *starvation*. That is, some jobs may be delayed to an unbounded extent. Inducing starvation is an inherent property of the total flow time metric. In particular, there exists inputs where any optimal schedule for total flow time forces the starvation of some job (see Lemma 2.1 in [2]). This property is undesirable.

From the discussion above, it is natural to conclude that in order to avoid starvation, one should bound the flow time of each job. This motivates the study of the minimization of the *maximum flow time*.

Problems: We address three basic types of parallel machine models. In each there are n jobs $J_1, ..., J_n$ to be scheduled on m machines $M_1, ..., M_m$. Each machine can process at most one job at a time, and each job must be processed in an uninterrupted fashion on one of the machines. We will also consider the *preemptive* case, in which a job may be interrupted on one machine and continued later (possibly on another machine) without penalty. Job J_j ($j = 1, ..., n$) is released at time $r_j \geq 0$ and cannot start processing before that time. In the most general setting, the machines are *unrelated*: job J_j takes $p_{ij} = p_j/s_{ij}$ time units when processed by machine M_i, where p_j is the processing requirement of job J_j and s_{ij} is the speed of machine M_i for job J_j. If the machines are *uniformly related*, then each machine M_i runs at a given speed s_i for all jobs. Finally, for *identical* machines, we assume that $s_i = 1$ for each machine M_i.

We denote the completion time of job J_j in a schedule S by C_j^s or C_j, if no confusion is possible. The flow time of job J_j is defined as $F_j = C_j - r_j$, and the *maximum flow time* $F_{\max}$ is $\max_{j=1,...,n} F_j$. We seek to minimize the maximum flow time.

In the *off-line* version of the problem, it is assumed that the scheduler has full information of the problem instance. By contrast, in the *on-line* version of the problem, jobs are introduced to the algorithm at their release times. Thus, the algorithm bases its decision only upon information related to already released jobs. In the on-line paradigm, we distinguish between the *clairvoyant* and *non-clairvoyant* model. In the clairvoyant model we assume that once a job is known to the scheduler, its processing time is also known. In the non-clairvoyant model the processing time of a job is unknown until its processing is completed.

Previous Work: To the best of our knowledge, the only known result about the non-preemptive max flow time scheduling problem is due to Bender et al. [2]. They address the on-line non-preemptive problem with identical parallel machines (in the notation of Graham et al. [6], this problem is noted $P|$on-line; $r_j|F_{\max}$). In [2] they claim that the *First In First Out* (FIFO) heuristic (that is, scheduling jobs in the order they arrive to the machine on which they will finish first) is a $(3 - 2/m)$-competitive algorithm[1].

When preemption is allowed, in each of the three types of parallel models, we observe that there are polynomial-time off-line algorithms for finding optimal

---

[1] A ρ-competitive algorithm is an on-line algorithm that finds a solution within a ρ factor of the optimum.

preemptive solutions: these are obtained by adapting the approaches proposed in [14,15] for the preemptive parallel machines problems with release times and deadlines. In [14,15] the objective function is the minimization of the maximum lateness $L_{max} = \max L_j$, where L_j is the lateness of job J_j, that is the completion time of J_j minus the its deadline (the time by which job J_j must be completed). We can use the algorithms in [14,15] for the preemptive maximum flow time minimization by setting the deadline of each job equal to its release time.

When the jobs release times are identical, the problem reduces to the classical makespan minimization problem. In this case the three types of parallel machine models have been studied extensively (see [3,6,12,18] for surveys). Here, we only mention that these related scheduling problems are all strongly NP-hard [5], and polynomial time approximation schemes[2] (PTAS) are known when the machines are either identical or uniformly related [7,8]. For unrelated machines, Lenstra, Shmoys and Tardos [16] gave a polynomial-time 2-approximation algorithm for this problem; and this is the currently known best approximation ratio achieved in polynomial time. They also proved that for any positive $\varepsilon < 1/2$, no polynomial-time $(1+\varepsilon)$-approximation algorithm exists, unless P=NP. Since the problem is NP-hard even for $m = 2$, it is natural to ask how well the optimum can be approximated when there is only a constant number of machines. In contrast to the previously mentioned inapproximability result for the general case, there exists a fully polynomial-time approximation scheme for the problem when m is fixed. Horowitz and Sahni [10] proved that for any $\varepsilon > 0$, an ε-approximate solution can be computed in $O(nm(nm/\varepsilon)^{m-1})$ time, which is polynomial in both n and $1/\varepsilon$ if m is constant. Recently, Jansen and Porkolab [11], and later improved by Fishkin, Jansen and Mastrolilli [4], presented a fully polynomial time approximation scheme for the problem whose running time is linear in the number of jobs.

Note that, as the makespan problem is a special case of the max flow time problem, all the mentioned negative results hold also for the problems addressed in this paper.

Our Results: In this paper, we investigate the max flow time problem in the off-line and on-line setting. We prove positive and negative theoretical results.

In the off-line setting, we address the unrelated parallel machines model (Section 2.1) and present, when the number m of machines is fixed, the first known fully polynomial time approximation scheme (FPTAS). Observe that no polynomial time approximation scheme is possible when the number of machines is part of the input [16], unless P=NP. Therefore, for fixed m obtaining a FPTAS is to some extent the strongest possible result.

In the on-line setting and when the machines are identical, we analyze the (non-preemptive) FIFO heuristic when preemption is allowed (noted as $P|$on-line; $pmtn; r_j|F_{max}$ according to Graham et al. [6]). Bender et al. [2] claimed that

---

[2] Algorithms that, for any fixed $\varepsilon > 0$, find a solution within a $(1 + \varepsilon)$ factor of the optimum in polynomial time. If the running time is bounded by a polynomial in the input size and $1/\varepsilon$, then these algorithms are called *fully polynomail time approximation schemes* (FPTAS).

this strategy is a $(3 - 2/m)$-competitive algorithm for the non-preemptive problem. We show (Section 3.1) that FIFO comes within the same bound of the optimal preemptive schedule length. Since FIFO does not depend on the sizes of the jobs, it is also an on-line non-clairvoyant algorithm with a $(3-2/m)$-competitive ratio. In Section 3.2 we show that no 1-competitive (optimal) on-line algorithm is possible for the preemptive problem $(P|\text{on-line}; pmtn; r_j|F_{\max})$. This result should be contrasted with the related problem $P|\text{on-line}; pmtn; r_j|C_{\max}$ (i.e., the same problem with makespan as objective function) that admits an optimal on-line algorithm [9]. In Section 3.3, we show that in the non-clairvoyant model the competitive ratio cannot be better than 2. This proves that the competitive ratio of FIFO matches the lower bound when $m = 2$. Finally, in Section 3.4 we address the problem with uniformly related parallel machines and identical processing times (noted as $Q|\text{on-line}; p_j = p; r_j|F_{\max}$ according to [6]). We show that in this case FIFO is 1-competitive (optimal).

Due to page limit, several proofs had to be omitted from this version of the paper. A complete version of the paper is available (http://www.idsia.ch/~monaldo /research_papers.html).

2   Offline Max Flow Time

2.1   A FPTAS for Unrelated Parallel Machines

In this section we consider the off-line problem of scheduling a set $\mathcal{J} = \{J_1, ..., J_n\}$ of n independent jobs on a set $M = \{M_1, ..., M_m\}$ of m unrelated parallel machines. We present a FPTAS when the number m of machines is a constant. Our approach consists of partitioning the set of jobs into *blocks* $B(1), B(2), ...$, such that jobs belonging to any block can be scheduled regardless of jobs belonging to other blocks (Separation Property). The FPTAS follows by presenting a $(1 + \varepsilon)$-approximation algorithm for each block of jobs.

Separation Property. Let $p_j = \min_{i=1,...,m} p_{ij}$ denote the smallest processing time of job J_j. Let $R = \{r(1), r(2), ..., r(\rho)\}$ be the set of all release dates ($\rho \leq n$ is the number of different release values). Assume, without loss of generality, that $r(1) < r(2) < ... < r(\rho)$. Set $r(\rho+1) = \infty$. Partition jobs according to their release times and let $N(i) = \{J_j : r_j = r(i)\}$, $i = 1, ..., \rho$, denote the set of jobs released at time $r(i)$. Finally, let $P_{N(i)}$ be the sum of the smallest processing times of jobs from $N(i)$, i.e., $P_{N(i)} = \sum_{J_j \in N(i)} p_j$.

Block Definition. The first block $B(1)$ is defined as follows. If $r(1)+P_{N(1)} \leq r(2)$ then $B(1) = N(1)$. Otherwise, if $r(1) + P_{N(1)} + P_{N(2)} \leq r(3)$ then $B(1) = N(1) \cup N(2)$, else continue similarly. More formally,

$$B(1) = \bigcup_{i=1,..,b_1} N(i)$$

where b_1 is the smallest positive integer such that

$$r(1) + \sum_{i=1,..,b_1} P_{N(i)} \leq r(b_1 + 1).$$

Therefore if a job belongs to $B(1)$ then it could be completed not later than time $r(b_1 + 1)$ (by assigning jobs to the machines with the smallest processing requirements).

Other possible blocks are obtained in a similar way: if $r(b_1 + 1) \leq r(\rho)$ then discard all jobs from $B(1)$ and apply a similar procedure to obtain the next block $B(2)$. More formally, for $w = 2, 3, ...$, the w-th block is defined as

$$B(w) = \bigcup_{i=b_{w-1}+1,..,b_w} N(i)$$

where b_w is the smallest positive integer such that

$$r(b_{w-1} + 1) + \sum_{i=b_{w-1}+1,..,b_w} P_{N(i)} \leq r(b_w + 1).$$

In the following, let us use β to denote the number of blocks. By definition, observe that $b_\beta = \rho$.

Block Property. Let $r_{B(i)}$ be the earliest release time of jobs from block $B(i)$, i.e., $r_{B(i)} = \min_{J_j \in B(i)} r_j$, and $P_{B(i)} = \sum_{J_j \in B(i)} p_j$. Formerly, we claim that jobs belonging to any block can be scheduled regardless of jobs belonging to other blocks. A sufficient condition to have this separation property would be that in *any* 'good' (optimal or approximate) solution all jobs from block $B(i)$ ($i = 1, ..., \beta$) could be scheduled between time $r_{B(i)}$ and $r_{B(i)} + P_{B(i)}$. However, this is not always true for this problem, as Example 1 shows.

Example 1. Consider an instance with 3 jobs and 2 machines. The data are reported in the table of Figure 1.

In this example we have only one block $B(1)$ and $r_{B(1)} + P_{B(1)} = 5$. In Figure 1 it is shown an optimal solution ($F^*_{\max} = 3$) in which the last job completes at time 6 ($> r_{B(1)} + P_{B(1)}$).

j	r_j	p_{1j}	p_{2j}
1	0	3	10
2	2	1	3
3	3	3	1

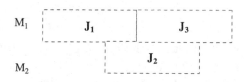

Fig. 1. Block example

We overcome the previous problem by showing that there exists always at least one 'good' (optimal or approximate) solution in which all jobs from block $B(i)$ $(i = 1, ..., \beta)$ are scheduled between time $r_{B(i)}$ and $r_{B(i)} + P_{B(i)}$. We prove this by exhibiting an algorithm which transforms any solution into another solution with the desired separation property. Moreover, the objective function value of the new solution is not worse than the previous one.

Separation Algorithm. Assume that we have a solution SOL of value $F_{\max}$ in which jobs from different blocks are not scheduled separately. Then there exists at least one block, say $B(w)$, in which the last job of $B(w)$ completes after time $r_{B(w)} + P_{B(w)}$. For those blocks $B(w)$, and starting with the block with the lowest index w, we show how to reschedule jobs from $B(w)$ such that they are completed within time $r_{B(w)} + P_{B(w)}$, and without worsening the solution value.

Let $C(i)$ denote the time all jobs from $N(i)$ are completed according to solution SOL, i.e., the time the last job from $N(i)$ completes. Observe that $F_{\max} = \max_i(C(i) - r(i))$. Recall the block definition $B(w) = \bigcup_{i=b_{w-1}+1,..,b_w} N(i)$, and let $N(l) \subseteq B(w)$ be the last released group of jobs such that

$$C(l) \leq r_{B(w)} + \sum_{i=b_{w-1}+1,...,l} P_{N(i)}.$$

By construction we have

$$C(x) > r_{B(w)} + \sum_{i=b_{w-1}+1,...,x} P_{N(i)}, \text{ for } x = l+1, ..., b_w.$$

Now remove from SOL all jobs belonging to $N(l+1) \cup ... \cup N(b_w)$ and reschedule them in order of non-decreasing release times and on the machine requiring the lowest processing time. We claim that according to the new solution SOL' the completion time $C'(i)$ of every class $N(i)$ is not increased, i.e. $C'(i) \leq C(i)$ for $i = l+1, ..., b_w$, and all jobs from $B(w)$ are completed by time $r_{B(w)} + P_{B(w)}$. Indeed, the new completion time $C'(l+1)$ of jobs from $N(l+1)$ is bounded by $C(l) + P_{N(l+1)}$ that is at most $r_{B(w)} + \sum_{i=b_{w-1}+1,...,l+1} P_{N(i)}$, and by construction less than $C(l+1)$. More generally, this property holds for every set $N(x+1)$ with $x = l+1, ..., b_w$, i.e.

$$C'(x+1) \leq C'(x) + P_{N(x+1)}$$
$$\leq r_{B(w)} + \sum_{i=b_{w-1}+1,...,x+1} P_{N(i)} < C(x+1).$$

It follows that in solution SOL' every job from $N(x)$ $(\subseteq B(w))$ is completed within time $r_{B(w)} + \sum_{i=b_{w-1}+1,...,x} P_{N(i)}$ and therefore every job from $B(w)$ is completed by time $r_{B(w)} + P_{B(w)}$. Moreover the maximum flow time $F'_{\max}$ of the new solution is not increased since $F'_{\max} = \max_i(C'(i) - r(i)) \leq \max_i(C(i) - r(i)) = F_{\max}$.

Lemma 1. *Without increasing the maximum flow time, any given solution can be transformed into a new feasible solution having all jobs from block $B(w)$ ($w = 1, ..., \beta$) scheduled between time $r_{B(w)}$ and $r_{B(w)} + P_{B(w)}$.*

Block Approximation. By Lemma 1 a $(1 + \varepsilon)$-approximate solution can be obtained as follows: starting from the first block, compute a $(1 + \varepsilon)$-approximate schedule for each block $B(w)$ that starts at time $r_{B(w)}$ and completes by time $r_{B(w)} + P_{B(w)}$, i.e., not later than the earliest starting time of the next block $B(w+1)$ of jobs. A $(1 + \varepsilon)$-approximate solution can be computed in polynomial time if there exists a polynomial time $(1 + \varepsilon)$-approximation algorithm for each block of jobs.

By previous arguments, we focus our attention on a single block of jobs and assume, without loss of generality, that the input instance is given by this set of jobs. For simplicity of notation we again use n to denote the number of jobs in the block instance and $\{J_1, ..., J_n\}$ the set of jobs. Moreover, we assume, without loss of generality, that the earliest release date is zero, i.e., $\min_j r_j = 0$.

Observe that $p_{\max} = \max_j p_j$ is a lower bound for the minimum objective value $F_{\max}^*$, i.e., $F_{\max}^* \geq p_{\max}$. By block definition, Lemma 1 and since $\min_j r_j = 0$, all jobs can be completed by time $\sum_{j=1}^n p_j \leq np_{\max}$. Moreover, any solution that completes within time $np_{\max}$ has a maximum flow time that cannot be larger than $np_{\max}$. Therefore, the optimal objective value $F_{\max}^*$ can be bounded as follows: $p_{\max} \leq F_{\max}^* \leq np_{\max}$. Without loss of generality, we restrict our attention to finding those solutions with maximum flow time at most $np_{\max}$. Therefore we can discard all solutions whose last job completes later than $2np_{\max}$, since all solutions with greater length have a maximum flow time larger than $np_{\max}$. Similarly, we will implicitly assume that job J_j cannot be scheduled on those machines M_i with $p_{ij} > np_{\max}$, since otherwise the resulting schedule would have a maximum flow time larger than $np_{\max}$.

In the following we show how to compute a $(1 + \varepsilon)$-approximate solution in which the last job completes not later than $2np_{\max}$. This solution can be always transformed into a $(1 + \varepsilon)$-approximate solution with the last job completing not later than $\sum_{j=1}^n p_j$ by Lemma 1.

The $(1 + \varepsilon)$-approximation algorithm is structured in the following three steps.

1. Round input values.
2. Find an optimal solution of the rounded instance.
3. Unround values.

We will first describe step 2, then step 1 with its "inverse" step 3.

An Optimal Algorithm. We start making some observations regarding the maximum flow time of a schedule. First renumber the jobs such that $r_1 \leq r_2 \leq ... \leq r_n$ holds. A simple job interchange argument shows that for a single machine, the maximum flow time is minimized if the jobs are processed in a non-decreasing order of release times. This property was first observed by Bender et al. [2].

We may view any m-machine schedule as an assignment of the set of jobs to machines with jobs assigned to machine M_i being processed in increasing order of index. Consequently given an assignment the max flow time is easily computed. We are interested in obtaining an assignment which minimizes $F_{\max}$. Thus we may regard assignment and schedule as synonymous.

A *completion configuration* $\mathbf{c}$ is a m-dimensional vector $\mathbf{c} = (c_1, ..., c_m)$: c_i denotes the completion time of machine M_i, for $i = 1, ..., m$. A *partial schedule*, σ_k is an assignment of the jobs $J_1, ..., J_k$ to machines. A *completion schedule* ω_k is an assignment of the remaining jobs $J_{k+1}, ..., J_n$ to machines. Consider two partial schedules σ_k^1 and σ_k^2 such that according to σ_k^1 the last job on machine M_i (for $i = 1, ..., m$) completes not later than the last job scheduled on the same machine M_i according to σ_k^2; moreover the maximum flow time of σ_k^1 is not larger than that of σ_k^2. If this happens we say that σ_k^1 *dominates* σ_k^2. It is easy to check that whatever is the completion schedule ω_k, the schedule obtained considering the assignment of jobs as in σ_k^1 and ω_k cannot be worse that attainable with σ_k^2 and ω_k. Therefore, with no loss, we can discard all dominated partial schedules. The reason is that by adding the remaining jobs $J_{k+1}, ..., J_n$ in order of increasing index, the completion time of the current job J_j ($j = k + 1, ..., n$) is a monotone non-decreasing function of the completion times of machines before scheduling J_j (and does not depend on how $J_1, ..., J_{j-1}$ are really scheduled). Therefore if jobs $J_1, ..., J_k$ are scheduled according to σ_k^1 then the maximum flow time of jobs $J_{k+1}, ..., J_n$, when scheduled according to any ω_k, cannot be larger than the maximum flow time of the same set of jobs when $J_1, ..., J_k$ are scheduled according to σ_k^2.

We encode a feasible schedule $\mathbf{s}$ by a $(m + 1)$-dimensional vector $\mathbf{s} = (c_1, ..., c_m, F)$, where $(c_1, ..., c_m)$ is a completion configuration and F is the maximum flow time in $\mathbf{s}$. We say that schedule $\mathbf{s}_1 = (c_1', ..., c_m', F')$ *dominates* $\mathbf{s}_2 = (c_1'', ..., c_m'', F'')$ if $c_i' \leq c_i''$, for $i = 1, ..., m$, and $F' \leq F''$. Moreover, since $F_{\max}^* \leq np_{\max}$ we classify as dominated all those schedule $\mathbf{s} = (c_1, ..., c_m, F)$ with $F > np_{\max}$. The latter implies $c_i \leq 2np_{\max}$ ($i = 1, ..., m$) in any not dominated schedule.

For every $\mathbf{s} = (c_1, ..., c_m, F)$, let us define the operator $\oplus$ as follows:

$$\mathbf{s} \oplus p_{ij} = (c_1, ..., c_i', ..., c_m, F')$$

where

$$c_i' = \begin{cases} c_i + p_{ij} & \text{if } r_j \leq c_i \\ r_j + p_{ij} & \text{otherwise} \end{cases}$$

and $F' = \max\{F; c_i' - r_j\}$.

The following dynamic programming algorithm computes the optimal solution:

Algorithm OPT-$F_{\max}$

---

1. **Initialization:** $L_0 \leftarrow \{(c_1 = 0, ..., c_m = 0, 0)\}$
2. **For** $j = 1$ **to** n
3.   **For** $i = 1$ **to** m
4.   For every vector $\mathbf{s} \in L_{j-1}$ put vector $\mathbf{s} \oplus p_{ij}$ in L_j
5.   Discard from L_j all dominated schedules
6. **Output:** return the vector $(c_1, ..., c_m, F) \in L_n$ with minimum F

At line 4, the algorithm schedules job J_j at the end of machine M_i. At line 5, all dominated partial schedules are discarded.

The total running time of the dynamic program is $O(nmD)$, where D is the maximum number of not dominated schedules at steps 4 and 5. Let δ be the maximum number of different values that each machine completion time c_i can take in any not dominated schedule. The reader should have no difficulty to bound D by $O(\delta^m)$. Therefore, the described algorithm is, for every fixed m, a polynomial time algorithm iff δ is polynomial in n and $1/\varepsilon$. The next subsection shows how to transform any given instance such that the latter happens.

Rounding and Unrounding Jobs. Let $\varepsilon > 0$ be an arbitrary small rational number and assume, for simplicity, that $1/\varepsilon$ is an integral value. The first step is to round down every processing and release time to the nearest lower value of $\frac{\varepsilon p_{\max}}{2n} i$, for $i = 0, 1, \ldots, 2n^2/\varepsilon$; clearly this does not increase the objective function value. Note that the largest release time r_n is not greater than $np_{\max}$ since all jobs can be completed by that time. Then, find the optimal solution SOL of the resulting instance by using the dynamic programming approach described in the previous subsection. Observe that, since in every not dominated schedule the completion time c_i of any machine M_i cannot be larger than $2np_{\max}$, then the maximum number δ of different values of c_i is now bounded by $1 + (2np_{\max})/(\frac{\varepsilon p_{\max}}{2n}) = 1 + 4n^2/\varepsilon$, i.e., polynomial in n and $1/\varepsilon$.

Solution SOL can be easily modified to be a feasible solution also for the original instance. First, delay the starting time of each job by $\frac{\varepsilon p_{\max}}{2n}$ (this is sufficient to guarantee that all jobs do not start before their original release date); the completion time of each job may increase by at most $\frac{\varepsilon p_{\max}}{2n}$. Second, replace the rounded processing values with the originals; now the completion time of each job may increase by at most $\varepsilon p_{\max}/2$ (here we are using the assumption that each processing time cannot be larger than $np_{\max}$, and that each processing time may increase by at most $\frac{\varepsilon p_{\max}}{2n}$). Therefore, we may potentially increase the maximum flow time of SOL by at most $\frac{\varepsilon p_{\max}}{2} + \frac{\varepsilon p_{\max}}{2n} \leq \varepsilon F^*_{\max}$. This results in a $(1 + \varepsilon)$-approximate solution for the block instance.

The total running time of the described FPTAS is determined by the dynamic programming algorithm, that is $O(nm(n^2/\varepsilon)^m)$.

Theorem 1. *For the problem of minimizing the maximum flow time in scheduling n jobs on m unrelated machines (m fixed), there exists a fully polynomial time approximation scheme that runs in $O(nm(n^2/\varepsilon)^m)$ time.*

3   Online Max Flow Time

3.1   Analysis of FIFO for P|on-line; pmtn; r_j|F_{max}

In this section we will analyze the FIFO heuristic when preemption is allowed and in the identical machines model. Bender et al. [2] claimed that this strategy is a $(3 - 2/m)$-competitive algorithm for nonpreemptive scheduling. We show that FIFO (that is non-preemptive) comes within the same bound of the optimal preemptive schedule length. Since FIFO does not depend on the sizes of the jobs, it is also an on-line non-clairvoyant algorithm with a $(3-2/m)$-competitive ratio. In Section 3.2 we will show that no 1-competitive (optimal) on-line algorithm is possible.

Lower Bounds. First observe that $p_{max} = \max_j p_j$ is a lower bound for the minimum objective value F_{max}^*, i.e., $F_{max}^* \geq p_{max}$. In the following we provide a second lower bound.

Consider a relaxed version of the problem in which a job J_j can be processed by more that one machine simultaneously and without changing the total processing time p_j that J_j spends on machines. Let us call this relaxed version of the problem as the *fractional problem*. Clearly the optimal value F_{max}^o of the fractional problem cannot be larger than F_{max}^*, i.e. the optimal preemptive max flow time.

Now, recall the definitions given in subsection 2.1, and without loss of generality, let us renumber the jobs $J_1, J_2, ..., J_n$ such that $r_1 \leq r_2 \leq ... \leq r_n$. Consider the following rule that we call *fractional FIFO*: schedule jobs in order of increasing index and assigning p_j/m time units of job J_j $(j = 1, ..., n)$ to each machine.

Lemma 2. *The optimal solution of the fractional problem can be obtained by using the fractional FIFO.*

Now according to the fractional FIFO, let the *fractional load* $\ell(i)$ at time $r(i)$ be defined as the total sum of processing times of jobs that at time $r(i)$ have been released but not yet finished. More formally, we have

$$\ell(1) = P_{N(1)},$$
$$\ell(i + 1) = P_{N(i+1)} + \max\{\ell(i) - m(r(i + 1) - r(i)); 0\}.$$

By Lemma 2, the maximum flow time $F_{N(i)}$ of jobs from $N(i)$ is the time required to process all jobs that at time $r(i)$ have been released but not yet finished, i.e. $F_{N(i)} = \ell(i)/m$. The optimal solution value F_{max}^o of the fractional solution is therefore equal to $\frac{\ell_{max}}{m} = \frac{1}{m} \max_{i=1,...,\rho} \ell(i)$. We will refer to this value ℓ_{max} as the *maximal fractional load* over time. Since the optimal solution value F_{max}^* of our original preemptive problem cannot be smaller than F_{max}^o, we have the following lower bounds

$$F_{max}^* \geq \max\{\frac{\ell_{max}}{m}; p_{max}\}. \tag{1}$$

Analysis of FIFO. We start showing that FIFO delivers a schedule whose maximum flow time is within $\frac{\ell_{\max}}{m} + 2(1 - \frac{1}{m})p_{\max}$.

Lemma 3. *FIFO returns a solution with maximum flow bounded by*

$$\frac{\ell_{\max}}{m} + 2(1 - \frac{1}{m})p_{\max}.$$

By Lemma 3 and inequality (1) it follows that FIFO is a $(3 - 2/m)$-competitive algorithm. Moreover, this bound is tight.

Theorem 2. *FIFO is $(3 - 2/m)$-competitive algorithm for $P|on\text{-}line; pmtn; r_j|F_{\max}$ and this bound is tight.*

3.2   A Lower bound for P|on-line; pmtn; r_j|F$_{\max}$

We show that no on-line preemptive algorithm can be 1-competitive.

Theorem 3. *The competitive ratio of any deterministic algorithm for $P|on\text{-}line; pmtn; r_j|F_{\max}$ is at least $1 + \frac{1}{14}$.*

This result should be contrasted with the related problem $P|on\text{-}line; pmtn; r_j|C_{\max}$ (i.e., the same problem with makespan as objective function) that admits an optimal on-line algorithm [9]. Moreover, we already observed that in the off-line setting the problem can be solved optimally in polynomial time by adapting the algorithm described in [14,15].

3.3   A Lower bound for P|on-line-nclv; r_j|F$_{\max}$

When jobs processing times are known at their arrival dates (clairvoyant model), Bender et al. [2] observed a simple lower bound of 3/2 on the competitive ratio of any on-line deterministic algorithm. In the following we show that in the non-clairvoyant model the competitive ratio cannot be better than 2. This shows that the competitive ratio of FIFO matches the lower bound when $m = 2$.

Theorem 4. *The competitive ratio of any deterministic algorithm for $P|on\text{-}line\text{-}nclv; r_j|F_{\max}$ is at least 2.*

3.4   Analysis of FIFO for Q|on-line; p_j=p; r_j|F$_{\max}$

We address the problem with identical and uniformly related parallel machines. We assume that the processing times of jobs are identical. Simple analysis shows that FIFO is optimal.

Theorem 5. *FIFO is 1-competitive for $Q|on\text{-}line; p_j = p; r_j|F_{\max}$.*

References

1. B. Awerbuch, Y. Azar, S. Leonardi, and O. Regev. Minimizing the flow time without migration. In *In Proceedings of the 31st Annual ACM Symposium on Theory of Computing (STOC'99)*, pages 198–205, 1999.
2. M. A. Bender, S. Chakrabarti, and S. Muthukrishnan. Flow and stretch metrics for scheduling continuous job streams. In *Proceedings of the 9th Annual ACM-SIAM Symposium on Discrete Algorithms (SODA'98)*, pages 270–279, 1998.
3. B. Chen, C. Potts, and G. Woeginger. A review of machine scheduling: Complexity, algorithms and approximability. *Handbook of Combinatorial Optimization*, 3:21–169, 1998.
4. A. Fishkin, K. Jansen, and M. Mastrolilli. Grouping techniques for scheduling problems: simpler and faster. In *9th Annual European Symposium on Algorithms (ESA'01)*, volume LNCS 2161, pages 206–217, 2001.
5. M. R. Garey and D. S. Johnson. *Computers and intractability; a guide to the theory of NP-completeness*. W.H. Freeman, 1979.
6. R. Graham, E. Lawler, J. Lenstra, and A. R. Kan. Optimization and approximation in deterministic sequencing and scheduling: A survey. volume 5, pages 287–326. North–Holland, 1979.
7. D. Hochbaum and D. Shmoys. Using dual approximation algorithms for scheduling problems: theoretical and practical results. *Journal of the ACM*, 34:144–162, 1987.
8. D. Hochbaum and D. Shmoys. A polynomial approximation scheme for machine scheduling on uniform processors: Using the dual approximation approach. *SIAM J. on Computing*, 17:539–551, 1988.
9. K. S. Hong and J. Y.-T. Leung. On-line scheduling of real-time tasks. *IEEE Transactions on Computing*, 41:1326–1331, 1992.
10. E. Horowitz and S. Sahni. Exact and approximate algorithms for scheduling nonidentical processors. *Journal of the ACM*, 23(2):317–327, 1976.
11. K. Jansen and L. Porkolab. Improved approximation schemes for scheduling unrelated parallel machines. In *Proceedings of the 31st Annual ACM Symposium on the Theory of Computing*, pages 408–417, 1999.
12. D. Karger, C. Stein, and J. Wein. Scheduling algorithms. In M. J. Atallah, editor, *Handbook of Algorithms and Theory of Computation*. CRC Press, 1997.
13. H. Kellerer, T. Tautenhahn, and G. J. Woeginger. Approximability and nonapproximability results for minimizing total flow time on a single machine. In *In Proceedings of the 28th Annual ACM Symposium on Theory of Computing (STOC'96)*, pages 418–426, 1996.
14. J. Labetoulle, E. L. Lawler, J. K. Lenstra, and A. H. G. R. Kan. Preemptive scheduling of uniform machines subject to release dates. In W. R. Pulleyblank, editor, *Progress in Combinatorial Optimization*, pages 245–261. Academic Press, 1984.
15. E. Lawler and J. Labetoulle. On preemptive scheduling of unrelated parallel processors by linear programming. *Journal of the ACM*, 25:612–619, 1978.
16. J. K. Lenstra, D. B. Shmoys, and E. Tardos. Approximation algorithms for scheduling unrelated parallel machines. *Mathematical Programming*, 46:259–271, 1990.
17. S. Leonardi and D. Raz. Approximating total flow time on parallel machines. In *Proc. 28th Annual ACM Symposium on the Theory of Computing (STOC'96)*, pages 110–119, 1997.
18. J. Sgall. On-line scheduling – a survey. In *A. Fiat and G. Woeginger, editors, On-Line Algorithms, Lecture Notes in Computer Science. Springer-Verlag, Berlin.*, 1997.

Linear Time Algorithms for Some NP-Complete Problems on (P_5,Gem)-Free Graphs

(Extended Abstract)

Hans Bodlaender[1], Andreas Brandstädt[2], Dieter Kratsch[3],
Michaël Rao[3], and Jeremy Spinrad[4]

[1] Institute of Information and Computing Sciences, Utrecht University
P.O. Box 80.089, 3508 TB Utrecht, The Netherlands
hansb@cs.uu.nl
[2] Fachbereich Informatik, Universität Rostock
A.-Einstein-Str. 21, 18051 Rostock, Germany
ab@informatik.uni-rostock.de
[3] Université de Metz, Laboratoire d'Informatique Théorique et Appliquée
57045 Metz Cedex 01, France
fax: ++ 00 33 387315309
{kratsch,rao}@sciences.univ-metz.fr
[4] Department of Electrical Engineering and Computer Science
Vanderbilt University, Nashville TN 37235, U.S.A.
spin@vuse.vanderbilt.edu

Abstract. A graph is (P_5,gem)-free, when it does not contain P_5 (an induced path with five vertices) or a gem (a graph formed by making an universal vertex adjacent to each of the four vertices of the induced path P_4) as an induced subgraph.

Using a characterization of (P_5,gem)-free graphs by their prime graphs with respect to modular decomposition and their modular decomposition trees [6], we obtain linear time algorithms for the following NP-complete problems on (P_5,gem)-free graphs: Minimum Coloring, Maximum Weight Stable Set, Maximum Weight Clique, and Minimum Clique Cover.

Keywords: algorithms, graph algorithms, NP-complete problems, modular decomposition, (P_5,gem)-free graphs.

1 Introduction

Graph decompositions play an important role in graph theory. The central role of decompositions in the recent proof of one of the major open conjectures in Graph Theory, the so-called Strong Perfect Graph Conjecture of C. Berge, is an exciting example [9]. Furthermore various decompositions of graphs such as decomposition by clique cutsets, tree-decomposition and clique-width are often used to design efficient graph algorithms. There are even beautiful general results stating that a variety of NP-complete graph problems can be solved in linear time for graphs of bounded treewidth and bounded clique-width, respectively [1,12].

A. Lingas and B.J. Nilsson (Eds.): FCT 2003, LNCS 2751, pp. 61–72, 2003.
© Springer-Verlag Berlin Heidelberg 2003

Despite the fact that modular decomposition is a well-known decomposition in graph theory having algorithmic uses that seem to be simple and obvious, there is relatively few research concerning non-trivial uses of modular decomposition such as designing polynomial time algorithms for NP-complete problems on special graph classes. An important exception are the many linear and polynomial time algorithms for cographs [10, 11] i.e. P_4-free graphs which are known to have a cotree representation which allows to solve various NP-complete problems in linear time when restricted to cographs, among them the problems Maximum (Weight) Stable Set, Maximum (Weight) Clique, Minimum Coloring and Minimum Clique Cover [10,11].

The original motivation to study (P_5,gem)-free graphs, as a natural generalization of cographs, by the authors of [6] had been to construct a faster, possibly linear time algorithm for the Maximum Stable Set problem on (P_5,gem)-free graphs. They established a characterization of the (P_5,gem)-free graphs by their prime induced subgraphs called the Structure Theorem for (P_5,gem)-free graphs. We show in this paper that the Structure Theorem is a powerful tool to design efficient algorithms for NP-complete problems on (P_5,gem)-free graphs. All our algorithms use the modular decomposition tree of the input graph and the structure of the prime (P_5,gem)-free graphs. We are convinced that efficient algorithms for other NP-complete graph problems (e.g. domination problems) on (P_5,gem)-free graphs can also be obtained by this approach.

It is remarkable that there are only few papers establishing efficient algorithms for NP-complete graph problems using modular decomposition and that most of them consider a single problem, namely Maximum (Weight) Stable Set. For work dealing with other problems we refer to [4,5,18]. Concerning the limits of modular decomposition it is known, for example, that Achromatic Number, List Coloring, and $\lambda_{2,1}$-Coloring with pre-assigned colors remain NP-complete on cographs [2,3,19]. This implies that these three problems are NP-complete on (P_5,gem)-free graphs.[1]

There is also a strong relation between modular decomposition and the clique-width of graphs. For example, if all prime graphs of a graph class have bounded size then this class has bounded clique-width. Problems definable in a certain logic, so-called LinEMSOL($\tau_{1,L}$)-definable problems, such as Maximum (Weight) Stable Set, Maximum (Weight) Clique and Minimum (Weight) Dominating Set, can be solved in linear time on any graph class of bounded clique-width, assuming a k-expression describing the graph is part of the input [12]. Many other NP-complete problems which are not LinEMSOL($\tau_{1,L}$)-definable can be solved in polynomial time on graph classes of bounded clique-width [15,20].

Brandstädt et al. have shown that (P_5,gem)-free graphs have clique-width at most five [7]. However this does not yet imply linear time algorithms for LinEMSOL($\tau_{1,L}$)-definable problems on (P_5,gem)-free graphs, since their approach does not provide a linear time algorithm to compute a suitable k-expression.

We present a linear time algorithm to solve the NP-complete Minimum Coloring problem on (P_5,gem)-free graphs using modular decomposition in Section 5. The NP-

---

[1] A proof, similarly to the one in [3] shows that $\lambda_{2,1}$-Coloring is NP-complete for graphs with at most one prime induced subgraph, the P_4, and hence for (P_5,gem)-free graphs.

complete problems Maximum Weight Stable Set, Maximum Weight Clique and Minimum Clique Cover can also be solved by linear time algorithms using modular decomposition for (P_5,gem)-free graphs. Due to space constraints, these algorithms are not shown in this extended abstract.

2 Preliminaries

We assume the reader to be familiar with standard graph theoretic notations. In this paper, $G = (V, E)$ is a finite undirected graph, and $|V| = n$ and $|E| = m$. $N(v) := \{u : u \in V, u \neq v, uv \in E\}$ denotes the *open neighborhood* of v and $N[v] := N(v) \cup \{v\}$ the *closed neighborhood* of v. The *complement graph* of G is denoted $\overline{G} = (V, \overline{E})$. For $U \subseteq V$ let $G[U]$ denote the subgraph of G induced by U. A graph is *co-connected* if its complement $\overline{G}$ is connected. If for $U \subset V$, a vertex not in U is adjacent to exactly k vertices in U then it is called a *k-vertex* for U.

A function $f : V \rightarrow \mathbb{N}$ is a *(proper) coloring* of the graph $G = (V, E)$, if $\{u, v\} \in E$ implies $f(u) \neq f(v)$. The *chromatic number* of G, denoted $\chi(G)$, is the smallest k such that the graph G has a k-coloring $f : V \rightarrow \{1, 2, \ldots, k\}$.

Let $G = (V, E)$ be a graph with vertex weight function $w : V \rightarrow \mathbb{N}$. The weight of a vertex set $U \subseteq V$ is defined to be $w(U) := \sum_{u \in U} w(u)$. We let $\alpha_w(G)$ denote the maximum weight of a stable set of G and $\omega_w(G)$ denote the maximum weight of a clique of G. A *weighted k-coloring* of (G, w) assigns to each vertex v of G $w(v)$ different colors, i.e. integers of $\{1, 2, \ldots, k\}$, such that $\{x, y\} \in E$ implies that no color assigned to x is equal to a color assigned to y. $\chi_w(G)$ denotes the smallest k such that the graph G with weight function w has a weighted k-coloring. Note that each weighted k-coloring of (G, w) corresponds to a multiset $S_1, S_2, \ldots, S_k$ of stable sets of G where $S_i, i \in \{1, 2, \ldots, k\}$, is the set of all vertices of G to which color i is assigned.

3 Modular Decomposition

Modular decomposition is a fundamental decomposition technique that can be applied to graphs, partially ordered sets, hypergraphs and other structures. It has been described and used under different names and it has been rediscovered various times. Gallai introduced and studied modular decomposition in his seminal 1967 paper [17] where it is used to decompose comparability graphs.

A vertex set $M \subseteq V$ is a *module* in G if for all vertices $x \in V \setminus M$, x is either adjacent to all vertices in M, or non-adjacent to all vertices in M. The *trivial modules* of G are $\emptyset, V$ and the singletons. A *homogeneous set* in G is a nontrivial module in G. A graph containing no homogeneous set is called *prime*. Note that the smallest prime graph is the P_4. A homogeneous set M is *maximal* if no other homogeneous set properly contains M.

Modular decomposition of graphs is based on the following decomposition theorem.

Theorem 1 ([17]). *Let $G = (V, E)$ be a graph with at least two vertices. Then exactly one of the following conditions holds:*

(i) G *is not connected: it can be decomposed into its connected components;*

(ii) $\overline{G}$ *is not connected: G can be decomposed into the connected components of $\overline{G}$;*

(iii) G *is connected and co-connected. There is some $U \subseteq V$ and a unique partition P of V such that*

    (a) $|U| > 3$,

    (b) $G[U]$ *is a maximal prime induced subgraph of G, and*

    (c) *for every class S of the partition P, S is a module of G and $|S \cap U| = 1$.*

Consequently there are three decomposition operations.

0-Operation: If G is disconnected then decompose it into its connected components $G_1, G_2, \ldots, G_r$.

1-Operation: If $\overline{G}$ is disconnected then decompose G into $G_1, G_2, \ldots G_s$, where $\overline{G}_1$, $\overline{G}_2, \ldots \overline{G}_s$ are the connected components of $\overline{G}$.

2-Operation: If $G = (V, E)$ is connected and co-connected then its maximal homogeneous sets are pairwise disjoint and they form the partition P of V. The graph $G[U]$ is obtained from G by contracting every maximal homogeneous set of G to a single vertex; it is called the *characteristic graph* of G and denoted by G^*. (Note that the characteristic graph of a connected and co-connected graph G is prime.)

The decomposition theorem and the above mentioned operations lead to the uniquely determined *modular decomposition tree* T of G. The leaves of the modular decomposition tree are the vertices of G. The interior nodes of T are labeled 0, 1 or 2 according to the operation corresponding to the node. Thus we call them 0-node (parallel node), 1-node (series node) and 2-node (prime node). Any interior node x of T corresponds to the subgraph of G induced by the set of all leaves in the subtree of T rooted at x, denoted by $G(x)$.

0-node. The children of a 0-node x correspond to the components obtained by a 0-operation applied to the disconnected graph $G(x)$.

1-node. The children of a 1-node x correspond to the components obtained by a 1-operation applied to the not co-connected graph $G(x)$.

2-node The children of a 2-node x correspond to the subgraphs induced by the maximal homogeneous sets or single vertices of the connected and co-connected graph $G(x)$. Additionally, the characteristic graph of $G(x)$ is assigned to the 2-node x.

The modular decomposition tree is of basic importance for many algorithmic applications, and in [22,13,14], linear time algorithms are given for determining the modular decomposition tree of an input graph.

Often, algorithms exploiting the modular decomposition have the following structure. Let Π be a graph problem to be solved on some graph class $\mathcal{G}$, e.g., Maximum Stable Set on (P_5,gem)-free graphs. First the algorithm computes the modular decomposition tree T of the input graph G using one of the linear time algorithms. Then in a bottom up fashion the algorithm computes for each node x of T the optimal value for the subgraph $G(x)$ of G induced by the set of all leaves of the subtree of T rooted at x. Thus the computation starts assigning the optimal value to the leaves. Then the algorithm computes the optimal value of an interior node x by using the optimal values of all children of x depending on the type of the node. Finally the optimal value of the root is the optimal value of Π for the input graph G. (Note that various more complicated variants of this method can be useful.)

Thus to specify such a modular decomposition based algorithm we only have to describe how to obtain the value for the leaves, and which formula to evaluate or which subproblem to solve on 0-nodes, 1-nodes and 2-nodes, using the values of all children as input. It is well-known how to do this for 0-nodes and 1-nodes for the NP-complete graph problems Maximum Weight Stable Set, Maximum Weight Clique, Minimum Coloring and Minimum Clique Cover from the corresponding cograph algorithm [10,11]. On the other hand to find out the algorithmic problem to solve on 2-nodes, called the 2-node subproblem, for solving problem Π using modular decomposition can be quite challenging.

4   The Structure Theorem for (P_5,Gem)-Free Graphs

To state the Structure Theorem of (P_5,gem)-free graphs we need to define three classes of (P_5,gem)-free graphs which together contain all prime (P_5,gem)-free graphs.

Definition 1. *A graph $G = (V, E)$ is called* matched cobipartite *if its vertex set V is partionable into two cliques C_1, C_2 with $|C_1| = |C_2|$ or $|C_1| = |C_2| - 1$ such that the edges between C_1 and C_2 form a matching and at most one vertex in C_1 and at most one vertex in C_2 are not covered by the matching.*

Definition 2. *A graph G is called* specific *if it is the complement of a prime induced subgraph of one of the three graphs in Figure 1.*

Fig. 1.

To establish a definition of the third class of prime graphs, we do need some more notions. A graph is *chordal* if it contains no induced cycles C_k, $k \geq 4$. See e.g. [8] for properties of chordal graphs. A graph is *cochordal* if its complement graph is chordal. A vertex v is *simplicial* in G if its neighborhood $N(v)$ in G is a clique. A vertex v is *cosimplicial* in G if it is simplicial in $\overline{G}$. It is well-known that every chordal graph has a simplicial vertex and that such a vertex can be found in linear time.

We also need the following kind of substituting a C_5 into a vertex: For a graph G and a vertex v in G, let the result of the extension operation $ext(G, v)$ denote the graph G' resulting from G by replacing v with a C_5 (v_1, v_2, v_3, v_4, v_5) of new vertices such that v_2, v_4 and v_5 have the same neighborhood in G as v, and v_1, v_3 have only their C_5

neighbors, i.e. have degree 2 in G'. For a vertex set $U \subseteq V$ of G, let $ext(G, U)$ denote the result of applying repeatedly the extension operation to all vertices of U. Note that the resulting graph does not depend on the order of replacing U vertices.

Definition 3. *For $k \geq 0$, let C_k be the class of prime graphs $G' = ext(G, Q)$ resulting from a (not necessarily prime) cochordal gem-free graph G by extending a clique Q of exactly k cosimplicial vertices of G. Thus, C_0 is the class of prime cochordal gem-free graphs.*

Clearly each graph in C_k contains k C_5's which are vertex-disjoint. It is also known that each graph in C_k has neither $\overline{C_4}$ nor $\overline{C_6}$ as an induced subgraph [6].

Lemma 1. *Let $G = (V, E)$ be a graph of C_k, $k \geq 1$. Then for every C_5 $C = (v_1, v_2, v_3, v_4, v_5)$ of G, the vertex set V has a partition into $\{v_1, v_2, v_3, v_4, v_5\}$, the stable set A of 0-vertices for C and the set B of 3-vertices for C such that all vertices of B have the same non consecutive neighbors in C, say v_2, v_4, v_5, and $G[B]$ is a cograph.*

Theorem 2 (Structure Theorem [6]). *A connected and co-connected graph G is (P_5, gem)-free if and only if the following conditions hold:*

(1) *The homogeneous sets of G are P_4-free (i.e., induce a cograph);*
(2) *For the characteristic graph G^* of G, one of the following conditions holds:*
   (2.1) *G^* is a matched co-bipartite graph;*
   (2.2) *G^* is a specific graph;*
   (2.3) *there is a $k \geq 0$ such that G^* is in C_k.*

Consequently, the modular decomposition tree T of any connected (P_5, gem)-free G contains at most one 2-node. If G is a cograph then T has no 2-node. If G is not a cograph then the only 2-node of T is its root.

5   An Algorithm for Minimum Coloring on (P_5, Gem)-Free Graphs

In this section we present a linear time algorithm for the Minimum Coloring problem on (P_5, gem)-free graphs. That is we are given a (P_5, gem)-free graph G, and want to determine $\chi(G)$.

Minimum Coloring is not LinEMSOL($\tau_{1,L}$) definable. Nevertheless there is a polynomial time algorithm for graphs of bounded clique-width [20]. However this algorithm is only of theoretical interest. For graphs of clique-width at most five (which is the best known upper bound for the clique-width of (P_5, gem)-free graphs [7]), the exponent r of the running time $O(n^r)$ of this algorithm is larger than 2000.

5.1   The Subproblems

We use the approach discussed in Section 3. Thus, we start by computing (in linear time) the modular decomposition tree T of G. For each node x of T, we compute $\chi(G(x))$. Suppose $x_1, x_2, \ldots, x_r$ are the children of x. For leaves, 0-nodes, and 1-nodes x, the steps of the linear time algorithm for Minimum Coloring on cographs can be used: If x is a **leaf** of T then $\chi(G(x)) := 1$. If x is a **0-node**, then $\chi(G(x)) := \max_{i=1,\ldots,r} \chi(G(x_i))$.

If x is a **1-node**, then $\chi(G(x)) := \sum_{i=1}^{r} \chi(G(x_i))$.

Suppose x is a **2-node** of T. Let $G^* = (V^*, E^*)$ be the characteristic graph assigned to x. We assign to the vertex set V^* of G^* the weight function $w^* : V^* \to \mathbb{N}$ such that $w^*(v_i) := \chi(G(x_i))$. We have that $\chi(G(x)) := \chi_{w^*}(G^*)$.

Thus, the Minimum Coloring problem on (P_5,gem)-free graphs becomes the problem of computing the minimum number of colors for a weighted coloring of (G^*, w^*), where G^* is a prime (P_5,gem)-free graph. The remainder of this section is devoted to this problem. The Structure Theorem tells us that G^* either is a matched co-bipartite graph, a specific graph, or there is a $k \geq 0$ with G^* in $\mathcal{C}_k$. In three subsections, each of these cases will be dealt with. We also use the following notation and lemma.

Let $N = \sum_{v \in V^*} w^*(v)$ be the total weight. Observe that N is at most the number of vertices of the original (P_5,gem)-free graph.

Lemma 2. *Let G be a perfect graph and w be a vertex weight function of G. Then $\chi_w(G) = \omega_w(G)$ and $\kappa_w(G) = \alpha_w(G)$.*

Proof. Let G' be the graph obtained from G by substituting each vertex v of G by a clique of cardinality $w(v)$. As any weighted coloring of (G, w) corresponds to a coloring of G' and vice versa, we have $\chi_w(G) = \chi(G')$. Similarly, $\omega_w(G) = \omega(G')$.

Let G be perfect. Then $\overline{G}$ is perfect by Lovasz's Perfect Graph Theorem [21]. $\overline{G'}$ is obtained from the perfect graph $\overline{G}$ by vertex multiplication, and thus it is perfect [21]. As G' is the complement of a perfect graph $(\overline{G'})$, it is perfect. Since G' is perfect we have $\chi(G') = \omega(G')$ and thus $\chi_w(G) = \omega_w(G)$. Similarly, since $\overline{G'}$ is perfect we obtain $\chi_w(\overline{G}) = \omega_w(\overline{G})$. Hence $\kappa_w(G) = \alpha_w(G)$. $\qquad\square$

We now discuss how to solve the weighted coloring problem for each of the three classes of prime (P_5,gem)-free graphs.

5.2 Matched Cobipartite Graphs

The graph G^* is cobipartite and thus perfect. By Lemma 2 we obtain $\chi_{w^*}(G^*) = \omega_{w^*}(G^*)$. One easily finds in linear time a partition of the vertex set of G^* into two cliques, C_1, and C_2. Now, as each maximal clique of G^* is either C_1, C_2, or an edge of G^*, $\omega_{w^*}(G^*) = \chi_{w^*}(G^*)$ can be computed by a linear time algorithm.

5.3 Specific Graphs

Each specific graph G^* is a prime induced subgraph of the complement of one of the three graphs in Figure 1. To solve the weighted coloring problem on specific graphs, we formulate this problem as an integer linear programming problem, and then argue that this ILP can be solved in constant time.

Consider the specific graph G^* with weights w^*. Let $\mathcal{S}$ be the collection of all maximal stable sets of G^*. We build an integer linear programming with for each $S \in \mathcal{S}$ a variable x_S, as follows.

$$\text{minimize} \sum_{S \in \mathcal{S}} x_S \text{ such that} \tag{1}$$

$$\sum_{v \in S, S \in \mathcal{S}} x_S \geq w(v) \quad \text{for all } v \in V \tag{2}$$

$$x_S \geq 0 \quad \text{for all } S \in \mathcal{S} \tag{3}$$

$$x_S \text{ integer} \quad \text{for all } S \in \mathcal{S} \tag{4}$$

With x we denote a vector containing for each $S \in \mathcal{S}$ a value x_S.

Let z be the optimal value of this ILP. z equals the minimum number of colors needed for (G^*, w^*). If we have a coloring of (G^*, w^*) with a minimum number of colors, then assign to each color one maximal stable set $S \in \mathcal{S}$, such that this color is given to (a subset of) all vertices in S. Let x_S be the number of colors assigned to S. Clearly, x_S is a non-negative integer. For each $v \in V$, as v has $w(v)$ colors, we have $\sum_{v \in S, S \in \mathcal{S}} x_S \geq w(v)$. $\sum_{S \in \mathcal{S}} x_S$ equals the number of colors. Conversely, suppose we have an optimal solution x_S of the ILP. For each $S \in \mathcal{S}$, we can take a set of x_S unique colors, and use these colors to color the vertices in x_S. As S is stable, this gives a proper coloring, and as $\sum_{v \in S, S \in \mathcal{S}} x_S \geq w(v)$, each vertex has sufficiently many colors available. So, this gives a coloring of (G^*, w^*) with z colors.

The relaxation of the ILP is the linear program, obtained by dropping the integer condition (4):

$$\text{minimize} \sum_{S \in \mathcal{S}} x_S \text{ such that} \tag{5}$$

$$\sum_{v \in S, S \in \mathcal{S}} x_S \geq w(v) \quad \text{for all } v \in V \tag{6}$$

$$x_S \geq 0 \quad \text{for all } S \in \mathcal{S} \tag{7}$$

Let x' be an optimal solution of this relaxation, with value $z' = \sum_{S \in \mathcal{S}} x'_S$.

As G^* is a specific graph, the linear program has a constant number of variables (namely, the number of maximal stable sets of G^*) and a constant number of constraints (at most nine, one per vertex of G^*), and hence can be solved in constant time. (E.g., enumerate all corners of the polyhedron spanned by program, and take the optimal one.) Note that we can write the linear program in the form $\max\{cx \mid Ax \leq b\}$, such that each element of A is either 0 or 1. Let Δ be the maximum value of a subdeterminant of this matrix A. It follows that Δ is bounded by a constant. Write $s = |\mathcal{S}|$.

Now we can use a result of Cook, Gerards, Schrijver, and Tardos, see Theorem 17.2 from [23]. This theorem tells us that the ILP has an optimal solution x'', such that for each $S \in \mathcal{S}$, $|x'_S - x''_S| \leq s\Delta$.

Thus, the following is an algorithm that finds the optimal solution to the ILP (and hence the number of colors needed for (G^*, w^*)) in constant time. First, find an optimal solution x' of the relaxation. Then, enumerate all integer vectors x'' with for all $S \in \mathcal{S}$,

$|x'_S - x''_S| \leq s\Delta$. For each such x'', check if it fulfils condition (2), and select the solution vector that fulfils the conditions with the minimum value. By Theorem 17.2 from [23], this is an optimal solution of the ILP. This method takes constant time, as s and Δ are bounded by constants, and thus 'only' a constant number of vectors have to be checked, and each is of constant size.[2]

A straightforward implementation of this procedure would not be practical, as more than $(s\Delta)^s$ vectors are checked, with s the number of maximal stable sets in one of the specific graphs. In a practical setting, one could first solve the linear program, and use that value as starting point in a branch and bound procedure.

Remark 1. The method not only works for the specific graphs, but for any constant size graph. This implies that Minimum Coloring can be solved in linear time for graphs whose modular decomposition has a constant upper bound on the size of the characteristic graphs.

Remark 2. In the full version we present an $O(N^3)$ time algorithm to solve the weighted coloring of the specific graphs, that has no large hidden constant in the running time.

5.4 $\bigcup_{k=0}^{\infty} C_k$

Let $G^* \in C_k$, for some $k \geq 0$, and w^* the weight function G^*. All C_5's of G^* can be computed by a linear time algorithm that first computes all vertices of degree two.

If $G^* = C_5$ then with the technique applied to specific graphs $\chi_{w^*}(G^*)$ can be computed in constant time. If $G^* \in C_0$ then it is cochordal and thus perfect. Hence $\chi_{w^*}(G^*) = \omega_{w^*}(G^*)$ by Lemma 2.

Lemma 3. *The Maximum Weight Clique problem and the weighted coloring problem can be solved by a linear time algorithm for cochordal graphs.*

Proof. Frank [16] gave a linear time algorithm to compute the maximum weight of a stable set of a chordal graph G. This implies that there is an $O(n^2)$ algorithm to compute the maximum weight of a clique in a cochordal graph $\overline{G}$ since $\omega_w(\overline{G}) = \alpha_w(G)$. To get a linear time algorithm, we must avoid the complementation; thus, we simulate Frank's algorithm applied to $\overline{G}$. This is Frank's algorithm: First it computes a perfect elimination ordering $v_1, \ldots, v_n$ of the input chordal graph $G = (V, E)$. Then a maximum weight stable set is constructed as follows. Initially, let $c\_w(v_i) = w(v_i)$, for all $1 < i \leq n$. For each i from 1 to n, if $c\_w(v_i) > 0$ then colour v_i red, and subtract $c\_w(v_i)$ from $c\_w(v_j)$ for all $v_j \in \{v_i\} \cup (N(v_i) \cap \{v_{i+1}, \ldots, v_n\})$. After all vertices have been processed, set $I = \emptyset$ and, for each i from n down to 1, if v_i is red and not adjacent to any vertex of I then $I = I \cup \{v_i\}$. When all vertices have been processed again, the algorithm terminates and outputs the maximum weight stable set I of (G, w).

We now describe our simulation of this algorithm. First a perfect elimination ordering $v_1, v_2, \ldots, v_n$ of $\overline{G}$ is computed in linear time (see e.g. [22]).

The maximum weight of a clique of G is constructed as follows. Initially, let $W' = 0$ and $s(v_i) = 0$ for all i ($1 \leq i \leq n$). For each i from 1 to n, if $w(v_i) - W' + s(v_i) > 0$

---

[2] Computer computation shows that $\Delta \leq 3$ for specific graphs.

then colour v_i red, set $W' = w(v_i) + s(v_i)$ and add $w(v_i) - W' + s(v_i)$ to $s(v_j)$ for all $v_j \in (N(v_i) \cap \{v_{i+1}, \ldots, v_n\})$.

After all vertices have been processed, set $K = \emptyset$ and, for each i from n down to 1, if v_i is red and adjacent to all vertices of K then $K = K \cup \{v_i\}$. Finally the algorithm outputs the maximum weight clique K of (G, w).

Clearly our algorithm runs in linear time. Its correctness follows from the fact that when treating the vertex v_i, the difference $W' - s(v_i)$ is precisely the value the original Frank algorithm applied to the complement of G would have subtracted from $c\_w(v_i)$ up to the point when it treats v_i. Thus our algorithm simulates Frank's algorithm on $\overline{G}$, and thus it is correct.   □

In the remaining case, we consider a prime graph $G^* \in \mathcal{C}_k$, $k \geq 1$ such that $G^* \neq C_5$.

Lemma 4. *Let $k \geq 1$, $G^* \in \mathcal{C}_k$ and $G^* \neq C_5$. Let $C = (v_1, v_2, v_3, v_4, v_5)$ be a C_5 in G^* and v_1 and v_3 its vertices of degree two. Let w^* be the vertex weight function of G^*. Then there is a minimum weight coloring S^* of (G^*, w^*) with precisely $\max(w^*(v_2), w^*(v_4) + w^*(v_5))$ stable sets containing at least one of the vertices of $\{v_2, v_4, v_5\}$.*

Proof. By Lemma 1, the set A of 0-vertices for $C = (v_1, v_2, v_3, v_4, v_5)$ is a stable set, $B = V^* \setminus (C \cup A) = N(v_2) \setminus C = N(v_4) \setminus C = N(v_5) \setminus C$, and $G^*[B]$ is a cograph.

Let S be any minimum weight coloring of (G^*, w^*). Since $N(v_1) \setminus C = N(v_3) \setminus C = \emptyset$ and $N(v_2) \setminus C = N(v_4) \setminus C = N(v_5) \setminus C = B$ we may assume that every stable set of S contains either none or two vertices of C. Therefore we study weighted colorings of a C_5 $C = (v_1, v_2, v_3, v_4, v_5)$ of G^* with vertex weights w^*, where all stable sets are non edges of C and call them partial weight colorings (abbr. pwc) of C. Clearly any pwc of $C = (v_1, v_2, v_3, v_4, v_5)$ contains at least $w^*(v_2)$ stable sets containing v_2, and it contains at least $w^*(v_4) + w^*(v_5)$ stable sets containing v_4 or v_5.

Let S' be a weighted coloring of G^* containing the smallest possible number of stable sets S with $S \cap \{v_2, v_4, v_5\} \neq \emptyset$. Let t be the number of stable sets S of S' satisfying $S \cap \{v_2, v_4, v_5\} \neq \emptyset$ and suppose that, contrary to the statement of the lemma, $t > \max(w^*(v_2), w^*(v_4) + w^*(v_5))$. Let $s(v)$ be the number of stable sets of S' containing the vertex v. Then $t > w^*(v_4) + w^*(v_5)$ implies $s(v_4) > w^*(v_4)$ or $s(v_5) > w^*(v_5)$. W.l.o.g. we may assume $s(v_4) > w^*(v_4)$. Hence there is a stable set $S' \in S'$ containing v_4. Consequently either $S' \subseteq \{v_2, v_4\} \cup A$ or $S' \subseteq \{v_1, v_4\} \cup A$. In both cases we replace the stable set S' of S' by $\{v_1, v_3\} \cup A$. Thus the replacement decrements the number of stable sets containing v_4 and possibly the number of stable sets containing v_2. Thus we obtain a new weighted coloring S'' of G^* with $t - 1$ stable sets S with $S \cap \{v_2, v_4, v_5\} \neq \emptyset$. This contradicts the choice of t. Consequently $t = \max(w^*(v_2), w^*(v_4) + w^*(v_5))$.   □

To extend any pwc of a C_5 C to G^* only two parameters are important: the number a of stable sets $\{v_1, v_3\}$ in the pwc of C, and the number b of non edges in the pwc of C different from $\{v_1, v_3\}$. Each of the a stable sets $\{v_1, v_3\}$ in the pwc of C, can be extended to a maximal stable set $\{v_1, v_3\} \cup A'$ of G^*, where A' is some maximal stable set of $G^* - C$. Each of the b non edges S, $S \neq \{v_1, v_3\}$, in the pwc of C has the unique extension to the maximal stable set $S \cup A$ of G^*.

By Lemma 4, for each C_5 of G^* there is a minimum weight coloring of G^* extending a pwc of the C_5 C with $b = \max(w^*(v_2), w^*(v_4) + w^*(v_5))$. Taking such a minimum weight coloring we can clearly remove vertices v_1 and v_3 from stable sets containing both until we obtain the smallest possible value of a in a pwc of C with $b = \max(w^*(v_2), w^*(v_4) + w^*(v_5))$.

Finally given a C_5 C, the smallest possible value of a in a pwc of C with $b = \max(w^*(v_2), w^*(v_4) + w^*(v_5))$ can be computed in constant time. (Details omitted.)

Now we are ready to present our coloring algorithm that computes a minimum weight coloring of (G^*, w^*) for a graph G^* of $\mathcal{C}_k$, $k \geq 1$. It removes at most k times the precomputed C_5 from the current graph until the remaining graph has no C_5 and is therefore a cochordal graph. Then by Lemma 3 there is an algorithm to solve the weighted coloring problem for the cochordal graph in linear time.

In each round, i.e. when removing one C_5 $C = (v_1, v_2, v_3, v_4, v_5)$ from the current graph G' with current weight function w', the algorithm proceeds as follows: First it computes in constant time a pwc of C such that $b = \max(w'(v_2), w'(v_4) + w'(v_5))$ and a as small as possible. Then the algorithm removes all vertices of C and obtains the graph $G'' = G' - C$. Furthermore it removes all vertices of the stable set A of 0-vertices for C in G' with weight at most a and decrements the weight of all other vertices in A by a. Recursively the algorithm solves the minimum weight coloring problem on the obtained graph G'' with weight function w''. Finally the minimum number of stable sets in a weighted coloring of (G', w') is obtained using the formula $\chi_{w'}(G') = a + \max(b, \chi_{w''}(G''))$.

Thus the algorithm removes at most $k \leq n$ times a C_5. Each pwc of a C_5 can be computed in constant time. For the final cochordal graph the minimum weight coloring can be solved in linear time. Hence the overall running time of the algorithm is linear. We have given a linear time algorithm for the weighted coloring problem for $\bigcup_{k \geq 0} \mathcal{C}_k$. We can finally conclude:

Theorem 3. *There is a linear time algorithm to solve the Minimum Coloring problem on (P_5,gem)-free graphs.*

6   Conclusion

We have shown how modular decomposition and the Structure Theorem for (P_5,gem)-free graphs can used to obtain a linear time algorithm to solve the Minimum Coloring problem. In a quite similar way one can construct a linear time algorithm to solve the Minimum Clique Cover problem on (P_5,gem)-free graphs. Modular decomposition can also be used to obtain linear time algorithms for the LinEMSOL($\tau_{1,L}$) definable NP-complete graph problems Maximum Weight Stable Set and Maximum Weight Clique on (P_5,gem)-free graphs. These algorithms are given in the full version of this paper.

Acknowledgement. Thanks are due to Alexander Schrijver for pointing towards Theorem 17.2 from his book [23].

References

1. S. ARNBORG, J. LAGERGREN, D. SEESE, Easy problems for tree-decomposable graphs, *J. Algorithms* 12 (1991), 308–340.
2. H. BODLAENDER, Achromatic number is NP-complete for cographs and interval graphs, *Inform. Process. Lett.* 31 (1989) 135–138
3. H.L. BODLAENDER, H.J. BROERSMA, F.V. FOMIN, A.V. PYATKIN, G.J. WOEGINGER, Radio labeling with pre-assigned frequencies, *Proceedings of the 10th European Symposium on Algorithms (ESA'2002)*, LNCS 2461 (2002) 211–222
4. H.L. BODLAENDER, K. JANSEN, On the complexity of the maximum cut problem, *Nord. J. Comput.* 7 (2000) 14–31
5. H.L. BODLAENDER, U. ROTICS, Computing the treewidth and the minimum fill-in with the modular decomposition, *Proceedings of the 8th Scandinavian Workshop on Algorithm Theory (SWAT'2002)*, LNCS 1851 (2002) 388–397
6. A. BRANDSTÄDT, D. KRATSCH, On the structure of (P_5,gem)-free graphs, *Manuscript* 2002
7. A. BRANDSTÄDT, H.-O. LE, R. MOSCA, Chordal co-gem-free graphs have bounded clique width, *Manuscript* 2002
8. A. BRANDSTÄDT, V.B. LE, J. SPINRAD, Graph Classes: A Survey, *SIAM Monographs on Discrete Math. Appl., Vol.* 3, SIAM, Philadelphia (1999)
9. M. CHUDNOVSKY, N. ROBERTSON, P.D.SEYMOUR, R.THOMAS, The Strong Perfect Graph Theorem, *Manuscript* 2002
10. D.G. CORNEIL, H. LERCHS, L. STEWART-BURLINGHAM, Complement reducible graphs, *Discrete Applied Math.* 3 (1981) 163–174
11. D.G. CORNEIL, Y. PERL, L.K. STEWART, Cographs: recognition, applications, and algorithms, *Congressus Numer.* 43 (1984) 249–258
12. B. COURCELLE, J.A. MAKOWSKY, U. ROTICS, Linear time solvable optimization problems on graphs of bounded clique-width, *Theory of Computing Systems* 33 (2000) 125–150
13. A. COURNIER, M. HABIB, A new linear algorithm for modular decomposition, *Trees in Algebra and Programming - CAAP '94*, LNCS 787 (1994) 68–84
14. E. DAHLHAUS, J. GUSTEDT, R.M. MCCONNELL, Efficient and practical algorithms for sequential modular decomposition, *J. Algorithms* 41 (2001) 360–387
15. W. ESPELAGE, F. GURSKI, E. WANKE, How to solve NP-hard graph problems on clique-width bounded graphs in polynomial time, *Proceedings of the 27th Workshop on Graph-Theoretic Concepts in Computer Science (WG 2001)*, LNCS 2204 (2001) 117–128
16. A. FRANK, Some polynomial algorithms for certain graphs and hypergraphs, *Proceedings of the Fifth British Combinatorial Conference (Univ. Aberdeen, Aberdeen, 1975)* 211–226, *Congressus Numerantium* No. XV, Utilitas Math., Winnipeg, Man. (1976)
17. T. GALLAI, Transitiv orientierbare Graphen, *Acta Mathematica Academiae Scientiarum Hungaricae* 18 (1967) 25–66
18. V. GIAKOUMAKIS, I. RUSU, Weighted parameters in ($P5$, $\overline{P5}$)-free graphs, *Discrete Appl. Math.* 80 (1997) 255–261
19. K. JANSEN, P. SCHEFFLER, Generalized coloring for tree-like graphs, *Discrete Appl. Math.* 75 (1997) 135–155
20. D. KOBLER, U. ROTICS, Edge dominating set and colorings on graphs with fixed clique-width, *Discrete Appl. Math.* 126 (2003) 197–221
21. L. LOVÁSZ, Normal hypergraphs and the perfect graph conjecture, *Discrete Math.* 2 (1972) 253–267
22. R.M. MCCONNELL, J. SPINRAD, Modular decomposition and transitive orientation, *Discrete Math.* 201 (1999) 189–241
23. A. SCHRIJVER, *Theory of Linear and Integer Programming*, John Wiley & Sons, Chichester, 1986.

Graph Searching, Elimination Trees, and a Generalization of Bandwidth

Fedor V. Fomin, Pinar Heggernes, and Jan Arne Telle

Department of Informatics, University of Bergen, N-5020 Bergen, Norway
{fomin,pinar,telle}@ii.uib.no

Abstract. The bandwidth minimization problem has a long history and a number of practical applications. In this paper we introduce a generalization of bandwidth to partially ordered layouts. We consider this generalization from two main viewpoints: graph searching and tree decompositions. The three graph parameters pathwidth, profile and bandwidth related to linear layouts can be defined by variants of graph searching using a standard fugitive. Switching to an inert fugitive, the two former parameters are generalized to treewidth and fill-in, and our first viewpoint considers the analogous tree-like generalization that arises from the bandwidth variant. Bandwidth also has a definition in terms of ordered path decompositions, and our second viewpoint generalizes this in a natural way to ordered tree decompositions. In showing that both generalizations are equivalent we employ the third viewpoint of elimination trees, as used in the field of sparse matrix computations. We call the resulting parameter the treespan of a graph and prove some of its combinatorial and algorithmic properties.

1 Motivation through Graph Searching Games

Different versions of graph searching has been attracting the attention of researchers from Discrete Mathematics and Computer Science for a variety of elegant and unexpected applications in different and seemingly unrelated fields. There is a strong resemblance of graph searching to certain pebble games [15] that model sequential computation. Other applications of graph searching can be found in VLSI theory since this game-theoretic approach to some important parameters of graph layouts such as the cutwidth [19], the topological bandwidth [18], the bandwidth [9], the profile [10], and the vertex separation number [8] is very useful for the design of efficient algorithms. There is also a connection between graph searching, pathwidth and treewidth, parameters that play an important role in the theory of graph minors developed by Robertson & Seymour [3,7,22]. Furthermore, some search problems have applications in problems of privacy in distributed environments with mobile eavesdroppers ('bugs') [11].

In the standard node-search version of searching, a single searcher is placed at a vertex of a graph G at every move, while from other vertices searchers are removed (see *e.g.* [15]). The purpose of searching is to capture an invisible fugitive moving fast along paths in G. The fugitive is not allowed to run through

A. Lingas and B.J. Nilsson (Eds.): FCT 2003, LNCS 2751, pp. 73–85, 2003.
© Springer-Verlag Berlin Heidelberg 2003

the vertices currently occupied by searchers. So the fugitive is caught when a searcher is placed on the vertex it occupies, and it has no possibility to leave the vertex because all the neighbors are occupied (guarded) by searchers. The goal of search games is to find a search strategy to *guarantee* the fugitive's capture while minimizing some resource usage.

Because the fugitive is invisible, the only information the searchers possess are the previous search moves that may give knowledge about subgraphs where the fugitive cannot possibly be present. This brings us to the interesting interpretation of the search problem [3] as the problem of fighting against damage spread in complex systems, *e.g.* the spread of a mobile computer virus in networks. Initially all vertices are viewed as contaminated (infected by a virus or damaged) and a contaminated vertex is cleared once it is occupied by a searcher (checked by an anti-virus program). A clear vertex v is recontaminated if there is a path without searchers leading from v to a contaminated vertex. In some applications it is required that recontamination should never occur and in this case we are interested in the so-called 'monotone' searching. For most of the search game variants considered in the literature it can be shown, sometimes by very clever techniques, that the resource usage does not increase in spite of this constraint [15,16,4,7]. The 'classical' goal of the search problem is to find the search program such that the maximum number of searchers in use at any move is minimized. The minimum number of searchers needed to clear the graph is related to the parameter called pathwidth. Dendris et al. [7] studied a variation of the node-search problem with *inert*, or lazy, fugitive. In this version of the game the fugitive is allowed to move only just before a searcher is placed on the vertex it occupies. The smallest number of searchers needed to find the fugitive in this version of searching is related to the parameter called treewidth [7].

Another criteria of optimality in node-searching, namely *search cost* was studied in [10]. Here the goal is to minimize the sum of the number of searchers in use over all moves of the search program. The search cost of a graph is equal to the interval completion number, or profile, which is the smallest number of edges in any interval supergraph of the given graph. Looking at the monotone search cost version but now with an inert fugitive, it is easy to see that this parameter is equal to the smallest number of edges in the chordal supergraph of a given graph, so called *fill-in*. (It is not clear if in this version of searching recontamination can help and this is an interesting open question.) We thus have the following elegant relation: the parameters related to standard node searching (pathwidth, profile) expressible in terms of interval completion problems, correspond in inert fugitive searching to chordal completion problems (treewidth, fill-in).

In this paper we want to minimize the maximum length of time (number of intermediate moves) during which a searcher occupies a vertex. A similar problem for pebbling games (that can be transferred into search terms) was studied by Rosenberg & Sudborough [23]. In terms of monotone pebbling (i.e., no recontamination allowed) this becomes the maximum lifetime of any pebble in the game. It turned out that this parameter is related to the bandwidth of a graph G, which is the minimum over all linear layouts of vertices in G of the maximum distance between images of adjacent vertices. The following table summarizes the knowledge about known relations between graph monotone searching and graph parameters.

	Number of Searchers	Cost of Searching	Occupation Time
Standard Search	pathwidth [15]	profile [10]	bandwidth [23]
Inert Search	treewidth [7]	fill-in	???

One of the main questions answered in this paper concerns the entry labeled ??? above: What kind of graph parameter corresponds to the minimum occupation time (*mot*) for monotone inert fugitive search? In section 2 we introduce a generalization of bandwidth to tree-like layouts, called *treespan*, based on what we call ordered tree decompositions. In section 3 we give the formal definition of the parameter $mot(G)$, and then in section 4 we show that it is equivalent to a parameter arising from elimination trees, as used in the sparse matrix computation community. In section 5 we show the equivalence also between this elimination tree parameter and treespan, thereby providing evidence that the entry labeled ??? above indeed corresponds to a natural generalization of bandwidth to partially ordered (tree) layouts. Finally in section 6 we obtain some algorithmic and complexity results on the treespan parameter.

2   Motivation through Tree Decompositions

We assume simple, undirected, connected graphs $G = (V, E)$, where $|V| = n$. We let $N(v)$ denote the neighbors of vertex v, and $d(v) = |N(v)|$ is the *degree* of v. The maximum degree of any vertex in G is denoted by $\Delta(G)$. For a set of vertices $U \subseteq V$, $N(U) = \{v \notin U \mid uv \in E \text{ and } u \in U\}$. $H \subset G$ means that H is a subgraph of G. For a rooted tree T and a vertex v in T, we let $T[v]$ denote the subtree of T with root in v.

A *chord* of a cycle C in a graph is an edge that connects two non-consecutive vertices of C. A graph G is *chordal* if there are no induced chordless cycles of length ≥ 4 in G. Given any graph $G = (V, E)$, a *triangulation* $G^+ = (V, E^+)$ of G is a chordal graph such that $E \subseteq E^+$.

A *tree decomposition* of a graph $G = (V, E)$ is a pair (X, T), where $T = (I, M)$ is a tree and $X = \{X_i \mid i \in I\}$ is a collection of subsets of V called *bags*, such that:

1. $\bigcup_{i \in I} X_i = V$
2. $uv \in E \Rightarrow \exists i \in I$ with $u, v \in X_i$
3. For all vertices $v \in V$, the set $\{i \in I \mid v \in X_i\}$ induces a connected subtree of T.

The *width* of a tree decomposition (X, T) is $tw(X, T) = \max_{i \in I} |X_i| - 1$. The *treewidth* of a graph G is the minimum width over all tree decompositions of G. A *path decomposition* is a tree decomposition (X, T) such that T is a path. The *pathwidth* of a graph G is the minimum width over all path decompositions of G. We refer to Bodlaender's survey [5] for further information on treewidth.

For a chordal graph G, the treewidth is one less than the size of the largest clique in G. For a non-chordal graph G, the treewidth is the minimum treewidth over all triangulations of G. This is due to the fact that a tree decomposition (X, T) of G actually corresponds to a triangulation of the given graph G: simply add edges to G such that each bag of X becomes a clique. The resulting graph, which we will call $tri(X, T)$ is a chordal graph of which G is a subgraph. In addition, any triangulation G^+ of G is equal to $tri(X, T)$ for some tree decomposition (X, T) of G.

Another reason why tree decompositions and chordal graphs are closely related is that chordal graphs are exactly the intersection graphs of subtrees of a tree [14]. Analogously, interval graphs are related to path decompositions, and they are the intersection graphs of subpaths of a path. A graph is *interval* if there is a mapping f of its vertices into sets of consecutive integers such that for each pair of vertices v, w the following is true: vw is an edge $\Leftrightarrow f(v) \cap f(w) \neq \emptyset$. Interval graphs form a subclass of chordal graphs. Similar to treewidth, the pathwidth of a graph G is one less than the smallest clique number over all triangulations of G into interval graphs.

The bandwidth of G, $bw(G)$, is defined as the minimum, over all linear orders of the vertices of G, maximum difference between labels of two adjacent vertices. Similar to pathwidth and treewidth, bandwidth can be defined in terms of triangulations as follows. A graph isomorphic to $K_{1,3}$ is referred to as a *claw*, and a graph that does not contain an induced claw is said to be *claw-free*. An interval graph G is a *proper interval graph* if it is claw-free [21]. As it was observed by Parra & Scheffler [20], the bandwidth of a graph G is one less than the smallest clique number over all triangulations of G into proper interval graphs. One can define bandwidth in terms of *ordered path decompositions*. In an ordered path decomposition, the bags are numbered $1, 2, ..., n$ from left to right. The first bag X_1 contains only one vertex of G, and for $1 \leq i \leq n - 1$ we have $|X_{i+1} \setminus X_i| = 1$, meaning that exactly one new graph vertex is introduced in each new bag. The number of bags a vertex v belongs to is denoted by $l(v)$. It is easy to show that $bw(G)$ is the minimum, over all ordered path decompositions, $\max\{l(v) - 1 \mid v \in V\}$.

The natural question here is, what kind of parameter corresponds to bandwidth when, instead of path decompositions, we switch to tree decompositions? This brings us to the definition of ordered tree decomposition and treespan.

Definition 1. *An* ordered tree decomposition (X, T, r) *of a graph* $G = (V, E)$ *is a tree decomposition* (X, T) *of* G *where* $T = (I, M)$ *is a rooted tree with root* $r \in I$, *such that:*

$|X_r| = 1$, *and if* i *is the parent of* j *in* T, *then* $|X_j \setminus X_i| = 1$.

Definition 2. *Given a graph* $G = (V, E)$ *and an ordered tree decomposition* (X, T, r) *of* G, *we define:*

$l(v) = |\{i \in I \mid v \in X_i\}|$ *(number of bags that contain* v*), for each* $v \in V$.
$ts(X, T, r) = \max\{l(v) \mid v \in V\} - 1$.

The treespan *of a graph* G *is* $ts(G) = \min\{ts(X, T, r) \mid (X, T, r)$ *is an ordered tree decomposition of* $G\}$.

Since every ordered path decomposition is an ordered tree decomposition, it is clear that for every graph G, $ts(G) \le bw(G)$.

3 Search Minimizing Occupation Time with Inert Fugitive

In this section we give a formal definition of minimum occupation time for inert fugitive searching. A *search program* Π on a graph $G = (V, E)$ is the sequence of pairs

$$(A_0, Z_0), (A_1, Z_1), \ldots, (A_m, Z_m)$$

such that

I. For $i \in \{0, \ldots, m\}$, $A_i \subseteq V$ and $Z_i \subseteq V$. We say that vertices A_i are *cleared*, vertices $V - A_i$ are *contaminated* and vertices Z_i are *occupied by searchers* at the ith step.

II. (*Initial state.*) $A_0 = \emptyset$ and $Z_0 = \emptyset$. All vertices are contaminated.

III. (*Final state.*) $A_0 = V$ and $Z_0 = \emptyset$. All vertices are cleared.

IV. (*Placing-removing searchers and clearing vertices.*) For $i \in \{1, \ldots, m\}$ there exists $v \in V$ and $Y_i \subseteq A_{i-1}$ such that $A_i - A_{i-1} = v$ and $Z_i = Y_i \cup \{v\}$. Thus at every step one of the searchers is placed on a contaminated vertex v while the others are placed on cleared vertices Y_i. The searchers are removed from vertices $Z_{i-1} - Y_i$. Note that Y_i is not necessarily a subset of Z_{i-1}.

V. (*Possible recontamination.*) For $i \in \{1, \ldots, m\}$ $A_i - \{v\}$ is the set of vertices $u \in A_{i-1}$ such that *every* uv-path has an internal vertex in Z_i. This means that the fugitive awakening in v can run to a cleared vertex u if there is a uv-path unguarded by searchers.

Dendris, Thilikos & Kirousis [7] initiated the study of inert search problem, where the problem is to find a search program Π with the smallest $\max_{i \in \{0, \ldots, m\}} |Z_i|$ (this maximum can be treated as the maximum number of searchers used in one step). It turns out that this number is equal to the treewidth of a graph. We find an alternative measure of search to be interesting as well. For a search program $\Pi = (A_0, Z_0), (A_1, Z_1), \ldots, (A_m, Z_m)$ on a graph $G = (V, E)$ and vertex $v \in V$ we define

$$\delta_i(v) = \begin{cases} 1, v \in Z_i \\ 0, v \notin Z_i \end{cases}$$

Then the number $\sum_{i=0}^{m} \delta_i(v)$ is the number of steps at which vertex v was occupied by searchers. For a program Π we define the *maximum vertex occupation time* to be $ot(\Pi, G) = \max_{v \in V} \sum_{i=0}^{m} \delta_i(v)$. The *vertex occupation time* of a graph G, denoted by $ot(G)$, is the minimum maximum vertex occupation time over all search programs on G.

A search program $(A_0, Z_0), (A_1, Z_1), \ldots, (A_m, Z_m)$ is *monotone* if $A_{i-1} \subseteq A_i$ for each $i \in \{1, \ldots, m\}$. Note that recontamination does not occur when a searcher is placed on a contaminated vertex thus awaking the fugitive.

Finally, for a graph G we define $mot(G)$ to be the minimum maximum vertex occupation time over all *monotone* search programs on G. We do not know whether $mot(G) = ot(G)$ for every graph G, and leave it as an interesting open question.

4 Searching and Elimination Trees

In this section we discuss a relation between $mot(G)$ and elimination trees of G. This relation is not only interesting in its own but also serves as a tool in further proofs.

For a graph $G = (V, E)$, an elimination order $\alpha : \{1, 2, ..., n\} \to V$ is a linear order of the vertices of G. For each given order α, a unique triangulation G_α^+ of G can be computed from the following procedure: starting with vertex $\alpha(1)$, at each step i, make the higher numbered neighbors of vertex $\alpha(i)$ in the transitory graph into a clique by adding edges. The resulting graph, which is denoted G_α^+, is chordal [12], and the given elimination ordering decides the quality of this resulting triangulation. The following lemma follows from the definition of G_α^+.

Lemma 1. *uv is an edge of G_α^+ $\Leftrightarrow$ uv is an edge of G or there is a path $u, x_1, x_2, ..., x_k, v$ in G with $k \geq 1$ such that all x_i are ordered before u and v by α (in other words, $\max\{\alpha^{-1}(x_i) \mid 1 \leq i \leq k\} < \min\{\alpha^{-1}(u), \alpha^{-1}(v)\}$).*

Definition 3. *For a vertex $v \in V$ we define $madj^+(v)$ to be the set of vertices $u \in V$ such that $\alpha(u) \geq \alpha(v)$ and uv is an edge of G_α^+. (The higher numbered neighbors of v in G_α^+.)*

Given a graph G, and an elimination order α on G, the corresponding *elimination tree* is a rooted tree $ET = (V, P)$, where the edges in P are defined by the following *parent* function: $parent(\alpha(i)) = \alpha(j)$ where $j = \min\{k \mid \alpha(k) \in madj^+(\alpha(i))\}$, for $i = 1, 2, ..., n$. Hence the elimination tree is a tree on the vertices of G, and vertex $\alpha(n)$ is always the root. The *height* of the elimination tree is the longest path from a leaf to the root. *Minimum elimination tree height* of a graph G, $mh(G)$ is the minimum height of an elimination tree corresponding to any triangulation of G. For a vertex $u \in V$ we denote by $ET[u]$ the subtree of ET rooted in u and containing all descendants (in ET) of u. It is important to note that, for two vertices u and v such that $ET[u]$ and $ET[v]$ are disjoint subtrees of ET, no vertex belonging to $ET[u]$ is adjacent to any vertex belonging to $ET[v]$ in G or G_α^+. In addition, $N(ET[v])$ is a clique in G_α^+, and a minimal vertex separator in both G_α^+ and G when v is not the only child of its parent in ET.

Let α be an elimination order of the vertices of a graph $G = (V, E)$ and let ET be the corresponding elimination tree of G. Observe that the elimination tree ET gives enough information about the chordal completion G^+ of G that ET corresponds to. It is important to understand that *any post order α of the vertices of ET is an elimination order on G that results in the same chordal completion $G_\alpha^+ = G^+$*. Thus given G and ET, we have all the information we need on the corresponding triangulation.

Definition 4. *Given an elimination tree ET of G, the* pruned subtree with root in x, $ET_p[x]$, *is the subtree obtained from $ET[x]$ by deleting all descendants of every vertex $y \in ET[x]$ such that $xy \in E(G)$ but no descendant of y is a neighbor of x in G.*

Thus, the leaves of $ET_p[x]$ are neighbors of x in G, and all lower numbered neighbors in G^+ of x are also included in $ET_p[x]$. In addition, there might clearly appear vertices in $ET_p[x]$ that are not neighbors of x in G. However, every neighbor of x in G^+ appears in $ET_p[x]$, as we prove in the following lemma.

Lemma 2. *Let α be an elimination order of graph $G = (V, E)$ and let ET be a corresponding elimination tree. Then for any $u, v \in V$, $u \in ET_p[v]$ if and only if $v \in madj^+(u)$.*

Proof. Let $u \in ET_p[v]$ and let w be a neighbor of v in G such that u is on a vw-path in ET. By the definition of pruned tree such a vertex w always exists. Because ET is an elimination tree, there is a uw-path P^+ in G_α^+ such that for any vertex x of P^+, $\alpha^{-1}(x) \leq \alpha^{-1}(u)$. By Lemma 1, this implies that there is also an uw-path P in G such that for any vertex x of P, $\alpha^{-1}(x) \leq \alpha^{-1}(u)$. Since w is adjacent to v in G, we conclude that $v \in madj^+(u)$.

Let $v \in madj^+(u)$. Then there is an uv-path P in G (and hence in G_α^+) such that all inner vertices of the path are ordered before u in α. Let w be the vertex of P adjacent to v. Because ET is elimination tree, we have that u is on vw-path in ET. Thus $u \in ET_p[v]$.

We define a parameter called *elimination span, es,* as follows:

Definition 5. *Given an elimination tree ET of a graph $G = (V, E)$, for each vertex $v \in V$ we define $s(v) = |ET_p[v]|$ and $es(ET) = \max\{s(v) \mid v \in V\} - 1$. The* elimination span *of a graph G is $es(G) = \min\{es(ET) \mid ET$ is an elimination tree of $G\}$.*

Theorem 1. *For any graph $G = (V, E)$, $es(G) = mot(G) - 1$.*

Proof. Let us prove $es(G) \leq mot(G) - 1$ first. Let $\Pi = (A_0, Z_0), (A_1, Z_1), \ldots, (A_m, Z_n)$ be a monotone search program. At every step of the program exactly one new vertex $A_i - A_{i-1}$ is cleared. Thus we can define the vertex ordering α by putting for $1 \leq i \leq n$ $\alpha(A_i - A_{i-1}) = n - i + 1$. At the ith step, when a searcher is placed at a vertex $u = A_i - A_{i-1}$ every vertex $v \in A_i$ such that there is a uv-path with no inner vertices in A_i should be occupied by a searcher (otherwise v would be recontaminated). Therefore, $v \in madj^+(u)$ and the number of steps when a vertex v is occupied by searchers, is $|\{u \mid v \in madj^+(u)\}|$. By Lemma 2, $|\{u \mid v \in madj^+(u)\}| = s(v)$ and we arrive at $es(ET) \leq mot(\Pi, G) - 1$.

We now show that $es(G) \geq mot(G) - 1$. Let ET be an elimination tree and let α be a corresponding elimination vertex ordering. We consider a search program Π where at the ith step of the program, $1 \leq i \leq n$, the searchers occupy the set of vertices $madj^+(v)$, where v is a vertex with $\alpha(v) = n - i + 1$. Let us first prove that Π is recontamination free. Suppose, on the contrary, that a vertex u is recontaminated at the ith step after placing a searcher on a vertex v. Then there is a uv-path P such that no vertex of P except v contains a searcher at the ith step. On the other hand, vertex u is after v in ordering α. Thus P should contain a vertex $w \in madj^{+}(u)$, $w \neq u$, occupied by a searcher. This is a contradiction. Since every vertex was occupied at least once and no recontamination occurs, we conclude that at the end of Π all vertices are cleared. Every vertex v was occupied by searchers during $|\{u \mid v \in madj^{+}(u)\}|$ steps and using Lemma 2 we conclude that $es(ET) \geq mot(\Pi, G) - 1$.

5   Ordered Tree Decompositions and Elimination Trees

In this section we discuss a relation between the treespan $ts(G)$ and elimination trees of G, establishing that $ts(G) = mot(G)$. We first give a simplified view of ordered tree decompositions and then proceed to prove some of their properties.

There are exactly n bags in X of an ordered tree decomposition (X, T, r) of G. Thus, the index set I for $X_i, i \in I$ can be chosen so that $I = V$, with $r \in V$. Then T is a tree on the vertices of G. To identify the bags and to define the correspondence between I and V uniquely, name the bags so that X_r is the bag corresponding to the root r of T. Regarding the bags in a top down fashion according to T, name the bag in which vertex v appears for the first time X_v and the corresponding tree node v. Thus if y is the parent of v in T then $X_v \setminus X_y = \{v\}$. This explains how to rename the bags and the vertices of T with elements from V given a tree decomposition based on I. However, if we replace i with v and I with V in Conditions 1 - 3 of the definition of a tree decomposition, and change condition in the definition of ordered tree decompositions to "$X_r = \{r\}$, and if y is the parent of v in T then $X_v \setminus X_y = \{v\}$", then this will automatically give a tree T on the vertices of G as we have explained above. For the remainder of this paper, when we mention an ordered tree decomposition (X, T, r), we will assume that T is a tree on the vertices of G as explained here. The following lemma will make the role of T even clearer.

Lemma 3. *Given a graph $G = (V, E)$ and a rooted tree $T = (V, P)$, there exists an ordered tree decomposition (X, T, r) of $G \Leftrightarrow$ for every edge $uv \in E$, u and v have an ancestor-descendant relationship in T.*

Proof. Assume that T corresponds to a valid ordered tree decomposition of G, but there is an edge uv in G such that $T[u]$ and $T[v]$ are disjoint subtrees of T. X_u is the first bag in which u appears and X_v is the first bag in which v appears, thus u and v do not appear in any bag X_w where w is on the path from u to the root or from v to the root in T. Thus if u and v appear together in any other bag X_y where y belongs to $T[u]$ or $T[v]$ or any other disjoint subtree in T, this would violate Condition 3 of a tree decomposition. Therefore, u and v cannot appear together in any bag, and there cannot exist a valid decomposition (X, T, r) of G.

For the reverse direction, assume that for every edge uv in G, u and v have an ancestor-descendant relationship in T. Assume without loss of generality that v is an ancestor of u. Then the bags can be defined so that 1) X_v contains v, 2) no bag X_y contains v where y is an ancestor of v, 3) for every vertex w on the path from v to u in T, X_w contains v (and w of course), and 4) X_u contains both u and v. We can see that all the conditions of an ordered tree decomposition are satisfied.

Lemma 4. *Let (X, T, r) be an ordered tree decomposition of a given graph. For every edge uv in $tri(X, T)$, u and v have an ancestor-descendant relationship in T.*

Proof. As we have seen in the proof of Lemma 3, if u and v belong to disjunct subtrees of T, then they cannot appear together in the same bag. Since only the bags are made into cliques, u and v cannot belong to the same clique in $tri(X, T)$, which means that the edge uv does not exist in $tri(X, T)$.

Lemma 5. *Let (X, T, r) be an ordered tree decomposition of a given graph. Let uv be an edge of $tri(X, T)$ such that v is an ancestor of u in T. Then v belongs to bag X_w for every w on the path from v to u including X_v and X_u.*

Proof. Vertex v appears for the first time in X_v on the path from the root, and u appears for the first time in X_u. For every vertex w on the path from v to u, exactly vertex w is introduced in X_w. Thus X_u is the first bag in which u and v both can belong to. In order for this to be possible, v must belong to bag X_w for every vertex w on the path from v to u in T.

Lemma 6. *For each graph G, there exists an ordered tree decomposition (X, T, r) of G of minimum treespan such that if u is a child of v in T then $v \in X_u$.*

Proof. Assume that u is a child of v in T and $v \notin X_u$. Clearly, uv is not an edge of G. Since v does not belong to any bag X_y for a descendant y of u, we can move u up to be a child of a node w in T where uw is an edge of G and where w is the first node on the path from v to the root that is a neighbor of u.

Lemma 7. *Let (X, T, r) be an ordered tree decomposition of G, and let $\alpha : \{1, ..., n\} \to V$ be a post order of T. Then $G_\alpha^+ \subseteq tri(X, T)$.*

Proof. Let uv be an edge of G_α^+, and assume without loss of generality that u has a lower number than v according to α. If uv is an edge of of G, then we are done. Otherwise, due to Lemma 1, there must exist a path $u, x_1, x_2, ..., x_k, v$ in G with $k \geq 1$ such that all x_i are ordered before u. Since α is a post order of T, none of the vertices $x_i, i = 1, ..., k$, can lie on the path from u to the root in T. Consequently and due to Lemma 3, since ux_1 is an edge of G, x_1 belongs to

$T[u]$. With the same argument, since $x_1, x_2, ..., x_k$ is a path in G, all the vertices $x_1, x_2, ..., x_k$ must belong to $T[u]$. Now, since vx_k is an edge in G, v must be an ancestor of x_k and thus of u in T, where u lies on the path from v to x_k. By Lemma 5, vertex v must be present in all bags X_w where w lies on the path from v to x_k, and consequently also in bag X_u. Therefore, u and v are both present in bag X_u and are neighbors in $tri(X, T)$.

Lemma 8. *Let (X, T, r) be an ordered tree decomposition of G, and let α be a post order of T. Let ET be the elimination tree of G_α^+. Then for any vertex u, if v is the parent of u in ET, then v lies on the path from u to the root in T.*

Proof. Since v is the parent of u in ET, uv is an edge of G_α^+. By Lemma 7, uv is also an edge of $tri(X, T)$. By Lemma 4, u and v must have an ancestor-descendant relationship in T. Since α is a post order of T, and $\alpha^{-1}(u) < \alpha^{-1}(v)$, v must be an ancestor of u in T.

Theorem 2. *For any graph G, $ts(G) = es(G)$.*

Proof. First we prove that $ts(G) \leq es(G)$. Let $ET = (V, P)$ be an elimination tree of G such that $es(G) = es(ET)$, and let r be the root vertex of ET. We define an ordered tree decomposition $(X = \{X_v \mid v \in V\}, T = ET, r)$ of G in the following way. For each vertex v in ET, put v in exactly the bags X_u such that $u \in ET_p[v]$. Regarding ET top down, each vertex u will appear for the first time in bag X_u, and clearly $|X_u \setminus X_v| = 1$ whenever v is the parent of u. It remains to show that (X, ET) is a tree decomposition of G. Conditions 1 and 3 of a tree decomposition are trivially satisfied since $ET_p[v]$ is connected and includes u for every vertex v. For Condition 2, if uv is an edge of G, then the lower numbered of v and u is a descendant of the other in ET. Let us say u is a descendant of v, then $u \in ET_p[v]$, and v and u will both appear in bag X_u. Thus (X, ET) is an ordered tree decomposition of G, and clearly, $ts(X, ET) = es(G)$. Consequently, $ts(G) \leq es(G)$.

Now we show that $es(G) \leq ts(G)$. Let (X, T, r) be an ordered tree decomposition of G with $ts(X, T, r) = ts(G)$. Let α be a post order on T, and let ET be the elimination tree of G_α^+. For any two adjacent vertices u and v in G, u and v must have an ancestor-descendant relationship both in T and in ET. Moreover, due to Lemma 8, all vertices that are on the path between u and v in ET must also be present on the path between u and v in T. Assume, without loss of generality, that u is numbered lower than v. By Lemma 5, v must belong to all the bags corresponding to the vertices on the path from v to u in T. Thus for each vertex v, $s(v)$ in ET is at most $l(v)$ in (X, T, r). Consequently, $es(G) \leq ts(G)$, and the proof is complete.

Theorems 1 and 2 imply the main combinatorial result of this paper.

Corollary 1. *For any graph G, $ts(G) = es(G) = mot(G)$.*

6   Treespan of Some Special Graph Classes

The *diameter* of a graph G, $diam(G)$, is the maximum length of a shortest path between any two vertices of G. The *density* of a graph G is defined as $dens(G) = (n - 1)/diam(G)$. The following result is well known

Lemma 9. [6] For any graph G, $bw(G) \geq \max\{dens(H) \mid H \subseteq G\}$.

A *caterpillar* is a tree consisting of a main path of vertices of degree at least two with some leaves attached to this main path.

Theorem 3. *For any graph G, $ts(G) \geq \max\{dens(H) \mid H \subseteq G$ and H is a caterpillar\}.*

Proof. Let the caterpillar H be a subgraph of G consisting of the following main path: $c_1, c_2, ..., c_{diam(H)-1}$. We view the bags of an ordered tree decomposition as labeled by vertices of G in the natural manner (as described before Lemma 3). Let (X, T, r) be an ordered tree decomposition of G with (X', T', r') being the topologically induced ordered tree decomposition on H, i.e. containing only bags labeled by a vertex from H, where we contract edges of T going to vertices labeled by vertices not in H to get T'. Let X_{c_i} be the 'highest' bag in (X', T', r') labeled by a vertex from the main path, so that only the subtree of (X', T', r') rooted at X_{c_i} contains any vertices from the main path. Let there be $h + 1$ bags on the path from X_{c_i} to the root $X_{r'}$ of (X', T', r'). Since vertex r' of H (a leaf unless $r' = c_i$) is adjacent to a vertex on the main path it appears in at least $h + 1$ bags, giving $ts(G) \geq h$. Moreover, by applying Lemma 3 we get that T' between its root $X_{r'}$ and X_{c_i} consists simply of a path without further children, so that the subtree rooted at X_{c_i} has $|V(H)| - h$ bags. Each of these bags contain a vertex from the main path since every leaf of H is adjacent in H only to a vertex on the main path, and by the pigeonhole principle we thus have that some main path vertex lives in at least $\lceil (|V(H)| - h)/(diam(H) - 1) \rceil$ bags. If $(|V(H)| - h)/(diam(H) - 1)$ is not an integer, then immediately we have the bound $ts(G) \geq \lfloor (|V(H)| - h)/(diam(H) - 1) \rfloor$. If $(diam(H) - 1)$ on the other hand does divide $(|V(H)| - h)$ then we apply the fact that at least $diam(H) - 2$ bags must contain at least two vertices from the main path, to account for edges between them, and for $diam(H) \geq 3$ (which holds except for the trivial case H a star) this increases the span of at least one main path vertex and we again get $ts(G) \geq \lfloor (|V(H)| - h)/(diam(H) - 1) \rfloor$.

Thus $ts(G) \geq \max\{h, \lfloor (|V(H)| - h)/(diam(H) - 1) \rfloor\}$. If $h \leq dens(H)$ we have that $\lfloor (|V(H)| - h)/(diam(H) - 1) \rfloor \geq (|V(H)| - 1)/diam(H)$ and therefore $\lfloor (|V(H)| - h)/(diam(H) - 1) \rfloor \geq dens(H)$. We conclude that $ts(G) \geq dens(H)$ and the lemma follows.

With this theorem, in connection with the following result from [2], we can conclude that $bw(G) = ts(G)$ for a caterpillar graph G.

Lemma 10. [2] For a caterpillar graph G, $bw(G) \leq \max\{dens(H) \mid H \subseteq G\}$.

Lemma 11. *For a caterpillar graph G, $bw(G) = ts(G) = \max\{dens(H) \mid H \subseteq G\}$.*

Proof. Let G be a caterpillar graph. Then, $bw(G) \geq ts(G) \geq \max\{dens(H) \mid H \subseteq G\} \geq bw(G)$. The first inequality was mentioned in Section 5, the second inequality is due to Theorem 3, and the last inequality is due to Lemma 10 since G is a caterpillar. Thus all of the mentioned parameters on G are equal.

A set of three vertices x, y, z of a graph G is called an *asteroidal triple* (AT) if for any two of these vertices there exists a path joining them that avoids the (closed) neighborhood of the third. A graph G is called an *asteroidal triple-free* (AT-free) graph if G does not contain an asteroidal triple. This notion was introduced by Lekkerkerker an Boland [17] for the following characterization of interval graphs: G is an interval graph if and only if it is chordal and AT-free.

A graph G is said to be *cobipartite* if it is the complement of a bipartite graph. Notice that cobipartite graphs form a subclass of AT-free claw-free graphs. Another subclass of AT-free claw-free graphs are the proper interval graphs, which were mentioned earlier. Thus G is a proper interval graph if and only if it is chordal and AT-free claw-free. A *minimal* triangulation of G is a triangulation H such that no proper subgraph of H is a triangulation of G. The following result is due to Parra and Scheffler.

Theorem 4. [20] *Let G be an AT-free claw-free graph. Then every minimal triangulation of G is a proper interval graph, and hence, $bw(G) = pw(G) = tw(G)$.*

Theorem 5. *For an AT-free claw-free graph G, $ts(G) = bw(G) = pw(G) = tw(G)$.*

Proof. Let G be AT-free claw-free and let H be its minimal triangulation such that $ts(G) = ts(H)$. Such a graph H must exist, since for an optimal ordered tree decomposition (X, T, r), the graph $tri(X, T)$ is chordal and $ts(tri(X, T)) = ts(G)$. Thus any minimal graph from the set of chordal graphs 'sandwiched' between $tri(X, T)$ and G can be chosen as H. By Theorem 4, H is a proper interval graph. Thus $\omega(H) - 1 = bw(H) \geq bw(G)$. Since $ts(H) \geq \omega(H) - 1$, we have that $ts(G) = ts(H) \geq \omega(H) - 1 \geq bw(G) \geq ts(G)$.

By the celebrated result of Arnborg, Corneil & Proskurowski [1], tree-width (and hence path-width and bandwidth) is NP-hard even for cobipartite graphs. Thus Theorem 5 yields the following corollary.

Corollary 2. *Computing treespan is NP-hard for cobipartite graphs.*

We conclude with an open question. For any graph G, $ts(G) \geq \lceil \Delta(G)/2 \rceil$. For trees of maximum degree at most 3 it is easy to prove that $ts(G) \leq \lceil \Delta(G)/2 \rceil$. It is an interesting question whether treespan can be computed in polynomial time for trees of larger max degree. Notice that bandwidth remains NP-complete on trees of max degree 3 [13].

References

1. S. ARNBORG, D.G. CORNEIL, AND A. PROSKUROWSKI, *Complexity of finding embeddings in a k-tree*, SIAM J. Alg. Disc. Meth., 8 (1987), pp. 277–284.
2. S.F. ASSMAN, G.W. PECK, M.M. SYSŁO, AND J.ZAK, *The bandwidth of caterpillars with hairs of length 1 and 2*, SIAM J. Alg. Disc. Meth., 2 (1981), pp. 387–392.
3. D. BIENSTOCK, *Graph searching, path-width, tree-width and related problems (a survey)*, DIMACS Ser. in Discrete Mathematics and Theoretical Computer Science, 5 (1991), pp. 33–49.
4. D. BIENSTOCK AND P. SEYMOUR, *Monotonicity in graph searching*, J. Algorithms, 12 (1991), pp. 239–245.
5. H.L. BODLAENDER, *A partial k-arboretum of graphs with bounded treewidth*, Theor. Comp. Sc., 209 (1998), pp. 1–45.
6. P.Z. CHINN, J. CHVÁTALOVÁ, A.K. DEWDNEY, AND N.E. GIBBS, *The bandwidth problem for graphs and matrices – a survey*, J. Graph Theory, 6 (1982), pp. 223–254.
7. N.D. DENDRIS, L.M. KIROUSIS, AND D.M. THILIKOS, *Fugitive-search games on graphs and related parameters*, Theor. Comp. Sc., 172 (1997), pp. 233–254.
8. J.A. ELLIS, I.H. SUDBOROUGH, AND J. TURNER, *The vertex separation and search number of a graph*, Information and Computation, 113 (1994), pp. 50–79.
9. F. FOMIN, *Helicopter search problems, bandwidth and pathwidth*, Discrete Appl. Math., 85 (1998), pp. 59–71.
10. F.V. FOMIN AND P.A. GOLOVACH, *Graph searching and interval completion*, SIAM J. Discrete Math., 13 (2000), pp. 454–464 (electronic).
11. M. FRANKLIN, Z. GALIL, AND M. YUNG, *Eavesdropping games: A graph-theoretic approach to privacy in distributed systems*, J. ACM, 47 (2000), pp. 225–243.
12. D. FULKERSON AND O. GROSS, *Incidence matrices and interval graphs*, Pacific Journal of Math., 15 (1965), pp. 835–855.
13. M.R. GAREY, R.L. GRAHAM, D.S. JOHNSON, AND D.E. KNUTH, *Complexity results for bandwidth minimization*, SIAM J. Appl. Math., 34 (1978), pp. 477–495.
14. F. GAVRIL, *The intersection graphs of subtrees in trees are exactly the chordal graphs*, J. Combin. Theory Ser. B, 16 (1974), pp. 47–56.
15. L.M. KIROUSIS AND C.H. PAPADIMITRIOU, *Searching and pebbling*, Theor. Comp. Sc., 47 (1986), pp. 205–218.
16. A.S. LAPAUGH, *Recontamination does not help to search a graph*, J. ACM, 40 (1993), pp. 224–245.
17. C.G. LEKKERKERKER AND J.C. BOLAND, *Representation of a finite graph by a set of intervals on the real line*, Fund. Math, 51 (1962), pp. 45–64.
18. F.S. MAKEDON, C.H. PAPADIMITRIOU, AND I.H. SUDBOROUGH, *Topological bandwidth*, SIAM J. Alg. Disc. Meth., 6 (1985), pp. 418–444.
19. F.S. MAKEDON AND I.H. SUDBOROUGH, *On minimizing width in linear layouts*, Disc. Appl. Math., 23 (1989), pp. 201–298.
20. A. PARRA AND P. SCHEFFLER, *Treewidth equals bandwidth for AT-free claw-free graphs*, Technical Report 436/1995, Technische Universität Berlin, Fachbereich Mathematik, Berlin, Germany, 1995.
21. F.S. ROBERTS, *Indifference graphs*, in Proof Techniques in Graph Theory, F. Harary, ed., Academic Press, 1969, pp. 139–146.
22. N. ROBERTSON AND P.D. SEYMOUR, *Graph minors – a survey*, in Surveys in Combinatorics, I. Anderson, ed., Cambridge Univ. Press, 1985, pp. 153–171.
23. A.L. ROSENBERG AND I.H. SUDBOROUGH, *Bandwidth and pebbling*, Computing, 31 (1983), pp. 115–139.

Constructing Sparse *t*-Spanners with Small Separators

Joachim Gudmundsson*

Department of Mathematics and Computing Science, TU Eindhoven
5600 MB Eindhoven, The Netherlands.

Abstract. Given a set of n points $\mathcal{S}$ in the plane and a real value $t > 1$ we show how to construct in time $\mathcal{O}(n \log n)$ a t-spanner $\mathcal{G}$ of $\mathcal{S}$ such that there exists a set of vertices $\mathcal{S}'$ of size $\mathcal{O}(\sqrt{n} \log n)$ whose removal leaves two disconnected sets $\mathcal{A}$ and $\mathcal{B}$ where neither is of size greater than $2/3 \cdot n$. The spanner also has some additional properties; low weight and constant degree.

1 Introduction

Complete graphs represent ideal communication networks but they are expensive to build; sparse spanners represent low cost alternatives. The weight of the spanner network is a measure of its sparseness; other sparseness measures include the number of edges, maximum degree and the number of Steiner points. Spanners for complete Euclidean graphs as well as for arbitrary weighted graphs find applications in robotics, network topology design, distributed systems, design of parallel machines, and many other areas, and have been subject to considerable research [1,2,5,8,14]. Consider a set $\mathcal{S}$ of n points in the plane. A network on $\mathcal{S}$ can be modeled as an undirected graph G with vertex set $\mathcal{S}$ and with edges $e = (u,v)$ of weight $wt(e)$. In this paper we will study Euclidean networks, a Euclidean network is a geometric network where the weight of the edge $e = (u,v)$ is equal to the Euclidean distance $d(u,v)$ between its two endpoints u and v. Let $t > 1$ be a real number. We say that G is a *t-spanner* for $\mathcal{S}$, if for every pair of points $u,v \in \mathcal{S}$, there exists a path in G of weight at most t times the Euclidean distance between u and v. A *sparse t-spanner* is defined to be a t-spanner with a linear number of edges and total weight (sum of edge weights) $\mathcal{O}(wt(MST(\mathcal{S})))$, where $wt(MST(\mathcal{S}))$ is the total weight of a minimal spanning tree of $\mathcal{S}$.

Many algorithms are known that compute t-spanners with $\mathcal{O}(n)$ edges that have additional properties such as bounded degree, small spanner diameter (i.e., any two points are connected by a t-spanner path consisting of only a small number of edges), low weight (i.e., the total length of all edges is proportional to the weight of a minimum spanning tree of $\mathcal{S}$), and fault-tolerance; see, e.g., [1,2,3,5,7,8,9,11,12,14,19], and the surveys [10,20]. All these algorithms compute t-spanners for any given constant $t > 1$.

---

* Supported by The Netherlands Organisation for Scientific Research (NWO).

A. Lingas and B.J. Nilsson (Eds.): FCT 2003, LNCS 2751, pp. 86–97, 2003.
© Springer-Verlag Berlin Heidelberg 2003

In this paper, we consider the construction of a sparse t-spanner with constant degree and with a provable balanced separator. Finding small separators in a graph is a problem that has been studied extensively within theoretical computer science for the last three decades, and a survey of the area can be found in the book by Rosenberg and Heath [17]. Spanners with good separators have, for example, applications in the construction of external memory data structures [16]. It is well-known that planar graphs have small separators and, hence any planar spanner has a small separator. Bose et al. [4] showed how to construct a planar t-spanner for $t \approx 10$ with constant degree and low weight. Also, it is known that the Delaunay triangulation is a t-spanner for $t = 2\pi/(3\cos(\pi/6))$ [13]. For arbitrary values of $t > 1$ this article is, to the best of the author's knowledge, the first time that separators have been considered.

Definition 1. *Given a graph $G = (V, E)$, a separator is a set of vertices $C \subset V$ whose removal leaves two disconnected sets $\mathcal{A}$ and $\mathcal{B}$. A separator C is said to be balanced if the size of both $\mathcal{A}$ and $\mathcal{B}$ is at most $2/3 \cdot |V|$.*

The main result of this paper is the following theorem.

Theorem 1. *Given a set $\mathcal{S}$ of n points in the plane and a constant $t > 1$, there is an $\mathcal{O}(n \log n)$-time algorithm that constructs a graph $\mathcal{G} = (\mathcal{S}, \mathcal{E})$*

1. *that it is a t-spanner of $\mathcal{S}$,*
2. *that has a linear number of edges,*
3. *that has weight $O(wt(MST(\mathcal{S})))$,*
4. *that has a balanced separator of size $\mathcal{O}(\sqrt{n}\log n)$,*
5. *and, in which each node has constant degree.*

The paper is organised as follows. First we present an algorithm that produces a t-spanner $\mathcal{G}$. Then, in Section 3, we prove that $\mathcal{G}$ has all the properties stated in Theorem 1.

2   Constructing a t-Spanner

In this section we first show an algorithm that, given a set $\mathcal{S}$ of n points in the plane together with a real value $t > 1$, produces a t-spanner $\mathcal{G}$. The algorithm works in two steps: first it produces a modified approximate θ-graph [6,12,18], denoted $\mathcal{G}_\theta$, which is then pruned using a greedy approach [1,5,8,11]. We show that the resulting graph, denoted $\mathcal{G}$, has two basic properties that will be used to prove that it is a sparse spanner with a balanced separator.

2.1   The Algorithm

It has long been known that for any constant $t > 1$, every point set $\mathcal{S}$ in the plane has a t-spanner with $\mathcal{O}(n)$ edges. One such construction is the θ-graph of $\mathcal{S}$. Let $\theta < \pi/4$ be a value such that $k_\theta = 2\pi/\theta$ is a positive integer. The θ-graph of $\mathcal{S}$ is obtained by drawing k_θ non-overlapping cones around each point

$p \in \mathcal{S}$, each spanning an angle of θ, and connecting p to the point in each cone closest to p. For each of these edges, p is said to be the *source* while the other endpoint is said to be the *sink*. The result is a t_θ-spanner with at most nk_θ edges. Here $t_\theta = (\cos(\theta) - \sin(\theta))^{-1}$. The time needed to construct the θ-graph for any constant θ is $\mathcal{O}(n \log n)$ [12].

Approximate the θ-Graph. Here we will build an approximate version of the θ-graph, which we denote a ϕ-graph $\mathcal{G}_\phi = (\mathcal{S}, \mathcal{E}_\phi)$. First build a θ'-graph $(\mathcal{S}, \mathcal{E}_{\theta'})$ with $\theta' = \epsilon\theta$, for some small constant ϵ, as shown in Fig. 1a. A point $v \in \mathcal{S}$ belongs to $\mathcal{S}_p$ if and only if $(p, v) \in \mathcal{E}_{\theta'}$ and p is the source of (p, v). Process each point $p \in \mathcal{S}$ iteratively as follows until $\mathcal{S}_p$ is empty. Let v be the point in $\mathcal{S}_p$ closest to p. Add the edge (p, v) to $\mathcal{E}_\phi$ and remove every point u from $\mathcal{S}_p$ for which it holds that $\angle vpu < (\theta/2)$, as illustrated in Fig. 1b. Continue until $\mathcal{S}_p$ is empty.

$\mathcal{G}_{\phi'}$ is a $t_{\phi'}$-spanner with $t_{\phi'} = (\cos(\phi') - \sin(\phi'))^{-1}$ and, since two adjacent cones may overlap, the number of outgoing edges is bounded by $4\pi/\theta$. Arya et al. [2] showed that a θ-graph can be pruned such that each point has constant degree. Applying this result to $\mathcal{G}_{\phi'}$ gives a t_ϕ-spanner $\mathcal{G}_\phi$ where each point has degree bounded by $O(\frac{t_{\phi'}}{\theta(t_\phi - t_\phi)})$. Note that the value of ϕ' is $\theta(1 + 2\epsilon)$.

Remove "long" Edges Intersecting "short" Edges. The remaining two steps of the construction algorithms are both pruning the graph. Prune $\mathcal{G}_\phi = (\mathcal{S}, \mathcal{E}_\phi)$ to obtain a graph $\mathcal{G}_\theta = (\mathcal{S}, \mathcal{E}_\theta)$ as follows. Build the minimum spanning tree $\mathcal{T}_{mst} = (\mathcal{S}, \mathcal{E}_{mst})$ of $\mathcal{S}$. Sort the edges in $\mathcal{E}_\phi$ and in $\mathcal{E}_{mst}$ with respect to their lengths. We obtain the two ordered sets $\mathcal{E}_\phi = \{e_1, \ldots, e_{O(n)}\}$ and $\mathcal{E}_{mst} = \{e'_1, \ldots, e'_{n-1}\}$ respectively. The idea is to process the edges in $\mathcal{E}_\phi$ in order, while maintaining a graph $\mathcal{T}$ that will cluster vertices that lie within distance l from each other, where $l = |e_i|/n^2$ and e_i is the edge just about to be processed. The graph will also contain information about the convex hull of each cluster and we will show that this can be done in linear time if the minimum spanning tree is given.

Initially $\mathcal{T}$ contains n clusters where every cluster is a single point. Assume that we are about to process an edge $e_i = (u, v) \in \mathcal{E}_\phi$. The first step is to merge all clusters in $\mathcal{T}$ that are connected by an edge of length at most $l = |e_i|/n^2$. This is done by extracting the shortest edge, $e'_j = (u'_j, v'_j)$, in $\mathcal{E}_{mst}$ and merging the two clusters C_1 and C_2 containing u'_j respectively v'_j. This is done until there are no more edges in $\mathcal{E}_{mst}$ of length less than $l = |e_i|/n^2$. At the same time we also compute the convex hull, denoted C, of C_1 and C_2, note that this can be done in linear time with respect to the decrease in complexity from C_1 and C_2, to C. Hence, in total, it will require linear time to update the convex hulls of the clusters. Now we are ready to process $e_i = (u, v)$. Let $m(u, l)$ and $m(v, l)$ denote the clusters in $\mathcal{T}$ containing u and v respectively. If e_i intersects the convex hull of either $m(u, l)$ or $m(v, l)$ then e_i is discarded, otherwise it is added to $\mathcal{E}_\theta$, as shown in Fig. 1c. Since the original graph is a ϕ-graph it is not hard to see that between every pair of clusters, C_1 and C_2, there is at least one edge $(u, v) \in \mathcal{E}_\phi$ such that u and v lies on the convex hull of C_1 and C_2 respectively. This finishes the second part of the algorithm and we sum it up by stating the following straight-forward observation.

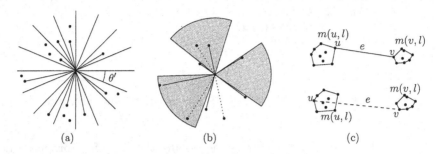

(a)   (b)   (c)

Fig. 1. (a) Constructing a θ'-graph, which is then (b) pruned to obtain a ϕ-graph. (c) Every edge is tested to see if it intersects the convex hulls of the clusters containing u and v.

Observation 1 *The above algorithm produces a graph $\mathcal{G}_\theta$ in time $\mathcal{O}(n \log n)$ which is a t_θ-spanner, where $t_\theta \leq (\frac{1}{\cos(\phi) - \sin(\phi)} + \frac{1}{n})$.*

Greedily Pruning the Graph. We are given a modified approximate θ-graph $\mathcal{G}_\theta$ for $t_\theta = t/(1 + \varepsilon)$. The final step is to run the greedy t_g-spanner algorithm with $\mathcal{G}_\theta$ and $t_g = (1 + \varepsilon)$ as input. The basic idea of the standard greedy algorithm is sorting the edges (by increasing weight) and then processing them in order. Greedy processing of an edge $e = (u, v)$ entails a shortest path query, i.e., checking whether the shortest path in the graph built so far has length at most $t \cdot d(u, v)$. If the answer to the query is no, then edge e is added to the spanner G, else it is discarded, see Fig. 2. The greedy algorithm was first considered by Althöfer et al. [1] and later variants of the greedy algorithm using clustering techniques improved the analysis [5,8,11]. In [8] it was observed that shortest path queries need not be answered precisely. Instead, approximate shortest path queries suffice, of course, this meant that the greedy algorithm, too, was only approximately simulated by the algorithm. The most efficient algorithm was recently presented by Gudmundsson et al. [11], where they show an $\mathcal{O}(n \log n)$-time variant of the greedy algorithm. In the approximate greedy algorithm an approximate shortest path query checks if the path is longer than $\tau \cdot d(u, v)$, where $1 < \tau < t$.

2.2 Two Basic Properties

The final result is a t-spanner $\mathcal{G} = (\mathcal{S}, \mathcal{E})$ with several nice properties, among them the following two simple and fundamental properties that will be used in

---

Algorithm STANDARD-GREEDY($G = (\mathcal{S}, E), t$)
1.   sort the edges in E by increasing weight
2. $E' := \emptyset$
3. $G' := (\mathcal{S}, E')$
4.   **for** each edge $(u, v) \in E$ **do**
5.   **if** SHORTESTPATH(G', u, v) $> t \cdot d(u, v)$ **then**
6. $E' := E' \cup \{(u, v)\}$
7. $G' := (\mathcal{S}, E')$
8.   output G'

Fig. 2. The naive $\mathcal{O}(|E|^2 \cdot |\mathcal{S}| \log |\mathcal{S}|)$-time greedy spanner algorithm

the analysis: the obtuse Empty-cone property, and the Leap-frog property. Let $\mathcal{C}(u, v, \theta)$ denote the (unbounded) cone with apex at u, spanning an angle of θ such that (u, v) splits the angle at u into two equal angles. An edge set E is said to have the Empty-cone property if for every edge $e = (u, v) \in E$ it holds that v is the point closest to u within $\mathcal{C}(u, v, \theta)$.

From the definition of θ-graphs it is obvious that $\mathcal{G}_\theta$ satisfies the Empty-cone property, actually we can see that the property can be somewhat strengthen to what we call an obtuse Empty-cone property. Assume w.l.o.g. that (u, v) is vertical, u lies below v and u is the source of e. Since u and v lies on the convex hull of $m(u, l)$ and $m(v, l)$ (otherwise e would have been discarded in the pruning step) it holds that there are two half disks intersecting (u, v) with radii $l = |e|/n^2$ and centers at u and v, see Fig 3a. Thus, the union of the half disks and the part of the cone $\mathcal{C}(u, v, \theta)$ within distance $|uv|$ from u is said to be an obtuse cone, and is denoted $\mathcal{C}_o(u, v, \theta)$. The following observation is straight-forward.

Observation 2 *The shortest edge that intersects an edge $e = (u, v) \in E$ satisfying the obtuse Empty-cone property must be longer than $\frac{2|e| \sin \theta/2}{n^2}$.*

Next we consider the Leap-frog property. Let $t \geq \tau > 1$. An edge set E satisfies the (t, τ)-*leapfrog* property if the following is true for every possible $E' = \{(u_1, v_1), \ldots, (u_m, v_m)\}$, which is a subset of E:

$$\tau \cdot wt(u_1, v_1) < \sum_{i=2}^{m} wt(u_i, v_i) + t \cdot \left(\sum_{i=1}^{m-1} wt(v_i, u_{i+1}) + wt(v_m, u_1) \right).$$

Informally, this definition says that if there exists an edge between u_1 and v_1 then any path, not including (u_1, v_1) must have length greater than $\tau \cdot wt(u_1, v_1)$, as illustrated in Fig. 3b.

Lemma 1. *Given a set of points in the plane and a real value $t > 1$ the above algorithm produces a t-spanner $\mathcal{G} = (\mathcal{S}, \mathcal{E})$ that satisfies the obtuse Empty-cone property, and the Leap-frog property.*

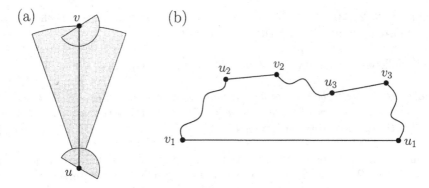

Fig. 3. (a) The shaded area, denoted $C_o(u, v, \theta)$, is empty if e satisfies the obtuse Empty-cone property. (b) Illustrating the Leap-frog property.

Proof. Since $\mathcal{E}$ is a subset of the edges in the approximate θ-graph $\mathcal{G}_\theta$ it immediately follows that $\mathcal{E}$ has the obtuse Empty-cone property.

Now, let C be the shortest simple cycle in $\mathcal{G}$ containing an arbitrary edge $e = (u, v)$. To prove that $\mathcal{G}$ satisfies the leapfrog property we have to estimate $wt(C) - wt(u, v)$. Let $e' = (u', v')$ be the longest edge of C. Among the cycle edges e' is examined last by the algorithm. What happens while the algorithm is examining e'? In [11] it was shown that if the algorithm adds an edge e' to the graph the shortest path between u' and v' must be longer than $\tau \cdot d(u', v')$ in the partial graph constructed so far. Hence, $wt(C) - d(u, v) \geq wt(C) - d(u', v') > \tau \cdot d(u', v') \geq \tau \cdot d(u, v)$. The lemma follows. $\qquad\square$

The obtuse Empty-cone property will be used to prove that $\mathcal{G}$ has a balanced separator and, the Leap-frog property will mainly be used to prove that the total weight of $\mathcal{G}$ is small, as will be shown in Section 3.2.

3   The Analysis

In this section we will perform a close analysis of the graph constructed by the algorithm presented in the previous section. First we study the separator property and then, in Section 3.2, we take a closer look at the remaining properties claimed in Theorem 1.

3.1   A Balanced Separator

In this subsection we prove that the graph $\mathcal{G} = (\mathcal{S}, \mathcal{E})$ has a balanced separator of size $O(\sqrt{n} \log n)$, by using the famous Planar Separator Theorem by Lipton and Tarjan [15].

Fact 1 *(Planar Separator Theorem [15]) Every planar graph G with n vertices can be partitioned into three parts $\mathcal{A}$, $\mathcal{B}$ and $\mathcal{C}$ such that $\mathcal{C}$ is a separator of*

G and $|\mathcal{A}| \leq 2n/3$, $|\mathcal{B}| \leq 2n/3$ and $|\mathcal{C}| \leq 2\sqrt{2}\sqrt{n}$. Furthermore, there is an algorithm to compute this partition in time $\mathcal{O}(n)$.

The following corollary is a straight-forward consequence of Fact 1.

Corollary 1. *Let G be a graph in the plane such that every edge of G intersects at most N other edges of G. It can be partitioned into three parts $\mathcal{A}$, $\mathcal{B}$ and $\mathcal{C}$ such that $\mathcal{C}$ is a separator of G and $|\mathcal{A}| \leq 2n/3$, $|\mathcal{B}| \leq 2n/3$ and $|\mathcal{C}| \leq 2\sqrt{2}\sqrt{n} \cdot N$.*

This corollary immediately suggests a way prove that $\mathcal{G}$ has a balanced separator of size $\mathcal{O}(N\sqrt{n})$, namely prove that every edge in $\mathcal{E}$ intersects at most N other edges in $\mathcal{E}$. It should be noted that it is not enough to prove that the intersection graph $\mathcal{I}$ of $\mathcal{G}$ has low complexity since finding a balanced separator in $\mathcal{I}$ does not imply a balanced separator of $\mathcal{G}$.

The first step is to partition the edge set $\mathcal{E}$ into a constant number of groups, each having the three nice properties listed below. The idea of partitioning the edge set into groups is borrowed from [7].

The edge set $\mathcal{E}$ can be partitioned into a constant number of groups such that the following three properties are satisfied for each subset:

1. **Near-parallel property:** Associate to each edge $e = (u, v)$ a slope as follows. Let h be a horisontal segment with left endpoint at the source of e. The of e is now the counter-clockwise angle between h and e. An edge e in $\mathcal{E}$ belongs to the subgroup $\mathcal{E}_i$ if the slope of e is between $(i-1)\beta$ and $i\beta$, for some small angle $\beta \ll \theta$.

2. **Length-grouping property:** Let $\gamma > 0$ be a small constant. The length of any two edges in $\mathcal{E}_{i,j}$ differ by at most a factor $\delta = (1 - \gamma)$ or by at least a factor $x\delta^{c-1}$.

   Consider a group $\mathcal{E}_i$ of near-parallel edges. Let the length of the longest edge in $\mathcal{E}_i$ be ℓ. Partition the interval $[0, \ell]$ into an infinite number of intervals $\{[\ell\delta, \ell], [\ell\delta^2, \ell\delta], [\ell\delta^3, \ell\delta^2], \ldots\}$. Define the subgroup $\mathcal{E}_{i,j}$ as containing the edges whose lengths lie in intervals $\{[\ell\delta^{j+1}, \ell\delta^j], [\ell\delta^{j+c+1}, \ell\delta^{j+c}], \ldots\}$. There is obviously only a constant number of such groups.

3. **Empty-region property:** Any two edges e_1 and e_2 in $\mathcal{E}_{i,j,k}$ that are near-parallel and almost of equal length are separated by a distance which is a large multiple of $|e_1|$. Hence, two "near-equal" edges cannot be close to each other.

   To achieve this grouping [7], construct a graph H where the nodes are edges of $\mathcal{E}_{i,j}$, and two "near-equal" nodes in H, say e_1 and e_2, are connected by an edge if e_1 intersects a large cylinder of radius $\alpha|e_2|$ and height $\alpha|e_2|$ centered at the center of the edge e_2 in $\mathcal{E}_{i,j}$, for some large constant α. This graph has constant degree, because by the Leap-frog property, there can be only a constant number of similar "near-equal" edges whose endpoints can be packed into the cylinder. Thus this graph has a constant chromatic number, and consequently a constant number of independent sets. Hence, $\mathcal{E}_{i,j}$ is subdivided into a constant number of groups, denoted $\mathcal{E}_{i,j,k}$.

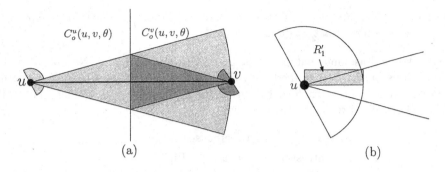

Fig. 4. (a) Illustrating the split of $C_o(u, v, \theta)$ into $C_o^u(u, v, \theta)$ and $C_o^v(u, v, \theta)$. (b) $\mathcal{R}_1'$ lies inside $\mathcal{R}_1$.

Let $e = (u, v)$ be an arbitrary edge in $\mathcal{E}$. Next we will prove that the number of edges in $\mathcal{D} = \mathcal{E}_{i,j,k}$, for any i, j and k, that may intersect e is bounded by $\mathcal{O}(\log n)$ and since there is only a constant number of groups this implies that e is intersected by at most a logarithmic number of edges of $\mathcal{E}$. For simplicity we will assume that e is horisontal.

To simplify the analysis we partition $C_o(u, v, \theta)$ into two regions, $C_o^u(u, v, \theta)$ and $C_o^v(u, v, \theta)$, where every point in $C_o^u(u, v, \theta)$ lies closer to u than to v, see Fig. 4a. We will prove that the number of edges intersecting (u, v) within the region $C_o^u(u, v, \theta)$ is bounded by $\mathcal{O}(\log n)$. By symmetry the proof also holds for the region $C_o^v(u, v, \theta)$ since a cone of size and shape as described by the region $C_o^u(u, v, \theta)$ can be placed within $C_o^v(u, v, \theta)$, see Fig. 4a. Hence, for the rest of this section we will only consider the region $C_o^u(u, v, \theta)$.

Let $\mathcal{D}' = \{e_1, e_2, \dots, e_r\}$ be the edges in $\mathcal{D}$ intersecting the part of e within $C_o^u(u, v, \theta)$, ordered from left to right with respect to their intersection with e. Let q_i denote the intersection point between e_i and e and let y_i denote the length of the intersection between a vertical line through q_i and $C_o^u(u, v, \theta)$.

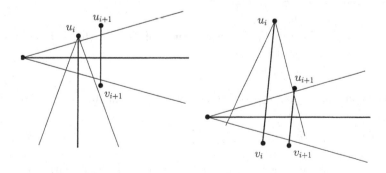

Fig. 5. Illustrating the proof of Lemma 2

Lemma 2. *The distance between any pair of consecutive points q_i and q_{i+1} along e is greater than $\frac{y_i}{2}\sin(\theta/2)$.*

Proof. We will assume that u_i and u_{i+1} lie above v_i and v_{i+1}. Note that in the calculations below we assumed that the edges in $\mathcal{D}$ are parallel but since the final bound is far from the exact solution the bound stated in the lemma is still valid. There are three cases to consider.

1. $|e_{i+1}| < \delta^c \cdot |e_i|$. We will have two subcases:
   a) e_{i+1} does not intersect $\mathcal{C}(u_i, v_i, \theta)$, see Fig. 5a.
      The distance between e_i and e_{i+1} is minimised when v_{i+1} is the intersection between the lower side of $\mathcal{C}(u, v, \theta)$ and the right side of $\mathcal{C}(u_i, v_i, \theta)$, and u_i lies on the top side of $\mathcal{C}(u, v, \theta)$. Now, straight-forward trigonometry shows that the horisontal distance between q_i and q_{i+1} is greater than $y_i \sin(\theta/2) > \frac{y_i}{2}\sin(\theta/2)$.
   b) e_{i+1} intersects $\mathcal{C}(u_i, v_i, \theta)$, see Fig. 5b.
      The distance between q_i and q_{i+1} is minimised when u_{i+1} lies on the right side of $\mathcal{C}(u_i, v_i, \theta)$ in a leftmost position. Again, using straight-forward trigonometry we obtain that the distance between q_i and q_{i+1} is greater than $(e_i(1 - \delta^{c-1})\sin(\theta/2) > y_i(1 - \delta^{c-1})\sin(\theta/2) > \frac{y_i}{2}\sin(\theta/2)$.
2. $|e_i| \leq \delta^c \cdot e_{i+1}$. We will have two subcases
   a) e_i does not intersect $\mathcal{C}(u_{i+1}, v_{i+1}, \theta)$, see Fig. 6a.
      The proof is almost identical to case 1a. The distance between q_i and q_{i+1} is minimised when v_{i+1} is the intersection between the lower side of $\mathcal{C}(u, v, \theta)$ and the right side of $\mathcal{C}(u_i, v_i, \theta)$, and u_i lies on the top side of $\mathcal{C}(u, v, \theta)$. Simple calculations show that the distance between q_i and q_{i+1} is greater than $\frac{y_i}{2}\sin(\theta/2)$.
   b) e_i intersects $\mathcal{C}(u_{i+1}, v_{i+1}, \theta)$, see Fig. 6b.
      The proof is similar to case 1b. The distance between q_i and q_{i+1} is minimised when u_i lies on the left side of $\mathcal{C}(u_{i+1}, v_{i+1}, \theta)$ in a rightmost position. Again, using straight-forward trigonometry we obtain that the distance between e_i and e_{i+1} is at least $(e_i(1 - \delta^{c-1})\sin(\theta/2) > y_i(1 - \delta^{c-1})\sin(\theta/2) > \frac{y_i}{2}\sin(\theta/2)$.
3. $\delta^c \cdot |e_i| \leq |e_{i+1}| \leq (1/\delta^c) \cdot |e_i|$.
   It follows from the Empty-region property of $\mathcal{D}$ that the distance between e_i and e_{i+1} is at least $(\alpha \cdot \max(|e_i|, |e_{i+1}|))$.

$\square$

We need one more lemma before we can state the main theorem of this section.

Lemma 3. *e intersects $\mathcal{O}(\log n)$ edges in $\mathcal{G}$.*

Proof. As above we assume w.l.o.g. that e is horisontal. Partition $\mathcal{C}_o^u(u, v, \theta)$ into two regions, the region $\mathcal{R}_1$ containing all points in $\mathcal{C}_o^u(u, v, \theta)$ with horisontal distance at most $(|e|/n^2)$ from u, and the region $\mathcal{R}_2$ containing the remaining

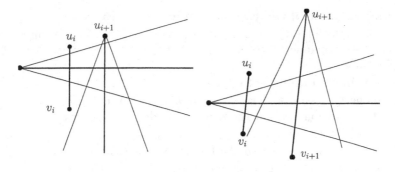

Fig. 6. Illustrating the proof of Lemma 2

region. Consider the disk D_u of radius $(|e|/n^2)$ with center at u. From the construction of $\mathcal{G}$ it holds that there is a half-disk centered at u and with radius $(|c|/n^2)$ that is empty. We may assume w.l.o.g. that the half-disk covers the upper right quadrant of D_u (otherwise it must cover the lower right quadrant of D_u), Fig. 4b.

Let us first consider the region $\mathcal{R}_1$. Let $\mathcal{R}_1'$ be the rectilinear box inside $\mathcal{R}_1$ with bottom left corner at u, width $(|e|/n^2)$ and height $(|e|/n^2 \cdot \sin(\theta/2))$, as illustrated in Fig. 4b. Every edge intersecting e within $\mathcal{R}_1$ must also intersect $\mathcal{R}_1'$, hence we may consider $\mathcal{R}_1'$ instead of $\mathcal{R}_1$. According to Lemma 2, the distance between q_i and q_{i+1}, is at least $\frac{|e|\cdot\sin(\theta/2)}{n^2} \cdot \frac{\sin(\theta/2)}{2}$, which implies that the total number of edges that may intersect e within $\mathcal{R}_1'$ is $\frac{|e|}{n^2} / \frac{|e|\cdot\sin^2(\theta/2)}{n^2} = \frac{1}{\sin^2(\theta/2)}$ which is constant since θ is a constant.

Next we consider the part of e within $\mathcal{R}_2$. The width of $\mathcal{R}_2$ is less than $(|e|/2)$, its left side has height at least $(|e|/n^2 \cdot \sin(\theta/2))$ and its right side has height at most $(|e|/2 \cdot \sin\theta)$. From Lemma 2 it holds that $y_{i+1} \geq y_i(1 + \frac{\sin^2(\theta/2)}{2\cos(\theta/2)})$ since the distance between q_i and q_{i+1} is at least $y_i/2 \cdot \sin(\theta/2)$. Set $\lambda = \frac{\sin^2(\theta/2)}{2\cos(\theta/2)}$. The length of the shortest edge $\ell_{\min}$ is $\Omega(1/n^2)$ according to Observation 2, and the value of y_i is at least $(1 + \lambda)^{i-1} \cdot \ell_{\min}$. The largest y-value is obtained for the rightmost intersection point q_b. Obviously y_b is bounded by $(|e|/2 \cdot \sin\theta)$, hence it holds that $(1 + \lambda)^b \cdot \ell_{\min} = \mathcal{O}(|e|)$ which is true for $b = \mathcal{O}(\log n)$. $\qquad\square$

Now we are ready to state the main theorem of this section, which is obtained by putting together Corollary 1 and Lemma 3 .

Theorem 2. $\mathcal{G}$ *has a balanced separator of size* $\mathcal{O}(\sqrt{n}\log n)$.

3.2   Other Properties

Theorem 1 claims that $\mathcal{G}$ has five properties which we will discuss below, one by one:

1. **$\mathcal{G}$ is a t-spanner of the complete Euclidean graph.**
   Since $\mathcal{G}_\theta$ is a $(t/(1+\varepsilon))$-spanner of the complete Euclidean graph and since $\mathcal{G}$ is a $(1+\varepsilon)$-spanner of $\mathcal{G}_\theta$ it follows that $\mathcal{G}$ is a t-spanner of the complete Euclidean graph.

2. **$\mathcal{G}$ has a linear number of edges.**
   This property is straight-forward since $\mathcal{G}$ is a subgraph of $\mathcal{G}_\theta$ and we already know from Section 2.1 that the number of edges in $\mathcal{G}_\theta$ is less than $n \cdot \frac{4\pi}{\theta}$.

3. **$\mathcal{G}$ has weight $O(wt|MST|)$.**
   Das and Narasimhan showed the following fact about the weight of graphs that satisfy the Leap-frog property.

   Fact 2 [Theorem 3 in [8]] There exists a constant $0 < \phi < 1$ such that the following holds: if a set of line segments E in d-dimensional space satisfies the (t,τ)-leapfrog property, where $t \geq \tau \geq \phi t + 1 - \phi > 1$, then $wt(E) = O(wt(MST))$, where MST is a minimum spanning tree connecting the endpoints of E. The constant implicit in the O-notation depends on t and d.

   The low weight property now follows from the above fact together with Lemma 1 and the fact that $\mathcal{G}_\theta$ is a $(t/(1+\varepsilon))$-spanner of the complete Euclidean graph of $\mathcal{S}$, hence it also includes a spanning tree of weight $\mathcal{O}(wt(MST(\mathcal{S})))$.

4. **$\mathcal{G}$ has a balanced separator**
   Follows from Theorem 2.

5. **$\mathcal{G}$ has constant degree.**
   This property is straight-forward since $\mathcal{G}$ is a subgraph of $\mathcal{G}_\phi$, constructed in Section 2.1, which has constant degree.

   This concludes the proof of Theorem 1.

4   Conclusions and Further Research

We have shown the first algorithm that given a set of points in the plane and a real value $t > 1$ constructs in time $\mathcal{O}(n \log n)$ a sparse t-spanner with constant degree and with a provably balanced separator. There are two obvious questions: (1) Is there a separator of size $\mathcal{O}(\sqrt{n})$, and (2) will the algorithm produce a t-spanner with similar properties in higher dimensions. Another interesting question to answer is if the greedy algorithm by itself produces a t-spanner with a balanced separator.

Acknowledgements. I am grateful to Anil Maheswari for introducing me to the problem, and to Mark de Berg, Otfried Cheong and Andrzej Lingas for stimulating and helpful discussions during the preparation of this article.

References

1. I. Althöfer, G. Das, D. P. Dobkin, D. Joseph, and J. Soares. On sparse spanners of weighted graphs. *Discrete Computational Geometry, 9.*
2. S. Arya, G. Das, D. M. Mount, J. S. Salowe, and M. Smid. Euclidean spanners: short, thin, and lanky. In *Proc. 27th Annual ACM Symposium on Theory of Computing*, pages 489–498, 1995.
3. J. Bose, J. Gudmundsson, and P. Morin. Ordered theta graphs. In *Proc. 14th Canadian Conference on Computational Geometry*, 2002.
4. J. Bose, J. Gudmundsson, and M. Smid. Constructing plane spanners of bounded degree and low weight. In *Proc. 10th European Symposium on Algorithms*, 2002.
5. B. Chandra, G. Das, G. Narasimhan, and J. Soares. New sparseness results on graph spanners. *International Journal of Computational Geometry and Applications*, 5:124–144, 1995.
6. K. L. Clarkson. Approximation algorithms for shortest path motion planning. In *Proc. 19th ACM Symposium on Computational Geometry*, pages 56–65, 1987.
7. G. Das, P. Heffernan, and G. Narasimhan. Optimally sparse spanners in 3-dimensional Euclidean space. In *Proc. 9th Annual ACM Symposium on Computational Geometry*, pages 53–62, 1993.
8. G. Das and G. Narasimhan. A fast algorithm for constructing sparse Euclidean spanners. *International Journal of Computational Geometry and Applications*, 7:297–315, 1997.
9. G. Das, G. Narasimhan, and J. Salowe. A new way to weigh malnourished Euclidean graphs. In *Proc. 6th ACM-SIAM Sympos. Discrete Algorithms*, pages 215–222, 1995.
10. D. Eppstein. Spanning trees and spanners. In J.-R. Sack and J. Urrutia, editors, *Handbook of Computational Geometry*, pages 425–461. Elsevier Science Publishers, Amsterdam, 2000.
11. J. Gudmundsson, C. Levcopoulos, and G. Narasimhan. Improved greedy algorithms for constructing sparse geometric spanners. *SIAM Journal of Computing*, 31(5):1479–1500, 2002.
12. J. M. Keil. Approximating the complete Euclidean graph. In *Proc. 1st Scandinavian Workshop on Algorithmic Theory*, pages 208–213, 1988.
13. J. M. Keil and C. A. Gutwin. Classes of graphs which approximate the complete Euclidean graph. *Discrete and Computational Geometry*, 7:13–28, 1992.
14. C. Levcopoulos, G. Narasimhan, and M. Smid. Improved algorithms for constructing fault-tolerant spanners. *Algorithmica*, 32:144–156, 2002.
15. R. J. Lipton and R. E. Tarjan. A separator theorem for planar graphs. *SIAM Journal of Applied Mathematics*, 36:177–189, 1979.
16. A. Maheswari. Personal communication, 2002.
17. A. L. Rosenberg and L. S. Heath. *Graph separators, with applications*. Kluwer Academic/Plenum Publishers, Dordrecht, the Netherlands, 2001.
18. J. Ruppert and R. Seidel. Approximating the d-dimensional complete Euclidean graph. In *Proc. 3rd Canadian Conference on Computational Geometry*, pages 207–210, 1991.
19. J. S. Salowe. Construction of multidimensional spanner graphs with applications to minimum spanning trees. In *Proc. 7th Annual ACM Symposium on Computational Geometry*, pages 256–261, 1991.
20. M. Smid. Closest point problems in computational geometry. In J.-R. Sack and J. Urrutia, editors, *Handbook of Computational Geometry*, pages 877–935. Elsevier Science Publishers, Amsterdam, 2000.

Composing Equipotent Teams

Mark Cieliebak[1], Stephan Eidenbenz[2], and Aris Pagourtzis[3]

[1] Institute of Theoretical Computer Science, ETH Zürich
cieliebak@inf.ethz.ch
[2] Basic and Applied Simulation Science (CCS-5), Los Alamos National Laboratory[†]
eidenben@lanl.gov
[3] Department of Computer Science, School of ECE,
National Technical University of Athens, Greece[‡]
pagour@cs.ntua.gr

Abstract. We study the computational complexity of k EQUAL SUM SUBSETS, in which we need to find k disjoint subsets of a given set of numbers such that the elements in each subset add up to the same sum. This problem is known to be NP-complete. We obtain several variations by considering different requirements as to how to compose teams of equal strength to play a tournament. We present:

- A pseudo-polynomial time algorithm for k EQUAL SUM SUBSETS with $k = O(1)$ and a proof of strong NP-completeness for $k = \Omega(n)$.
- A polynomial-time algorithm under the additional requirement that the subsets should be of equal cardinality $c = O(1)$, and a pseudo-polynomial time algorithm for the variation where the common cardinality is part of the input or not specified at all, which we proof NP-complete.
- A pseudo-polynomial time algorithm for the variation where we look for two equal sum subsets such that certain pairs of numbers are not allowed to appear in the same subset.

Our results are a first step towards determining the dividing lines between polynomial time solvability, pseudo-polynomial time solvability, and strong NP-completeness of subset-sum related problems; we leave an interesting set of questions that need to be answered in order to obtain the complete picture.

1   Introduction

The problem of identifying subsets of equal value among the elements of a given set is constantly attracting the interest of various research communities due to its numerous applications, such as production planning and scheduling, parallel processing, load balancing, cryptography, and multi-way partitioning in VLSI design, to name only a few. Most research has so far focused on the version where

---

[†] LA–UR–03:1158; work done while at ETH Zürich.
[‡] Work partially done while at ETH Zürich, supported by the Human Potential Programme of EU, contract no HPRN-CT-1999-00104 (AMORE).

A. Lingas and B.J. Nilsson (Eds.): FCT 2003, LNCS 2751, pp. 98–108, 2003.
© Springer-Verlag Berlin Heidelberg 2003

the subsets must form a partition of the given set; however, the variant where we skip this restriction is interesting as well. For example, the TWO EQUAL SUM SUBSETS problem can be used to show NP-hardness for a minimization version of PARTIAL DIGEST (one of the central problems in computational biology whose exact complexity is unknown) [4]. Further applications may include: forming similar groups of people for medical experiments or market analysis, web clustering (finding groups of pages of similar content), or fair allocation of resources.

Here, we look at the problem from the point of view of a tournament organizer: Suppose that you and your friends would like to organize a soccer tournament (you may replace soccer with the game of your ·choice) with a certain number of teams that will play against each other. Each team should be composed of some of your friends and – in order to make the tournament more interesting – you would like all teams to be of equal strength. Since you know your friends quite well, you also know how well each of them plays. More formally, you are given a set of n numbers $A = \{a_1, \ldots, a_n\}$, where the value a_i represents the excellence of your i-th friend in the chosen game, and you need to find k teams (disjoint subsets[1] of A) such that the values of the players of each team add up to the same number.

This problem can be seen as a variation of BIN PACKING with fixed number of bins. In this new variation we require that all bins should be filled to the same level while it is not necessary to use all the elements. For any set A of numbers, let $\mathrm{sum}(A) := \sum_{a \in A} a$ denote the sum of its elements. We call our problem k EQUAL SUM SUBSETS, where k is a fixed constant:

Definition 1 (k EQUAL SUM SUBSETS). *Given is a set of n numbers $A = \{a_1, \ldots, a_n\}$. Are there k disjoint subsets $S_1, \ldots, S_k \subseteq A$ such that $\mathrm{sum}(S_1) = \ldots = \mathrm{sum}(S_k)$?*

The problem k EQUAL SUM SUBSETS has been recently shown to be NP-complete for any constant $k \geq 3$ [3]. The NP-completeness of the particular case where $k = 2$ has been shown earlier by Woeginger and Yu [8]. To the best of our knowledge, the variations of k EQUAL SUM SUBSETS that we study in this paper have not been investigated before in the literature.

We have introduced parameter k for the number of equal size subsets as a fixed constant that is part of the problem definition. An interesting variation is to allow k to be a fixed function of the number of elements n, e.g. $k = \frac{n}{q}$ for some constant q. In the sequel, we will always consider k as a function of n; whenever k is a constant we simply write $k = O(1)$.

The definition of k EQUAL SUM SUBSETS corresponds to the situation in which it is allowed to form subsets that do not have the same number of elements. In some cases this makes sense; however, we may want to have the same

---

[1] Under a strict formalism we should define A as a set of elements which have *values* $\{a_1, \ldots, a_n\}$. For convenience, we prefer to identify elements with their values. Moreover, the term "disjoint subsets" refers to subsets that contain elements of A with different indices.

number of elements in each subset (this would be especially useful in composing teams for a tournament). We thus define k EQUAL SUM SUBSETS OF SPECIFIED CARDINALITY as follows:

Definition 2 (k EQUAL SUM SUBSETS OF SPECIFIED CARDINALITY). *Given are a set of n numbers $A = \{a_1, \dots, a_n\}$ and a cardinality c. Are there k disjoint subsets $S_1, \dots, S_k \subseteq A$ with $sum(S_1) = \dots = sum(S_k)$ such that each S_i has cardinality c?*

There are two nice variations of this problem, depending on the parameter c. The first is to require c to be a fixed constant; this corresponds to always playing a specific game (e.g. if you always play soccer then it is $c = 11$). We call this problem k EQUAL SUM SUBSETS OF CARDINALITY c. The second variation is to require only that all teams should have an equal number of players, without specifying this number; this indeed happens in several "unofficial" tournaments, e.g. when composing groups of people for medical experiments, or in online computer games. We call the second problem k EQUAL SUM SUBSETS OF EQUAL CARDINALITY.

Let us now consider another aspect of the problem. Your teams would be more efficient and happy if they consisted of players that like each other or, at least, that do not hate each other. Each of your friends has a list of people that she/he prefers as team-mates or, equivalently, a list of people that she/he would not like to have as team-mates. In order to compose k equipotent teams respecting such preferences/exclusions, you should be able to solve the following problem:

Definition 3 (k EQUAL SUM SUBSETS WITH EXCLUSIONS). *Given are a set of n numbers $A = \{a_1, \dots, a_n\}$, and an exclusion graph $G_{ex} = (A, E_{ex})$ with vertex set A and edge set $E_{ex} \subseteq A \times A$. Are there k disjoint subsets $S_1, \dots, S_k \subseteq A$ with $sum(S_1) = \dots = sum(S_k)$ such that each set S_i is an independent set in G_{ex}, i.e., there is no edge between any two vertices in S_i?*

An overview of the results presented in this paper is given below. In Section 2, we propose a dynamic programming algorithm for k EQUAL SUM SUBSETS with running time $O(\frac{nS^k}{k^{k-1}})$, where n is the cardinality of the input set and S is the sum of all numbers in the input set; the algorithm runs in pseudo-polynomial time[2] for $k = O(1)$. For k EQUAL SUM SUBSETS with $k = \Omega(n)$, we show strong NP-completeness[3] in Section 3 by proposing a reduction from 3-PARTITION.

---

[2] That is, the running time of the algorithm is polynomial in (n, m), where n denotes the cardinality of the input set and m denotes the largest number of the input, but it is not necessarily polynomial in the length of the representation of the input (which is $O(n \log m)$).

[3] This means that the problem remains NP-hard even when restricted to instances where all input numbers are polynomially bounded in the cardinality of the input set. In this case, no pseudo-polynomial time algorithm can exist for the problem (unless P = NP). For formal definitions and a detailed introduction to the theory of NP-completeness the reader is referred to [5].

In Section 4, we propose a polynomial-time algorithm for k EQUAL SUM SUBSETS OF CARDINALITY c. The algorithm uses exhaustive search and runs in time $O(n^{kc})$, which is polynomial in n as the two parameters k and c are fixed constants. For k EQUAL SUM SUBSETS OF SPECIFIED CARDINALITY, we show NP-completeness; the result holds also for k EQUAL SUM SUBSETS OF EQUAL CARDINALITY. However, we show that none of these problems is strongly NP-complete, by presenting an algorithm that can solve them in pseudo-polynomial time.

In Section 5, we study k EQUAL SUM SUBSETS WITH EXCLUSIONS, which is NP-complete since it is a generalization of k EQUAL SUM SUBSETS. We present a pseudo-polynomial time algorithm for the case where $k = 2$. We also give a modification of this algorithm that additionally guarantees that the two sets will have an equal (specified or not) cardinality.

We conclude in Section 6 presenting a set of open questions and problems.

1.1 Number Representation

In many of our proofs, we use numbers that are expressed in the number system of some base B. We denote by $\langle a_1, \dots, a_n \rangle$ the number $\sum_{1 \leq i \leq n} a_i B^{n-i}$; we say that a_i is the i-th digit of this number. In our proofs, we will choose base B large enough such that even adding up all numbers occurring in the reduction will not lead to carry-digits from one digit to the next. Therefore, we can add numbers digit by digit. The same holds for scalar products. For example, having base $B = 27$ and numbers $\alpha = \langle 3, 5, 1 \rangle, \beta = \langle 2, 1, 0 \rangle$, then $\alpha + \beta = \langle 5, 6, 1 \rangle$ and $3 \cdot \alpha = \langle 9, 15, 3 \rangle$.

We will generally make liberal use of the notation and allow different bases for each digit. We define the concatenation of two numbers by $\langle a_1, \dots, a_n \rangle \parallel \langle b_1, \dots, b_m \rangle := \langle a_1, \dots, a_n, b_1, \dots, b_m \rangle$, i.e., $\alpha \parallel \beta = \alpha B^m + \beta$, where m is the number of digits in β. We will use $\Delta_n(i) := \langle 0, \dots, 0, 1, 0, \dots, 0 \rangle$ for the number that has n digits, all 0's except for the i-th position where the digit is 1. Furthermore, $\mathbf{1}_n := \langle 1, \dots, 1 \rangle$ is the number that has n digits, all 1's, and $\mathbf{0}_n := \langle 0, \dots, 0 \rangle$ has n zeros. Notice that $\mathbf{1}_n = B^n - 1$.

2 A Pseudo-Polynomial Time Algorithm for k EQUAL SUM SUBSETS with $k = O(1)$

We present a dynamic programming algorithm for k EQUAL SUM SUBSETS that uses basic ideas of well-known dynamic programming algorithms for BIN PACKING with fixed number of bins [5]. For constant k, this algorithm runs in pseudo-polynomial time.

For an instance $A = \{a_1, \dots, a_n\}$ of k EQUAL SUM SUBSETS, let $S = \mathrm{sum}(A)$. We define boolean variables $F(i, s_1, \dots, s_k)$, where $i \in \{1, \dots, n\}$ and $s_j \in \{0, \dots, \lfloor \frac{S}{k} \rfloor\}$ for $1 \leq j \leq k$. Variable $F(i, s_1, \dots, s_k)$ will be TRUE if there are k disjoint subsets $X_1, \dots, X_k \subseteq \{a_1, \dots, a_i\}$ with $\mathrm{sum}(X_j) = s_j$, for $1 \leq j \leq k$.

There are k sets of equal sum if and only if there exists a value $s \in \{1, \ldots, \lfloor \frac{S}{k} \rfloor\}$ such that $F(n, s, \ldots, s) = \text{TRUE}$.

Clearly, $F(1, s_1, \ldots, s_k)$ is TRUE if and only if either $s_i = 0$ for $1 \leq i \leq k$ or there exists index j such that $s_j = a_1$ and $s_i = 0$ for all $1 \leq i \leq k, i \neq j$.

For $i \in \{2, \ldots, n\}$ and $s_j \in \{0, \ldots, \lfloor \frac{S}{k} \rfloor\}$, variable $F(i, s_1, \ldots, s_k)$ can be expressed recursively as follows:

$$F(i, s_1, \ldots, s_k) = F(i-1, s_1, \ldots, s_k) \vee$$
$$\bigvee_{\substack{1 \leq j \leq k \\ s_j - a_i \geq 0}} F(i-1, s_1, \ldots, s_{j-1}, s_j - a_i, s_{j+1}, \ldots, s_k).$$

The value of all variables can be determined in time $O(\frac{nS^k}{k^{k-1}})$, since there are $n \lfloor \frac{S}{k} \rfloor^k$ variables, and computing each variable takes at most time $O(k)$. This yields the following

Theorem 1. *There is a dynamic programming algorithm that solves k EQUAL SUM SUBSETS for input $A = \{a_1, \ldots, a_n\}$ in time $O(\frac{n \cdot S^k}{k^{k-1}})$, where $S = sum(A)$. For $k = O(1)$ this algorithm runs in pseudo-polynomial time.*

3   Strong NP-Completeness of k EQUAL SUM SUBSETS with $k = \Omega(n)$

In Section 2 we gave a pseudo-polynomial time algorithm for k EQUAL SUM SUBSETS assuming that k is a fixed constant. We will now show that this is unlikely if k is a fixed *function* of the cardinality n of the input set. In particular, we will prove that k EQUAL SUM SUBSETS is strongly NP-complete if $k = \Omega(n)$.

Let $k = \frac{n}{q}$ for some fixed integer $q \geq 2$. We provide a polynomial reduction from 3-PARTITION, which is defined as follows: Given a multiset of $n = 3m$ numbers $P = \{p_1, \ldots, p_n\}$ and a number h with $\frac{h}{4} < p_i < \frac{h}{2}$, for $1 \leq i \leq n$, are there m pairwise disjoint sets $T_1, \ldots, T_m$ such that $sum(T_j) = h$, for $1 \leq j \leq m$? Observe that in a solution for 3-PARTITION, there are exactly three elements in each set T_j.

Lemma 1. *If $k = \frac{n}{q}$ for some fixed integer $q \geq 2$, then 3-PARTITION can be reduced to k EQUAL SUM SUBSETS.*

Proof. Let $P = \{p_1, \ldots, p_n\}$ and h be an instance of 3-PARTITION. If all elements in P are equal, then there is a trivial solution. Otherwise, let $r = 3 \cdot (q-2) + 1$ and

$$a_i = \langle p_i \rangle \parallel \mathbf{0}_r, \text{ for } 1 \leq i \leq n$$
$$b_j = \langle h \rangle \parallel \mathbf{0}_r, \text{ for } 1 \leq j \leq \frac{2n}{3}$$
$$d_{k,\ell} = \langle 0 \rangle \parallel \Delta_r(k), \text{ for } 1 \leq k \leq r, 1 \leq \ell \leq \frac{n}{3}$$

Here, we use base $B = 2nh$ for all numbers. Let A be the set containing all numbers a_i, b_j and $d_{k,\ell}$. We will use A as an instance of k EQUAL SUM SUBSETS. The size of A is $n' = n + \frac{2n}{3} + r \cdot \frac{n}{3} = n + \frac{2n}{3} + (3 \cdot (q-2) + 1) \cdot \frac{n}{3} = q \cdot n$. We prove that there is a solution for the 3-PARTITION instance P and h if and only if there are $\frac{n'}{q}$ disjoint subsets of A with equal sum.

"only if": Let $T_1, \ldots, T_m$ be a solution for the 3-PARTITION instance. This induces m subsets of A with sum $\langle h \rangle \parallel \mathbf{0}_r$, namely $S_i = \{a_i \mid p_i \in T_i\}$. Together with the $\frac{2n}{3}$ subsets that contain exactly one of the b_j's each, we have $n = \frac{n'}{q}$ subsets of equal sum $\langle h \rangle \parallel \mathbf{0}_r$.

"if": Assume there is a solution $S_1, \ldots, S_n$ for the k EQUAL SUM SUBSETS instance. Let S_j be any set in this solution. Then $\text{sum}(S_j)$ will have a zero in the r rightmost digits, since for each of these digits, there are only $\frac{n}{3}$ numbers in A for which this digit is non-zero (which are not enough to have one of them in each of the n sets S_j). Thus, only numbers a_i and b_j can occur in the solution; moreover, we only need to consider the first digit of these numbers (as the other are zeros).

Since not all numbers a_i are equal, and the solution consists of $\frac{n'}{q} = n$ disjoint sets, there must be at least one b_j in one of the subsets in the solution. Thus, for all j we have $\text{sum}(S_j) \geq h$. On the other hand, the sum of all a_i's and of all b_j's is exactly $n \cdot h$, therefore $\text{sum}(S_j) = h$, which means that all a_i's and all b_j's appear in the solution. More specifically, there are $\frac{2n}{3}$ sets in the solution such that each of them contains exactly one of the b_j's, and each of the remaining $\frac{n}{3}$ sets in the solution consists only of a_i's, such that the corresponding a_i's add up to h. Therefore, the latter sets immediately yield a solution for the 3-PARTITION instance. $\square$

In the previous proof, r is a constant, therefore numbers a_i and b_j are polynomial in h and numbers $d_{k,\ell}$ are bounded by a constant. Since 3-PARTITION is strongly NP-complete [5], k EQUAL SUM SUBSETS is strongly NP-hard for $k = \frac{n}{q}$ as well. Obviously, k EQUAL SUM SUBSETS is in NP even if $k = \frac{n}{q}$ for some fixed integer $q \geq 2$, thus we have the following

Theorem 2. k EQUAL SUM SUBSETS *is* NP-*complete in the strong sense for* $k = \frac{n}{q}$, *for any fixed integer* $q \geq 2$.

4   Restriction to Equal Cardinalities

In this section we study the setting where we do not only require the teams to be of equal strength, but to be of equal cardinality as well. If we are interested in a specific type of game, e.g. soccer, then the size of the teams is also fixed, say $c = 11$, and we have k EQUAL SUM SUBSETS OF CARDINALITY c. This problem is solvable in time polynomial in n by exhaustive search as follows: compute all $N = \binom{n}{c}$ subsets of the input set A that have cardinality c; consider all $\binom{N}{k}$ possible sets of k subsets, and for each one check if it consists of disjoint subsets of equal sum. This algorithm needs time $O(n^{ck})$, which is polynomial in n, since c and k are constants.

On the other hand, if the size of the teams is not fixed, but given as part of the input, then we have k EQUAL SUM SUBSETS OF SPECIFIED CARDINALITY. We show that this problem is NP-hard by modifying a reduction used in [3] to show NP-completeness of k EQUAL SUM SUBSETS. The reduction is from ALTERNATING PARTITION, which is the following NP-complete [5] variation of PARTITION: Given n pairs of numbers $(u_1, v_1), \ldots, (u_n, v_n)$, are there two disjoint sets of indices I and J with $I \cup J = \{1, \ldots, n\}$ such that $\sum_{i \in I} u_i + \sum_{j \in J} v_j = \sum_{i \in I} v_i + \sum_{j \in J} u_j$ (equivalently, $\sum_{i \in I} u_i + \sum_{j \in J} v_j = \sum_{i \notin I} u_i + \sum_{j \notin J} v_j$)?

Lemma 2. ALTERNATING PARTITION *can be reduced to* k EQUAL SUM SUBSETS OF SPECIFIED CARDINALITY *for any* $k \geq 2$.

Proof. We transform a given ALTERNATING PARTITION instance with pairs $(u_1, v_1), \ldots, (u_n, v_n)$ into a k EQUAL SUM SUBSETS OF SPECIFIED CARDINALITY instance as follows: Let $S = \sum_{i=1}^{n} (u_i + v_i)$. For each pair (u_i, v_i) we create two numbers $u_i' = \langle u_i \rangle \parallel \Delta_n(i)$ and $v_i' = \langle v_i \rangle \parallel \Delta_n(i)$. In addition, we create $k - 2$ (equal) numbers $b_1, \ldots, b_{k-2}$ with $b_i = \langle \frac{S}{2} \rangle \parallel \Delta_n(n)$. Finally, for each b_i we create $n - 1$ numbers $d_{i,j} = \langle 0 \rangle \parallel \Delta_n(j)$, for $1 \leq j \leq n - 1$. While we set the base of the first digit to $k \cdot S$, for all other digits it suffices to use base $n + 1$, in order to ensure that no carry-digits can occur in any addition in the following proof. The set A that contains all u_i''s, v_i''s, b_i's, and d_{ij}'s, together with chosen cardinality $c = n$, is our instance of k EQUAL SUM SUBSETS OF SPECIFIED CARDINALITY.

Assume first that we are given a solution for the ALTERNATING PARTITION instance, i.e., two indices sets I and J. We create k equal sum subsets $S_1, \ldots, S_k$ as follows: for $i = 1, \ldots, k - 2$, we have $S_i = \{b_i, d_{i,1}, \ldots, d_{i,n-1}\}$; for the remaining two subsets, we let $u_i' \in S_{k-1}$, if $i \in I$, and $v_j' \in S_{k-1}$, if $j \in J$, and we let $u_j' \in S_k$, if $j \in J$, and $v_i' \in S_k$, if $i \in I$. Clearly, all these sets have n elements, and their sum is $\langle \frac{S}{2} \rangle \parallel \mathbf{1}_n$.

Now assume we are given a solution for the k EQUAL SUM SUBSETS OF SPECIFIED CARDINALITY instance, i.e., k equal sum subsets $S_1, \ldots, S_k$ of cardinality n; in this case, all numbers participate in the sets S_i, and the elements in each S_i sum up to $\langle \frac{S}{2} \rangle \parallel \mathbf{1}_n$. Since the first digit of each b_i equals $\frac{S}{2}$, we may assume w.l.o.g. that for each $1 \leq i \leq k - 2$, set S_i contains b_i and does not contain any number with non-zero first digit (i.e., it does not contain any u_j' or any v_j'). Therefore, all u_i''s and v_i''s (and only these numbers) are in the remaining two subsets; this yields an alternating partition for the original instance, as u_i' and v_i' can never be in the same subset since both have the $(i + 1)$-th digit non-zero. □

Since the problem k EQUAL SUM SUBSETS OF SPECIFIED CARDINALITY is obviously in NP, we get the following:

Theorem 3. *For any* $k \geq 2$, k EQUAL SUM SUBSETS OF SPECIFIED CARDINALITY *is NP-complete*.

Remark: Note that the above reduction, hence also the theorem, holds also for the variation k EQUAL SUM SUBSETS OF EQUAL CARDINALITY. This requires to employ a method where additional extra digits are used in order to force the equal sum subsets to include all augmented numbers that correspond to numbers in the ALTERNATING PARTITION instance; a similar method has been used in [8] to establish the NP-completeness of TWO EQUAL SUM SUBSETS (called EQUAL-SUBSET-SUM there).

However, these problems are not strongly NP-complete for fixed constant k. We will now describe how to convert the dynamic programming algorithm of Section 2 to a dynamic programming algorithm for k EQUAL SUM SUBSETS OF SPECIFIED CARDINALITY and for k EQUAL SUM SUBSETS OF EQUAL CARDINALITY.

It suffices to add to our variables k more dimensions corresponding to cardinalities of the subsets. We define boolean variables $F(i, s_1, \ldots, s_k, c_1, \ldots, c_k)$, where $i \in \{1, \ldots, n\}$, $s_j \in \{0, \ldots, \lfloor \frac{S}{k} \rfloor\}$ for $1 \leq j \leq k$, and $c_j \in \{0, \ldots, \lfloor \frac{n}{k} \rfloor\}$ for $1 \leq j \leq k$. Variable $F(i, s_1, \ldots, s_k, c_1, \ldots, c_k)$ will be TRUE if there are k disjoint subsets $X_1, \ldots, X_k \subseteq \{a_1, \ldots, a_i\}$ with $\text{sum}(X_j) = s_j$ and the cardinality of X_j is c_j, for $1 \leq j \leq k$.

There are k subsets of equal sum and equal cardinality c if and only if there exists a value $s \in \{1, \ldots, \lfloor \frac{S}{k} \rfloor\}$ such that $F(n, s, \ldots, s, c, \ldots, c) = \text{TRUE}$. Also, there are k subsets of equal sum and equal (non-specified) cardinality if and only if there exists a value $s \in \{1, \ldots, \lfloor \frac{S}{k} \rfloor\}$ and a value $d \in \{1, \ldots, \lfloor \frac{n}{k} \rfloor\}$ such that $F(n, s, \ldots, s, d, \ldots, d) = \text{TRUE}$.

Clearly, $F(1, s_1, \ldots, s_k, c_1, \ldots, c_k) = \text{TRUE}$ if and only if either $s_i = 0, c_i = 0$ for $1 \leq i \leq k$, or there exists index j such that $s_j = a_1, c_j = 1$, and $s_i = 0$ and $c_i = 0$ for all $1 \leq i \leq k$, $i \neq j$.

Each variable $F(i, s_1, \ldots, s_k, c_1, \ldots, c_k)$, for $i \in \{2, ..., n\}$, $s_j \in \{0, \ldots, \lfloor \frac{S}{k} \rfloor\}$, and $c_j \in \{0, \ldots, \lfloor \frac{n}{k} \rfloor\}$, can be expressed recursively as follows:

$$F(i, s_1, \ldots, s_k, c_1, \ldots, c_k) = F(i-1, s_1, \ldots, s_k, c_1, \ldots, c_k) \vee$$
$$\bigvee_{\substack{1 \leq j \leq k \\ s_j - a_i \geq 0 \\ c_j > 0}} F(i-1, s_1, \ldots, s_j - a_i, \ldots, s_k, c_1, \ldots, c_j - 1, \ldots, c_k).$$

The boolean value of all variables can be determined in time $O(\frac{S^k \cdot n^{k+1}}{k^{2k-1}})$, since there are $n \lfloor \frac{S}{k} \rfloor^k \lfloor \frac{n}{k} \rfloor^k$ variables, and computing each variable takes at most time $O(k)$. This yields the following:

Theorem 4. *There is a dynamic programming algorithm that solves k EQUAL SUM SUBSETS OF SPECIFIED CARDINALITY and k EQUAL SUM SUBSETS OF EQUAL CARDINALITY for input $A = \{a_1, \ldots, a_n\}$ in running time $O(\frac{S^k \cdot n^{k+1}}{k^{2k-1}})$, where $S = \text{sum}(A)$. For $k = O(1)$ this algorithm runs in pseudo-polynomial time.*

5   Adding Exclusion Constraints

In this section we study the problem k EQUAL SUM SUBSETS WITH EXCLU-
SIONS where we are additionally given an *exclusion graph* (or its complement: a
preference graph) and ask for teams that take this graph into account.

Obviously, k EQUAL SUM SUBSETS WITH EXCLUSIONS is NP-complete, since
k EQUAL SUM SUBSETS (shown NP-complete in [3]) is the special case where
the exclusion graph is empty ($E_{ex} = \emptyset$). Here, we present a pseudo-polynomial
algorithm for the case $k = 2$, using a dynamic programming approach similar-
in-spirit to the one used for finding two equal sum subsets (without exclusions)
[1].

Let $A = \{a_1, \dots, a_n\}$ and $G_{ex} = (A, E_{ex})$ be an instance of k EQUAL SUM
SUBSETS WITH EXCLUSIONS. We assume w.l.o.g. that the input values are ordered, i.e., $a_1 \leq \dots \leq a_n$. Let $S = \sum_{i=1}^{n} a_i$.

We define boolean variables $F(k, t)$ for $k \in \{1, \dots, n\}$ and $t \in \{1, \dots, S\}$.
Variable $F(k, t)$ will be TRUE if there exists a set $X \subseteq A$ such that $X \subseteq
\{a_1, \dots, a_k\}$, $a_k \in X$, sum$(X) = t$, and X is independent in G_{ex}. For a TRUE
entry $F(k, t)$ we store the corresponding set in a second variable $X(k, t)$.

We compute the value of all variables $F(k, t)$ by iterating over t and k. The
algorithm runs until it finds the smallest $t \in \{1, \dots, S\}$ for which there are
indices $k, \ell \in \{1, \dots, n\}$ such that $F(k, t) = F(\ell, t) = $ TRUE; in this case, sets
$X(k, t)$ and $X(\ell, t)$ constitute a solution: sum$(X(k, t)) = $ sum$(X(\ell, t)) = t$, both
sets are disjoint due to minimality of t, and both sets are independent in G_{ex}.

We initialize the variables as follows. For all $1 \leq k \leq n$, we set $F(k, t) =$
FALSE for $1 \leq t < a_k$ and for $\sum_{i=1}^{k} a_i < t \leq S$; moreover, we set $F(k, a_k) =$
TRUE and $X(k, a_k) = \{a_k\}$. Observe that these equations already define $F(1, t)$
for $1 \leq t \leq S$, and $F(k, 1)$ for $1 \leq k \leq n$.

After initialization, the table entries for $k > 1$ and $a_k \leq t \leq \sum_{i=1}^{k} a_i$ can be
computed recursively: $F(k, t)$ is TRUE if there exists an index $\ell \in \{1, \dots, k-1\}$
such that $F(\ell, t - a_k)$ is TRUE and the subset $X(\ell, t - a_k)$ remains independent
in G_{ex} when adding a_k. The recursive computation is as follows.

$$F(k, t) = \bigvee_{\ell=1}^{k-1} [F(\ell, t - a_k) \wedge \forall a \in X(\ell, t - a_k), (a, a_k) \notin E_{ex}].$$

If $F(k, t)$ is set to TRUE due to $F(\ell, t - a_k)$, then we set $X(k, t) = X(\ell, t -
a_k) \cup \{a_k\}$. The key observation for showing correctness is that for each $F(k, t)$
considered by the algorithm there is at most one $F(\ell, t - a_k)$ that is TRUE, for
$1 \leq \ell \leq k - 1$; if there were two, say ℓ_1, ℓ_2, then $X(\ell_1, t - a_k)$ and $X(\ell_2, t - a_k)$
would be a solution to the problem and the algorithm would have stopped earlier
– a contradiction. This means that all subsets considered are constructed in a
unique way, and therefore no information can be lost.

In order to determine the value $F(k, t)$, the algorithm considers $k - 1$ table
entries. As shown above, only one of them may be TRUE; for such an entry, say
$F(\ell, t - a_k)$, the (at most ℓ) elements of $X(\ell, t - a_k)$ are checked to see if they

exclude a_k. Hence, computation of $F(k,t)$ takes time $O(n)$ and the total time
complexity of the algorithm is $O(n^2 \cdot S)$. Therefore, we have the following

Theorem 5. TWO EQUAL SUM SUBSETS WITH EXCLUSIONS *can be solved for
input* $A = \{a_1, \ldots, a_n\}$ *and* $G_{ex} = (A, E_{ex})$ *in pseudo-polynomial time* $O(n^2 \cdot S)$,
where $S = sum(A)$.

Remarks: Observe that the problem k EQUAL SUM SUBSETS OF CARDINALITY
c WITH EXCLUSIONS, where cardinality c is constant, and an exclusion graph is
given, can be solved by exhaustive search in time $O(n^{kc})$ in the same way as the
problem k EQUAL SUM SUBSETS OF CARDINALITY c is solved (see Section 4).

Moreover, we can have a pseudo-polynomial time algorithm for k EQUAL SUM
SUBSETS OF EQUAL CARDINALITY WITH EXCLUSIONS, where the cardinality is
part of the input, if $k = 2$, by modifying the dynamic programming algorithm
for TWO EQUAL SUM SUBSETS WITH EXCLUSIONS as follows. We introduce a
further dimension in our table F, the cardinality, and set $F(k,t,c)$ to TRUE
if there is a set X with $\text{sum}(X) = t$ (and all other conditions as before), and
such that the cardinality of X equals c. Again, we can fill the table recursively,
and we stop as soon as we find values $k, \ell \in \{1, \ldots, n\}, t \in \{1, \ldots, S\}$ and
$c \in \{1, \ldots, n\}$ such that $F(k,t,c) = F(\ell,t,c) = \text{TRUE}$, which yields a solution.
Notice that the corresponding two sets must be disjoint, since otherwise removing
their intersection would yield two subsets of smaller equal cardinality that are
independent in G_{ex}; thus, the algorithm - which constructs two sets of minimal
cardinality - would have stopped earlier. Table F now has $n^2 \cdot S$ entries, thus we
can solve TWO EQUAL SUM SUBSETS WITH EXCLUSIONS in time $O(n^3 \cdot S)$.

Note that the above sketched algorithm does not work for specified cardinalities, because there may be exponentially many ways to construct a subset of
the correct cardinality.

6   Conclusion – Open Problems

In this work we studied the problem k EQUAL SUM SUBSETS and some of its
variations. We presented a pseudo-polynomial time algorithm for constant k, and
proved strong NP-completeness for non-constant k, namely for the case in which
we want to find $\frac{n}{q}$ subsets of equal sum, where n is the cardinality of the input
set and q a constant. We also gave pseudo-polynomial time algorithms for the k
EQUAL SUM SUBSETS OF SPECIFIED CARDINALITY problem and for the TWO
EQUAL SUM SUBSETS WITH EXCLUSIONS problem, as well as for variations of
them.

Several questions remain open. Some of them are: determine the exact borderline between pseudo-polynomial time solvability and strong NP-completeness
for k EQUAL SUM SUBSETS, for k being a function different than $\frac{n}{q}$, for example
$k = \frac{\log n}{q}$; find faster dynamic programming algorithms for k EQUAL SUM SUB-
SETS OF SPECIFIED CARDINALITY; and, finally, determine the complexity of k
EQUAL SUM SUBSETS WITH EXCLUSIONS, i.e. is it solvable in pseudo-polynomial
time or strongly NP-complete?

Another promising direction is to investigate approximation versions related to the above problems, for example "given a set of numbers A, find k subsets of A with sums that are as similar as possible". For $k = 2$, the problem has been studied by Bazgan et al. [1] and Woeginger [8]; an FPTAS was presented in [1]. We would like to find out whether there is an FPTAS for any constant k. Finally, it would be interesting to study phase transitions of these problems with respect to their parameters, in a spirit similar to the work of Borgs, Chayes and Pittel [2], where they analyzed the phase transition of TWO EQUAL SUM SUBSETS.

Acknowledgments. We would like to thank Peter Widmayer for several fruitful discussions and ideas in the context of this work.

References

1. C. Bazgan, M. Santha, and Zs. Tuza; *Efficient approximation algorithms for the Subset-Sum Equality problem*; Proc. ICALP'98, pp. 387–396.
2. C. Borgs, J.T. Chayes, and B. Pittel; *Sharp Threshold and Scaling Window for the Integer Partitioning Problem*; Proc. STOC'01, pp. 330–336.
3. M. Cieliebak, S. Eidenbenz, A. Pagourtzis, and K. Schlude; *Equal Sum Subsets: Complexity of Variations*; Technical Report 370, ETH Zürich, Department of Computer Science, 2003.
4. M. Cieliebak, S. Eidenbenz, and P. Penna; *Noisy Data Make the Partial Digest Problem NP-hard*; Technical Report 381, ETH Zürich, Department of Computer Science, 2002.
5. M.R. Garey and D.S. Johnson; *Computers and Intractability: A Guide to the Theory of NP-completeness*; Freeman, San Francisco, 1979.
6. R.M. Karp; *Reducibility among combinatorial problems*; in R.E. Miller and J.W. Thatcher (eds.), Complexity of Computer Computations, Plenum Press, New York, pp. 85 – 103, 1972.
7. S. Martello and P. Toth; *Knapsack Problems*; John Wiley & Sons, Chichester, 1990.
8. G.J. Woeginger and Z.L. Yu; *On the equal-subset-sum problem*; Information Processing Letters, 42(6), pp. 299–302, 1992.

Efficient Algorithms for GCD and Cubic Residuosity in the Ring of Eisenstein Integers*

Ivan Bjerre Damgård and Gudmund Skovbjerg Frandsen

BRICS**
Department of Computer Science
University of Aarhus
Ny Munkegade
DK-8000 Aarhus C, Denmark
{ivan,gudmund}@daimi.au.dk

Abstract. We present simple and efficient algorithms for computing gcd and cubic residuosity in the ring of Eisenstein integers, $\mathbf{Z}[\zeta]$, i.e. the integers extended with ζ, a complex primitive third root of unity. The algorithms are similar and may be seen as generalisations of the binary integer gcd and derived Jacobi symbol algorithms. Our algorithms take time $O(n^2)$ for n bit input. This is an improvement from the known results based on the Euclidean algorithm, and taking time $O(n \cdot M(n))$, where $M(n)$ denotes the complexity of multiplying n bit integers. The new algorithms have applications in practical primality tests and the implementation of cryptographic protocols.

1 Introduction

The Eisenstein integers, $\mathbf{Z}[\zeta] = \{a + b\zeta \mid a, b \in \mathbf{Z}\}$, is the ring of integers extended with a complex primitive third root of unity, i.e. ζ is root of $x^2 + x + 1$. Since the ring $\mathbf{Z}[\zeta]$ is a unique factorisation domain, a greatest common divisor (gcd) of two numbers is well-defined (up to multiplication by a unit). The gcd of two numbers may be found using the classic Euclidean algorithm, since $\mathbf{Z}[\zeta]$ is an Euclidean domain, i.e. there is a norm $N(\cdot) : \mathbf{Z}[\zeta] \setminus \{0\} \mapsto \mathbf{N}$ such that for $a, b \in \mathbf{Z}[\zeta] \setminus \{0\}$ there is $q, r \in \mathbf{Z}[\zeta]$ such that $a = qb + r$ with $r = 0$ or $N(r) < N(b)$.

When a gcd algorithm is directly based on the Euclidean property, it requires a subroutine for division with remainder. For integers there is a very efficient alternative in the form of the binary gcd, that only requires addition/subtraction and division by two [12]. A corresponding Jacobi symbol algorithm has been analysed as well [11].

It turns out that there are natural generalisations of these binary algorithms over the integers to algorithms over the Eisenstein integers for computing the

---

* Partially supported by the IST Programme of the EU under contract number IST-1999-14186 (ALCOM-FT).
** Basic Research in Computer Science, Centre of the Danish National Research Foundation.

A. Lingas and B.J. Nilsson (Eds.): FCT 2003, LNCS 2751, pp. 109–117, 2003.
© Springer-Verlag Berlin Heidelberg 2003

gcd and the cubic residuosity symbol. The role of 2 is taken by the number $1-\zeta$, which is a prime of norm 3 in $\mathbf{Z}[\zeta]$.

We present and analyse these new algorithms. It turns out that they both have bit complexity $O(n^2)$, which is an improvement over the so far best known algorithms by Scheidler and Williams [8], Williams [16], Williams and Holte [17]. Their algorithms have complexity $O(nM(n))$, where $M(n)$ is the complexity of integer multiplication and the best upper bound on $M(n)$ is $O(n \log n \log \log n)$ [10].

1.1   Related Work

The asymptotically fastest algorithm for integer gcd takes time $O(n \log n \log \log n)$ and is due to Schönhage [9]. There is a derived algorithm for the Jacobi symbol of complexity $O(n(\log n)^2 \log \log n)$. For practical input sizes the most efficient algorithms seems to be variants of the binary gcd and derived Jacobi symbol algorithms [11,7].

If ω_n is a complex primitive nth root of unity, say $\omega_n = e^{2\pi/n}$ then the ring $\mathbf{Z}[\omega_n]$ is known to be norm-Euclidean for only finitely many n and the smallest unresolved case is $n = 17$ [6,4].

Weilert have generalised both the "binary" and the asymptotically fast gcd algorithms to $\mathbf{Z}[\omega_4] = \mathbf{Z}[i]$, the ring of Gaussian integers [13,14]. In the latter case Weilert has also described a derived algorithm for computing the quartic residue symbol [15], and in all cases the complexity is identical to the complexity of the corresponding algorithm over $\mathbf{Z}$.

Williams [16], Williams and Holte [17] both describe algorithms for computing gcd and cubic residue symbols in $\mathbf{Z}[\omega_3]$, the Eisenstein integers. Scheidler and Williams describe algorithms for computing gcd and nth power residue symbol in $\mathbf{Z}[\omega_n]$ for $n = 3, 5, 7$ [8]. Their algorithms all have complexity $O(nM(n))$ for $M(n)$ being the complexity of integer multiplication.

Weilert suggests that his binary (i.e. $(1+i)$-ary) gcd algorithm for the Gaussian integers may generalise to other norm-Euclidean rings of algebraic integers [13]. Our gcd algorithm for the Eisenstein integers was obtained independently, but it may nevertheless be seen as a confirmation of this suggestion in a specific case. It is an open problem whether the "binary" approach to gcd computation may be further generalised to $\mathbf{Z}[\omega_5]$.

Weilert gives an algorithm for the quartic residue symbol that is derived from the asymptotically fast gcd algorithm over $\mathbf{Z}[i]$. For practical purposes, however, it would be more interesting to have a version derived from the "binary" approach. In the last section of this paper, we sketch how one can obtain such an algorithm.

1.2   Applications

Our algorithms may be used for the efficient computation of cubic residuosity in other rings than $\mathbf{Z}[\zeta]$ when using an appropriate homomorphism. As an example, consider the finite field $GF(p)$ for prime $p \equiv 1 \bmod 3$. A number $z \in \{1, \ldots, p-1\}$ is a cubic residue precisely when $z^{(p-1)/3} \equiv 1 \bmod p$, implying that (non)residuosity may be decided by a (slow) modular exponentiation. However, it is possible to decide cubic residuosity much faster provided we make some preprocessing depending only on p. The preprocessing consists in factoring p over $\mathbf{Z}[\zeta]$, i.e. finding a prime $\pi \in \mathbf{Z}[\zeta]$ such that $p = \pi\bar{\pi}$. A suitable π may be found as $\pi = \gcd(p, r - \zeta)$, where $r \in \mathbf{Z}$ is constructed as a solution to the quadratic equation $x^2 + x + 1 = 0 \bmod p$. Following this preprocessing cubic residuosity of any z is decided using that $z^{(p-1)/3} \equiv 1 \bmod p$ if and only if $[z/\pi] = 1$, where $[\cdot/\cdot]$ denotes the cubic residuosity symbol.

When the order of the multiplicative group in question is unknown, modular exponentiation cannot be used, but it may still be possible to identify some nonresidues by computing residue symbols. In particular, the primality test of Damgård and Frandsen [2] uses our algorithms for finding cubic nonresidues in a more general ring.

Computation of gcd and cubic residuosity is also used for the implementation of cryptosystems by Scheidler and Williams [8], and by Williams [16].

2   Preliminary Facts about $\mathbf{Z}[\zeta]$

$\mathbf{Z}[\zeta]$ is the ring of integers extended with a primitive third root of unity ζ (complex root of $z^2 + z + 1$). We will be using the following definitions and facts (see f.x. [3]).

Define the two conjugate mappings $\sigma_i : \mathbf{Z}[\zeta] \mapsto \mathbf{Z}[\zeta]$ by $\sigma_i(\zeta) = \zeta^i$ for $i = 1, 2$. The rational integer $N(\alpha) = \sigma_1(\alpha)\sigma_2(\alpha) \geq 0$ is called the norm of $\alpha \in \mathbf{Z}[\zeta]$, and $N(a + b\zeta) = a^2 + b^2 - ab$. (Note that $\sigma_2(\cdot)$ and $N(\cdot)$ coincides with complex conjugation and complex norm, respectively).

A unit in $\mathbf{Z}[\zeta]$ is an element of norm 1. There are 6 units in $\mathbf{Z}[\zeta]$: $\pm 1, \pm\zeta, \pm\zeta^2$. Two elements $\alpha, \beta \in \mathbf{Z}[\zeta]$ are said to be associates if there exists a unit ϵ such that $\alpha = \epsilon\beta$.

A prime π in $\mathbf{Z}[\zeta]$ is a non-unit such that for any $\alpha, \beta \in \mathbf{Z}[\zeta]$, if $\pi|\alpha\beta$, then $\pi|\alpha$ or $\pi|\beta$.

$1 - \zeta$ is a prime in $\mathbf{Z}[\zeta]$ and $N(1 - \zeta) = 3$. A primary number has the form $1 + 3\beta$ for some $\beta \in \mathbf{Z}[\zeta]$. If $\alpha \in \mathbf{Z}[\zeta]$ is not divisible by $1 - \zeta$ then α is associated to a primary number. (The definition of *primary* seems to vary in that some authors require the alternate forms $\pm 1 + 3\beta$ [5] and $-1 + 3\beta$ [3], but our definition is more convenient in the present context).

A simple computation reveals that the norm of a primary number has residue 1 modulo 3, and since the norm is a multiplicative homomorpism it follows that every $\alpha \in \mathbf{Z}[\zeta]$ that is not divisible by $1 - \zeta$ has $N(\alpha) \equiv 1(\bmod 3)$.

3   Computing GCD in $\mathbf{Z}[\zeta]$

It turns out that the well-known binary integer gcd algorithm has a natural generalisation to a gcd algorithm for the Eisenstein integers. The generalised algorithm is best understood by relating it to the binary algorithm in a nonstandard version. The authors are not aware of any description of the latter in the literature (for the standard version see f.x. [1]).

A slightly nonstandard version of the binary gcd is the following. Every integer can be represented as $(-1)^i \cdot 2^j \cdot (4m+1)$, where $i \in \{0, 1\}$, $j \geq 0$ and $m \in \mathbf{Z}$. Without loss of generality, we may therefore assume that the numbers in question are of the form $(4m+1)$. One iteration consists in replacing the numerically larger of the two numbers by their difference. If it is nonzero then the dividing 2-power (at least 2^2) may be removed without changing the gcd. If necessary the resulting odd number is multiplied with -1 to get a number of the form $4m+1$ and we are ready for the next iteration. It is fairly obvious that the product of the numeric values of the two numbers decreases by a factor at least 2 in each step until the gcd is found, and hence the gcd of two numbers a, b can be computed in time $(\log^2 |ab|)$.

To make the analogue, we recall that any element of $\mathbf{Z}[\zeta]$ that is not divisible by $1 - \zeta$ is associated to a (unique) primary number, i.e. a number of the form $1 + 3\alpha$. This implies that any element in $\mathbf{Z}[\zeta] \setminus \{0\}$ has a (unique) representation on the form $(-\zeta)^i \cdot (1-\zeta)^j \cdot (1+3\alpha)$ where $0 \leq i < 6$, $0 \leq j$ and $\alpha \in \mathbf{Z}[\zeta]$. In addition, the difference of two primary numbers is divisible by $(1-\zeta)^2$, since $3 = -\zeta^2(1-\zeta)^2$. Now a gcd algorithm for the Eisenstein integers may be formulated as an analogue to the binary integer gcd algorithm. We may assume without loss of generality that the two input numbers are primary. Replace the (normwise) larger of the two numbers with their difference. If it is nonzero, we may divide out any powers of $(1 - \zeta)$ that divide the difference (at least $(1 - \zeta)^2$) and convert the remaining factor to primary form by multiplying with a unit. We have again two primary numbers and the process may be continued. In each step we are required to identify the (normwise) larger of two numbers. Unfortunately it would be too costly to compute the relevant norm, but it suffices to choose the large number based on an approximation that we can afford to compute. By a slightly nontrivial argument one may prove that the product of the norms of the two numbers decreases by a factor at least 2 in each step until the gcd is found, and hence the gcd of two numbers α, β can be computed in time $O(\log^2 N(\alpha\beta))$.

Algorithm 1 describes the details including a start-up to bring the two numbers on primary form.

Theorem 1. *Algorithm 1 takes time $O(\log^2 N(\alpha\beta))$ to compute the gcd of α, β, or formulated alternatively, the algorithm has bit complexity $O(n^2)$.*

Proof. Let us assume that a number $\alpha = a + b\zeta \in \mathbf{Z}[\zeta]$ is represented by the integer pair (a, b). Observe that since $N(\alpha) = a^2 + b^2 - ab$, we have that $\log |a| + \log |b| \leq \log N(\alpha) \leq 2(\log |a| + \log |b|)$ for $a, b \neq 0$, i.e. the logarithm of the norm is proportional to the number of bits in the representation of a number.

We may do addition, subtraction on general numbers and multiplication by units in linear time. Since $(1 - \zeta)^{-1} = (2 + \zeta)/3$, division by (and check for divisibility by) $(1 - \zeta)$ may also be done in linear time.

Algorithm 1 Compute gcd in $\mathbf{Z}[\zeta]$

Require: $\alpha, \beta \in \mathbf{Z}[\zeta] \setminus \{0\}$
Ensure: $g = \gcd(\alpha, \beta)$
 1: Let primary $\gamma, \delta \in \mathbf{Z}[\zeta]$ be defined by $\alpha = (-\zeta)^{i_1} \cdot (1 - \zeta)^{j_1} \cdot \gamma$ and $\beta = (-\zeta)^{i_2} \cdot (1 - \zeta)^{j_2} \cdot \delta$.
 2: $g \leftarrow (1 - \zeta)^{\min\{j_1, j_2\}}$
 3: Replace α, β with γ, δ.
 4: **while** $\alpha \neq \beta$ **do**
 5:     LOOP INVARIANT: α, β are primary
 6:     Let primary γ be defined by $\alpha - \beta = (-\zeta)^i \cdot (1 - \zeta)^j \cdot \gamma$
 7:     Replace "approximately" larger of α, β with γ.
 8: **end while**
 9: $g \leftarrow g \cdot \alpha$

Clearly, the startup part of the algorithm that brings the two numbers on primary form can be done in time $O(\log^2 N(\alpha\beta))$. Hence, we need only worry about the while loop.

We want to prove that the norm of the numbers decrease for each iteration. The challenge is to see that forming the number $\alpha - \beta$ does not increase the norm too much. In fact $N(\alpha - \beta) \leq 4 \cdot \max\{N(\alpha), N(\beta)\}$. This follows trivially from the fact that the norm is non-negative combined with the equation $N(\alpha + \beta) + N(\alpha - \beta) = 2(N(\alpha) + N(\beta))$ that may be proven by an elementary computation. Hence, for the γ computed in the loop of the algorithm, we get $N(\gamma) = 3^{-j}N(\alpha - \beta) \leq 3^{-2}4 \cdot \max\{N(\alpha), N(\beta)\}$. In each iteration, γ ideally replaces the one of α and β with the larger norm. However, we can not afford to actually compute the norms to find out which one is the larger. Fortunately, by Lemma 1, it is possible in linear time to compute an approximate norm that may be slightly smaller than the exact norm, namely up to a factor $9/8$. When γ replaces the one of α and β with the larger approximate norm, we know that $N(\alpha\beta)$ decreases by a factor at least $9/4 \cdot 8/9 = 2$ in each iteration, i.e. the total number of iterations is $O(\log N(\alpha\beta))$.

Each loop iteration takes time $O(\log N(\alpha\beta))$ except possibly for finding the exponent of $(1 - \zeta)$ that divides $\alpha - \beta$. Assume that $(1 - \zeta)^{t_i}$ is the maximal power of $(1 - \zeta)$ that divides $\alpha - \beta$ in the ith iteration. Then the combined time complexity of all loop iterations is $O((\sum_i t_i) \log N(\alpha\beta))$. We also know that the norm decreases by a factor at least $3^{t_i - 2} \cdot 2$ in the ith iteration, i.e. $\prod_i (3^{t_i - 2} \cdot 2) \leq N(\alpha\beta)$. Since there is only $O(\log N(\alpha\beta))$ iterations it follows that $\prod_i 3^{t_i} \leq (9/2)^{O(\log N(\alpha\beta))} N(\alpha\beta)$ and hence $\sum_i t_i = O(\log N(\alpha\beta))$.

Lemma 1. *Given $\alpha = a + b\zeta \in \mathbf{Z}[\zeta]$ it is possible to compute an approximate norm $\tilde{N}(\alpha)$ such that*

$$\frac{8}{9}N(\alpha) \leq \tilde{N}(\alpha) \leq N(\alpha)$$

in linear time, i.e. in time $O(\log N(\alpha))$.

Proof. Note that

$$N(a + b\zeta) = \frac{(a - b)^2 + a^2 + b^2}{2}.$$

Given $\epsilon > 0$, we let $\tilde{d}$ denote some approximation to integer d satisfying that $(1 - \epsilon)|d| \le \tilde{d} \le |d|$. Note that

$$(1 - \epsilon)^2 N(a + b\zeta) \le \frac{\widetilde{(a - b)}^2 + \tilde{a}^2 + \tilde{b}^2}{2} \le N(a + b\zeta)$$

Since we may compute $a - b$ in linear time it suffices to compute ~-approximations and square them in linear time for some $\epsilon < 1/18$. Given d in the usual binary representation, we take $\tilde{d}$ to be $|d|$ with all but the 6 most significant bits replaced with zeroes, in which case

$$(1 - \frac{1}{32})|d| \le \tilde{d} \le |d|$$

and we can compute $\tilde{d}^2$ from d in linear time.

4   Computing Cubic Residuosity in Z[ζ]

Just as the usual integer gcd algorithms may be used for constructing algorithms for the Jacobi symbol, so can our earlier strategy for computing the gcd in $\mathbf{Z}[\zeta]$ be used as the basis for an algorithm for computing the cubic residuosity symbol.

We start by recalling the definition of the cubic residuosity symbol.

$$[\cdot/\cdot] : \mathbf{Z}[\zeta] \times (\mathbf{Z}[\zeta] - (1 - \zeta)\mathbf{Z}[\zeta]) \mapsto \{0, 1, \zeta, \zeta^{-1}\}$$

is defined as follows:

- For prime $\pi \in \mathbf{Z}[\zeta]$ where π is not associated to $1 - \zeta$:

$$[\alpha/\pi] = (\alpha^{\frac{N(\pi)-1}{3}}) \bmod \pi$$

- For number $\beta = \prod_{i=1}^{t} \pi_i^{m_i} \in \mathbf{Z}[\zeta]$ where β is not divisible by $1 - \zeta$:

$$[\alpha/\beta] = \prod_{i=1}^{t} [\alpha/\pi_i]^{m_i}$$

Note that these rules imply $[\alpha/\epsilon] = 1$ for a unit ϵ and $[\alpha/\beta] = 0$ when $\gcd(\alpha, \beta) \ne 1$. In addition, we will need the following laws satisfied by the cubic residuosity symbol (recall that β is primary when it has the form $\beta = 1 + 3\gamma$ for $\gamma \in \mathbf{Z}[\zeta]$) [5]:

- Modularity:

$$[\alpha/\beta] = [\alpha'/\beta], \quad \text{when } \alpha \equiv \alpha' (\bmod \beta).$$

– Multiplicativity:
$$[\alpha\alpha'/\beta] = [\alpha/\beta] \cdot [\alpha'/\beta].$$

– The cubic reciprocity law:

$$[\alpha/\beta] = [\beta/\alpha], \quad \text{when } \alpha \text{ and } \beta \text{ are both primary.}$$

– The complementary laws (for primary $\beta = 1 + 3(m + n\zeta)$, where $m, n \in \mathbf{Z}$)

$$[1 - \zeta/\beta] = \zeta^m,$$
$$[\zeta/\beta] = \zeta^{-(m+n)},$$
$$[-1/\beta] = 1.$$

The cubic residuosity algorithm will follow the gcd algorithm closely. In each iteration we will assume the two numbers α, β to be primary with $\tilde{N}(\alpha) \geq \tilde{N}(\beta)$. We write their difference on the form $\alpha - \beta = (-\zeta)^i(1 - \zeta)^j\gamma$, for primary $\gamma = 1+3(m+n\zeta)$. By the above laws, $[\alpha/\beta] = \zeta^{mj-(m+n)i}[\gamma/\beta]$. If $\tilde{N}(\alpha) < \tilde{N}(\beta)$, we use the reciprocity law to swap γ and β before being ready to a new iteration. The algorithm stops, when the two primary numbers are identical. If the identical value (the gcd) is not 1 then the residuosity symbol evaluates to 0.

Algorithm 2 describes the entire procedure including a start-up to ensure that the numbers are primary.

Algorithm 2 Compute cubic residuosity in $\mathbf{Z}[\zeta]$

Require: $\alpha, \beta \in \mathbf{Z}[\zeta] \setminus \{0\}$, and β is not divisible by $(1 - \zeta)$
Ensure: $c = [\alpha/\beta]$
 1: Let primary $\gamma, \delta \in \mathbf{Z}[\zeta]$ be defined by $\alpha = (-\zeta)^{i_1} \cdot (1 - \zeta)^{j_1} \cdot \gamma$ and $\beta = (-\zeta)^{i_2} \cdot \delta$.
 2: Let $m, n \in \mathbf{Z}$ be defined by $\delta = 1 + 3m + 3n\zeta$.
 3: $t \leftarrow mj_1 - (m + n)i_1 \bmod 3$
 4: Replace α, β by γ, δ.
 5: If $\tilde{N}(\alpha) < \tilde{N}(\beta)$ then interchange α, β.
 6: **while** $\alpha \neq \beta$ **do**
 7:     LOOP INVARIANT: α, β are primary and $\tilde{N}(\alpha) \geq \tilde{N}(\beta)$
 8:     Let primary γ be defined by $\alpha - \beta = (-\zeta)^i \cdot (1 - \zeta)^j \cdot \gamma$
 9:     Let $m, n \in \mathbf{Z}$ be defined by $\beta = 1 + 3m + 3n\zeta$.
10:     $t \leftarrow t + mj - (m + n)i \bmod 3$
11:     Replace α with γ.
12:     If $\tilde{N}(\alpha) < \tilde{N}(\beta)$ then interchange α, β.
13: **end while**
14: If $\alpha \neq 1$ then $c \leftarrow 0$ else $c \leftarrow \zeta^t$

Theorem 2. *Algorithm 2 takes time $O(\log^2 N(\alpha\beta))$ to compute $[\alpha/\beta]$, or formulated alternatively, the algorithm has bit complexity $O(n^2)$.*

Proof. The complexity analysis from the gcd algorithm carries over without essential changes.

5 Computing GCD and Quartic Residuosity in the Ring of Gaussian Integers

We may construct fast algorithms for gcd and quartic residuosity in the ring of Gaussian integers, $\mathbf{Z}[i] = \{a+bi \mid a, b \in \mathbf{Z}\}$, in a completely analogous way to the algorithms over the Eisenstein integers. In the case of the gcd, this was essentially done by Weilert [13]. However, the case of the quartic residue symbol may be of independent interest since such an algorithm is likely to be more efficient for practical input values than the asymptically ultrafast algorithm [15].

Here is a sketch of the necessary facts (see [5]). There are 4 units in $\mathbf{Z}[i]$: $\pm 1, \pm i$. $1 + i$ is a prime in $\mathbf{Z}[i]$ and $N(1 + i) = 2$. A primary number has the form $1 + (2 + 2i)\beta$ for some $\beta \in \mathbf{Z}[i]$. If $\alpha \in \mathbf{Z}[i]$ is not divisible by $1 + i$ then α is associated to a primary number.

In particular, any element in $\mathbf{Z}[i] \setminus \{0\}$ has a (unique) representation on the form $i^j \cdot (1 + i)^k \cdot (1 + (2 + 2i)\alpha)$ where $0 \le j < 4$, $0 \le k$ and $\alpha \in \mathbf{Z}[i]$. In addition, the difference of two primary numbers is divisible by $(1 + i)^3$, since $(2 + 2i) = -i(1 + i)^3$. This is the basis for obtaining an algorithm for computing gcd over the Gaussian integers analogous to Algorithm 1. This new algorithm has also bit complexity $O(n^2)$ as one may prove when using that $N((1+i)^3) = 8$ and $N(\alpha - \beta) \le 4 \cdot \max\{N(\alpha), N(\beta)\}$.

For computing quartic residuosity, we need more facts [5]. If π is a prime in $\mathbf{Z}[i]$ and π is not associated to $1 + i$ then $N(\pi) \equiv 1(\bmod 4)$, and the quartic residue symbol $[\cdot/\cdot] : \mathbf{Z}[i] \times (\mathbf{Z}[i] - (1 + i)\mathbf{Z}[i]) \mapsto \{0, 1, -1, i, -i\}$ is defined as follows:

- For prime $\pi \in \mathbf{Z}[i]$ where π is not associated to $1 + i$:

$$[\alpha/\pi] = (\alpha^{\frac{N(\pi)-1}{4}}) \bmod \pi$$

- For number $\beta = \prod_{j=1}^{t} \pi_j^{m_j} \in \mathbf{Z}[i]$ where β is not divisible by $1 + i$:

$$[\alpha/\beta] = \prod_{j=1}^{t} [\alpha/\pi_j]^{m_j}$$

The quartic residuosity symbol satisfies in addition

- Modularity:
$$[\alpha/\beta] = [\alpha'/\beta], \quad \text{when } \alpha \equiv \alpha'(\bmod \beta).$$

- Multiplicativity:
$$[\alpha\alpha'/\beta] = [\alpha/\beta] \cdot [\alpha'/\beta].$$

- The quartic reciprocity law:
$$[\alpha/\beta] = [\beta/\alpha] \cdot (-1)^{\frac{N(\alpha)-1}{4} \cdot \frac{N(\beta)-1}{4}}, \quad \text{when } \alpha \text{ and } \beta \text{ are both primary.}$$

- The complementary laws (for primary $\beta = 1 + (2 + 2i)(m + ni)$, where $m, n \in \mathbf{Z}$)

$$[1 + i/\beta] = i^{-n-(n+m)^2},$$
$$[i/\beta] = i^{n-m}.$$

This is the basis for obtaining an algorithm for computing quartic residuosity analogous to Algorithm 2. This new algorithm has also bit complexity $O(n^2)$.

References

1. Eric Bach and Jeffrey Shallit. *Algorithmic number theory. Vol. 1.* Foundations of Computing Series. MIT Press, Cambridge, MA, 1996. Efficient algorithms.
2. Ivan B. Damgård and Gudmund Skovbjerg Frandsen. An extended quadratic Frobenius primality test with average and worst case error estimates. Research Series RS-03-9, BRICS, Department of Computer Science, University of Aarhus, February 2003. Extended abstract in these proceedings.
3. Kenneth Ireland and Michael Rosen. *A classical introduction to modern number theory*, Vol. 84 of *Graduate Texts in Mathematics*. Springer-Verlag, New York, second edition, 1990.
4. Franz Lemmermeyer. The Euclidean algorithm in algebraic number fields. *Exposition. Math.* **13**(5) (1995), 385–416.
5. Franz Lemmermeyer. *Reciprocity laws.* Springer Monographs in Mathematics. Springer-Verlag, Berlin, 2000. From Euler to Eisenstein.
6. Hendrik W. Lenstra, Jr. Euclidean number fields. I. *Math. Intelligencer* **2**(1) (1979/80), 6–15.
7. Shawna Meyer Eikenberry and Jonathan P. Sorenson. Efficient algorithms for computing the Jacobi symbol. *J. Symbolic Comput.* **26**(4) (1998), 509–523.
8. Renate Scheidler and Hugh C. Williams. A public-key cryptosystem utilizing cyclotomic fields. *Des. Codes Cryptogr.* **6**(2) (1995), 117–131.
9. A. Schönhage. Schnelle Berechnung von Kettenbruchentwicklungen. *Acta Informat.* **1** (1971), 139–144.
10. A. Schönhage and V. Strassen. Schnelle Multiplikation grosser Zahlen. *Computing (Arch. Elektron. Rechnen)* **7** (1971), 281–292.
11. Jeffrey Shallit and Jonathan Sorenson. A binary algorithm for the jacobi symbol. *ACM SIGSAM Bull.* **27**(1) (1993), 4–11.
12. J. Stein. Computationals problems associated with Racah algebra. *J. Comput. Phys.* **1** (1967), 397–405.
13. André Weilert. $(1 + i)$-ary GCD computation in $\mathbf{Z}[i]$ is an analogue to the binary GCD algorithm. *J. Symbolic Comput.* **30**(5) (2000), 605–617.
14. André Weilert. Asymptotically fast GCD computation in $\mathbf{Z}[\mathbf{i}]$. In *Algorithmic number theory (Leiden, 2000)*, Vol. 1838 of *Lecture Notes in Comput. Sci.*, pp. 595–613. Springer, Berlin, 2000.
15. André Weilert. Fast computation of the biquadratic residue symbol. *J. Number Theory* **96**(1) (2002), 133–151.
16. H. C. Williams. An M^3 public-key encryption scheme. In *Advances in cryptology— CRYPTO '85 (Santa Barbara, Calif., 1985)*, Vol. 218 of *Lecture Notes in Comput. Sci.*, pp. 358–368. Springer, Berlin, 1986.
17. H. C. Williams and R. Holte. Computation of the solution of $x^3 + Dy^3 = 1$. *Math. Comp.* **31**(139) (1977), 778–785.

An Extended Quadratic Frobenius Primality Test with Average and Worst Case Error Estimates* **

Ivan Bjerre Damgård and Gudmund Skovbjerg Frandsen

BRICS* * *
Department of Computer Science, University of Aarhus.
{ivan,gudmund}@daimi.au.dk

Abstract. We present an Extended Quadratic Frobenius Primality Test (EQFT), which is related to an extends the Miller-Rabin test and the Quadratic Frobenius test (QFT) by Grantham. EQFT takes time about equivalent to 2 Miller-Rabin tests, but has much smaller error probability, namely $256/331776^t$ for t iterations of the test in the worst case. We give bounds on the average-case behaviour of the test: consider the algorithm that repeatedly chooses random odd k bit numbers, subjects them to t iterations of our test and outputs the first one found that passes all tests. We obtain numeric upper bounds for the error probability of this algorithm as well as a general closed expression bounding the error. For instance, it is at most 2^{-143} for $k = 500, t = 2$. Compared to earlier similar results for the Miller-Rabin test, the results indicates that our test in the average case has the effect of 9 Miller-Rabin tests, while only taking time equivalent to about 2 such tests. We also give bounds for the error in case a prime is sought by incremental search from a random starting point.

1 Introduction

Efficient methods for primality testing are important, in theory as well as in practice. Tests that always return correct results exist see for instance [1], but all known tests of this type are only of theoretical interest because they are much too inefficient to be useful in practice. In contrast, tests that accept composite numbers with bounded probability are typically much more efficient. This paper presents and analyses one such test. Primality tests are used, for instance, in public-key cryptography, where efficient methods for generating large, random primes are indispensable tools. Here, it is important to know how the test behaves in the average case. But there are also scenarios (e.g., in connection with Diffie-Hellman key exchange) where one needs to test if a number n is prime and where

* Partially supported by the IST Programme of the EU under contract number IST-1999-14186 (ALCOM-FT).
** Full paper is available at http://www.brics.dk/RS/03/9/index.html
* * * Basic Research in Computer Science, Centre of the Danish National Research Foundation.

A. Lingas and B.J. Nilsson (Eds.): FCT 2003, LNCS 2751, pp. 118–131, 2003.
© Springer-Verlag Berlin Heidelberg 2003

n may have been chosen by an adversary. Here the worst case performance of the test is important.

Virtually all known probabilistic tests are built on the same basic principle: from the input number n, one defines an Abelian group and then tests if the group structure we expect to see if n is prime, is actually present. The well-known Miller-Rabin test uses the group Z_n^* in exactly this way. A natural alternative is to try a quadratic extension of Z_n, that is, we look at the ring $Z_n[x]/(f(x))$ where $f(x)$ is a degree 2 polynomial chosen such that it is guaranteed to be irreducible if n is prime. In that case the ring is isomorphic to the finite field with n^2 elements, $GF(n^2)$. This approach was used successfully by Grantham [6], who proposed the Quadratic Frobenius Test (QFT), and showed that it accepts a composite with probability at most $1/7710$, i.e. a better bound than may be achieved using 6 independent Miller-Rabin tests, while asymptotically taking time approximately equivalent to only 3 such tests. Müller proposes a different approach based on computation of square roots, the MQFT [7,8] which takes the same time as QFT and has error probability essentially[1] $1/131040$. Just as for the Miller-Rabin test, however, it seems that most composites would be accepted with probability much smaller than the worst-case numbers. A precise result quantifying this intuition would allow us to give better results on the average case behaviour of the test, i.e., when it is used to test numbers chosen at random, say, from some interval. Such an analysis has been done by Damgård, Landrock and Pomerance for the Miller-Rabin test, but no corresponding result for QFT or MQFT is known.

In this paper, we propose a new test that can be seen as an extension of QFT. We call this the Extended Quadratic Frobenius test (EQFT). EQFT comes in two variants, EQFTac which works well in an average case analysis and EQFTwc, which is better for applications where the worst case behavior is important.

For the average case analysis: consider an algorithm that repeatedly chooses random odd k-bit numbers, subject each number to t iterations of EQFTac, and outputs the first number found that passes all t tests. Under the ERH, each iteration takes expected time equivalent to about 2 Miller-Rabin tests, or $2/3$ of the time for QFT/MQFT (the ERH is only used to bound the run time and does not affect the error probability). Let $q_{k,t}$ be the probability that a composite is output. We derive numeric upper bounds for $q_{k,t}$, e.g., we show $q_{500,2} \le 2^{-143}$, and also show a general upper bound, namely for $2 \le t \le k-1$, $q_{k,t}$ is $O(k^{3/2}2^{(\sigma_t+1)t}t^{-1/2}4^{-\sqrt{2\sigma_t tk}})$ with an easily computable big-O constant, where $\sigma_t = \log_2 24 - 2/t$. Comparison to the similar analysis by Damgård et al. for the MR test indicates that for $t \ge 2$, our test in the average case roughly speaking has the effect of 9 Miller-Rabin tests, while only taking time equivalent to 2 such tests. We also analyze the error probability when a random k-bit prime is instead generated using incremental search from a random starting point, still using (up to) t iterations of our test to distinguish primes from composites.

---

[1] The test and analysis results are a bit different, depending on whether the input is 3 or 1 modulo 4, see [7,8] for details

Concerning worst case analysis, we show that t iterations of EQFTwc err with probability at most $256/331776^t$ except for an explicit finite set of numbers[2]. The same worst case error probability can be shown for EQFTac, but this variant is up to 4 times slower on worst case inputs than in the average case, namely on numbers n where very large powers of 2 and 3 divide $n^2 - 1$. For EQFTwc, on the other hand, t iterations take time equivalent to about $2t + 2$ MR tests on all inputs (still assuming ERH). For comparison with EQFT/MQFT, assume that we are willing to spend the same fixed amount of time testing an input number. Then EQFTwc gives asymptotically a better bound on the error probability: using time approximately corresponding to $6t$ Miller-Rabin tests, we get error probability $1/7710^{2t} \approx 1/19.8^{6t}$ using QFT, $1/131040^{2t} \approx 1/50.8^{6t}$ using MQFT, and $256/331776^{3t-1} \approx 1/576^{6t}$ using EQFTwc.

2   The Intuition behind EQFT

2.1   A Simple Initial Idea

Given the number n to be tested, we start by constructing a quadratic extension $Z_n[X]/(f(X))$, which is kept fixed during the entire test (across all iterations). We let H be the multiplicative group in this extension ring. If n is prime, the quadratic extension is a field, and so H is cyclic of order $n^2 - 1$. We may of course assume that n is not divisible by 2 or 3, which implies that $n^2 - 1$ is always divisible by 24. Let H_{24} be the subgroup of elements of order dividing 24. If H is cyclic, then clearly $|H_{24}| = 24$. On the other hand, if n is not prime, H is the direct product of a number of subgroups, one for each distinct prime factor in n, and we may have $|H_{24}| \gg 24$.

Now, suppose we are already given an element $r \in H$ of order 24. Then a very simple approach to a primality test could be the following: Choose a random element z in H, and verify that $z^n = \bar{z}$, where $\bar{z}$ refers to the standard conjugate (explained later). This implies $z^{n^2-1} = 1$ for any invertible z and so is similar to the classical Fermat test. It is, however, in general a much stronger test than just checking the order of z. Then, from z construct an element z' chosen from H_{24} with some "suitable" distribution. For this intuitive explanation, just think of z' as being uniform in H_{24}. Now check that $z' \in< r >$, i.e. is a power of r. This must be the case if n is prime, but may fail if n is composite. This is similar to the part of the MR test that checks for existence of elements of order 2 different from -1.

To estimate the error probability, let ω be the number of distinct prime factors in n. Since H is the direct product of ω subgroups, H_{24} is typically of order 24^ω [3]. As one might then expect, it can be shown that the error probability of the test is at most $24/24^\omega$ times the probability that $z^n = \bar{z}$. The factor $24^{1-\omega}$ corresponds to the factor of $2^{1-\omega}$ one obtains for the MR test.

---

[2] namely if n has no prime factors less than 118, or if $n \geq 2^{42}$

[3] it may be smaller, but then the Fermat-like part of the test is stronger than otherwise, so we only consider the maximal case in this section

2.2   Some Problems and Two Ways to Solve Them

It is not clear how to construct an element of order 24 (if it exists at all), and we have not specified how to construct z' from z. We present two different approaches to these problems.

EQFTwc. In this approach, we run a start-up procedure that may discover that n is composite. But if not, it constructs an element of order 24 and also guarantees that H contains ω distinct subgroups, each of order divisible by $2^u 3^v$, where $2^u, 3^v$ are the maximal 2- and 3-powers dividing $n^2 - 1$. This procedure runs in expected time $O(1)$ Miller-Rabin tests. Details on the idea behind it are given in Section 5. Having run the start-up procedure, we construct z' as $z' = z^{(n^2-1)/24}$. Note that without the condition on the subgroups of H, we could have $z' = 1$ always which would clearly be bad. Each z can be tested in time approximately 2 MR tests, for any n. This leads to the test we call EQFTwc (since it works well in a worst case analysis).

EQFTac. The other approach avoids spending time on the start-up. This comes at the cost that the test becomes slower on n's where u, v are very large. But this only affects a small fraction of the potential inputs and is not important when testing randomly chosen n, since then the expected values of u, v are constant.

The basic idea is the following: we start choosing random z's immediately, and instead of trying to produce an element in H_{24} from z, we look separately for an element of order dividing 3 and one of order dividing 8. For order 3, we compute $z^{(n^2-1)/3^v}$ and repeatedly cube this value at most v times. This is guaranteed to produce an element of order 3, if 3 divides the order of z. If we already know an element ξ_3 of order 3, we can check that the new element we produce is in the group generated by ξ_3, and if not, n is composite. Of course, we do not know an element of order 3 from the start, but note that the computations we do on each z may produce such an element. So if we do several iterations of the test, as soon as an iteration produces an element of order 3, this can be used as ξ_3 by subsequent iterations. A similar idea can be applied to elements of order 8.

This leads to a test of strength comparable to EQFTwc, except for one problem: the iterations we do before finding elements of the right order may have larger error probability than the others. This can be compensated for by a number of further tricks: rather than choosing z uniformly, we require that $N(z)$ has Jacobi symbol 1, where $N()$ is a fixed homomorphism from H to Z_n^* defined below. This means we can expect z to have order a factor 2 smaller than otherwise[4], and this turns out to improve the error probability of the Fermat-like part of the test by a factor of $2^{1-\omega}$. Moreover, some partial testing of the elements we produce is always possible: for instance, we know n is composite if we see an element of order 2 different from -1. These tricks imply that the test, up to

---

[4] This also means that we should look for an element ξ_4 of order 4 (and not 8) in the part of the test that produces elements of order a 2-power

a small constant factor on the error probability, is as good as if we had known ξ_3, ξ_4 from the start. This version of the test is called EQFTac (since it works well in an average case analysis). We show that it satisfies the same upper bound on the error probability as we have for EQFTwc.

2.3   Comparison to Other Tests

We give some comments on the similarities and difference between EQFT and Grantham's QFT. In QFT the quadratic extension, that is, the polynomial $f(x)$, is randomly chosen, whereas the element corresponding to our z is chosen deterministically, given $f(x)$. This seems to simplify the error analysis for EQFT. Other than that, the Fermat part of QFT is transplanted almost directly to EQFT. For the test for roots of 1, QFT does something directly corresponding to the square root of 1 test from Miller-Rabin, but does nothing relating to elements of higher order. In fact, several of our ideas cannot be directly applied to QFT since there, $f(x)$ changes between iterations. As for the running time, since our error analysis works for any (i.e. a worst case) quadratic extension, we can pick one that has a particularly fast implementation of arithmetic, and this is the basis for the earlier mentioned difference in running time between EQFT and QFT.

A final comment relates to the comparison in running times between Miller-Rabin, Grantham's and our test. Using the standard way to state running times in the literature, the Miller-Rabin, resp. Grantham's, resp. our test run in time $\log n + o(\log n)$ resp. $3 \log n + o(\log n)$ resp. $2 \log n + o(\log n)$) multiplications in Z_n. However, the running time of Miller-Rabin is actually $\log n$ *squarings* $+o(\log n)$ multiplications in Z_n, while the $3 \log n$ ($2 \log n$) multiplications mentioned for the other tests are a mix of squarings and multiplications. So we should also compare the times for modular multiplications and squarings. On a standard, say, 32 bit architecture, a modular multiplication takes time about 1.25 times that of a modular squaring if the numbers involved are very large. However, if we use the fastest known modular multiplication method (which is Montgomery's in this case, where n stays constant over many multiplications), the factor is smaller for numbers in the range of practical interest. Concrete measurements using highly optimized C code shows that it is between 1 and 1.08 for numbers of length 500-1000 bits. Finally, when using dedicated hardware the factor is exactly 1 in most cases. So we conclude that the comparisons we stated are quite accurate also for practical purposes.

2.4   The Ring $R(n, c)$ and EQFTac

Definition 1. *Let n be an odd integer and let c be a unit modulo n.*
   Let $R(n, c)$ denote the ring $\mathbf{Z}[x]/(n, x^2 - c)$.

More concretely, an element $z \in R(n, c)$ can be thought of as a degree 1 polynomial $z = ax + b$, where $a, b \in \mathbf{Z}_n$, and arithmetic on polynomials is modulo $x^2 - c$ where coefficients are computed on modulo n.

Let p be an odd prime. If c is not a square modulo p, i.e. $(c/p) = -1$, then the polynomial $x^2 - c$ is irreducible modulo p and $R(p,c)$ is isomorphic to $GF(p^2)$.

Definition 2. *Define the following multiplicative homomorphisms on* $R(n,c)$ *(assume* $z = ax + b$*):*

$$\bar{\ } : R(n,c) \mapsto R(n,c), \ \bar{z} = -ax + b \tag{1}$$

$$N(\cdot) : \quad R(n,c) \mapsto \mathbf{Z}_n, \quad N(z) = \bar{z} \cdot z = b^2 - ca^2 \tag{2}$$

and define the map $(\cdot/\cdot) : \mathbf{Z} \times \mathbf{Z} \mapsto \{-1, 0, 1\}$ *to be the Jacobi symbol.*

The maps $\bar{\ }$ and $N(\cdot)$ are both multiplicative homomorphisms whether n is composite or n is a prime. The primality test will be based on some additional properties that are satisfied when p is a prime and $(c/p) = -1$, in which case $R(p,c) \simeq GF(p^2)$:

Frobenius property / generalised Fermat property: Conjugation, $z \mapsto \bar{z}$, is a field automorphism on $GF(p^2)$. In characteristic p, the Frobenius map that raises to the p'th power is also an automorphism, using this it follows easily that

$$\bar{z} = z^p \tag{3}$$

Quadratic residue property / generalised Solovay-Strassen property: The norm, $z \mapsto N(z)$, is a surjective multiplicative homomorphism from $GF(p^2)$ to the subfield $GF(p)$. As such the norm maps squares to squares and non-squares to non-squares, it follows from the definition of the norm and (3) that

$$z^{(p^2-1)/2} = N(z)^{(p-1)/2} = (N(z)/p) \tag{4}$$

4'th-root-of-1-test / generalised Miller-Rabin property: Since $GF(p^2)$ is a field there are only four possible 4th roots of 1 namely 1, -1 and ξ_4, $-\xi_4$, the two roots of the cyclotomic polynomial $\Phi_4(x) = x^2 + 1$. In particular, this implies for $p^2 - 1 = 2^u 3^v q$ where $(q, 6) = 1$ that if $z \in GF(p^2) \setminus \{0\}$ is a square then

$$z^{3^v q} = \pm 1, \quad \text{or} \quad z^{2^i 3^v q} = \pm \xi_4 \text{ for some } i = 0, \dots, u-3 \tag{5}$$

3'rd-root-of-1-test: Since $GF(p^2)$ is a field there is only three possible 3rd roots of 1 namely 1 and ξ_3, ξ_3^{-1}, the two roots of the cyclotomic polynomial $\Phi_3(x) = x^2 + x + 1$. In particular, this implies for $p^2 - 1 = 2^u 3^v q$ where $(q, 6) = 1$ that if $z \in GF(p^2) \setminus \{0\}$ then

$$z^{2^u q} = 1, \quad \text{or} \quad z^{2^u 3^i q} = \xi_3^{\pm 1} \text{ for some } i = 0, \dots, v-1 \tag{6}$$

The actual test will have two parts (see algorithm 1). In the first part, a specific quadratic extension is chosen, i.e. $R(n,c)$ for an explicit c. In the second part, the above properties of $R(n,c)$ are tested for a random choice of z. When the EQFTac is run several times on the same n, only the second part is executed multiple times. The second part receives two extra inputs, a 3rd and a 4th root of 1. On the first execution of the second part these are both 1. During later

---

Algorithm 1 Extended Quadratic Frobenius Test (EQFTac).

First part (construct quadratic extension):

Require: input is odd number $n \geq 13$
Ensure: output is "composite" or c satisfying $(c/n) = -1$
 1: **if** n is divisible by a prime less than 13 **return** "composite"
 2: **if** n is a perfect square **return** "composite"
 3: choose a small c with $(c/n) = -1$; **return** c

Second part (make actual test):

Require: input is n, c, r_3, r_4, where $n \geq 5$ not divisible by 2 or 3, $(c/n) = -1$, $r_3 \in \{1\} \cup \{\xi \in R(n,c) \mid \Phi_3(\xi) = 0\}$ and $r_4 \in \{1, -1\} \cup \{\xi \in R(n,c) \mid \Phi_4(\xi) = 0\}$
  Let u, v be defined by $n^2 - 1 = 2^u 3^v q$ for $(q, 6) = 1$.
Ensure: output is "composite", or "probable prime", s_3, s_4, where $s_3 \in \{1\} \cup \{\xi \in R(n,c) \mid \Phi_3(\xi) = 0\}$ and $s_4 \in \{1, -1\} \cup \{\xi \in R(n,c) \mid \Phi_4(\xi) = 0\}$
 4: select random $z \in R(n,c)^*$ with $(N(z)/n) = 1$
 5: **if** $\bar{z} \neq z^n$ or $z^{(n^2-1)/2} \neq 1$ **return** "composite"
 6: **if** $z^{3^v q} \neq 1$ and $z^{2^i 3^v q} \neq -1$ for all $i = 0, \ldots, u - 2$ **return** "composite"
 7: **if** we found $i_0 \geq 1$ with $z^{2^{i_0} 3^v q} = -1$ (there can be at most one such value) then
    let $R_4(z) = z^{2^{i_0-1} 3^v q}$. Else let $R_4(z) = z^{3^v q} (= \pm 1)$;
    if $(r_4 \neq \pm 1$ and $R_4(z) \notin \{\pm 1, \pm r_4\})$ **return** "composite"
 8: **if** $z^{2^u q} \neq 1$ and $\Phi_3(z^{2^u 3^i q}) \neq 0$ for all $i = 0, \ldots, v - 1$ **return** "composite"
 9: **if** we found $i_0 \geq 0$ with $\Phi_3(z^{2^u 3^{i_0} q}) = 0$ (there can be at most one such value)
    then let $R_3(z) = z^{2^u 3^{i_0} q}$ else let $R_3(z) = 1$;
    if $(r_3 \neq 1$ and $R_3(z) \notin \{1, r_3^{\pm 1}\})$ **return** "composite"
10: **if** $r_3 = 1$ and $R_3(z) \neq 1$ then let $s_3 = R_3(z)$ else let $s_3 = r_3$;
    if $r_4 = \pm 1$ and $R_4(z) \neq \pm 1$ then let $s_4 = R_4(z)$ else let $s_4 = r_4$;
    return "probable prime", s_3, s_4

---

executions of the second part some nontrivial roots are possibly constructed. If so they are transferred to all subsequent executions of the second part.

Here follows some more detailed comments to algorithm 1:

Line 1 ensures that $24 \mid n^2 - 1$. In addition, we will use that n has no small prime factors in the later error analysis.

Line 2 of the algorithm is necessary, since no c with $(c/n) = -1$ exists when n is a perfect square.

Line 3 of the algorithm ensures that $R(n, c) \simeq GF(n^2)$ when n is a prime. Lemma 2 defines more precisely what "small" means.

Line 4 makes sure that z is a square, when n is a prime.

Line 5 checks equations (3) and (4), the latter in accordance with the condition enforced in line 4.

Line 6 checks equation (5) to the extent possible without having knowledge of ξ_4, a primitive 4th root of 1.

Line 7f continues the check of equation (5) by using any ξ_4 given on the input.

Line 8 checks equation (6) to the extent possible without having knowledge of ξ_3, a primitive 3rd root of 1.

Line 9f continues the check of equation (6) by using any ξ_3 given on the input.

2.5   Implementation of the Test

High powers of elements in $R(n, c)$ may be computed efficiently when c is (numerically) small. Represent $z \in R(n, c)$ in the natural way by $((A_z, B_z) \in \mathbf{Z}_n \times \mathbf{Z}_n$, i.e. $z = A_z x + B_z$.

Lemma 1. *Let $z, w \in R(n, c)$:*

1. *$z \cdot w$ may be computed from z and w using 3 multiplications and $O(\log c)$ additions in $\mathbf{Z}_n$*
2. *z^2 may be computed from z using 2 multiplications and $O(\log c)$ additions in $\mathbf{Z}_n$*

Proof. For 1, we use the equations $A_{zw} = m_1 + m_2$ and $B_{zw} = (cA_z + B_z)(A_w + B_w) - (cm_1 + m_2)$ with $m_1 = A_z B_w$ and $m_2 = B_z A_w$. For 2, we need only observe that in the proof of 1, $z = w$ implies that $m_1 = m_2$.

We also need to argue that it is easy to find a small c with $(c/n) = -1$. One may note that if $n = 3 \bmod 4$, then $c = -1$ can always be used, and if $n = 5 \bmod 8$, then $c = 2$ will work. In general, we have the following:

Lemma 2. *Let n be an odd composite number that is not a perfect square. Let $\pi_-(x, n)$ denote the number of primes $p \leq x$ such that $(p/n) = -1$, and, as usual, let $\pi(x)$ denote the total number of primes $p \leq x$. Assuming the Extended Riemann Hypothesis (ERH), there exists a constant C (independent of n) such that*

$$\frac{\pi_-(x, n)}{\pi(x)} > \frac{1}{3} \quad \text{for all } x \geq C(\log n \log \log n)^2$$

Proof. We refer to the full paper for the proof that is based on [2, th.8.4.6].

Theorem 1. *Let n be a number that is not divisible by 2 or 3, and let $u \geq 3$ and $v \geq 1$ be maximal such that $n^2 - 1 = 2^u 3^v q$. There is an implementation of algorithm 1 that on input n takes expected time equivalent to $2 \log n + O(u + v) + o(\log n)$ multiplications in $\mathbf{Z}_n$, when assuming the ERH.*

Remark 1. We can only prove a bound on the expected time, due to the random selection of an element z (in line 4) having a property that is only satisfied by half the elements, and to the selection of a suitable c (line 3), where at least a third of the candidates are usable. Although there is in principle no bound on the maximal time needed, the variance around the expectation is small because the probability of failing to find a useful z and c drops exponentially with the number of attempts. We emphasize that the ERH is only used to bound the running time (of line 3) and does not affect the error probability, as is the case with the original Miller test.

The detailed implementation of algorithm 1 may be optimized in various ways. The implementation given in the proof that follows this remark has focused on simplicity more than saving a few multiplications. However, we are not aware of any implementation that avoids the $O(u+v)$ term in the complexity analysis.

Proof. We will first argue that only lines 5-9 in the algorithm have any significance in the complexity analysis.

line 2. By Newton iteration the square root of n may be computed using $O(\log \log n)$ multiplications.

line 3. By lemma 2, we expect to find a c of size $O((\log n \log \log n)^2)$ such that $(c/n) = -1$ after three attempts (or discover that n is composite).

line 4. z is selected randomly from $R(n, c) \setminus \{0\}$. We expect to find z with $(N(z)/n) = 1$ after two attempts (or discover that n is composite).

line 5-9. Here we need to explain how it is possible to simultaneously verify that $\bar{z} = z^n$, and do both a 4'th-root-of-1-test and a 3'rd-root-of-1-test without using too many multiplications. We refer to lemma 1 for the implementation of arithmetic in $R(n, c)$.

Define s, r by $n = 2^u 3^v s + r$ for $0 < r < 2^u 3^v$. A simple calculation confirms that

$$q = ns + rs + (r^2 - 1)/(2^u 3^v), \tag{7}$$

where the last fraction is integral. Go through the following computational steps using the z selected in line 4 of the algorithm:

1. compute z^s.
   This uses $2 \log n + o(\log n)$ multiplications in $\mathbf{Z}_n$.
2. compute z^n.
   Starting from step 1 this requires $O(v + u)$ multiplications in $\mathbf{Z}_n$.
3. verify $z^n = \bar{z}$.
4. compute z^q.
   One may compute z^q from step 1 using $O(v+u)$ multiplications in $\mathbf{Z}_n$, when using (7) and the shortcut $z^{ns} = \overline{z^s}$, where the shortcut is implied by step 3 and exponentiation and conjugation being commuting maps.
5. compute $z^{3^v q}, z^{2 \cdot 3^v q}, z^{2^2 3^v q}, \ldots, z^{2^{u-2} 3^v q}$.
   Starting from step 4 this requires $O(v + u)$ multiplications in $\mathbf{Z}_n$.
6. verify that $z^{3^v q} = 1$ or $z^{2^i 3^v q} = -1$ for some $0 \le i \le u - 2$. If there is $i_0 \ge 1$ with $z^{2^{i_0} 3^v q} = -1$ and if ξ_4 is present, verify that $z^{2^{i_0-1} 3^v q} = \pm \xi_4$.
7. compute $z^{2^u q}, z^{2^u 3q}, z^{2^u 3^2 q}, \ldots, z^{2^u 3^{v-1} q}$.
   Starting from step 4 this requires $O(v + u)$ multiplications in $\mathbf{Z}_n$.
8. By step 6 there must be an i ($0 \le i \le v$) such that $z^{2^u 3^i q} = 1$. Let i_0 be the smallest such i. If $i_0 \ge 1$ verify that $z^{2^u 3^{i_0-1} q}$ is a root of $x^2 + x + 1$. If ξ_3 is present, verify in addition that $z^{2^u 3^{i_0-1} q} = \xi_3^{\pm 1}$

3 An Expression Bounding the Error Probability

Theorem 2 assumes that the auxiliary inputs r_3, r_4 are "good", which should be taken to mean that they are non-trivial third and fourth roots of 1, and are roots in the third and fourth cyclotomic polynomial (provided such roots exist in $R(n, c)$. When EQFT is executed as described earlier, we cannot be sure that r_3, r_4 are good. However, the probability that they are indeed good is sufficiently large that the theorem can still be used to bound the actual error probability as shown in Theorem 3 (for proofs, see the full paper):

Theorem 2. *Let n be an odd composite number with prime power factorisation $n = \prod_{i=1}^{\omega} p_i^{m_i}$, let $\Omega = \sum_{i=0}^{\omega} m_i$, and let c satisfy that $(c/n) = -1$. Given good values of the inputs r_3, r_4, the error probability of a single iteration of the second part of the EQFTac (algorithm 1) is bounded by*

$$\beta(n, c) \leq 24^{1-\omega} \prod_{i=1}^{\omega} p_i^{2(1-m_i)} \operatorname{sel}[(c/p_i), \frac{(n/p_i - 1, (p_i^2 - 1)/24)}{(p_i^2 - 1)/24}, \frac{12}{p_i - 1}] \leq 24^{1-\Omega}$$

where, we have adopted the notation $\operatorname{sel}[\pm 1, E_1, E_2]$ for a conditional expression with the semantics $\operatorname{sel}[-1, E_1, E_2] = E_1$ and $\operatorname{sel}[1, E_1, E_2] = E_2$.

Theorem 3. *Let n be an odd composite number with ω distinct prime factors.*
For any $t \geq 1$, the error probability $\beta_t(n)$ of t iterations of EQFTac (algorithm 1) is bounded by

$$\beta_t(n) \leq \max_{(c/n)=-1} 4^{\omega-1} \beta(n, c)^t$$

4 EQFTac: Average Case Behaviour

4.1 Uniform Choice of Candidates

Let M_k be the set of odd k-bit integers $(2^{k-1} < n < 2^k)$. Consider the algorithm that repeatedly chooses random numbers in M_k, until one is found that passes t iterations of EQFTac, and outputs this number.

The expected time to find a "probable prime" with this method is at most tT_k/p_k, where T_k is the expected time for running the test on a random number from M_k, and p_k is the probability that such a number is prime. Suppose we choose n at random and let $n^2 - 1 = 2^u 3^v q$, where q is prime to 2 and 3. It is easy to see that the expected values of u and v are constant, and so it follows from Theorem 1 that T_k is $2k + o(k)$ multiplications modulo a k bit number. This gives approximately the same time needed to generate a probable prime, as if we had used $2t$ iterations of the Miller-Rabin test in place of t iterations of EQFTac. But, as we shall see, the error probability is much smaller than with $2t$ MR tests.

Let $q_{k,t}$ be the probability that the algorithm above outputs a composite number. When running t iterations of our test on input n, it follows from Theorem 3 and Theorem 2 that the probability $\beta_t(n)$ of accepting n satisfies

$$\beta_t(n) \leq 4^{\omega-1}24^{t(1-\Omega)} \max\{\frac{(n/p - 1, (p^2 - 1)/24)}{(p^2 - 1)/24}, \frac{12}{p-1}\}^t$$

where p is the largest prime factor in n and Ω is the number of prime factors in n, counted with multiplicity. This expression is extremely similar to the one for the Rabin test found in [5]. Therefore we can find bounds for $q_{k,t}$ in essentially the same way as there. Details can be found in the full paper. We obtain numerical estimates for $q_{k,t}$, some sample results are shown in the table 1, which contains $-\log_2$ of the estimates, so we assert that, e.g., $q_{500,2} \leq 2^{-143}$.

Table 1. Lower bounds on $-\log_2 q_{k,t}$

$k \setminus t$	1	2	3	4
300	42	105	139	165
400	49	125	165	195
500	57	143	187	221
600	64	159	208	245
1000	86	212	276	325

We also get a closed expression (with an easily computable big-O constant):

Theorem 4. *For* $2 \leq t \leq k-1$, *we have that* $q_{k,t}$ *is* $O(k^{3/2}2^{(\sigma_t+1)t}t^{-1/2}4^{-\sqrt{2\sigma_t tk}})$

Comparing to corresponding results in [5] for the Miller-Rabin test one finds that if several iteration of EQFTac are performed, then roughly speaking each iteration has the effect of 9 Miller-Rabin tests, while only taking time equivalent to about 2 M-R tests.

4.2   Incremental Search

The algorithm we have just analysed is in fact seldom used in practice. Most real implementations will not want to choose candidates for primes uniformly at random. Instead one will choose a random starting point n_0 in M_k and then test $n_0, n_0 + 2, n_0 + 4, \ldots$ for primality until one is found that passes t iterations of the test. Many variations are possible, such as other step sizes, various types of sieving, but the basic principle remains the same. The reason for applying such an algorithm is that test division by small primes can be implemented much more efficiently (see for instance [4]). On the other hand, the analysis we did above depends on the assumption that candidates are independent. In [3], a way to get around this problem for the Miller-Rabin test was suggested. We apply an improvement of that technique here.

We will analyse the following example algorithm which depends on parameters t and s: choose n_0 uniformly in M_k and test $n_0, n_0 + 2, .., n_0 + 2(s-1)$ using t iterations of EQFTac. If no probable prime is found, start over with a new independently chosen value of n_0. Output the first number found that passes all t iterations of EQFTac.

We argue in the full paper that the expected time to find a probable prime by the above algorithm is at most $O(tk^2)$ multiplications modulo k bit numbers, if s is $\theta(k)$. Practice shows that for $s = 10 \ln 2^k$, we need almost all the time only one value of n_0, and so $st(2k + o(k))$ multiplications is an upper bound[5]. Let $Q_{k,t,s}$ be the probability that the above algorithm outputs a composite number. Table 2 shows sample numeric results of our estimates of $Q_{k,t,s}$.

Table 2. Estimates of the overall error probability with incremental search, lower bounds on $-\log_2 Q_{k,t,s}$ using $s = c \cdot \ln(2^k)$ and $c = 10$.

$k \setminus t$	1	2	3	4
300	18	74	107	133
400	26	93	132	162
500	34	109	153	186
600	40	125	174	210
1000	62	176	239	288

5   EQFTwc: Worst Case Analysis

We present in this section the version of our test (EQFTwc) which is fast for all n and has essentially the same error probability bound as EQFTac. The price for this is an expected start up cost of $\leq 2 \log n + o(\log n)$ multiplications in $\mathbf{Z}_n$ for the first iteration of the test. For comparison of our test with the earlier tests of Grantham, Müller and Miller-Rabin, assume that we are willing to spend some fixed amount of time testing an input number, say, approximately corresponding to the time for t Miller-Rabin tests. Then, using our test, we get asymptotically a better bound on the error probability: using Miller-Rabin, Grantham[6], Müller [7,8], and EQFTwc, respectively, we get error bounds $4^{-t}, 19.8^{-t}, 50.8^{-t}$ and approximately 576^{-t}.

In Section 2, the general idea behind EQFTwc was explained. The only point left open was the following: we need to design a start-up procedure that can either discover that n is composite, or construct an element r_{24} of order 24, and also guarantee that all Sylow-2 and -3 subgroups of $R(n, c)^*$ have order at least $2^u, 3^v$

---

[5] Of course, this refers to the run time when only the EQFTac is used. In practice, one would use test division and other tricks to eliminate some of the non primes faster than EQFTac can do it. This may reduce the run time significantly. Any such method can be used without affecting the error estimates, as long as no primes are rejected.

Algorithm 2 Extended Quadratic Frobenius Test (EQFTwc).

First iteration:

Require: input is an odd number $n \geq 5$
Ensure: output is "composite", or "probable prime", $c \in \mathbf{Z}_n$, $r_{24} \in R(n,c)^*$, where
$(c/n) = -1$ and $\Phi_{24}(r_{24}) = 0$.
1: **if** n is divisible by 2 or 3 **return** "composite"
2: **if** n is a perfect square or a perfect cube **return** "composite"
3: choose a small c with $(c/n) = -1$
4: compute $r \in R(n,c)$ satisfying $r^2 + r + 1 = 0$ (may **return** "composite")
5: a: **if** $n \equiv 1 \bmod 3$ then select a random $z \in R(n,c)^*$ with $(N(z)/n) = -1$ and
   $res_3(z) \neq 1$.
   b: **if** $n \equiv 2 \bmod 3$ then **repeat**
      Make a Miller-Rabin primality test on n (may **return** "composite")
      select a random $z \in R(n,c)^*$ with $(N(z)/n) = -1$ and compute $res_3(z)$
   until either the Miller-Rabin test returns composite or the selected z satisfies
   that $res_3(z) \neq 1$
6: **if** $\bar{z} \neq z^n$ **return** "composite".
7: Let $r_{24} = z^{(n^2-1)/24}$. If $r_{24}^8 \neq r^{\pm 1}$ or $r_{24}^{12} \neq -1$ **return** "composite".
8: **return** "probable prime", c, r_{24}
Subsequent iterations:
Require: input is n, c, r_{24}, where $n \geq 5$ is not divisible by 2 or 3, $(c/n) = -1$, and
   $\Phi_{24}(r_{24}) = 0$
Ensure: output is "composite" or "probable prime"
9: select random $z \in R(n,c)^*$
10: **if** $\bar{z} \neq z^n$ **return** "composite"
11: **if** $z^{(n^2-1)/24} \notin \{r_{24}^i \mid i = 0, \ldots, 23\}$ **return** "composite"
12: **return** "probable prime"

(where as usual, $2^u, 3^v$ are the maximal 2- and 3-powers dividing $n^2 - 1$). We do
this by choosing $z \in R(n,c)^*$ in such a way that *if* n is prime, then z is both a
non-square and a non-cube. This means that we can expect that $z^{(n^2-1)/2} = -1$
and that $z^{(n^2-1)/3} = r^{\pm 1}$, where r is a primitive 3rd root of 1. If this is not the
case, n is composite. If it is, n may still be composite, but we have the required
condition on the Sylow-2 and -3 subgroups, and we can set $r_{24} = z^{(n^2-1)/24}$. The
subsequent iterations of the test are then very simple: take a random $z \in R(n,c)$
and check whether $\bar{z} = z^n$ and $z^{(n^2-1)/24} \in \{r_{24}^i \mid i = 0, \ldots, 23\}$

Before presenting the algorithm, we need to define a homomorphism res_3
from the ring $R(n,c)^*$ into the complex third roots of unity $\{1, \zeta, \zeta^2\}$. This
homomorphism will be used to recognize cubic nonresidues.

Definition 3. *For arbitrary $n \geq 5$ with $(n, 6) = 1$, for arbitrary c with $(c/n) = -1$, assume there exists an $r = gx + h \in R(n,c)$ with $r^2 + r + 1 = 0$, and if $n \equiv 1 \bmod 3$ assume in addition that $r \in \mathbf{Z}_n$, i.e. $g = 0$.*
Define $res_3 : R(n,c)^* \mapsto \{1, \zeta, \zeta^2\} \subseteq \mathbf{Z}[\zeta]$ by

$$res_3(ax + b) = \begin{cases} [b^2 - ca^2 \ / \ \gcd(n, r - \zeta)], & \text{if } n \equiv 1 \bmod 3 \\ [(b + a(\zeta - h)/g) \ / \ n], & \text{if } n \equiv 2 \bmod 3 \end{cases}$$

where $[\cdot / \cdot]$ denotes the cubic residuosity symbol.

To find the element z mentioned above, we note that computing the Jacobi symbol will let us recognize $1/2$ of all elements as nonsquares. One might expect that applying res_3 would let us recognize $2/3$ of all elements as noncubes. Unfortunately, all we can show is that res_3 is nontrivial except possibly when n is a perfect cube, or n is composite and $n \equiv 2 \mod 3$. To handle this problem, we take a pragmatic solution: Run a Miller-Rabin test and a search for noncubes in parallel. If n is prime then the search for a noncube will succeed, and if n is composite then the MR-test (or the noncube search) will succeed.

The following results are proved in the full paper:

Theorem 5. *There is an implementation of algorithm 2 that on input n takes expected time equivalent to at most $2 \log n + o(\log n)$ multiplications in $\mathbf{Z}_n$ per iteration, when assuming the ERH. The first iteration has an additional expected start up cost equivalent to at most $2 \log n + o(\log n)$ multiplications in $\mathbf{Z}_n$.*

Theorem 6. *Let n be an odd composite number with prime power factorisation $n = \prod_{i=1}^{\omega} p_i^{m_i}$, let $\Omega = \sum_{i=0}^{\omega} m_i$. If $\gamma_t(n)$ denotes the probability that n passes t iterations of the EQFTwc test (algorithm 2) then*

$$\gamma_t(n)$$

$$\leq \max_{(c/n)=-1} 4^{\omega-1}(24^{1-\omega} \prod_{i=1}^{\omega} p_i^{2(1-m_i)} \mathrm{sel}[(c/p_i), \frac{(n/p_i-1, p_i^2-1)}{p_i^2-1}, \frac{(n^2/p_i^2-1, p_i-1)}{(p_i-1)^2}])^t$$

$$\leq 4^{\omega-1}24^{t(1-\Omega)}$$

If n has no prime factor ≤ 118 or $n \geq 2^{42}$ then $\gamma_t(n) \leq 4^4 24^{-4t} \approx 2^{8-18.36t}$

References

1. Manindra Agrawal, Neeraj Kayal, and Nitin Saxena. PRIMES is in P. Preprint 2002. Department of Computer Science & Engineering, Indian Institute of Technology, Kanpur Kanpur-208016, INDIA, 2002.
2. Eric Bach and Jeffrey Shallit. *Algorithmic number theory. Vol. 1.* Foundations of Computing Series. MIT Press, Cambridge, MA, 1996. Efficient algorithms.
3. Jørgen Brandt and Ivan Damgård. On generation of probable primes by incremental search. In *Advances in cryptology—CRYPTO '92 (Santa Barbara, CA, 1992)*, Vol. 740 of *Lecture Notes in Comput. Sci.*, pp. 358–370. Springer, Berlin, 1993.
4. Jørgen Brandt, Ivan Damgård, and Peter Landrock. Speeding up prime number generation. In *Advances in cryptology—ASIACRYPT '91 (Fujiyoshida, 1991)*, Vol. 739 of *Lecture Notes in Comput. Sci.*, pp. 440–449. Springer, Berlin, 1993.
5. Ivan Damgård, Peter Landrock, and Carl Pomerance. Average case error estimates for the strong probable prime test. *Math. Comp.* **61**(203) (1993), 177–194.
6. Jon Grantham. A probable prime test with high confidence. *J. Number Theory* **72**(1) (1998), 32–47.
7. Siguna Müller. A probable prime test with very high confidence for $n \equiv 1 \mod 4$. In *Advances in cryptology—ASIACRYPT 2001 (Gold Coast)*, Vol. 2248 of *Lecture Notes in Comput. Sci.*, pp. 87–106. Springer, Berlin, 2001.
8. Siguna Müller. A probable prime test with very high confidence for $n \equiv 3 \mod 4$. *J. Cryptology* **16**(2) (2003), 117–139.

Periodic Multisorting Comparator Networks*

Marcin Kik

Institute of Mathematics, Wrocław University of Technology
ul. Wybrzeże Wyspiańskiego 27, 50-370 Wrocław, Poland
kik@im.pwr.wroc.pl

Abstract. We present a family of periodic comparator networks that
transform the input so that it consists of a few sorted subsequences.
The depths of the networks range from 4 to $2\log n$ while the number
of sorted subsequences ranges from $2\log n$ to 2. They work in time
$c\log^2 n + O(\log n)$ with $4 \leq c \leq 12$, and the remaining constants
are also suitable for practical applications. So far, known periodic
sorting networks of a constant depth that run in time $O(\log^2 n)$ (a
periodic version of AKS network [7]) are impractical because of complex
structure and very large constant factor hidden by big "Oh".

Keywords: sorting, comparator networks, parallel algorithms.

1 Introduction

Comparator is a simple device capable of sorting two elements. Many comparators can be connected together to form a comparator network. This way we get
the classical framework for sorting algorithms. Optimal arranging the comparators turned out to be a challenge. The main complexity measures of comparator
networks are time complexity (depth or number of steps) and the number of
comparators. The most famous sorting network is AKS network with asymptotically optimal depth $O(\log n)$ [1], however the big constant hidden by big "Oh"
makes it impractical. The Batcher networks of depth $\approx \frac{1}{2}\log^2 n$ [2], seem to be
very attractive for practical applications.

A *periodic* network is repeatedly used on the intermediate results until the
output becomes sorted, thus the same comparators are reused many times. In this
case, the time complexity is the depth of the network multiplied by the number
of iterations. The main advantage of periodicity is the reduction of the amount of
hardware (comparators) needed for the realization of the sorting algorithm, with
a very simple control mechanism providing the output of one iteration as the
input for the next iteration. Dowd et al, [3], reduced the number of comparators
from $\Omega(n\log^2 n)$ to $\frac{1}{2}n\log n$, while keeping the sorting time $\log^2 n$, by the
use of a periodic network of depth $\log n$. (The networks of depth d have at most
$dn/2$ comparators.) There are some periodic sorting networks of a constant depth
([10], [5], [7]). In [7], constant depth networks with time complexity $O(\log^2 n)$ are

---

* Research supported by KBN grant 7T11C 3220 in the years 2002, 2003.

A. Lingas and B.J. Nilsson (Eds.): FCT 2003, LNCS 2751, pp. 132–143, 2003.
© Springer-Verlag Berlin Heidelberg 2003

obtained by "periodification" of the AKS network, and more practical solutions with time complexity $O(\log^3 n)$, are obtained by "periodification" of the Batcher network. On the other hand there is not known any $\omega(\log n)$ lower bound on the time complexity of periodic sorting networks of constant depth. Closing the gap between the known upper bound of $O(\log^2 n)$ and the trivial general lower bound $\Omega(\log n)$ seems to be a very hard problem.

Periodic networks of constant depth can also be used for simpler tasks, such as merging sorted sequences [6], or resorting sequences with few values modified [4].

1.1 New Results

We assume that the values are stored in the registers and the only allowed operations are *compare-exchange* operations (*applications of comparators*) on the pairs of registers. Such an operation takes the two values stored in the pair of registers and stores the lower value in the first register and the greater value in the second register. (This interpretation differs from the one presented for instance in [8] but is more useful when periodic comparator networks are concerned.)

We present a family of periodic comparator networks $N_{m,k}$. The input size of $N_{m,k}$ is $n = 4m2^k$. The depth of $N_{m,k}$ is $2\lceil k/m \rceil + 2$. In Section 4 we prove the following theorem.

Theorem. *The periodic network $N_{m,k}$ transforms the input into $2m$ sorted subsequences of length $n/(2m)$ in time $4k^2 + 8km + O(k + m)$.*

For example, the network $N_{1,k}$ is a network of depth $\approx 2 \log n$ that produces 2 sorted sequences in time $\approx 4 \log^2 n + O(\log n)$. On the other hand, $N_{k,k}$ is a network of depth 4 that transforms the input into $\approx 2 \log n$ sorted sequences in time $\approx 12 \log^2 n + O(\log n)$. Due to the large constants in the known periodic constant depth networks sorting in time $O(\log^2 n)$, [7], it could be interesting alternative to use $N_{k,k}$ to produce very much ordered (although not completely sorted) output.

The output produced by $N_{m,k}$ can be finally sorted by a network merging $2m$ sequences. This can be performed by the very efficient multiway merge sorting networks [9]. It is an interesting problem to find efficient periodic network of constant depth that merges multiple sorted sequences. The periodic networks of constant depth that merge two sorted sequences in time $O(\log n)$ are already known [6].

As $N_{m,k}$ outputs multiple sorted sequences, we call it a *multisorting* network. Much simpler multisorting networks of constant depth exist if some additional operations are allowed (such as permutations of the elements in the registers between the iterations). However, we consider only the case restricted to the compare-exchange operations.

2   Preliminaries

By a *comparator network* we mean a set of *registers* $R_0, \ldots, R_{n-1}$ together with a finite sequence of *layers of comparators*. Every moment a register R_i contains a single value (denoted by $v(R_i)$) from some totally ordered set, say $\mathbb{N}$. We say that the network stores a sequence $v(R_0), \ldots, v(R_{n-1})$. A subset S of registers is *sorted* if for all R_i, R_j in S, $i < j$ implies that $v(R_i) \leq v(R_j)$. A *comparator* is denoted by an ordered pair of registers (R_i, R_j). If $v(R_i) = x$ and $v(R_j) = y$ before an *application* of the comparator (R_i, R_j), then $v(R_i) = \min\{x, y\}$ and $v(R_j) = \max\{x, y\}$ after the application of (R_i, R_j). A set of comparators L forms a *layer* if each register is contained in at most one of the comparators of L. So all the comparators of a layer can be applied simultaneously. We call such application a *step*. The *depth* of the network is the number of its layers. An *input* is the initial value of the sequence $v(R_0), \ldots, v(R_{n-1})$. An *output* of the network N is the sequence $v(R_0), \ldots, v(R_{n-1})$ obtained after application of all its layers (*application of N*) on some initial input sequence. We can iterate the network's application, by applying it to the output of its previous application. We call such network a *periodic network*. The *time* complexity of the periodic network is the number of steps performed in all iterations.

3   Definition of the Network $N_{m,k}$

We define a periodic network $N_{m,k}$ for positive integers m and k. For the sake of simplicity we fix the values m and k and denote $N_{m,k}$ by N. Network N contains n registers $R_0, \ldots, R_{n-1}$, where $n = 4m \cdot 2^k$. It will be useful to imagine that the registers are arranged in a three-dimensional matrix M of size $2 \times 2m \times 2^k$. For $0 \leq x \leq 1$, $0 \leq y \leq 2m - 1$ and $0 \leq z \leq 2^k - 1$, the element $M_{x,y,z}$ is a register R_i such that $i = x + 2y + 4mz$. For the intuitions, we assume that Z and Y coordinates are increasing downwards and rightwards respectively. By a *column* $C_{x,y}$ we mean a subset of registers $M_{x,y,z}$ with $0 \leq z < 2^k$. $P_y = C_{0,y} \cup C_{1,y}$ is a *pair of columns*. An *Z-slice* is a subset of registers with the same Z coordinate.

Let $d = \lceil k/m \rceil$. We define the sets of comparators X, Y_0, Y_1, and Z_i, for $0 \leq i < d$, as follows. (Comparators of X, Y_j and Z_i are called *X-comparators*, *Y-comparators* and *Z-comparators*, respectively.) The comparators of X, Y_0 and Y_1 act in each Z-slice separately (see Figure 1). Set X contains comparators $(M_{0,y,z}, M_{1,y,z})$, for all y and z. Let Y be an auxiliary set of all comparators $(M_{x,y,z}, M_{x,y',z})$ such that $y' = (y + 1) \bmod 2m$. Y_0 contains all comparators $(M_{x,y,z}, M_{x,y',z})$ from Y, such that y is even. Y_1 consists of these comparators from Y that are not in Y_0. Note that the layer Y_1 contains *nonstandard* comparators $(M_{x,2m-1,z}, M_{x,0,z})$ (i.e. comparators that place the greater value in the register with lower index).

In order to describe Z_i we define a matrix α of size $d \times 2m$ (with the rows indexed by the first coordinate) such that, for $0 \leq i < d$ and $0 \leq j < 2m$:

- if j is even then $\alpha_{i,j} = d \cdot j/2 + i$,
- if j is odd $\alpha_{i,j} = \alpha_{i,2m-1-j}$.

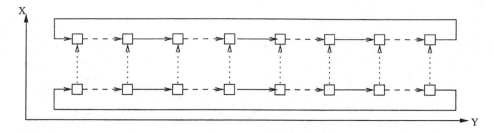

Fig. 1. Comparator connections within a single Z-slice. Dotted (respectively, dashed and solid) arrows represent comparators from X (respectively, Y_0 and Y_1).

For example, for $m = 4$ and $4 < k \leq 8$, α is the following matrix:

$$\begin{bmatrix} 0\,6\,2\,4\,4\,2\,6\,0 \\ 1\,7\,3\,5\,5\,3\,7\,1 \end{bmatrix}.$$

For $0 \leq i < d$, Z_i consists of comparators $(M_{1,y,z}, M_{0,y,z'})$ such that $0 \leq y < 2m$ and $z' = z + 2^{k-1-\alpha_{i,y}}$ provided that $0 \leq z$, $z' < 2^k$ and $k - 1 - \alpha_{i,y} \geq 0$. By a *height* of the comparator $(M_{x,y,z}, M_{x',y',z'})$ we mean $z' - z$. Note that each single Z-comparator is contained within a single pair of columns and all comparators of Z_i contained in the same pair of columns are are of the same height which is a power of two. All Z-comparators of height $2^{k-1}, 2^{k-2}, \ldots, 2^{k-d}$ (which are from $Z_0, Z_1, \ldots, Z_{d-1}$, respectively) are placed in the pairs of columns P_0 and P_{2m-1}. All Z-comparators of height $2^{k-1-d}, \ldots, 2^{k-2d}$ (from $Z_0, \ldots, Z_{d-1}$) are placed in P_2 and P_{2m-3}. And so on. Generally, for $0 \leq i < d$ and $0 \leq y < m$, the height of all comparators of Z_i contained in P_{2y} and in $P_{2m-1-2y}$ is $2^{k-1-dy-i}$.

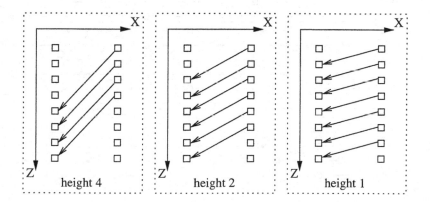

Fig. 2. Z-comparators of different heights within the pairs of columns, for $k = 3$.

The sequence of layers of the network N is $(L_0, \ldots, L_{2d+1})$ where $L_{2i} = X$, $L_{2i+1} = Z_i$, for $0 \leq i < d$, and $L_{2d} = Y_0$, $L_{2d+1} = Y_1$.

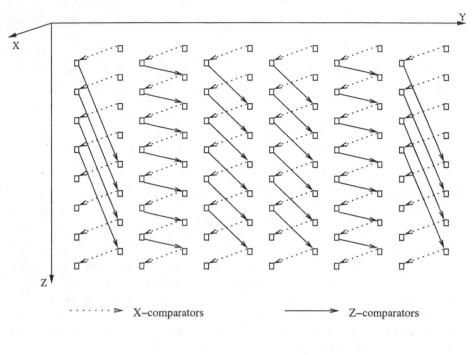

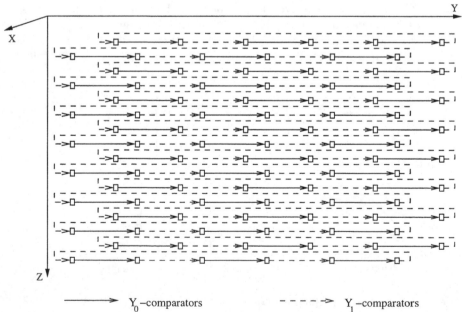

Fig. 3. Network $N_{3,3}$. For clarity, the Y-comparators are drawn separately.

A set of comparators K is *symmetric* if $(R_i, R_j) \in K$ implies $(R_{n-1-j}, R_{n-1-i}) \in K$. Note that all layers of N are symmetric.

Figure 3 shows a network $N_{k,m}$, for $k = m = 3$. As $m \geq k$, this network contains only one layer of Z-comparators Z_0.

4   Analysis of the Computation of $N_{m,k}$

The following theorem is a more detailed version of the theorem stated in the introduction.

Theorem 1. *After $T \leq 4k^2 + 8mk + 7k + 14m + 6\frac{k}{m} + 13$ steps of the periodic network $N_{m,k}$ all its pairs of columns are sorted.*

We denote $N_{m,k}$ by N. By the zero-one principle, [8], it is enough to show this property for the case when only zeroes and ones are stored in the registers. We replace zeroes by negative numbers and ones by positive numbers. These numbers can increase their absolute values between the applications of subsequent layers in periodic computation of N, but can not change their signs. We show that, after T steps, negative values preceed all positive values within each pair of columns.

Initially, let $v(R_0), \dots v(R_{n-1})$ be arbitrary sequence of the values from $\{-1, 1\}$. We apply N to this sequence as a periodic network. We call the application of the layer Y_i (respectively, X, Z_i) an Y-*step* (respectively, X-*step*, Z-*step*).

To make the analysis more intuitive, we assume that each register stores (besides the value) an unique *element*. The *value of an element* e stored in R_i, (denoted $v(e)$) is equal to $v(R_i)$. If $v(e) > 0$ then e is *positive*. Otherwise e is *negative*. If just before the application of comparator $c = (R_i, R_j)$ we have $v(R_i) > v(R_j)$ then during the application of c the elements are exchanged between R_i and R_j. If c is from Y_0 or Y_1 then the elements are exchanged also if $v(R_i) = v(R_j)$. If e is a positive (respectively, negative) element contained in R_i or R_j, before the application of c, then e *wins* in c if, after the application of c, it ends up in R_j (respectively, R_i). Otherwise e *loses* in c.

We call the elements that are stored during the X-steps and Z-steps in the pairs of columns P_{2i}, for $0 \leq i < m$, *right-running* elements. The remaining elements are called *left-running*.

Let $k' = md$. (Recall that $d = \lceil k/m \rceil$.) Let $\delta = 1/(4k')$. Note that $k'\delta < 1$. By *critical comparators* we mean the comparators between P_{2m-1} and P_0 from the layer Y_1. We modify the computation of N as follows:

- After each Z-step, we increase the values of the positive right-running elements and decrease the values of the negative left-running elements by δ. (We call it δ-*increase*.)
- When a positive right-running (respectively, negative left-running) element e wins in a critical comparator, we increase $v(e)$ to $\lfloor v(e) + 1 \rfloor$ (respectively, decrease $v(e)$ to $\lceil v(e) - 1 \rceil$).

Note that once a positive (respectively, negative) element becomes right-running (respectively, left-running) it remains right-running (respectively, left-running) for ever. All the positive left-running and negative right-running elements have absolute value 1.

Lemma 1. *If, during the Z-step t, $|v(e)| = l + y'\delta$, where l and y' are nonnegative integers such that $l \geq 2$ and $0 \leq y' < k'$, then, during t, e can be processed only by comparators with height $2^{k-1-y'}$.*

Let e be a positive element. (A negative element behaves symmetrically.) Since $v(e) > 1$, e is a right-running element during step t. At the moment when e started being right-running, its value was equal 1. A right-running element can be δ-increased at most k' times between its subsequent wins in the critical comparators, and $k'\delta < 1$. Thus e reached the value 2 when it entered P_0 for the first time. Then its value was being increased by δ, after each Z-step (d times in each P_{2j}), and rounded up to the next integer during its wins in critical comparators. The lemma follows from the definition of α and Z_i: The heights of the Z-comparators from the subsequent Z-layers Z_i, for $0 \leq i < d$, in the subsequent pairs of columns P_{2j}, for $0 \leq j < m$, are the decreasing powers of two. $\square$

We say that a register $M_{x,y,z}$ is *l-dense for v* if

- in the case $v > 0$: $v(M_{x,y,z+i\lceil 2^l \rceil}) \geq v$, for all $i \geq 0$ such that $z + i\lceil 2^l \rceil < 2^k$, and
- in the case $v < 0$: $v(M_{x,y,z-i\lceil 2^l \rceil}) \leq v$ for all $i \geq 0$ such that $z - i\lceil 2^l \rceil \geq 0$.

Note that, for $l < 0$, "*l-dense*" means "*0-dense*". An *element is l-dense for v* if it is stored in a register that is *l-dense for v*.

Lemma 2. *If $M_{x,y,z}$ is l-dense for $v > 0$ (respectively, $v < 0$), then, for $0 < v' \leq v$ (respectively, $v \leq v' < 0$), $M_{x,y,z}$ is l-dense for v'.*

If $M_{x,y,z}$ is l-dense for $v > 0$ (respectively, $v < 0$), then, for all $j \geq 0$ (respectively, $j \leq 0$), $M_{x,y,z+j\lceil 2^l \rceil}$ is l-dense for v.

If $M_{x,y,z}$ is l-dense for $v > 0$ (respectively, $v < 0$) and $M_{x,y,z+\lfloor 2^{l-1} \rfloor}$ (respectively, $M_{x,y,z-\lfloor 2^{l-1} \rfloor}$) is l-dense for v, then $M_{x,y,z}$ is $(l-1)$-dense for v.

The properties can be easily derived from the definition. $\square$

Lemma 3. *Let L be any layer of N and $(M_{x,y,z}, M_{x',y',z'}) \in L$.*

If $M_{x,y,z}$ or $M_{x',y',z'}$ is l-dense for $v > 0$ (respectively, $v < 0$), just before an application of L, then $M_{x',y',z'}$ (respectively, $M_{x,y,z}$) is l-dense for v just after the application of L.

If $M_{x,y,z}$ and $M_{x',y',z'}$ are l-dense for v, just before the application of L, then $M_{x,y,z}$ and $M_{x',y',z'}$ are l-dense for v just after the application of L.

Proof. The lemma follows from the fact that, for each integer i such that $0 \leq z + i\lceil 2^l \rceil$, $z' + i\lceil 2^l \rceil < 2^k$, the comparator $(M_{x,y,z+i\lceil 2^l \rceil}, M_{x',y',z'+i\lceil 2^l \rceil})$ is also in L. $\square$

Corollary 1. *If an element l-dense for v wins during an application of a layer L of N, then it remains l-dense for v. If it looses to another element l-dense for v, then it also remains l-dense for v. If it wins in critical comparator and $v > 0$ (respectively, $v < 0$), then it becomes l-dense for $\lfloor v + 1 \rfloor$ (respectively, $\lceil v - 1 \rceil$).*

If just before Z-step t, e is right-running positive (respectively, left-running negative) element l-dense for $v > 0$ (respectively, $v < 0$), and, during t, e looses to another element l-dense for v or wins, then it becomes l-dense for $v + \delta$ (respectively, $v - \delta$), after the δ-increase following t.

The following lemma states that each positive element e that was right-running for a long time is contained in a dense foot of the elements with the value $v(e)$ or greater, and an analogical property holds for left-running negative values.

Lemma 4. *Consider the configuration of N after the Z-step. For nonnegative integers l,s and y' such that $y' \leq k'$, for each element e:*

If $v(e) = l + 2 + s + y'\delta$, then e is $(k - l)$-dense for $l + 2 + y'\delta$ and, if $y' > l$, then e is $(k - l - 1)$-dense for $l + 2 + y'\delta$.

If $v(e) = -(l + 2 + s + y'\delta)$, then e is $(k - l)$-dense for $-(l + 2 + y'\delta)$ and, if $y' > l$, then e is $(k - l - 1)$-dense for $-(l + 2 + y'\delta)$.

Proof. We prove only the first part. The second part is analogical since all layers of N are symmetrical. The proof is on induction by l. Let $0 \leq l < k$. Let e be any element with $v(e) = l + 2 + s + y'\delta$, for some nonnegative integers s, y', where $y' \leq k'$. The element e was right-running during each of the last y' Z-steps. These steps were preceeded by a critical step t, that increased $v(e)$ to $l + 2 + s$. Let t_i (respectively, t'_i) be the $(i + 1)$-st X-step (respectively, Z-step) after step t. Let M_{x_i,y_i,z_i} (respectively, $M_{x'_i,y_i,z'_i}$) be the register that stored e just after t_i (respectively, t'_i). Let v_i denote the value $l + 2 + i\delta$. During each step t_i and t'_i, all elements e' with $v(e') \geq v(e)$, in the pair of columns containing e, are $(k - l)$-dense for v_i. (For $l = 0$ it is obvious, since the "height" of N is 2^k, and, for $l > 0$, it follows from the induction hypothesis and Corollary 1, since e' was $(k - l)$-dense for $l + 1$ already before t, and, hence, $(k - l)$-dense for v_0 just after t.)

Claim (Breaking Claim). For $0 \leq i \leq l$, just after the X-step t_i, the registers $M_{0,y_i,z_i + 2^{k-i}}$ and $M_{1,y_i,z_i + 2^{k-i}}$ are $(k - l)$-dense for v_i, if they exist.

We prove the claim by induction on i. For $i = 0$ it is obvious. ($M_{0,y_i,z_i + 2^k}$ and $M_{1,y_i,z_i + 2^k}$ do not exist.)

Let $0 < i \leq l$. Consider the configuration just after step t_{i-1}. (See Figure 4.) Since t_{i-1} was an X-step, $v(M_{1,y_{i-1},z_{i-1}}) \geq v(e)$ and, hence, $M_{1,y_{i-1},z_{i-1}}$ is $(k-l)$-dense for v_{i-1}. Thus, $M_{1,y_{i-1},z_{i-1}+2^{k-i}}$ is $(k-l)$-dense for v_{i-1}, since 2^{k-i} is multiple of 2^{k-l}. By the induction hypothesis of the claim, $M_{0,y_{i-1},z_{i-1}+2^{k-i+1}}$ and $M_{1,y_{i-1},z_{i-1}+2^{k-i+1}}$ are $(k - l)$-dense for v_{i-1}. Just after the step t'_{i-1}, $M_{1,y_{i-1},z_{i-1}+2^{k-i}}$, and $M_{1,y_{i-1},z_{i-1}+2^{k-i+1}}$ remain $(k-l)$-dense for v_{i-1}, since they were compared to the registers $M_{0,y_{i-1},z_{i-1}+2^{k-i+1}}$ and $M_{0,y_{i-1},z_{i-1}+2^{k-i+2}}$

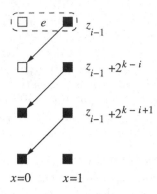

Fig. 4. The configuration after t_{i-1} in $P_{y_{i-1}}$ in the registers with Z-coordinates $z_{i-1} + j2^{k-i}$, for $0 \leq j < 4$. (Black registers are $(k-l)$-dense for v_{i-1}. Arrows denote the comparators from t'_{i-1}.)

that were $(k-l)$-dense for v_{i-1}. $M_{0,y_{i-1},z_{i-1}+2^{k-i+1}}$ remains $(k-l)$-dense for v_{i-1}. $M_{0,y_{i-1},z_{i-1}+2^{k-i}}$ also becomes (or remains) $(k-l)$-dense for v_{i-1}, since it was compared to $M_{1,y_{i-1},z_{i-1}}$. Thus, just after the Z-step t'_{i-1}, for $x \in \{0,1\}$, the registers $M'_x = M_{x,y_{i-1},z'_{i-1}+2^{k-i}}$ are $(k-l)$-dense for v_{i-1} (and for v_i, after the δ-increase). (Either $z'_{i-1} = z_{i-1}$ and $M'_x = M_{x,y_{i-1},z_{i-1}+2^{k-i}}$, or $z'_{i-1} = z_{i-1}+2^{k-i}$ and $M'_x = M_{x,y_{i-1},z_{i-1}+2^{k-i+1}}$.) If $i \bmod d = 0$ then, during the next two Y-steps, the elements from both M'_0 and M'_1 together with the element e are moved "horizontally" to $P_{2i/d}$ (wining by the way). Thus, by Corollary 1, just before and after the X-step t_i, for $x \in \{0,1\}$, the registers $M_{x,y_i,z_i+2^{k-i}}$ are $(k-l)$-dense for v_i. This completes the proof of the claim.

The next claim shows how the values v_l or greater form twice more condensed foot below e.

Claim (Condensing Claim). After the Z-step t'_l, e is $(k-l-1)$-dense for v_l (and for v_{l+1}, after the δ-increase).

Consider the configuration just after X-step t_l. The registers M_{x_l,y_l,z_l} and, by the Breaking Claim, $M_{0,y_l,z_l+2^{k-l}}$ and $M_{1,y_l,z_l+2^{k-l}}$ are $(k-l)$-dense for v_l. Since the last step was an X-step, M_{1,y_l,z_l} is $(k-l)$-dense for v_l.

Consider the following scenarios of the Z-step t'_l (see Figure 5):

1. e remains in M_{0,y_l,z_l}: Then the register $M_{0,y_l,z_l+2^{k-l-1}}$ becomes $(k-l)$-dense for v_l, by Lemma 3, since M_{1,y_l,z_l} was $(k-l)$-dense for v_l just before t'_l. Thus e becomes $(k-l-1)$-dense for v_l, by Lemma 2.

2. e is moved from M_{1,y_l,z_l} to $M_{0,y_l,z_l+2^{k-l-1}}$: Then by Corollary 1, e remains $(k-l)$-dense for v_l, and the register $M_{0,y_l,z_l+2^{k-l}}$ remains $(k-l)$-dense for v_l. Thus e becomes $(k-l-1)$-dense for v_l, by Lemma 2.

3. e remains in M_{1,y_l,z_l}: Then $v(e) \leq v(M_{0,y_l,z_l+2^{k-l-1}}) \leq v(M_{1,y_l,z_l+2^{k-l-1}})$ just before t'_l. (The second inequality is forced by the X-step t_l.) Hence, for $x \in \{0,1\}$, $R'_x = M_{x,y_l,z_l+2^{k-l-1}}$ was $(k-l)$-dense for v_l just before t'_l. During

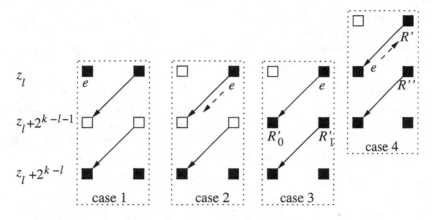

Fig. 5. The scenarios of t'_l.

t'_l the register R'_1 is compared to $M_{0,y_l,z_l+2^{k-l}}$. So R'_1 remains $(k-l)$-dense for v_l. Since e was compared to R'_0, it also remains $(k-l)$-dense for v_l. By Lemma 2, e is $(k-l-1)$-dense for v_l just after t'_l.

4. e is moved from M_{0,y_l,z_l} to $R' = M_{1,y_l,z_l-2^{k-l-1}}$: During t'_l, R' was compared to M_{x_l,y_l,z_l} and $R'' = M_{1,y_l,z_l}$ was compared to $M_{0,y_l,z_l+2^{k-l-1}}$ that was $(k-l)$-dense for v_l just before t'_l, by the Breaking Claim applied to the element in R'. Thus, by Lemma 3, the registers R' and R'' remain $(k-l)$-dense for v_l just after t'_l. By Lemma 2, R' is $(k-l-1)$-dense for v_l just after t'_l.

Since there are no other scenarios for c and the subsequent δ-increase is the same for all positive elements in P_{y_l}, the proof of the claim is completed.

By Corollary 1, the element e remains $(k-l-1)$-dense for v_i, for $i > l$, since other elements in its pair of columns with values $v(e)$ or greater are now also $(k-l-1)$-dense for v_i, and during Y-steps e is wining (right-running).

For $l \geq k$, "$(k-l)$-dense for v" means "0-dense for v". The element e with $v(e) = k+1+k\delta$ is 0-dense for $k+1+k\delta$. All the positive elements below it increase their values at the same rate as e. Thus, when $v(e)$ reaches $k+2$, it becomes 0-dense for $k+2$. By repeating this reasoning for the values $k+2$ and greater we complete the proof of the Lemma 4. $\square$

By Lemma 4, whenever any element e reaches the value $k+2$ (in the pair of columns P_0) it is 0-dense for $k+2$. Then, by the Breaking Claim, after the X-step after e reaches the value $k+2+k\delta$, e is stored in a register $M_{x,y,z}$ such that $M_{0,y,z+1}$ is also 0-dense for $k+2+k\delta$. Hence, all the elements following e in its pair of columns are 0-dense for $k+2+k\delta$. By Corollary 1, this property of e remains valid forever. Since the network is symmetric, we have the following corollary:

Corollary 2. *Consider a configuration in a pair of columns P_y just after an X-step.*

If, for some register $R_i \in P_y$, $v(R_i) \geq k + 2 + k\delta$, then, for all $R_j \in P_y$ such that $j \geq i$, we have $v(R_j) \geq k + 2 + k\delta$.

If, for some register $R_i \in P_y$, $v(R_i) \leq -(k + 2 + k\delta)$, then, for all $R_j \in P_y$ such that $j \leq i$, we have $v(R_j) \leq -(k + 2 + k\delta)$.

Now, it is enough to show that, after the last X-step of the first T steps, all right-running positive and all left-running negative elements have the absolute values $k+2+k\delta$ or greater. Then in each pair of columns containing right-running elements, the -1s are above the positive values, and in each pair of columns containing left-running elements, the 1s are below the negative elements.

Lemma 5. *If, after m Y-steps, and the next $k'(k+1) + k$ Z-steps, and the next X-step, e is a left-running positive (respectively, right-running negative) element, then e remains left-running (respectively, right-running) forever.*

Let e be positive. (The proof for e negative is analogical). During each of the first m Y-steps, e was compared with the positive right-running elements. For $t \geq 0$, let y_t be such that e was in P_{y_t} just after the $(t+1)$st Y-step. For $0 \leq i < m$, let S_i (respectively, S_i') denote the set of positive elements that were in P_{y_i} (respectively, $P_{(y_i+1) \bmod 2m}$) just after $(i+1)$st Y-step. Let S'' be the set of negative elements in $P_{y_{m-1}}$ just after the mth Y-step. For $0 \leq i < m$, $|S_{m-1}| = 2 \cdot 2^k - |S''| \leq |S_i'|$, since $S_{m-1} \subseteq S_i$ and $|S_i| \leq |S_i'|$. Note that, for all $t \geq m$, during the $(t+1)$st Y-step, the pair of columns containing (left-running) S'' is compared to the pair of columns containing (right-running) $S'_{t \bmod m}$.

After the next $k'(k+1) + k$ Z-steps all the elements of S'' have values $-(k + 2 + k\delta)$ or less, and, for $0 \leq i < k$, the elements of S_i' have values $k + 2 + k\delta$ or greater (they have walked at least $k + 1$ times through the critical comparators and then increased their values by δ at least k times during Z-steps). Let t' be the next X-step. Let t be any Y-step after t' such that e is still in the same pair of columns as S''. Before the step t, the elements in S'' and each S_i' were processed by an X-step after their absolute values had reached $k+2+k\delta$. Hence, by the Corollary 2, just before the Y-step t, all the final $|S_i'|$ registers of the pair of columns containing S_i' store the values $k + 2 + k\delta$ or greater and the pair of columns containing S'' has all the initial $|S''|$ registers filled with the values $-(k + 2 + k\delta)$ or less. Thus, e is stored in one of its remaining $2 \cdot 2^k - |S''|$ final registers and, during the Y-step t, e is compared with a value $k + 2 + k\delta$ or greater and it must remain left-running. □

The depth of N is $2d + 2$. Each iteration of N performs two Y-steps as its last steps. Thus the first m Y-steps are performed during the first $(2d+2)\lceil m/2 \rceil$ steps. Each iteration of N performs d Z-steps. Thus, the next $k'(k + 1) + k$ Z-steps are performed during the next $(2d + 2)\lceil (k'(k+1)+k)/d \rceil$ steps. After the next X-step, t', by Lemma 5, the set S of positive right-running and negative left-running elements remains fixed. After the next $\lceil (k'(k+1)+k)/d \rceil$ iterations absolute values of elements in S are $k + 2 + k\delta$ or greater. (t' was the first step of these iterations.) After the first X-step of the next iteration, by Corollary 2, in all pairs of columns the negative values preceed the positive values. We can now replace negative values with zeroes, positive values with ones, and, by the zero-one principle, we have all the pairs of columns sorted. (Note that, by the definition of N, once all the pairs of columns are sorted, they remain sorted for ever.)

We can estimate the number of steps by $T \leq (2d+2)(\lceil m/2 \rceil + 2\lceil (k'(k+1) + k)/d \rceil) + 1$. Recall that $d = \lceil k/m \rceil$. It can be verified that $T \leq 4k^2 + 8mk + 7k + 14m + 6\frac{k}{m} + 13$. This completes the proof of Theorem 1.

Remarks: Note that the network $N_{1,k}$ can be simplified to a periodic sorting network of depth $2\log n$, by removing the Y-steps and merging P_0 with P_1. However, better networks exist, [3], with depth $\log n$ that sort in $\log n$ iterations. Note also that the arrangement of the registers in the matrix M can be arbitrary. We can select the one that is most suitable for the subsequent merging.

Acknowledgments. I would like to thank Mirosław Kutyłowski for his useful suggestions and comments on this paper.

References

1. M. Ajtai, J. Komlós and E. Szemerédi. Sorting in $c\log n$ parallel steps. Combinatorica, Vol. 3, pages 1–19, 1983.
2. K. E. Batcher. Sorting networks and their applications. Proceedings of 32nd AFIPS, pages 307–314, 1968.
3. M. Dowd, Y. Perl, L. Rudolph, and M. Saks. The periodic balanced sorting network. Journal of the ACM, Vol. 36, pages 738–757, 1989.
4. M. Kik. Periodic correction networks. Proceedings of the Euro-Par 2000, Springer Verlag, LNCS 1900, pages 471–478, 2000.
5. M. Kik, M. Kutyłowski and G. Stachowiak. Periodic constant depth sorting network. Proceedings of the 11th STACS, Springer Verlag, LNCS 775, pages 201–212, 1994.
6. M. Kutyłowski, K. Loryś and B. Oesterdiekhoff. Periodic merging networks. Proceedings of the 7th ISAAC, pages 336–345, 1996.
7. M. Kutyłowski, K. Loryś, B. Oesterdiekhoff, and R. Wanka. Fast and feasible periodic sorting networks. Proceedings of the 55th IEEE-FOCS, 1994.
8. D. E. Knuth. The art of Computer Programming. Volume 3: Sorting and Searching. Addison-Wesley, 1973.
9. De-Lei Lee and K. E. Batcher. A multiway merge sorting network. IEEE Transactions on Parallel and Distributed Systems 6, pages 211–215, 1995.
10. U. Schwiegelshohn. A short-periodic two-dimensional systolic sorting algorithm. IEEE International Conference on Systolic Arrays, pages 257–264, 1988.

Fast Periodic Correction Networks

Grzegorz Stachowiak

Institute of Computer Science, University of Wrocław,
Przesmyckiego 20, 51-151 Wrocław, Poland
gst@ii.uni.wroc.pl

Abstract. We consider the problem of sorting N-element inputs differing from already sorted sequences on t entries. To perform this task we construct a comparator network that is applied periodically. The two constructions for this problem made by previous authors required $O(\log n + t)$ iterations of the network. Our construction requires $O(\log n + (\log \log N)^2 (\log t)^3)$ iterations which makes it faster for $t \gg \log N$.

Keywords: sorting network, comparator, periodic sorting network.

1 Introduction

Sorting is one of the most fundamental problems of computer science. A classical approach to sort a sequence of keys is to apply a comparator network. Apart from a long tradition, comparator networks are particularly interesting due to hardware implementations. They can be also implemented as sorting algorithms for parallel computers.

In our approach sorted elements are stored in registers $r_1, r_2, \ldots, r_N$. Registers are indexed with integers or elements of other linearly ordered sets. A *comparator* $[i : j]$ is a simple device connecting registers r_i and $r_j (i < j)$. It compares the keys they contain and if the key in r_i is bigger, it swaps the keys. The general problem is the following. At the beginning of the computations the input sequence of keys is placed in the registers. Our task is to sort the sequence of keys according to the linear order of register indices by applying a sequence of comparators. The sequence of comparators is the same for all possible inputs. We assume that comparators connecting disjoint pairs of registers can work in parallel. Thus we arrange the sequence of comparators into a series of *layers* which are sets of comparators connecting disjoint pairs of registers. The total time needed by such a network to sort a sequence is proportional to the number of layers called the network's *depth*.

Much research concerning sorting networks was done in the past. Most famous results are asymptotically optimal AKS [1] sorting network of depth $O(\log N)$ and more 'practical' Batcher [2] network of depth $\sim \frac{1}{2} \log^2 N$ (from now on all the logarithms are binary).

Some research was devoted to problems concerning periodic sorting networks. Such a comparator network is applied not once but many times in a series of iterations. The input of the first iteration is the sequence to be sorted. The input of $(i+1)$st iteration is the output of ith iteration. The output of the last iteration should always be sorted. The total time needed to sort an input sequence is the product of the number of iterations

A. Lingas and B.J. Nilsson (Eds.): FCT 2003, LNCS 2751, pp. 144–156, 2003.
© Springer-Verlag Berlin Heidelberg 2003

and the depth of the network. Constructing such networks especially of small constant depth gives hope to reduce the amount of hardware needed to build sorting comparator networks. It can be done by applying the same small chip many times to sort an input. We can also view such a network as a building block of a sorting network in which layers are repeated periodically. Main results concerning periodic sorting networks are presented in the table:

	depth	# iterations
DPS [3]	$\log N$	$\log N$
Schwiegelsohn [15]	8	$O(\sqrt{N} \log N)$
KKS [5]	$O(k)$	$O(N^{1/k})$
Loryś et al. [9]	3-5	$O(\log^2 N)$

Last row of this table requires some words of explanation. The paper [9] describes a network of depth 5, but a later paper [10] reduces this value to 3. The number of iterations $O(\log^2 N)$ is achieved by periodification of AKS sorting network for which the constant hidden behind big O is very big. Periodification of Batcher network requires less iterations for practical sizes of the input, though it requires the time $O(\log^3 N)$ asymptotically. It is not difficult to show that 3 is the minimal depth of a periodic sorting network which requires $o(N)$ iterations to sort an arbitrary input.

A sequence obtained from a sorted one by t changes being either swaps between pairs of elements or changes on single positions we call t-disturbed. We define t-correction network to be a specialized network sorting t-disturbed inputs. Such networks were designed to obtain a sorted sequence from an output produced by a sorting network having t faulty comparators [14,11,16]. There are also other potential applications in which we have to deal with sequences that differ not much from a sorted one. Let us consider a large sorted database with N entries. In some period of time we make t modifications of the database and want to have it sorted back. It can be more effective to use a specialized correction unit in such a case, than to apply a sorting network. Results concerning such correction networks are presented in [4,16].

There was some interest in constructing periodic comparator networks of a constant depth, that sort t-disturbed inputs. The reason is that the fastest known constant depth periodic sorting networks have running time $O(\log^2 N)$. On the other hand in some applications faster correction networks can replace sorting networks. Two periodic correction networks were already constructed by Kik and Piotrów [6,12]. The first of them has depth 8 and the other has depth 6. Both of them require $O(\log N + t)$ iterations for considered inputs where N is input size and t is the number of modifications. The running time is $O(\log N)$ for $t = O(\log N)$ and the constants hidden behind the big O are small. Unfortunately it is not known how fast these networks complete sorting if $t \gg \log N$.

In this paper we construct a periodic t-correction network to deal with $t : \log N \ll t \ll N$. The reason we assume that t is small in comparison to N is the following. If t is about the same as N, then the periodification scheme gives a practical periodic sorting network of depth 3 requiring $O(\log^3 N) = O(\log^3 t)$ iterations. Actually we do not hope to get better performance in such a case. Our network has depth 3 and running time: $O(\log N + (\log \log N)^2 (\log t)^3)$. We should mention that in our construction we do not use AKS sorting network. If this network was used (also in the auxiliary construction of a non periodic t-correction network) we would get the running time: $O(\log N + (\log\log N)(\log t)^2)$. In such case the AKS constant would stand in front of $(\log\log N)(\log t)^2$.

Now we remind of a couple of useful properties of comparator networks. The first of them is a general property of all comparator networks. Let us assume we have two inputs for a fixed comparator network. We say that we have relation $(x_1, x_2, \ldots, x_N) \leq (y_1, y_2, \ldots, y_N)$ between these inputs if for all i we have $x_i \leq y_i$.

Lemma 1.1. *If we apply the same comparator network to inputs for which we have* $(x_1, x_2, \ldots, x_N) \leq (y_1, y_2, \ldots, y_N)$ *then this relation is preserved for the outputs.*

The analysis of sorting networks is most often based on the following lemma [7]

Lemma 1.2 (zero–one principle). *A comparator network is a sorting network if and only if it can sort any input consisting only of 0s and 1s.*

This lemma is the reason, why from now on we consider inputs consisting only of 0s and 1s. Thus we consider only t-disturbed sequences consisting of 0s and 1s. We note, that 0-1 sequence $x_1, \ldots, x_N$ is t disturbed if for some index b called the *border* at most t entries in $x_1, \ldots, x_b$ are 1s and at most t entries in $x_{b+1}, \ldots, x_N$ are 0s. These 1s (0s) we call *displaced*.

Let us remind the proof of zero–one principle. The input consists of arbitrary elements. We prove that the comparator network sorts it. We consider an arbitrary a from this input and show it gets to the register corresponding to its rank in the sequence. We replace elements bigger than a by 1, and smaller by 0. Indeed the only difference between outputs for sequences where a is replaced by 0 or 1 respectively is the register with the index corresponding to rank(a).

Now we deal with an arbitrary t-disturbed input. We transform it to a t-disturbed 0-1 sequence as in the proof of zero–one principle. This gives us a useful analog of zero-one principle for t-correction networks.

Lemma 1.3. *A comparator network is a t-correction network if it can sort any t-disturbed input consisting of 0s and 1s.*

We define *dirty area* for 0-1 a sequence stored in the registers during computations of a comparator network. Dirty area is the smallest set of subsequent registers such that all registers with lower indices contain 0s and all registers with bigger indices contain 1s. A specialized comparator network that sorts any input having dirty area of a given size we call a *cleaning network*.

2   Periodic Sorting Networks

In this section we remind the periodification scheme in [9]. Actually what we present is closer to the version of this scheme described by Oesterdiekhoff [10] which produces a network of depth 3. In comparison to previous authors we change the construction of Schwiegelsohn edges and embed only a single copies of sorting and merging networks.

The analysis of the network is almost the same as in abovementioned papers and we do not show it.

The periodification scheme is a method to convert a non periodic sorting network having $T(p)$ layers for input size p into a periodic sorting network of depth 3. This periodic network sorts any input containing $\Theta(pT(p))$ items in $O(T(p)\log p)$ iterations. We take advantage of the fact, that for any sorting network $T(p) = \Omega(\log p)$. The periodification scheme applied to Batcher sorting network gives a periodic sorting network which needs $O(\log^3 N)$ iterations to sort an arbitrary input of size N. If we put AKS sorting network into this scheme, we get a periodic sorting network requiring $O(\log^2 N)$ iterations which is (due to very large constants in AKS) worse solution for practical N.

In the periodification scheme registers are indexed with pairs $(i,j), 1 \leq i \leq p, 1 \leq j \leq q$ ordered lexicographically. Thus we view these registers as arranged in rectangular matrix $p \times q$ of p rows and q columns. We have the rows with smallest indices i at the 'top' and those with biggest indices at the 'bottom' of the array. We also view columns with smallest indices j to be on the left hand side and those with biggest indices to be on the right hand side. The parameter $q = 10(T(p) + \log p)$ is an even number (for simplicity from now on we write $\log p$ instead of $\lceil \log p \rceil$).

The periodic sorting network consists of three subsequent layers A, B and C. The layers A and B which are layers of odd-even transposition sort network are called horizontal steps. They are sets of comparators:

$$A = \{[(i, 2j-1) : (i, 2j)] | i, j = 1, 2, \dots\}$$
$$B = \{[(i, 2j) : (i, 2j+1)] | i, j = 1, 2, \dots\} \cup \{[(i, q) : (i+1, 1)] | i = 1, 2, \dots\}$$

The edges of A and B connecting registers of the same row we call *horizontal*. The layers A, B alone sort any input but in general the time to do it is very long.

Defining layer C called vertical step is much more complicated. We first divide the columns of registers into six subsequent areas: S, M_L, X_L, Y, X_R, M_R. Each of the areas contains an even number of columns. First two columns form an area S where so called 'Schwiegelsohn' edges are located. So the columns with numbers $3, 4, \dots, 2\log p + 2$ are in the area M_L. Next $2T(p)$ columns form area X_L. Last $2\log p$ columns are contained in area M_R. Area X_R consists of $2T(p)$ columns directly preceding M_R. And the area Y contains all the columns between X_L and X_R. We now say where the comparators of layer C are in each area.

In area S the comparators form the set

$$\{[(2i-1, 1) : (2i, 2)] | i = 1, 2, \dots\}$$

Note that this way of defining "Schwiegelsohn" edges differs from one described in previous papers on this subject. Comparators of C in all other areas unlike those in S connect always registers in the same column. There are no comparators in area Y on layer C.

In each area M_L and M_R we embed a single copy of a network which merges two sorted sequences of length $p/2$. In this network's input of length p even indexed entries are one sequence and odd indexed entries are the other. We also assume, that the sequence in odd indexed entries does not have more 1s than one contained in even

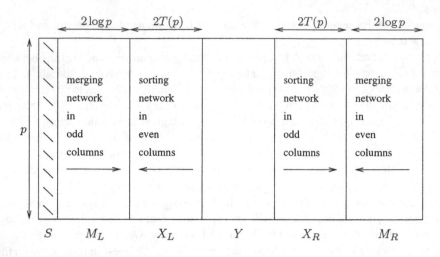

Fig. 1. Areas defined to embed C-layer. Arrows indicate the order of layers of embeded networks.

indexed entries. A comparator network merging two such sequences is the series of layers $L_1, L_2, \ldots, L_{\log p-1}$ where

$$L_i = \{[2j : 2j + 2^{\log p - i} - 1]|j = 1, 2, \ldots\}.$$

Thus the set of comparators in M_L is equal to

$$\{[(k, 2j+1) : (l, 2j+1)]|[k : l] \in L_j, j = 1, 2, \ldots\}.$$

The set of comparators in M_R is equal to

$$\{[(k-1, q-2j+2)) : (l-1, q-2j+2)]|[k : l] \in L_j, j = 1, 2, \ldots\}.$$

For technical reasons the network embedded in M_R is moved one row up.

Finally we define comparators in X_L and X_R. These comparators are embedding of a single sorting network in each area. Let this sorting network be the series of layers $L'_1, L'_2, \ldots, L'_{T(p)}$. Let $j_L = 2 + 2\log p + 2T(p)$ be the last column of X_L. The set of comparators in X_L is equal

$$\{[(k, j_L - 2(j-1)) : (l, j_L - 2(j-1))]|[k : l] \in L'_j, j = 1, 2, \ldots\}.$$

Analogously if $j_R = q - 2\log p - 2T(p) + 1$ is the first column of X_R, then the set of comparators in X_R is equal

$$\{[(k, j_R + 2(j-1)) : (l, j_R + 2(j-1))]|[k : l] \in L'_j, j = 1, 2, \ldots\}.$$

The edges connecting registers in the same column we call *vertical*. Almost all the edges of step C are vertical. Only the slanted edges in S are not vertical.

Our aim in the analysis of the network obtained in periodification scheme is to prove that it sorts any input in $O(T(p)\log p)$ steps. The proof easily follows from the key lemma

Lemma 2.1 (key lemma). *There exist constants c and d such that after $d \cdot q$ steps*

- *the bottom $c \cdot p$ rows contain only 1s if there are more 1s than 0s in the registers;*
- *the top $c \cdot p$ rows contain only 0s if there are more 0s then 1s in the registers.*

Indeed if we consider only the rows containing dirty area in the key lemma, then this area is guaranteed to be reduced by a constant factor within $O(q)$ steps. Thus applying the key lemma $O(\log p)$ times we reduce this area within $O(q \log p)$ steps to a constant number of rows. Next $O(q)$ steps sort such a reduced dirty area.

We do not describe the proof of key lemma, but define some notions from it to use them further in the paper. In this proof it is assumed, that two 1s or 0s compared by a horizontal edge are exchanged. In a given moment of computations we call an item (i.e. 0 or 1) right-running (left-running) if it is placed in the right (left) register by a horizontal edge of the recently executed horizontal step. We can extend this definition on wrap-around edges of layer B in a natural way saying that they put right-running items in the first column and left-running items in the last. A column containing right-running (left-running) items is called R-column (L-column). Analyzing the network we can observe 'movement' of R-columns of 1s to the right and L-columns of 0s to the left. Thus any column is alternately L-column and R-column and the change occurs during every horizontal step. The only exception are two columns of S. From the proof of key lemma it also follows, that we have the following property

Fact 2.2 *Assume we add any vertical edges to the layer C in area Y. For such a new network the key lemma still holds.*

Now we modify periodification scheme step by step to obtain at the end periodic t-correction network. First we introduce a construction of a *periodic cleaning network* sorting any N-element input with the dirty area of size $qt, q \geq 10(T(2t) + 2\log t)$. In this construction registers are arranged into q columns and dirty area is contained in t subsequent rows. This network needs $O(q \log t)$ iterations to do its job. The construction of periodic correction network is based on this cleaning network. We first build a simple non periodic cleaning network

Lemma 2.3. *Assume we have a sorting network of depth $T(t)$ for input size t. We can construct a comparator network of depth $T'(t) = T(2t) + \log t$ which sorts any input with dirty area of size t.*

Proof. We divide the set of all registers $r_1, r_2, \ldots, r_N$ into N/t disjoint parts each consisting of t subsequent registers. Thus we obtain part P_1 containing registers $r_1, \ldots, r_t$, P_2 containing registers $r_{t+1}, \ldots, r_{2t}$, P_3 containing registers $r_{2t+1}, \ldots, r_{3t}$, and so on. The cleaning network consists of two steps. First we have networks sorting keys in $P_{2i} \cup P_{2i+1}$ for each i. It requires $T(2t)$ layers. Then we have networks merging elements in P_{2i-1} with those in P_{2i} for each i. It requires $\log t$ layers of the network.
□

Now we can build a periodic cleaning network. We do it substituting sorting network in the periodification scheme with the cleaning network described above. This way we can reduce X_L and X_R to $2T'(t)$ columns. We also reduce M_L and M_L to $2 \log t$

columns, by embedding only $\log t$ last layers of merging network instead of the whole merging network applied in periodification scheme. These layers are (after relabeling) $L_1, L_2, \ldots, L_{\log t}$ where

$$L_i = \{[2j+1 : 2j+2^{\log t-i+1}] | j = 1, 2, \ldots\}.$$

They merge any two sequences that do not differ by more than $t/2$ 1s. So instead of a sorting network we use a cleaning one and we reduce the merging network. Such reduced sorting and merging networks are not distinguishable from original merging and sorting networks if we deal only with inputs having dirty areas of size at most qt. The analysis of such a periodification scheme for cleaning networks is the same as the original one for sorting networks and gives us the following fact

Lemma 2.4. *The periodic cleaning network described above has depth 3 and sorts any input with dirty area having t rows in $O(q \log t)$ iterations.*

One can notice that there are no edges of layer C in Y in this construction. If we add any vertical edges in Y or any other edges with the difference between row numbers of end registers bigger than t to layer C, then the network remains a cleaning network. Roughly speaking by adding such edges we are going to transform the periodic cleaning network into a periodic t-correction network.

3   Main Construction

In this section we define our periodic t-correction network. To do it we need another non periodic comparator network. We call it (t, Δ, δ)-*semi-correction* network. If a t-disturbed input with dirty area of size Δ is processed by such a network, then the dirty area size is guaranteed to be reduced to δ. Now we present quite unoptimal construction of (t, Δ, δ)-semi-correction network.

We divide the set of all registers $r_1, r_2, \ldots, r_N$ into N/Δ disjoint parts each consisting of Δ subsequent registers. Thus we obtain part P_1 containing registers $r_1, \ldots, r_\Delta$, P_2 containing registers $r_{\Delta+1}, \ldots, r_{2\Delta}$, P_3 containing registers $r_{2\Delta+1}, \ldots, r_{3\Delta}$, and so on. The construction consists of two steps. In step 1 we give new indices to the registers of each sum $P_{2k} \cup P_{2k+1}, k = 1, 2, \ldots$. These indices are lexicographically ordered pairs $(i, j), 1 \leq i \leq 2t\Delta/\delta, 1 \leq j \leq \delta/(2t)$. The ordering of new indices is the same as the main ordering of indices. We apply a t-correction network to each column j of each sum separately. This way we obtain dirty area of size at most δ in each sum. In step 2 we repeat the construction from step 1 for sums $P_{2k-1} \cup P_{2k}$. Because any dirty area of size Δ is contained in one of the sums $P_l \cup P_{l+1}$ from step 1 or 2, this dirty area is reduced to size δ. Thus we get the following lemma

Lemma 3.1. *Let $t \ll \delta$ and $t \ll \Delta/\delta$. There exists a (t, Δ, δ)-semi-correction network of depth $O\left(\log x + (\log t \log \log x)^2\right)$, where $x = \Delta/\delta$.*

Proof. Description of t-correction networks of depth $O\left(\log N + d(\log t \log \log N)^2\right)$ (N is the input size) can be found in [4,16]. We apply such a network in the construction presented above and obtain a semi-correction network with desired depth. Simple calculations are left to the reader.   □

Now at last we get to the main construction of this paper. We assume, that $\log N \ll t \ll N$ and want to construct an efficient periodic t-correction network. Without loss of generality we assume that t is even. Let $S(N,t) = O\left(\log N/\log t + (\log\log N)^2(\log t)^2\right)$ be the maximum depth of a (t,Δ,δ)-semi-correction network for $x = \Delta/\delta = N^{1/\log t}$. As before $T(t)$ is the depth of a sorting network. In our construction the registers are arranged into an array of q columns and N/q rows, where

$$q = \max\{10(T(4t+4)+2\log t), 4(T(4t+4)+2\log t)+2S(N,t)\}$$

The rows of this array are divided into N/pq *floors* which are sets of $p = 4t+4$ subsequent rows. So the floor 1 consists of rows $1,2,\ldots,p$, floor 2 of rows $p+1,p+2,\ldots,2p$ and so on. We use the notions of 'bottom' and 'top' registers from the proof of key lemma. Thus we divide each floor into two halves: top and bottom. They consist of $p/2 = 2t+2$ top and bottom rows of each floor respectively. We define a *family of floors* to be a the set of all floors whose indices differ by $i \cdot \log t$ for some integer i. Altogether we have $\log t$ families of floors. To each family of floors we assign the index of its first floor.

From now on we all the time deal only with t-disturbed 0-1 input sequences. Any such a sequence has a border index b. The b-th register we call the border register. Its row we call the border row. Its floor we call the border floor. In the analysis we take into account only behavior of displaced 1s. Due to symmetry of the network the analysis for displaced 0s is the same and can be omitted.

We begin with defining a particular kind of periodic cleaning network, which the whole construction is based on. By adding comparators to this network we finally obtain a periodic t-correction network. The periodic cleaning network is constructed in the similar way as one in the previous chapter.

Above all we want to have some relation between vertical edges in areas X_L and X_R and the division of rows into floors. These comparators are embeddings of a cleaning network for dirty area $p/2 - 2t + 2$ in each area. Note that such a network also sorts any input with dirty area of size t, so can be used in the construction of periodic cleaning network for t dirty rows. The cleaning network consists of three subsequent parts. The first part are sorting networks – each sorting a group of p subsequent registers corresponding to a single floor. This part has depth $T(p)$. The second part consists of merging networks which merge neighboring upper and lower halves of each pair of subsequent groups from the first part. It has depth $\log p$. The third part is the last layer which we can add arbitrarily, because any layer of comparators does nothing to a sorted sequence. This layer is defined a bit later in the paper.

Parts S, M_L, M_R are defined exactly the same way as earlier for a periodic cleaning network. So as we previously proved the periodic network we now defined is a cleaning network for dirty areas consisting of at most t rows and the following key lemma describing its running time holds

Lemma 3.2 (key lemma). *We consider $t', t' \leq t$ subsequent rows of above defined network, such that above (below) these rows there are only 0s (1s). Let we have majority of 0s (1s) in these rows. There exist constants c and d such that after $d \cdot q$ steps the top (bottom) $c \cdot t'$ of these t' rows contain only 0s (1s).*

Note that if we add to C any edges in Y or connecting rows whose difference is bigger than t, then the key lemma still holds and so all its consequences hold too. We prove the following lemma

Lemma 3.3. *The periodic cleaning network described above sorts considered inputs with dirty area having $a \cdot t$ rows in $O(qa + q\log t)$ iterations.*

Proof. If the number of rows in dirty area is smaller than t then a standard reasoning for periodic sorting networks works. We need only to consider what happens if the number of rows in dirty area is bigger than t. If there are at least highest $t/2$ rows of dirty area above the border row, then we can apply key lemma to these rows. Since the input is t-disturbed we have majority of 0s in these rows. So we obtain $ct/2$ top rows of 0s in time dq. Thus the dirty area is reduced by $ct/2$ rows. In the opposite case an analogous reasoning can be applied to $t/2$ lowest rows where we have majority of 1s.□

Now we add some comparators to layer C so that our network gains new properties. First we add in area S comparators

$$\{[(2i,1) : (2i+t+1,2)]|i\}.$$

To formulate the fact which follows from the presence of these comparators we must specify what exactly we mean by right-running items. In the proof of key lemma right-running items were those 0s and 1s which were on the right of a horizontal edge after step A or B. We redefine it saying that in area S right-running items go right in step C instead of step A that is just after this step. Analogously we can redefine left-running items. We assume that two diplaced 1s or two 0s are swapped by an edge if this is an edge of step B or a slanted edge of step C or an edge of step A not belonging to area S. Displaced 1s are not swapped with non-displaced 1s. We can now formulate a simple property of our network that is preserved when we add edges

Fact 3.4 *In the network defined above right-running displaced 1s remain right-running as long as they are more than $t+1$ rows above border row.*

Now for a while we assume, that we deal only with displaced 1s that are more that one floor above the border. We remind, that R-columns and L-columns after a given step are columns containing right-running and left-running items respectively. We can note that R-column which gets to the column j_R while moving through X_R is first sorted separately on each floor by the first part of the cleaning network. Next the displaced 1s from each floor go half a floor down by the second part. An analogous process is also performed for left-running 1s in X_L as long as they remain left-running.

Thus after the second part of their way through X_R right-running displaced 1s are located at the bottom of the top half of each floor above the border floor. Analogously left-running displaced 1s are also moved just before the last layer embedded in X_L to the bottom of the top half of each floor.

We now should specify what the additional layer in the third part of X_R does. Formally speaking this layer is the set of comparators

$$\{[(kp+p/2+2i) : (kp+p/2-2i+1)]|, 0 < i \le t/2\}.$$

It moves right-running displaced 1s that went through X_R to odd indexed rows in the middle of each floor. Analogously the last layer embedded in X_L is

$$\{[(kp+p/2+2i-1):(kp+p/2-2i)]|,0<i\leq\lceil t/2\rceil\}.$$

It moves left-running displaced 1s to even indexed rows in the middle of each floor.

Let us call these rows for a while starting rows of these 1s. We can see that these all right-running displaced 1s then pass M_R, S, M_L and X_L without being moved by vertical edges in M_R, M_L, X_L. Note, that they encounter vertical edges only in M_L and they are at the bottom of these edges. The same happens to left-running 1s when they pass M_L, S, M_R and X_R. After passing X_L each right-running 1 is $t+2$ rows below its starting (odd) row. After passing X_R each left-running 1 is 2 rows above its starting (even) row. These 1s are still on the same floors as their starting rows. Similar facts can be proved for displaced 0s below the border which are also moved by last layers of X_L and X_R described above.

Now we define the vertical edges added in area Y of layer C. These comparators are embeddings of four semi-correction networks in each family of floors. Now we describe the comparators embedded in r-th family of floors. Let $a_1, a_2, \ldots, a_{2N/(q\log t)}$ be the indices of odd rows in this family of floors. We can build a $(t, N^{1-(r-1)/\log t}, N^{1-r/\log t})$-semi-correction network on registers with these indices. The depth of this network is not bigger (from the assumption about q) than the number of odd indexed columns in Y. Let this network be the sequence of layers $L_1, L_2, L_3, \ldots$. The first set of comparators is

$$\{[(j_L+2j-1,k):(j_L+2j-1,l):[k,l]\in L_j]\}.$$

We assumed that after passing X_L right-running 1s are in odd rows. Assume that they can be present only in $N^{1-(r-1)/\log t}$ odd rows of rth family directly above the border. When they pass Y they can be present only in $N^{1-r/\log t}$ odd rows of rth family directly above the border. Passing Y in family $r = \log t$ finally causes these 1s get to some of t odd rows of this family directly above the border. We formulate this assertion as a fact later because we need some additional assumptions. Analogously we can embed the same network once again to deal with left-running 1s that are in even rows. Formally speaking we add to C the following set of comparators

$$\{[(j_R-2j+1,k+1):(j_R-2j+1,l+1):[k,l]\in L_j]\}.$$

This set of edges again causes left running 1s which are in $N^{1-(r-1)/\log t}$ even rows of rth family directly above the border reduce the number of these rows between these 1s and the border to at most $N^{1-r/\log t}$. Analogously we also embed two copies of $(t, N^{r/\log t}, N^{(r-1)/\log t})$-semi-correction network to deal with displaced 0s below the border row.

We described the whole network and the way it works informally. To make this analysis more formal we assign colors to displaced 1s. We use five colors: blue, black, red, yellow and green. Let β be the index of the border floor. We assume the following rules of coloring displaced 1s:

– At the beginning the color of all displaced 1s is blue.

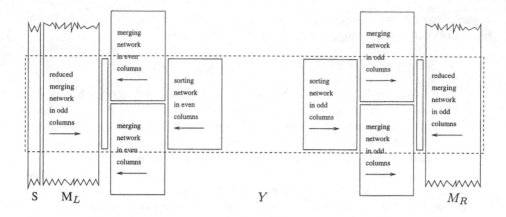

Fig. 2. Comparator networks embeded on a single floor.

- If a blue 1 is compared with a non-blue 1 by a vertical edge, then the blue 1 behaves like a 0.
- When any 1 gets to the floor with the index not smaller than $\beta - 1$, it changes its color to green.
- When a right-running non-blue 1 gets to the floor $\beta - 2$, it changes its color to green.
- When a non-green left-running 1 changes to be right-running it becomes blue.
- When a blue 1 gets from Y to outside of Y it changes its color to black.
- When a black 1 enters Y from outside of Y on the floor belonging to the family 1, then it changes its color to red.
- When a red 1 leaves Y on the floor in the last family of floors (family $\log t$), then it becomes yellow.

First we prove, that all green 1s stay close to the border. They prove to be all the time at the floors with indices not smaller than $\beta - 2$, so they are not more than $13t$ rows above the border row. We notice that right-running 1s can go only to the lower rows. Left-running 1s can go to the higher rows only in area S of layer C and by wrap-around edges of layer B. So only left-running can go up from the floor $\beta - 2$. Each q horizontal steps a left-running 1 can go up by maximum $t + 2$ rows. But on the other hand each q horizontal steps it passes X_L once. Passing X_L it goes to the row t-th or lower counting from the bottom of floor $\beta - 2$. Thus it cannot leave floor $\beta + 2$ going up not more than $t + 2$ rows. Moreover because our network is periodic correction network for dirty area of t rows we have the following fact

Fact 3.5 *If all displaced 1s are green, then the time to sort all the items above the border is $O(q \log t)$.*

Now we consider a right-running blue 1 or a left-running blue 1 assuming it stays left-running. From what we said before a right-running 1 stops to be right-running only when it is green. We want to see how quickly it becomes green. After $O(q)$ steps this 1 stops to be blue. The worst case is that it becomes black. The following fact can be viewed as a summary of what we said defining comparators of last column of X_L and X_R. We take advantage of the fact that right-running 1s that just changed from being left-running above floor $\beta - 1$ are blue. We also take advantage of the fact, that right-running 1s which are more than t rows above the border do not become left-running. Such 1s is the only factor that could disturb the 1s we are interested in to go one floor down.

Fact 3.6 *Any black, red or yellow right-running 1 on the floor higher than $\beta - 1$ passing areas X_R, M_R, S, M_L, X_L goes one floor down and ends up in an odd indexed row. Any black, red or yellow left-running 1 on the floor higher than β passing areas X_L, M_L, S, M_R and X_R goes one floor down and ends up in an even indexed row.*

The comparators in Y connect only the rows belonging to the same family of floors. So passing Y a displaced 1 does not change its family of floors. Thus we have the next fact.

Fact 3.7 *Every q horizontal steps a black or red 1 gets from a family r to the family $r + 1$. The exception is family $r = \log t$ from which it gets to family 1.*

So after at most $q \log t$ horizontal steps a black 1 becomes red, unless it starts to be green. We measure the *distance* of a red 1 that is in family r to the border as the number of rows that belong to the family r and are between this 1 and the floor $\beta - 2$. Passing Y in family 1 a red 1 reduces this distance from at most N to at most $2N^{1-1/\log t}$. Then it gets to families $2, 3, \ldots, \log t - 1$. Passing Y in family r a red 1 reduces this distance from $2N^{1-(r-1)/\log t}$ to $2N^{1-r/\log t}$. Then after passing Y in family $\log t$ a red 1 is in the distance at most $2t$. This way a red 1 becomes yellow after $q \log t$ horizontal steps. Now it is at most $\log t + 2$ floors above the border. A yellow right-running 1 goes at least 1 floor down each q horizontal steps, till it becomes green after at most $q \log t$ horizontal steps.

This whole process of color change from blue to green takes altogether $3q \log t$ horizontal steps. It always succeeds for right-running 1s. Left-running 1s can switch to be right-running before they become green. They have to do it before $3q \log t$ horizontal steps in which they have to become green if they are all the time left-running. In such a case they become inevitably green after next $3q \log t$ iterations as right-running 1s. Thus we have the following fact.

Fact 3.8 *All disturbed 1s start to be green after at most $6q \log t$ horizontal steps.*

Because inputs having only green 1s are quickly sorted we get the main result of the paper

Theorem 3.9. *The periodic t-correction network we defined in this paper sorts any t disturbed input in $O(q \log t)$ iterations, which is equal to*

$$O(\log N + (\log \log N)^2 (\log t)^3)$$

Acknowledgments. The author wishes to thank Marek Piotrów and other coworkers from algorithms and complexity group of his institute for helpful discussions.

References

1. M. Ajtai, J. Komolós, E. Szemerédi, Sorting in $c\log n$ parallel steps, *Combinatorica* 3 (1983), 1–19.
2. K.E. Batcher, Sorting networks and their applications, in *AFIPS Conf. Proc.* 32 (1968), 307–314.
3. M. Dowd, Y. Perl, M. Saks, L. Rudolph. The Periodic Balanced Sorting Network. *Journal of the ACM* 36 (1989), 738–757.
4. M. Kik, M. Kutyłowski, M. Piotrów, Correction Networks, in *Proc. of 1999 ICPP*, 40–47.
5. M. Kik, M. Kutyłowski, G. Stachowiak, Periodic constant depth sorting networks, *Proc. of the 11th STACS*, 1994, 201–212.
6. M. Kik, Periodic Correction Networks, *EUROPAR 2000 Proceedings*, LNCS 1900, 471–478
7. D. E. Knuth, *The Art of Computer Programming*, Vol 3, 2nd edition, Addison Wesley, Reading, MA, 1975.
8. J. Krammer, Lösung von Datentransportproblemen in integrierten Schaltungen. Dissertation, TU München 1991.
9. K. Loryś, M. Kutyłowski, B. Oesterdiekoff, R. Wanka, Fast and Feasible Periodic Sorting Networks of Constant Depth, *Proc of 35 IEEE-FOCS*, 1994, 369–380.
10. B. Oesterdiekoff, On the Minimal Period of Fast Periodic Sorting Networks, *Technical Report TR-RI-95-167*, University of Paderborn, 1995.
11. M. Piotrów, Depth Optimal Sorting Networks Resistant to k Passive Faults in *Proc. 7th SIAM Symposium on Discrete Algorithms* (1996), 242–251 (also accepted for *SIAM J. Comput.*).
12. M. Piotrów, Periodic Random-Fault-Tolerant Correction Networks, *Proceedings of 13th SPAA*, ACM 2001, 298–305.
13. L. Rudolph, A Robust Sorting Network, *IEEE Transactions on Computers* 34(1985), 326–336.
14. M. Schimmler, C. Starke, A Correction Network for N-Sorters, *SIAM J. Comput.* 18 (1989), 1179–1197.
15. U. Schwiegelsohn. A shortperiodic two-dimensional systolic sorting algorithm. In International Conference on Systolic Arrays, Computer Society Press, Baltimore 1988, 257–264.
16. G. Stachowiak, Fibonacci Correction Networks, in *Algorithm Theory – SWAT 2000* , M Halldórsson (Ed.) , LNCS 1851, Springer 2000, 535–548.

Games and Networks

Christos Papadimitriou

The Computer Science Division
University of California, Berkeley
Berkeley, CA 94720-1776
christos@cs.berkeley.edu

Abstract. Modern networks are the product of, and arena for, the complex interactions between selfish entities. This talk surveys recent work (with Alex Fabrikant, Eli Maneva, Milena Mihail, Amin Saberi, and Scott Shenker) on various instances in which the theory of games offers interesting insights to networks. We study the Nash equilibria of a simple and novel network creation game in which nodes/players add edges, at a cost, to improve communication delays. We point out that the heavy tails in the degree distribution of the Internet topology can be the result of a trade-off between connection costs and quality of service for each arriving node. We study an interesting class of games called network congestion games, and prove positive and negative complexity results on the problem of computing pure Nash equilibria in such games. And we show that shortest path auctions, which are known to involve huge overpayments in the worst case, are "frugal" in expectation in several random graph models appropriate for the Internet.

A. Lingas and B.J. Nilsson (Eds.): FCT 2003, LNCS 2751, p. 157, 2003.
© Springer-Verlag Berlin Heidelberg 2003

One-Way Communication Complexity of Symmetric Boolean Functions

Jan Arpe*, Andreas Jakoby**, and Maciej Liśkiewicz***

Institut für Theoretische Informatik, Universität zu Lübeck
{arpe,jakoby,liskiewi}@tcs.uni-luebeck.de

Abstract. We study deterministic one-way communication complexity of functions with Hankel communication matrices. In this paper some structural properties of such matrices are established and applied to the one-way two-party communication complexity of symmetric Boolean functions. It is shown that the number of required communication bits does not depend on the communication direction, provided that neither direction needs maximum complexity. Moreover, in order to obtain an optimal protocol, it is in any case sufficient to consider only the communication direction from the party with the shorter input to the other party. These facts do not hold for arbitrary Boolean functions in general. Next, gaps between one-way and two-way communication complexity for symmetric Boolean functions are discussed. Finally, we give some generalizations to the case of multiple parties.

1 Introduction

The communication complexity of two-party protocols was introduced by Yao in 1979 [15]. The theory of communication complexity evolved into an important branch of computational complexity (for a general survey of the theory see e.g. Kushilevitz and Nisan [9]).

In this paper we consider one-way communication, i.e. we restrict the communication to a single round. This simple model has been investigated by several authors for different types of communication such as fully deterministic, probabilistic, nondeterministic, and quantum (see e.g. [15,12,1,11,3,8,7]). We study the deterministic setting. One-way communication complexity finds application in a wide range of areas, e.g. it provides lower bounds on VLSI complexity and on the size of finite automata (cf. [5]). Moreover, one-way communication complexity of symmetric Boolean functions is connected to binary decision diagrams by the following observation due to Wegener [14]: The size of an optimal protocol coincides with the number of nodes at a certain level in a minimal OBDD.

We consider the standard two-party communication model: Initially the parties, called Alice and Bob, hold disjoint parts of input data x and y, respectively.

---

* Supported by DFG research grant Re 672/3.
** Part of this work was done while visiting International University Bremen, Germany.
*** On leave from Instytut Informatyki, Uniwersytet Wrocławski, Wrocław, Poland.

A. Lingas and B.J. Nilsson (Eds.): FCT 2003, LNCS 2751, pp. 158–170, 2003.
© Springer-Verlag Berlin Heidelberg 2003

In order to compute a function $f(x, y)$, they exchange messages between each other according to a communication protocol.

In a (deterministic) one-way protocol $\mathcal{P}$ for f, one of the parties sends a single message to the other party, and then the latter party computes the output $f(x, y)$. We call $\mathcal{P}$ a protocol of type $A \to B$ if Alice sends to Bob and of type $B \to A$ if Bob sends to Alice. The size of $\mathcal{P}$ is the number of different messages that can potentially be transmitted via the communication channel according to $\mathcal{P}$. The one-way communication size $S^{A \to B}(f)$ of f is the size of the best protocol of type $A \to B$. It is clear that the respective one-way communication complexity is $C^{A \to B}(f) = \lceil \log S^{A \to B}(f) \rceil$. For the case when Bob sends messages to Alice, we analogously use the notation $S^{B \to A}$ and $C^{B \to A}$. Note that throughout this paper, log always denotes the binary logarithm.

The main results of this paper deal with one-way communication complexity of symmetric Boolean functions – an important subclass of all Boolean functions. A Boolean function F is called symmetric, if permuting the input bits does not effect the function value. Some examples for symmetric functions are *and, or, parity, majority,* and arbitrary *threshold* functions. We assume that the input bits for a given F are partitioned into two parts, one part consisting of m bits held by Alice and the other part consisting of n bits only known to Bob. As the function value of a symmetric Boolean function only depends on the number of 1's in the input (cf. [13]), it is completely determined by the sum of the number of 1's in Alice's input part and the number of 1's in Bob's part. Hence for such functions, we are faced with the problem of determining the one-way communication complexity of a function $f : \{0, \dots, m\} \times \{0, \dots, n\} \to \{0, 1\}$ associated to F, where $f(x, y)$ only depends on the sum $x + y$. Note that $S^{A \to B}(F) \leq m+1$ is a trivial upper bound on the one-way communication size of F.

Let us assume that Alice's input part is at most as large as Bob's is (i.e. let $m \leq n$). While for arbitrary functions this property does not imply which communication direction admits the better one-way protocols, we show that the converse is true for symmetric Boolean functions F, namely in this case we have $C^{A \to B}(F) \leq C^{B \to A}(F)$. Moreover, we prove that if some protocol of type $A \to B$ does not require maximal size, i.e. if $S^{A \to B}(F) < m + 1$, then both directions yield the same complexities, i.e. $C^{A \to B}(F) = C^{B \to A}(F)$.

We also present a class of families of symmetric Boolean functions for which one-way communication is *almost* as powerful as two-way communication. More precisely, for any family of symmetric Boolean functions $F_1, F_2, F_3 \dots$ with $F_m : \{0, 1\}^{2m} \to \{0, 1\}$, let $f_m : \{0, \dots, m\} \times \{0, \dots, m\} \to \{0, 1\}$ denote the integer function associated to F_m. We prove that if $f_m \subseteq f_{m+1}$ for all $m \in \mathbb{N}$, then either the one-way communication complexities of $F_1, F_2, F_3 \dots$ are almost all equal to a constant c or the two-way communication complexities of $F_1, F_2, F_3 \dots$ are infinitely often maximal. We show that one can easily test whether the first or the second case occurs: The two-way communication complexities are infinitely often maximal if and only if the unary language $\{0^{k+\ell} \mid f_m(k, \ell) = 1, \ m, k, \ell \in \mathbb{N}\}$ is nonregular.

On the other hand, we construct an example of a symmetric Boolean function having one-way communication complexity exponentially larger than its two-way communication complexity. Finally, we generalize the two-party model to the case of multiple parties and extend our results to such a setting.

Our proofs are based on the fact that the communication matrix of the integer function f associated with a symmetric Boolean function F is a Hankel matrix. In general, the entries of the communication matrix M_f of f are defined by $m_{i,j} = f(i,j)$. A Hankel matrix is a matrix in which the entries on each anti-diagonal are constant (equivalently, $m_{i,j}$ only depends on $i + j$). Hankel matrices are completely determined by the entries of their first rows and their last columns. Thus with any $(m + 1) \times (n + 1)$-Hankel matrix H we associate a function f_H such that $f_H(0), f_H(1), \ldots, f_H(n)$ compose the first row of H and $f_H(n), f_H(n+1), \ldots, f_H(m+n)$ make up its last column. One of the main technical contributions of this paper is a theorem saying that if $m \leq n$ and H has less than $m + 1$ different rows, then f_H is periodic on a certain large interval. We apply this property to the one-way communication size using a known relationship between this measure and the number of different rows in communication matrices.

As a byproduct, we obtain a word combinatorial property: Let w be an arbitrary string over some alphabet Σ. Then, for $m \leq \lceil |w|/2 \rceil$ and $n = |w| - m + 1$, the number of different substrings of w of length n is at most as large as the number of different substrings of w of length m. Moreover, if the former number is strictly less than m (note that it can be at most m in general), then the number of different substrings of length n and the number of different substrings of length m coincide.

The paper is organized as follows: In Section 2, we introduce basic definitions and notation. Section 3 deals with the examination of the number of different rows and columns in Hankel matrices involving certain periodicity properties. In Section 4, we state some applications of these properties. Then, in Section 5, we present a class of symmetric Boolean functions with both maximal one-way and two-way communication complexity, and then we construct a symmetric Boolean function with an exponential gap between its one-way and its two-way communication complexity. Finally, in Section 6, we discuss natural extensions of our results to the case of multiple parties.

2   Preliminaries

For any integers $0 \leq k < k'$, let $[k..k']$ denote the set $\{k, k + 1, \ldots, k'\}$, and denote $[0..k]$ by $[k]$ for short. By $\mathbb{N}$ we denote the set of nonnegative integers. We consider deterministic one-way communication protocols between Alice and Bob for functions $f : [m] \times [n] \to \Sigma$, where Σ is an arbitrary (finite or infinite) nonempty set. More specifically, we assume that Alice holds a value $x \in [m]$, and Bob holds a value $y \in [n]$ for some fixed positive integers m and n. Their aim is to compute the value $f(x, y)$.

Let $\mathcal{M}(m, n)$ denote the set of all $(m + 1) \times (n + 1)$ matrices $M = (m_{i,j})$, with $m_{i,j} \in \Sigma$. It will be convenient for us to enumerate the rows from 0 to m and the columns from 0 to n. For a given function $f : [m] \times [n] \to \Sigma$, we denote by M_f the corresponding communication matrix in $\mathcal{M}(m, n)$.

Definition 1. *For a matrix $M \in \mathcal{M}(m, n)$, define $\#\mathrm{row}(M)$ to be the number of different rows of M, and similarly let $\#\mathrm{col}(M)$ be the number of different columns of M. Furthermore, for any $i, j \in [m]$, let $i \sim_M j$ denote that the rows i and j of M are equal.*

It is easy to characterize the one-way communication size by $\#\mathrm{row}$ and $\#\mathrm{col}$.

Fact 1. *For all $m, n \in \mathbb{N}$ and for every function $f : [m] \times [n] \to \Sigma$, it holds that*
$$S^{A \to B}(f) = \#\mathrm{row}(M_f) \quad \text{and} \quad S^{B \to A}(f) = \#\mathrm{col}(M_f).$$

In this paper we will restrict ourselves to functions f that only depend on the sum of the arguments. Note that for such functions f the communication matrix M_f is a Hankel matrix. The problem of finding protocols for such restricted f arises naturally when one considers symmetric Boolean functions.

Definition 2. *Let $f : [s] \to \mathbb{N}$, $\lambda \geq 1$ and $s_1, s_2 \in [s]$ with $s_1 \leq s_2 - \lambda$. We call f λ-periodic on $[s_1..s_2]$, if for all $x \in [s_1..s_2 - \lambda]$, $f(x) = f(x + \lambda)$.*

Obviously, f is λ-periodic on $[s_1..s_2]$ if and only if for all $x, x' \in [s_1..s_2]$ with $\lambda \mid (x - x')$, it holds that $f(x) = f(x')$.

3   Periodicity of Rows and Columns in Hankel Matrices

This section is devoted to examine the relationship between the number of different rows and the number of different columns in a Hankel matrix. Lemmas 1 through 3 are technical preparations for Theorem 1 which gives an explicit characterization of a certain periodic behaviour of the function associated with a Hankel matrix and of the Hankel matrix itself. Theorems 2 and 3 reveal all possible constellations of values for $\#\mathrm{row}(H)$ and $\#\mathrm{col}(H)$ for a Hankel matrix H. The results will be applied to the theory of one-way communication in Section 4.

Fact 2. *Let $f : [s] \to \mathbb{N}$ be λ-periodic on $[s_1..s_2] \subseteq [s]$ and on $[t_1..t_2] \subseteq [s]$ such that $s_1 \leq t_1$ and $t_1 + \lambda \leq s_2$. Then f is λ-periodic on $[s_1..t_2]$.*

Lemma 1. *Let $H \in \mathcal{M}(m, n)$ be a Hankel matrix, $m_0, m_1 \in [m]$ with $m_0 < m_1$, and $\lambda \in [1..m_1 - m_0]$. Then the following two statements are equivalent:*

(a) f_H is λ-periodic on $[m_0..m_1 + n]$.
(b) For all $x \in [m_0..m_1]$ and all $k \in \mathbb{N}$ such that $x + k\lambda \leq m_1$, $x \sim_H x + k\lambda$.

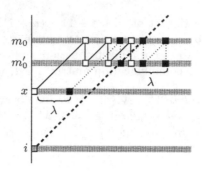

Fig. 1. An illustration of Case 1.

Proof. "(a)⇒(b)": Let $x \in [m_0..m_1]$ and $k \in \mathbb{N}$ such that $x + k\lambda \leq m_1$. For all $y \in [n]$, $x + y \geq m_0$ and $x + y + k\lambda \leq m_1 + n$. Since f_H is λ-periodic on $[m_0..m_1 + n]$, we have $f_H(x + y) = f_H(x + k\lambda + y)$.

"(b)⇒(a)": Let $x \in [m_0..m_1 + n - \lambda]$. We consider two cases. If $x \leq m_0 + n$, then $f_H(x) = f_H(m_0 + (x - m_0)) = f_H(m_0 + \lambda + (x - m_0)) = f_H(x + \lambda)$, because $m_0 \sim_H m_0 + \lambda$ by hypothesis. If on the other hand $x > m_0 + n$, then $x - n > m_0$ and $x - n + \lambda \leq m_1$. By hypothesis, $x - n \sim_H x - n + \lambda$, and thus $f_H(x) = f_H(x - n + n) = f_H(x - n + \lambda + n) = f_H(x + \lambda)$. □

Corollary 1. *Let $H \in \mathcal{M}(m, n)$ be a Hankel matrix and $i, j \in [m]$ with $i < j$. Then $i \sim_H j$ if and only if f_H is $(j - i)$-periodic on $[i..j + n]$.*

Corollary 2. *Let $H \in \mathcal{M}(m, n)$ be a Hankel matrix. If f_H is λ-periodic on $[m_0..m_1 + n]$ for some $m_0, m_1 \in [m]$ with $m_0 < m_1$ and some $\lambda \in [1..m_1 - m_0]$, then $\#\mathrm{row}(H) \leq m_0 + \lambda + m - m_1$, where equality holds if and only if all rows $0, \ldots, m_0 + \lambda - 1$ and $m_1 + 1, \ldots, m$ are pairwise different.*

Lemma 2. *Let $H \in \mathcal{M}(m, n)$ be a Hankel matrix and $m_0, m_0', i, j \in [m]$ such that $m_0 \leq i < j$, $m_0' - m_0 \leq n + 1$, $j - m_0 \leq n + 1$, $i \sim_H j$, and $m_0 \sim_H m_0'$. Then f_H is $(j - i)$-periodic on $[m_0..j + n]$.*

Proof. Choose $\lambda = j - i$ and $\mu_0 = m_0' - m_0$. By Corollary 1, f_H is

(i) μ_0-periodic on $[m_0..m_0' + n]$ and
(ii) λ-periodic on $[i..j + n]$.

Let $x \in [m_0..j + n - \lambda]$. In order to show that $f_H(x + \lambda) = f_H(x)$, we consider:
Case 1: $m_0 \leq x < i$: Let $k \in \mathbb{N}$ such that $i \leq x + k\mu_0 \leq i + \mu_0 - 1$. We need to show that

$$x, x + k\mu_0, x + k\mu_0 + \lambda, x + \lambda \in [m_0..m_0' + n] \quad \text{and} \tag{1}$$
$$x + k\mu_0, x + k\mu_0 + \lambda \in [i..j + n] \tag{2}$$

in order to apply properties (i) and (ii) to the corresponding elements. Property (1) follows from $m_0 \leq x$ and $x+k\mu_0+\lambda \leq i+\mu_0+\lambda-1 = j+m_0'-m_0-1 \leq m_0'+n$. Property (2) is due to $i \leq x+k\mu_0$ and $x+k\mu_0+\lambda \leq j-1+\mu_0 \leq j+n$. Now (cf. Fig. 1) $f_H(x) = f_H(x+k\mu_0) = f_H(x+k\mu_0+\lambda) = f_H(x+\lambda)$, where the first and the last equality follow from properties (1) and (i), and the middle equality is due to properties (2) and (ii).

Case 2: $i \leq x \leq j+n-\lambda$: In this case, $f_H(x) = f_H(x+\lambda)$ by Corollary 1. $\square$

The following lemma is symmetric to the previous one:

Lemma 3. *Let $H \in \mathcal{M}(m,n)$ be a Hankel matrix and $m_1, m_1', i, j \in [m]$ such that $i < j \leq m_1$, $m_1 - m_1' \leq n+1$, $m_1 - i \leq n+1$, $i \sim_H j$, and $m_1 \sim_H m_1'$. Then f_H is $(j-i)$-periodic on $[i..m_1+n]$.*

Proof. Let $H = (h_{i,j})$. We define $\lambda = j - i$ and $H' = (h_{\mu,\nu}') \in \mathcal{M}(m,n)$ by $h_{\mu,\nu}' = h_{m-\mu,n-\nu}$ for $(\mu,\nu) \in [m] \times [n]$, i.e. we rotate H by 180 degrees in the plane. Clearly, H' is again a Hankel matrix. Moreover, we have $f_H(z) = f_{H'}(m+n-z)$ for all $z \in [m+n]$. We set $m_0 = m - m_1$, $m_0' = m - m_1'$, $i' = m - j$, and $j' = m - i$. Now it is easy to check that H', i', j', m_0, and m_0' fulfill the preconditions of Lemma 2 and $m+n-x-\lambda \in [m_0..j'+n-\lambda]$, thus yielding $f_H(x+\lambda) = f_{H'}(m+n-x-\lambda) = f_{H'}(m+n-x) = f_H(x)$. $\square$

Theorem 1. *Let $m \leq n+1$ and $H \in \mathcal{M}(m,n)$ be a Hankel matrix with $\#\mathrm{row}(H) < m+1$. Then there exist $\lambda \in [1..n]$ and $m_0, m_1 \in [m]$ with $m_1 - m_0 \geq \lambda$ such that the following two properties hold:*

(a) The function f_H is λ-periodic on $[m_0..m_1+n]$.

(b) If $i, j \in [m]$ with $i < j$ and $i \sim_H j$, then $i, j \in [m_0..m_1]$ and $\lambda \mid (j-i)$.

Moreover, m_0, m_1 and λ can be explicitly determined as follows:

$$m_0 = \min\{k \in [m] \mid \exists k' \in [m] \text{ with } k' > k \text{ and } k \sim_H k'\},$$
$$m_1 = \max\{k \in [m] \mid \exists k' \in [m] \text{ with } k' < k \text{ and } k \sim_H k'\}, \text{ and}$$
$$\lambda = \min\{j - i \mid i, j \in [m] \text{ with } i \sim_H j \text{ and } i < j\}.$$

Proof. Since $\#\mathrm{row}(H) < m+1$, there exist $i, j \in [m]$ with $i < j$ such that $i \sim_H j$. Thus, m_0, m_1 and λ are well-defined. Clearly, $m_1 - m_0 \geq \lambda$. Choose $i_0, j_0 \in [m]$ such that $i_0 \sim_H j_0$ and $j_0 - i_0 = \lambda$. Since $m \leq n$, all preconditions of Lemma 2 and Lemma 3 are satisfied. Thus we conclude that f_H is λ-periodic on both discrete intervals $[m_0..j_0+n]$ and $[i_0..m_1+n]$. Fact 2 now yields property (a). Now let $i, j \in [m]$ with $i < j$ and $i \sim_H j$. Let $k \in \mathbb{N}$ such that $j - i = k\lambda + r$ with $0 \leq r < \lambda$. By property (a), f_H is λ-periodic on $[m_0..m_1+n]$, and so by Lemma 1 (note that $i+k\lambda = j - r \leq j \leq m_1$), we have $i+k\lambda \sim_H i \sim_H j$. As $r = j - i - k\lambda < \lambda$ and λ is the minimal difference between two equal rows of different indices, we have $r = 0$, so $\lambda \mid (j-i)$. $\square$

Using Corollary 2 we deduce two consequences of Theorem 1:

Corollary 3. *For H, m_0, m_1 and λ as in Theorem 1, $\#\mathrm{row}(H) = m_0 + \lambda + m - m_1$, i.e. H has exactly $m_0 + \lambda + m - m_1$ pairwise different rows.*

Corollary 4. *Let* $m \leq n+1$ *and* $H \in \mathcal{M}(m,n)$ *be a Hankel matrix with* $\#\mathrm{row}(H) < m+1$. *Then* $\#\mathrm{col}(H) \leq \#\mathrm{row}(H)$.

The next lemma states an "expansion property" of Hankel matrices with at least two equal rows.

Lemma 4. *For arbitrary* $m, n \in \mathbb{N}$ *let* $H \in \mathcal{M}(m,n)$ *be a Hankel matrix with* $\#\mathrm{row}(H) < m+1$. *Then there exist* $m' \geq n$ *and a Hankel matrix* $\tilde{H} \in \mathcal{M}(m',n)$ *such that* $\#\mathrm{row}(\tilde{H}) = \#\mathrm{row}(H)$ *and* $\#\mathrm{col}(\tilde{H}) = \#\mathrm{col}(H)$.

Sketch of proof. We duplicate the area between two equal rows until the total number of rows exceeds the total number of columns n. This process effects neither the number of different rows nor the number of different columns.   □

Theorem 2. *Let* $m \leq n+1$ *and* $H \in \mathcal{M}(m,n)$ *be a Hankel matrix with* $\#\mathrm{row}(H) < m+1$. *Then* $\#\mathrm{row}(H) = \#\mathrm{col}(H)$.

Proof. From Corollary 4, we have $\#\mathrm{row}(H) \geq \#\mathrm{col}(H)$. By Lemma 4, there exist $m' \geq n$ and a Hankel matrix $\tilde{H} \in \mathcal{M}(m',n)$ such that $\#\mathrm{row}(\tilde{H}) = \#\mathrm{row}(H)$ and $\#\mathrm{col}(\tilde{H}) = \#\mathrm{col}(H)$. Thus, again by Corollary 4, we obtain $\#\mathrm{row}(H) = \#\mathrm{row}(\tilde{H}) = \#\mathrm{col}(\tilde{H}^T) \leq \#\mathrm{row}(\tilde{H}^T) = \#\mathrm{col}(\tilde{H}) = \#\mathrm{col}(H)$. Consequently, we have $\#\mathrm{row}(H) = \#\mathrm{col}(H)$.   □

Theorem 3. *Let* $m \leq n$ *and* $H \in \mathcal{M}(m,n)$ *be a Hankel matrix with* $\#\mathrm{row}(H) = m+1$. *Then* $\#\mathrm{col}(H) \geq m+1$.

Proof. Induction on n: For $n = m$, we have $H = H^T$ and thus $\#\mathrm{col}(H) = \#\mathrm{row}(H^T) = \#\mathrm{row}(H) = m+1$. Now suppose that $n > m$. Let $H' \in \mathcal{M}(m, n-1)$ be the matrix H without its last column. We consider two cases:

Case 1: $n \sim_{H^T} n'$ for some $n' \in [n-1]$. Then $\#\mathrm{col}(H) = \#\mathrm{col}(H')$. In addition, $\#\mathrm{row}(H') = m+1$, because if $\#\mathrm{row}(H') \leq m$ was true, then we had $i \sim_{H'} j$ for some $0 \leq i < j \leq m$, and thus $i \sim_H j$, since $f_H(i+n) = f_H(i+n') = f_H(j+n') = f_H(j+n)$. Thus, we get $\#\mathrm{col}(H) = \#\mathrm{col}(H') \geq m+1$ by induction hypothesis.

Case 2: $n \not\sim_{H^T} n'$ for all $n' \in [n-1]$. Then $\#\mathrm{col}(H) = \#\mathrm{col}(H') + 1$. Once again, we have to consider two subcases:

Case 2a: $\#\mathrm{row}(H') = m+1$: Then $\#\mathrm{col}(H) = \#\mathrm{col}(H') + 1 = m+2 > m+1$ by induction hypothesis.

Case 2b: $\#\mathrm{row}(H') \leq m$: Assume that $\#\mathrm{row}(H') < m$, and let

$$m_0 = \min\{k \in [m] \mid \exists k' \in [m] \text{ with } k' > k \text{ and } k \sim_H k'\},$$
$$m_1 = \max\{k \in [m] \mid \exists k' \in [m] \text{ with } k' < k \text{ and } k \sim_H k'\},$$
$$\lambda = \min\{k' - k \mid k, k' \in [m] \text{ with } k < k' \text{ and } k \sim_H k'\},$$

where m_0', m_1' and λ' are the corresponding numbers for H'. By Corollary 3, $\#\mathrm{row}(H') = m_0' + m - m_1' + \lambda'$, and f is λ'-periodic on $[m_0'..m_1' + n - 1]$ by

Theorem 1. Since $\#\text{row}(H') < m$ by assumption, $\lambda' < m_1' - m_0'$. In particular, $m_0 \sim_H m_0 + \lambda'$, and thus $\lambda \mid \lambda'$ by Theorem 1. Consequently, $m_0 \leq m_0'$, $m_1 \geq m_1' - 1$ and $\lambda \leq \lambda'$. Hence again by Corollary 3,

$$\#\text{row}(H) = m_0 + m - m_1 + \lambda \leq m_0' + m - (m_1' - 1) + \lambda'$$
$$\leq m_0' + m - m_1' + \lambda' + 1 = \#\text{row}(H') + 1 < m + 1,$$

contradicting the precondition $\#\text{row}(H) = m + 1$. Thus, $\#\text{row}(H') = m$. By Theorem 2, $\#\text{col}(H') = \#\text{row}(H') = m$. Consequently, $\#\text{col}(H) = \#\text{col}(H') + 1 = m + 1$. $\qquad\square$

Note that for Hankel matrices over Σ with $|\Sigma| \geq m + n + 1$ we can say even more. Namely, if $m \leq n$, then for all $r \in [m + 1..n + 1]$, there exists a Hankel matrix $H \in \mathcal{M}(m, n)$ with $\#\text{row}(H) = m + 1$ and $\#\text{col}(H) = r$. To see this, define $f : [m] \times [n] \to \Sigma = \{a_0, \ldots, a_{m+n}\}$ by $f(x, y) = a_{(x+y) \bmod r}$. Then $H = M_f$ is a Hankel matrix fulfilling the requested properties.

4 Applications

Theorems 2 and 3 can be summarized in terms of one-way communication as follows.

Theorem 4. *Let $m \leq n$ and $f : [m] \times [n] \to \Sigma$ be a function for which the corresponding communication matrix M_f is a Hankel matrix. Then the following properties hold: (a) $S^{A \to B}(f) \leq S^{B \to A}(f)$. (b) If $S^{A \to B}(f) < m + 1$, then $S^{A \to B}(f) = S^{B \to A}(f)$.*

This result can immediately be applied to symmetric Boolean functions:

Corollary 5. *Let $m \leq n$ and $F : \{0,1\}^m \times \{0,1\}^n \to \{0,1\}$ be a symmetric Boolean function. Then the following properties hold: (a) $S^{A \to B}(F) \leq S^{B \to A}(F)$. (b) If $S^{A \to B}(F) < m + 1$, then $S^{A \to B}(F) = S^{B \to A}(F)$.*

The results of the last paragraph can also be applied to word combinatorics as follows:

Theorem 5. *Let w be an arbitrary string over some alphabet Σ, and let $N_w(i)$ denote the number of different subwords of w of length i. Then, for $m \leq \lceil |w|/2 \rceil$ and $n = |w| - m + 1$, we have $N_w(n) \leq N_w(m)$. Moreover, if $N_w(n) < m$ (note that $N_w(n) \leq m$ in general), then $N_w(n) = N_w(m)$.*

5 One-Way versus Two-Way Protocols

In this section we first present a class of families of functions for which one-way communication complexities are *almost* the same as two-way communication complexities. We denote the two-way complexity of F by $C(F)$. Let $F_1, F_2, F_3 \ldots$ with $F_m : \{0,1\}^{2m} \to \{0,1\}$ be a family of symmetric Boolean functions and let $f_m : [m] \times [m] \to \{0,1\}$ denote the integer function associated to F_m, i.e. $F(x_1, \ldots, x_{2m}) = 1$ if and only if $f(\sum_{i=1}^m x_i, \sum_{i=m+1}^{2m} x_i) = 1$.

Theorem 6. *Let $F_1, F_2, F_3 \ldots$ be a family of symmetric Boolean functions such that $f_m \subseteq f_{m+1}$ for all $m \in \mathbb{N}$. Then either*
(a) for almost all $m \in \mathbb{N}$, $C^{A \to B}(F_m) = c$ for some constant c or
(b) for infinitely many $m \in \mathbb{N}$, $C(F_m) = \lceil \log(m+1) \rceil$.
Moreover, (b) holds iff the language $L = \{0^{k+\ell} \mid f_m(k, \ell) = 1, \ m, k, \ell \in \mathbb{N}\}$ is nonregular.

Proof. First, Theorem 11.3 in [6] gives a nice characterization of (non)regular unary languages in terms of the rank of certain Hankel matrices. This characterization was first observed by Condon et al. in [2]. It says that the unary language L is nonregular if and only if for infinitely many $m \in \mathbb{N}$, $\mathrm{rank}(M_{f_m}) = m + 1$ (i.e. the communication matrix M_{f_m} has maximum rank). Second, Mehlhorn and Schmidt [10] showed that $C(f) \geq \log(\mathrm{rank}(M_f))$ for every f. Combining these facts we get that for nonregular L, $C(f_m) = \lceil \log(m+1) \rceil$ for infinitely many $m \in \mathbb{N}$.

On the other hand, if L is regular then by the Myhill-Nerode Theorem [4] the infinite matrix $M = (m_{i,j})_{i,j \in \mathbb{N}}$ defined by $m_{i,j} = 1$ iff $0^{i+j} \in L$, has constant number of different rows. Hence the theorem follows.   □

Example 1. Let $F_m(x_1, x_2, \ldots, x_{2m}) = 1$ iff the number of 1's in the sequence $x_1, x_2, \ldots, x_{2m}$ is the square of some integer. By Theorem 6 either for all $m \in \mathbb{N}$, $C(F_m), C^{A \to B}(F_m) \leq c$ for some constant c or for infinitely many $m \in \mathbb{N}$, $C^{A \to B}(F_m) = C(F_m) = \lceil \log(m+1) \rceil$. Since the language $\{0^n \mid n$ is the square of some integer$\}$ is nonregular, the (one-way) communication complexity of F_m is maximal for infinitely many $m \in \mathbb{N}$.

Next, we construct a symmetric Boolean function with an exponential difference between its one-way and its two-way communication complexity. Let $p_0, p_1, \ldots$ with $p_i < p_{i+1}$ for all $i \in \mathbb{N}$ be the sequence of all prime numbers. According to the Prime Number Theorem, there are at least $\frac{\ell}{\log \ell}$ prime numbers in the interval $[\ell]$ for all $\ell \geq 5$. For $k = \lceil \log \log m \rceil$ and $n = 2^k \cdot (1 + \prod_{i=0}^{2^k - 1} p_i)$, consider the function $f : [m] \times [n] \to \{0, 1\}$ defined by $f(x, y) = 1$ iff $\lfloor \frac{z}{2^k} \rfloor \bmod p_{z \bmod 2^k} = 0$, where $z = x + y$. Using the following two-way protocol, one can see that the two-way communication complexity of f is at most $4 \log \log m$: In the first round, Bob sends $y_0 = y \bmod 2^k$ to Alice. In the second round, Alice sends $z_0 = (x + y_0) \bmod 2^k$ and $z' = \lfloor \frac{x + y_0}{2^k} \rfloor \bmod p_{z_0}$ to Bob. Finally, Bob computes $f(x, y)$ by checking whether $(\lfloor \frac{y}{2^k} \rfloor + z') \bmod p_{z_0} = 0$.

Note that $z_0 = z \bmod 2^k$. The correctness of the protocol can be seen by investigating the addition of integers using a remainder representation.

Lemma 5. $C(f) \leq 4 \log \log m$.

For the one-way communication complexity of f we obtain:

Lemma 6. $\#\mathrm{row}(M_f) = m + 1$, *i.e.* $C^{A \to B}(f) = \lceil \log(m+1) \rceil$.

Theorem 7. *For the symmetric Boolean function $F : \{0, 1\}^m \times \{0, 1\}^n \to \{0, 1\}$ associated with f, we have $C(F) \in O(\log \log m)$ and $C^{A \to B}(F) \in \Theta(\log m)$.*

6   Multiparty Communication

So far we have analyzed the case that a fixed input partition for a function is given. However, sometimes it is also of interest to examine the communication complexity of a fixed function under *varying* the input partition. A typical question for this scenario is whether we can partition the input in such a way that the communication complexities for protocols of type $A \to B$ and $B \to A$ coincide. The main tool for these examinations is the *diversity* $\Delta(f)$ of f which we introduce below. For a function $f : [s] \to \Sigma$ and $m \in [s]$, define $f_m : [m] \times [s-m] \to \Sigma$ by $f_m(x,y) = f(x+y)$ for $x \in [m]$ and $y \in [s-m]$, and let $r_f(m) = \#\mathrm{row}(M_{f_m})$. We define $\Delta(f) = \max_{m \in [s]} r_f(m)$.

Lemma 7. *For every function $f : [s] \to \Sigma$, the following conditions hold:*

(a) $r_f(m) = m+1$ for all $m \in [\Delta(f) - 1]$,
(b) if $\Delta(f) \le \frac{s}{2}$, then $r_f(m) = \Delta(f)$ for all $m \in [\Delta(f) - 1 \,.. \, s - \Delta(f) + 1]$,
(c) $r_f(m) \ge r_f(m+1)$ for all $m \in [\Delta(f) - 1 \,.. \, s - 1]$.

It is an immediate consequence of Lemma 7 that $\Delta(f)$ equals the minimum m such that M_{f_m} has less than $m+1$ different rows, provided that such an m exists.

The diversity is helpful to analyze the case that more than two parties are involved. For such multiparty communication we assume that the input is distributed among d parties $P_1, \ldots, P_d$. Every party P_i knows a value $x_i \in [m_i]$. The goal is to compute a fixed function $f : [m_1] \times \ldots \times [m_d] \to \Sigma$. Analogously to communication matrices in the two-party case, we use multidimensional arrays to represent f.

Let $\mathcal{M}(m_1, \ldots, m_d)$ be the set of all d-dimensional $(m_1+1) \times \ldots \times (m_d+1)$ arrays M with entries $M(i_1, \ldots, i_d) \in \Sigma$ for $i_j \in [m_j]$, $j \in [1..d]$. M is called the *communication array* of a function f iff $M(i_1, \ldots, i_d) = f(i_1, \ldots, i_d)$. We denote the communication array of f by M_f.

Recall that in the two-party model the sender has to specify the row/column his input belongs to. In the multiparty case each party will have to specify the type of subarray determined by his input value. Therefore, for each $k \in [1..d]$ and each $x \in [m_k]$, we define the subarray $M_x^{(k)} \in \mathcal{M}(m_1, \ldots, m_{k-1}, m_{k+1}, \ldots, m_d)$ of M by $M_x^{(k)}(i_1, \ldots, i_{k-1}, i_{k+1}, \ldots, i_d) = M(i_1, \ldots, i_{k-1}, x, i_{k+1}, \ldots, i_d)$ for all $0 \le i_j \le m_j$, $j \in [1..d] \setminus \{k\}$. Finally, for $k \in [1..d]$ we define $\#\mathrm{sub}_k(M)$ as the number of different subarrays with fixed k^{th} dimension:

$$\#\mathrm{sub}_k(M) = |\{ \, M_x^{(k)} \mid x \in [m_k] \, \}| \,.$$

We call $M \in \mathcal{M}(m_1, \ldots, m_d)$ a *Hankel array*, if $M(i_1, \ldots, i_d) = M(j_1, \ldots, j_d)$ for every pair $(i_1, \ldots, i_d), (j_1, \ldots, j_d) \in [m_1] \times \ldots \times [m_d]$ with $i_1 + \ldots + i_d = j_1 + \ldots + j_d$. For a Hankel array $M \in \mathcal{M}(m_1, \ldots, m_d)$, let $f_M : [\sum_{i=1}^d m_i] \to \Sigma$ be defined by $f_M(x) = M(x_1, \ldots, x_d)$, if $x = x_1 + \ldots + x_d$. Note that f_M is well-defined since M is a Hankel array.

Lemma 8. *For a function f such that the corresponding communication array M is a Hankel array, we have $r_{f_M}(m_k) = \#\mathrm{sub}_k(M)$ for every $k \in [1..d]$.*

As a natural extension of two-party communication complexity we consider the case that the parties $P_1, \ldots, P_d$ are connected by a directed chain of the parties specified by a permutation $\pi : [1..d] \to [1..d]$, i.e. $P_{\pi(i)}$ can only send messages to $P_{\pi(i+1)}$ for $i \in [d-1]$. Let S^π be the size of an optimal protocol. More precisely, S^π is the number of possible communication sequences on the network in an optimal protocol.

We will now present a protocol of minimal size for a fixed chain network and functions f such that M_f is a Hankel array. During the computation the parties use the arrays $M_i \in \mathcal{M}(\sum_{j=1}^{i} m_{\pi(j)}, m_{\pi(i+1)}, \ldots, m_{\pi(d)})$, where M_i is the Hankel array defined by

$$M_i(y_i, \ldots, y_d) = M_f(z_1, \ldots, z_d)$$

for all $y_i \in [\sum_{j=1}^{i} m_{\pi(j)}], y_{i+1} \in [m_{\pi(i+1)}], \ldots, y_d \in [m_{\pi(d)}]$ and values $z_1 \in [m_1], \ldots, z_d \in [m_d]$ with $y_i = \sum_{j=1}^{i} z_{\pi(j)}$ and $y_j = z_{\pi(j)}$ for all $j \in [i+1..d]$. Furthermore, let $\Gamma_i(y_i)$ be the minimum value z such that $(M_i)_z^{(1)} = (M_i)_{y_i}^{(1)}$. The protocol works as follows: (1) $P_{\pi(1)}$ sends $\gamma_1 = \Gamma_1(x_{\pi(1)})$ to $P_{\pi(2)}$. (2) For $i \in [2..d-1]$, $P_{\pi(i)}$ receives γ_{i-1} from $P_{\pi(i-1)}$ and sends $\gamma_i = \Gamma_i(x_{\pi(i)} + \gamma_{i-1})$ to $P_{\pi(i+1)}$. (3) $P_{\pi(d)}$ receives γ_{d-1} from $P_{\pi(d-1)}$. Then $M_d(\gamma_{d-1} + x_{\pi(d)})$ gives the result of the function.

Theorem 8. *For a function f such that $M_f \in \mathcal{M}(m_1, \ldots, m_d)$ is a Hankel array the size of the protocol presented above is minimal.*

Note that the communication size S^π may depend on the order π of the parties on the chain. We will state that for $m_{\pi(i)} \leq m_{\pi(i+1)}$ for all $i \in [1..d-1]$ the ordering is optimal with respect to the communication size.

Theorem 9. *Let f be a function such that $M_f \in \mathcal{M}(m_1, \ldots, m_d)$ is a Hankel array and $\pi : [1..d] \to [1..d]$ be a permutation with $m_{\pi(i)} \leq m_{\pi(i+1)}$ for all $i \in [1..d-1]$. Then for every permutation $\pi' : [1..d] \to [1..d]$ $S^\pi(f) \leq S^{\pi'}(f)$.*

A second generalization of the two-party model is the simultaneous communication complexity $(C^{\|})$, where all parties can simultaneously write in a single round on a blackboard. This means that the messages send by each party do not depend on the messages send by the other parties. After finishing the communication round, each party has to be able to compute the result of the function (see e.g. [9]). For two-party communication it is well-known that $C^{\|}(f) = C^{A \to B}(f) + C^{B \to A}(f)$. Similarly, for the d-party case we have $C^{\|}(f) = \sum_{i \in [1..d]} \lceil \log \#\mathrm{sub}_i(M_f) \rceil$. Hence, if M_f is a Hankel array and if for some dimension $k \in [1..d]$ we have $\#\mathrm{sub}_k(M_f) \leq \min_{i \in [1..d]} m_i$, then by Lemmas 7 and 8 $C^{\|}(f) = d \cdot \lceil \log \Delta(f_{M_f}) \rceil$.

As a third generalization, we consider the case that in each round some party can write a message on a blackboard. The message may depend on messages that have been published on the board in previous rounds. We restrict the communication such that each party (except for the last one) publishes exactly one message on the blackboard, and in each round exactly one message is published. After finishing the communication rounds, at least one party has to be able to compute the result of the function. Let $S^{\square}$ be the corresponding size of an optimal protocol. Note that this model generalizes both of the previous models.

Theorem 10. *Let f be a function such that $M_f \in \mathcal{M}(m_1, \ldots, m_d)$ is a Hankel array and let $\pi : [1..d] \to [1..d]$ be a permutation such that $m_{\pi(i)} \leq m_{\pi(i+1)}$ for all $i \in [1..d-1]$. Then $S^{\pi}(f_M) = S^{\square}(f_M)$.*

7 Conclusions and Open Problems

In this paper we have investigated one-way communication complexity of functions for which the corresponding communication matrices are Hankel matrices. We have established some structural properties of such matrices. As a direct application, we have obtained a complete solution to the problem of how the communication direction in *deterministic* one-way communication protocols effects the communication complexity of symmetric Boolean functions. One possible direction of future research is to study other kinds of one-way communication such as *nondeterministic* and *randomized* for the class of symmetric functions.

Another interesting extension of the topic is to drop the restriction to *one-way* protocols and consider the deterministic two-way communication complexity of symmetric Boolean functions for both a bounded and an unbounded number of communication rounds. This particularly involves results about the computation of the rank of Hankel matrices. In addition, consequences for word combinatorics and OBDD theory may be of interest.

Acknowledgment. We would like to thank Ingo Wegener for his useful comment on the connection between one-way communication and OBDD theory.

References

1. F. Ablayev, Lower bounds for one-way probabilistic communication complexity and their application to space complexity. *Theoretical Comp. Sc.*, 157 (1996), 139–159.
2. A. Condon, L. Hellerstein, S. Pottle, and A. Wigderson, On the power of finite automata with both nondeterministic and probabilistic states. *SIAM J. Comput.*, 27 (1998), 739–762.
3. P. Ďuriš, J. Hromkovič, J.D.P. Rolim, and G. Schnitger, On the power of Las Vegas for one-way communication complexity, finite automata, and polynomial-time computations. Proc. *14th STACS*, Springer, 1997, 117–128.
4. J. E. Hopcroft and J. D. Ullman, *Formal Languages and Their Relation to Automata.* Addison-Wesley, Reading, Massachusetts, 1969.
5. J. Hromkovič, *Communication Complexity and Parallel Computing.* Springer, 1997.

6. I. S. Iohvidov, *Hankel and Toeplitz Matrices and Forms*. Birkhäuser, Boston, 1982
7. H. Klauck, On quantum and probabilistic communication: Las Vegas and one-way protocols. Proc. *32nd STOC*, 2000, 644–651.
8. I. Kremer, N. Nisan, and D. Ron, On randomized one-round communication complexity. *Computational Complexity*, 8 (1999), 21–49.
9. E. Kushilevitz and N. Nisan, *Communication Complexity*. Camb. Univ. Press, 1997.
10. K. Mehlhorn and E. M. Schmidt, Las Vegas is better than determinism in VLSI and distributed computing. Proc. *14th STOC*, 1982, 330–337.
11. I. Newman and M. Szegedy, Public vs. private coin flips in one round communication games. Proc. *28th STOC*, 1996, 561–570.
12. C. Papadimitriou and M. Sipser, Communication complexity. *J. Comput. System Sci.*, 28 (1984), 260–269.
13. I. Wegener, *The complexity of Boolean functions*. Wiley-Teubner, 1987.
14. I. Wegener, personal communication, April 2003.
15. A. C. Yao, Some complexity questions related to distributive computing. Proc. *11th STOC*, 1979, 209–213.

Circuits on Cylinders

Kristoffer Arnsfelt Hansen[1], Peter Bro Miltersen[1], and V. Vinay[2]

[1] Department of Computer Science, University of Aarhus, Denmark
{arnsfelt,bromille}@daimi.au.dk
[2] Indian Institute of Science, Bangalore, India.
vinay@csa.iisc.ernet.in

Abstract. We consider the computational power of constant width polynomial size cylindrical circuits and nondeterministic branching programs. We show that every function computed by a $\Pi_2 \circ \mathbf{MOD} \circ \mathbf{AC}^0$ circuit can also be computed by a constant width polynomial size cylindrical nondeterministic branching program (or cylindrical circuit) and that every function computed by a constant width polynomial size cylindrical circuit belongs to $\mathbf{ACC}^0$.

1 Introduction

In this paper we consider the computational power of constant width, polynomial size *cylindrical* branching programs and circuits.

It is well known that there is a rough similarity between the computational power of *width* restricted circuits and *depth* restricted circuits, but that this similarity is not a complete equivalence. For instance, the class of functions computed by a family of circuits of *quasi*-polynomial size and polylogarithmic depth is equal to the class of functions computed by a family of circuits of quasi-polynomial size and polylogarithmic width. On the other hand, the class of functions computed by a family of circuits of *polynomial* size and polylogarithmic width (non-uniform **SC**) is, in general, conjectured to be different from the class of functions computed by a family of circuits of polynomial size and polylogarithmic depth (non-uniform **NC**). For the case of *constant* depth and width, there is a provable difference in computational power; the class of functions computable by constant depth circuits of polynomial size, i.e. $\mathbf{AC}^0$, is a proper subset of the functions computable by constant width circuits (or branching programs) of polynomial size, the latter being, by Barrington's Theorem [1], the bigger class $\mathbf{NC}^1$. On the other hand, Vinay [9] and Barrington et al [2,3] showed that by putting a *geometric* restriction on the computation, the difference disappears: The class of functions computable by *upwards planar*, constant width, polynomial size circuits (or nondeterministic branching programs) is exactly $\mathbf{AC}^0$. Thus, both $\mathbf{AC}^0$ and $\mathbf{NC}^1$ can be captured by a constant width as well as by a constant depth circuit model. It is then natural to ask if one can similarly capture classes between $\mathbf{AC}^0$ and $\mathbf{NC}^1$ defined by various constant depth circuit models, such as $\mathbf{ACC}^0$ and $\mathbf{TC}^0$, by some natural constant width circuit or branching program model.

A. Lingas and B.J. Nilsson (Eds.): FCT 2003, LNCS 2751, pp. 171–182, 2003.
© Springer-Verlag Berlin Heidelberg 2003

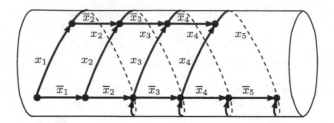

Fig. 1. A cylindrical branching program of width 2 computing PARITY.

Building upon the results in this paper, such a characterisation has recently been obtained for $\mathbf{ACC}^0$ [6]: The class of functions computed by *planar* constant width, polynomial size circuits is exactly $\mathbf{ACC}^0$.

In this paper we consider a slightly more relaxed geometric restriction than upwards planarity, yet more restrictive than planarity: We consider the functions computed by *cylindrical* polynomial size, constant width circuits (or nondeterministic branching programs). Informally (for formal definitions, see the next section), a layered circuit (branching program) is cylindrical if it can be embedded on the surface of a cylinder in such a way that each layer is embedded on a cross section of the cylinder (disjoint from the cross sections of the other layers), no wires intersect and all wires between two layers are embedded on the part of the cylinder between the two corresponding cross sections (see Fig. 1).

It is immediate that constant width polynomial size cylindrical branching programs have more computational power than constant width polynomial size upwards planar branching programs: The latter compute only functions in $\mathbf{AC}^0$ [2] while the former may compute PARITY (see Fig. 1). We ask what their exact computational power is and show that their power does not extend much beyond computing functions such as PARITY. Indeed, they can only compute functions in $\mathbf{ACC}^0$. To be precise, the first main result of this paper is the following *lower* bound on the power of cylindrical computation.

Theorem 1. *Every Boolean function computed by a polynomial size* $\mathbf{\Pi}_2 \circ \mathbf{MOD} \circ \mathbf{AC}^0$ *circuit is also computed by a constant width, polynomial size cylindrical nondeterministic branching program.*

By a $\mathbf{\Pi}_2 \circ \mathbf{MOD} \circ \mathbf{AC}^0$ circuit we mean a polynomial sized circuit with an AND gate at the output, a layer of OR gates feeding the AND gate, a layer of MOD_m gates (perhaps for many different constant-bounded values of m) feeding the OR gates and a (multi-output) $\mathbf{AC}^0$ circuit feeding the MOD gates. It is not known if the inclusion is proper. We prove Theorem 1 by a direct construction, generalising and extending the simple idea of Fig. 1.

Our second main result is the following *upper* bound on the power of cylindrical computation.

Theorem 2. *Every Boolean function computed by a constant width, polynomial size cylindrical circuit is in* $\mathbf{ACC}^0$.

Due to space constraints, the proof of Theorem 2 is omitted from this version of the paper. Instead we provide a proof of the weaker statement that cylindrical branching programs only compute functions in $\mathbf{ACC}^0$. We do however give an overview of a proof of Theorem 2. The full proof can be found in the technical report version of this paper [7].

The simulation is done (as were many previous results about constant width computation) by using the theory of finite monoids and the results of Barrington and Therien [4]. The notions of upwards planarity and of cylindricality share the property that all arcs flow along a common direction. This allows these notions to be captured by local constraints, which allows one to transfer the analysis of the restricted branching programs and circuits into an appropriate algebraic setting. Thus, we show the inclusion by relating the computation of cylindrical circuits to solving the word problem of a certain finite monoid and then show that this monoid is solvable.

A standard simulation shows that every Boolean function computed by a constant width, polynomial size cylindrical nondeterministic branching program is also computed by a constant width, polynomial size cylindrical circuit. For completeness, we describe this simulation in Proposition 3. Thus, one can exchange "cylindrical nondeterministic branching program" with "cylindrical circuit" and vice versa in our two main results.

Organisation of Paper. In Sect. 2, we formally define the notions of cylindrical branching program and circuits. We also give an overview of the algebraic tools we use. In Sect. 3, we show Theorem 1. In Sect. 4 we show the weaker version of Theorem 2 for cylindrical branching programs (instead of circuits), and in Sect. 5, we give an overview of the full proof of Theorem 2. We conclude with some discussions and open problems in Sect. 6.

2 Preliminaries

Bounded Depth Circuits. Let $A \subset \{0, \ldots, m-1\}$. Using the notation of Grolmusz and Tardos [5], a MOD_m^A gate takes n boolean inputs $x_1, \ldots, x_n$ and outputs 1 if $\sum_{i=1}^n x_i \in A$ (mod m) and 0 otherwise. We let **MOD** denote the family of MOD_m^A gates for all constant bounded m and all A. Similarly will **AND** and **OR** denote the family of unbounded fanin AND and OR gates.

If G is a family of boolean gates and $\mathcal{C}$ is a family of circuits we let $G \circ \mathcal{C}$ denote the class of polynomial size circuit families consisting of a G gate taking circuits from $\mathcal{C}$ as inputs.

$\mathbf{AC}^0$ is the class of functions computed by polynomial size bounded depth circuits consisting of NOT gates and unbounded fanin AND and OR gates. $\mathbf{ACC}^0$ is the class of functions computed when we also allow unbounded fanin MOD gates computing MOD_k for constants k. We will also use $\mathbf{AC}^0$ and $\mathbf{ACC}^0$ to denote the class of circuits computing the languages in the respective classes.

Cylindrical Branching Programs and Circuits. A digraph $D = (V, A)$ is called *layered* if there is a partition $V = V_0 \cup V_1 \cup \cdots \cup V_h$ such that all arcs of

A goes from *layer* V_i to the *next layer* V_{i+1} for some i. We call h the *depth* of D, $|V_i|$ the width of layer i and $k = \max |V_i|$ the width of D.

Let $[k]$ denote the integers $\{1, \ldots, k\}$. For $a, b \in [k]$ where $a \not\equiv b + 1$ (mod k) we define the (cyclic) interval $[a, b]$ to be the set $\{a, \ldots, b\}$ if $a \leq b$ and $\{a, \ldots, k\} \cup \{1, \ldots, b\}$ if $a > b$. Furthermore let $(a, b) = [a, b] \setminus \{a, b\}$, and let $(a, b) = [k] \setminus \{a, b\}$ if $a \equiv b + 1$ (mod k).

Let D be a layered digraph in which all layers have width k. We will assume the nodes in each layer numbered $1, \ldots, k$, and refer to nodes by these numbers. Then, D is called a *cylindrical* if the following property is satisfied: For every pair of arcs going from layer l to layer $l + 1$ connecting node a to node c and node b to node d the following must hold: Nodes in the interval (a, b) of layer l can only connect to nodes in the interval $[c, d]$ of layer $l + 1$ and nodes in the interval (b, a) of layer l can only connect to nodes in the interval $[d, c]$ of layer $l + 1$.

Notice this is equivalent of saying that nodes in the interval (c, d) of layer $l + 1$ can only connect to nodes in the interval $[a, b]$ of layer l and nodes in the interval (d, c) of layer $l + 1$ can only connect to nodes in the interval $[b, a]$ of layer l.

A *nondeterministic branching program*[1] is an acyclic digraph where all arcs are labelled by either a literal, i.e. a variable or a negated variable, or a boolean constant, and an initial and a terminal node. An input is accepted if and only if there is a path from the initial node to the terminal node in the digraph that results from substituting constants for the literals according to the input and then deleting arcs labelled by 0.

We will only consider branching programs in *layered form*, that is, viewed as a digraph it is layered. We can assume that the initial node is in the first layer and the terminal node in the last layer, and furthermore that these are the only nodes incident to arcs in these layers. We can also assume that all layers have the same number of nodes, by the addition of dummy nodes.

By a *cylindrical branching program* we will then mean a bounded-width nondeterministic branching program in layered form, which is cylindrical when viewed as a digraph.

A *cylindrical circuit* is a circuit consisting of fanin 2 AND and OR gates and fanin 1 COPY gates, which when viewed as a digraph is a cylindrical digraph. Inputs nodes can be literals or boolean constants. The output gate is in the last layer. We can assume that all layers have the same number of nodes by adding dummy input nodes to the first layer and dummy COPY gates to the other layers.

A standard simulation of nondeterministic branching programs by circuits extends to cylindrical branching programs and cylindrical circuits. We give the details for completeness.

---

[1] Our definition deviates slightly from the usual definition where nodes rather than edges are labelled by literals and unlabelled nodes serve as special nondeterministic "choice"-nodes, but it is easily seen to be polynomially equivalent - also in the cylindrical case - and it is more convenient for us.

Proposition 3. *Every function computed by a width k, depth d cylindrical branching program is also computed by a width $O(k)$, depth $O(d \log k)$ cylindrical circuit*

Proof. Replace every node in the branching program by an OR-gate. Replace each arc, going from, say, node u to node v and labelled with the literal x, with a new AND-gate taking two inputs, gate u and the literal x and with the output of the AND-gate feeding gate v.

This transformation clearly preserves the cylindricality of the graph. Also, the width of the circuit is linear in the width of the branching program. The resulting OR-gates may have fan-in bigger than two. We replace each such gate with a tree of fan-in two OR-gates, preserving the width and blowing up the depth by at most a factor of $O(\log k)$.   □

Monoids and Groups. Let x and y be elements of a group G. The *commutator* of x and y is the element $x^{-1}y^{-1}xy$. The subgroup $G^{(1)}$ of G generated by all of the commutators in G is called the *commutator subgroup* of G. In general, let $G^{(i+1)}$ denote the commutator subgroup of $G^{(i)}$. G is *solvable* if $G^{(n)}$ is the trivial group for some n. It follows that an Abelian group, and in particular a cyclic group, is solvable.

A *monoid* is a set M with an associative binary operation and a two sided identity. A subset G of M is a *group in* M if it is a group with respect to the operation of M. Note that a group G in M is not necessarily a submonoid of M as the identity element of G may not be equal to the identity element of M. M is called *solvable* if every group in M is solvable. The *word problem* for a finite monoid M is the computation of the product $x_1 x_2 \ldots x_n$ given $x_1, x_2, \ldots, x_n$ as input. A theorem by Barrington and Therien [4] states that the word problem for a solvable finite monoid is in **ACC^0**.

3   Simulation of Bounded Depth Circuits by Cylindrical Branching Programs

In this section, we prove Theorem 1. As a starting point, we shall use the "only if" part of the following correspondence established by Vinay [9] and Barrington et al [2]. We include here a proof of the "only if" part for completeness.

Theorem 4. *A language is in **AC^0** if and only if it is accepted by a polynomial size, constant width upwards planar branching program.*

Here an *upwards planar* branching program is a layered branching program satisfying, that for every pair of arcs going from layer l to layer $l+1$ connecting node a to node c and node b to node d, if $a < b$ then $c \le d$.

We need some simple observations. First observe that if we can simulate a class of circuits $\mathcal{C}$ with upwards planar (cylindrical) branching programs, then we can also simulate **AND** $\circ \, \mathcal{C}$ by upwards planar (cylindrical) branching programs by simply concatenating the appropriate branching programs.

Another way to combine branching programs is by *substitution* where we simply substitute a branching program for the edges corresponding to a particular literal. The effect of this is captured in the following lemma.

Lemma 5. *If $f(x_1, \ldots, x_n)$ is computed by an upwards planar (cylindrical) branching program of size s_1 and width w_1 and $g_1, \ldots, g_n$ and $\overline{g}_1, \ldots, \overline{g}_n$ are computed by upwards planar branching programs, each of size s_2 and width w_2 then $f(g_1, \ldots, g_n)$ is computed by an upwards planar (cylindrical) branching program of size $O(s_1 w_1 s_2)$ and width $O(w_1^2 w_2)$.*

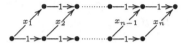

Fig. 2. An upwards planar branching program computing OR.

Combining the above observations with the construction in Fig. 2, simulating an OR gate, we have established the "only if" part of Theorem 4.

Simulation of a MOD_m^A gate can be done as shown in Fig. 3 if one disregards the top nodes in the first and last layers and modifies the connections between the second-to-last layer to take the set A into account. Thus, combining this construction with Lemma 5, the "only if" part of Theorem 4 and the closure of cylindrical branching programs under polynomial fan-in AND, we have established that we can simulate $\textbf{AND} \circ \textbf{MOD} \circ \textbf{AC}^0$ circuits by bounded width polynomial size cylindrical circuits.

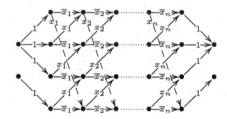

Fig. 3. A cylindrical branching program fragment for MOD_4.

The construction as shown in Fig. 3 has actually more use, by seeing it as computing elements of M_2, where M_2 is the monoid of binary relations on [2]. The general construction of a branching program fragment for MOD_m^A taking n inputs is as follows: Without loss of generality we can assume that $|A| = 1$ and in fact $A = \{0\}$ since we aim for simulating $\textbf{OR} \circ \textbf{MOD}$. The branching program fragment will have $n + 3$ layers. The first and last layer of width 2 and the middle layers of width m. The top node in the first layer has arcs to all nodes but node 1 and the bottom node has an arc to node 1. The top node in the last layer has arcs from all nodes but the one in A and the bottom node has an arc from this node. The nodes in the middle layers represent the sum of a prefix of the input modulo m in the obvious way. Consider now the elements of M_2 shown in Fig. 4. The branching program fragment just described corresponds to (a) and (b) for $m = 2$ and $m > 2$ respectively, when the simulated MOD gate evaluates to 0. In both cases, the fragment correspond to (c) when the simulated MOD gate evaluates to 1.

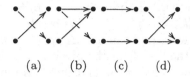

(a)   (b)   (c)   (d)

Fig. 4. Some elements of M_2.

We can now describe our construction for simulating $\mathbf{OR} \circ \mathbf{MOD}$ circuits. The construction interleaves branching program fragments for (d) between the branching program fragments for the MOD gates. This can be seen as a way of "short circuiting" the branching program in the case that one of the MOD gates evaluate to 1. Finally we add layers at both ends picking out the appropriate nodes for the simulation. The entire construction is shown in Fig. 5. The correctness can easily be verified.

The simulation of $\mathbf{OR} \circ \mathbf{MOD}$ circuits, the "only if" part of Theorem 4, Lemma 5, and the closure of cylindrical branching programs under polynomial fan-in AND, together completes the proof of Theorem 1.

Fig. 5. A cylindrical branching program computing $\mathrm{MOD} \vee \cdots \vee \mathrm{MOD}$.

4   Simulation of Cylindrical Branching Programs by Bounded Depth Circuits

In this section, we compensate for the omitted proof of Theorem 2 sketched in the next section, by giving a simpler (but similar) proof of the weaker result that constant width polynomial size cylindrical nondeterministic branching programs compute only functions in $\mathbf{ACC}^0$.

In fact, we shall prove that for fixed k the following "branching program value problem" BPV_k is in $\mathbf{ACC}^0$: Given a width k cylindrical branching program *and* a truth assignment to its variables, decide if the program accepts. As any function computed by width k cylindrical polynomial size branching program clearly is a Skyum-Valiant projection [8] of BPV_k, we will be done.

We shall prove that BPV_k is in $\mathbf{ACC}^0$ by showing that it reduces, by an $\mathbf{AC}^0$ reduction, to the word problem of the monoid M_k we define next. Then, we show that the monoid M_k is solvable, and since this implies, by the result of Barrington and Therien [4] that the word problem for M_k is in $\mathbf{ACC^0}$, our proof will be complete.

We define M_k to be the monoid of *binary relations* on $[k]$ which capture the calculation of width k branching programs embedded on a cylinder in the following sense: M_k is the monoid generated by all the relations which express how arcs can travel between two adjacent layers in an width k cylindrical digraph. The monoid operation is the usual composition operation of binary relations, i.e., if $A, B \in M_k$ and $x, y \in [k]$, $xABy \Leftrightarrow \exists z : xAz \wedge zBy$.

BPV_k reduces to the word problem for M_k by the following $\mathbf{AC^0}$ reduction: Substitute constants for the literals in the branching program according to the truth assignment. Consider now the cylindrical digraph D consisting only of arcs which have the constant 1 associated. Then, the branching program accepts the input given if and only if there is a path from the initial node in the first layer to the terminal node in the last layer of D. We can decide this by simply decomposing D into a sequence $A_1, A_2, \ldots, A_h$ of elements from M_k, computing the product $A = A_1 A_2 \cdots A_h$ and checking whether this is different from the zero element of M_k.

Thus, we just need to show that M_k is solvable. Our proof is finished by the following much stronger statement.

Proposition 6. *All groups in M_k are cyclic.*

Proof. Let $G \subseteq M_k$ be a group with identity E. Let $A \in G$ and let R be the set of all x such that xEx. As will be shown next it will be enough to consider elements of R to capture the structure of A.

Let $x \in R$. Since $AA^{-1} = E$ there exists z such that xAz and $zA^{-1}x$. Since $A^{-1}A = E$ it follows zEz, that is, $z \in R$. Hence there exists a function $\pi_A : R \to R$ such that

$$\forall x : xA\pi_A(x) \wedge \pi_A(x)A^{-1}x$$

To see that A is completely described by by π_A, we define a relation $\hat{A}$ on $[k]$ such that $x\hat{A}y \Leftrightarrow \pi_A(x) = y$. That is, $\hat{A}$ is just π_A viewed as a relation. Since $\hat{A} \subseteq A$ it follows $E\hat{A}E \subseteq EAE = A$. Conversely let xAy. Since $E^kA = A$ there exists $z \in R$ such that xEz and zAy. Since $\pi_A(z)A^{-1}z$ we get $\pi_A(z)Ey$. That is xEz, $z\hat{A}\pi_A(z)$ and $\pi_A(z)Ey$. Thus $xE\hat{A}Ey$. Hence we obtain that $A = E\hat{A}E$.

We would like to have both that π_A is a permutation and that $\{\pi_A | A \in G\}$ is a group. This is in general not true, since E can be any transitive relation in M_k.

To obtain this we will first simplify the structure of the elements of G using the following equivalence relation on [k] defined by

$$x \sim y \Leftrightarrow (xEy \wedge yEx) \vee x = y.$$

Let $A \in G$. If $x \sim x'$ and $y \sim y'$ then $xAy \Leftrightarrow x'Ay'$, since $EAE = A$. Thus A gives rise to a relation $\tilde{A}$ on $[k]/\sim$ where $xAy \Leftrightarrow [k]_x \tilde{A}[k]_y$ and it will follow that $\{\tilde{A} | A \in G\}$ is an isomorphic group of G.

For this we need to show that $\widetilde{AB} = \tilde{A}\tilde{B}$. This follows since $[k]_x\widetilde{AB}[k]_z \Leftrightarrow xABz \Leftrightarrow \exists y : xAy \wedge yBz \Leftrightarrow \exists y : [k]_x\tilde{A}[k]_y \wedge [k]_y\tilde{B}[k]_z \Leftrightarrow [k]_x\tilde{A}\tilde{B}[k]_z$

We can find an isomorphic copy of this group in M_k as follows. Choose for each equivalence class $[k]_x$ a representative $r([k]_x)$ in $[k]_x$. Define a relation C on $[k]$ such that $xCy \Leftrightarrow x = y = r([k]_x)$. Thus $\forall x : r([k]_x)Cr([k]_x)$. Let $\sigma : G \to M_k$ be given by $\sigma(A) = CAC$. Then $\sigma(G)$ is the desired isomorphic copy of G. We can thus assume that the equivalence classes with respect to $\sim$ are of size 1.

We now return to the study of π_A. The following property, that for $x, y \in R$ it holds that $xEy \Leftrightarrow \pi_A(x)E\pi_A(y)$, is satisfied:

If xEy then $\pi_A(x)A^{-1}y$ since $A^{-1}E = A^{-1}$. As $A^{-1}A = E$ it follows that $\pi_A(x)E\pi_A(y)$.

Conversely if $\pi_A(x)E\pi_A(y)$ then $xA\pi_A(y)$ since $xA\pi_A(x)$ and $AE = A$. As $\pi_A(y)A^{-1}y$ and $AA^{-1} = E$ it then follows that xEy.

We can now conclude that π_A is a permutation on R: If $\pi_A(x) = \pi_A(y)$ then $\pi_A(x) \sim \pi_A(y)$ so $x \sim y$, that is, $x = y$. Also π_A is uniquely defined : Assume $\hat{\pi}_A : R \to R$ satisfies

$$\forall x : xA\hat{\pi}_A(x) \wedge \hat{\pi}_A(x)A^{-1}x$$

Let $x \in R$. We then obtain $\pi_A(x) \sim \hat{\pi}_A(x)$ so $\pi_A(x) = \hat{\pi}_A(x)$. Hence $\pi_A = \hat{\pi}_A$.

Now we can conclude that $\{\pi_A | A \in G\}$ is a permutation group which is isomorphic to G. For this we need to show that $\pi_{AB} = \pi_B \circ \pi_A$.

Let $x \in R$. Since $xA\pi_A(x)$ and $\pi_A(x)B\pi_B \circ \pi_A(x)$ it follows $xAB\pi_B \circ \pi_A(x)$. Since $\pi_B \circ \pi_A(x)B^{-1}\pi_A(x)$ and $\pi_A(x)A^{-1}x$ it follows $\pi_B \circ \pi_A(x)B^{-1}A^{-1}x$, i.e. $\pi_B \circ \pi_A(x)(AB)^{-1}x$

Since π_{AB} is uniquely defined the result follows.

To show that $\{\pi_A | A \in G\}$ is cyclic we need the following fact, which easily follows from the definition of cylindricality

Fact: Let A be a relation which can be directly embedded on a cylinder. Let $p_1 < p_2 < \ldots p_m$ and $q_1 < q_2 < \cdots < q_m$ and π a permutation on $[m]$ such that $\forall i : p_iAq_{\pi(i)}$. Then π is in the cyclic group of permutations on $[m]$ generated by the cycle $(1\ 2\ldots m)$.

Now let $r_1 < r_2 < \cdots < r_m$ be the elements of R. Write $A \in G$ as $A = A_1A_2\ldots A_h$ where the A_i's can be directly embedded on the cylinder. Since $r_iA\pi_A(r_i)$ we have for all i, elements of $[k]$, $r_i = q_i^0, q_i^1, \ldots, q_i^h = \pi_A(r_i)$ such that $q_i^jA_{j+1}q_i^{j+1}$. For fixed j all the q_i^j's are distinct. If not we would have i_1 and i_2 such that $r_{i_1}A\pi_A(r_{i_2})$ and $r_{i_2}A\pi_A(r_{i_1})$. But then since $\pi_A(r_{i_1})A^{-1}r_{i_1}$ and $\pi_A(r_{i_2})A^{-1}r_{i_2}$ we then get $r_{i_1}Er_{i_2}$ and $r_{i_2}Er_{i_1}$. That is $r_{i_1} \sim r_{i_2}$ which implies $r_{i_1} = r_{i_2}$. Now by the fact and induction on h we have a permutation π in the cyclic group generated by the cycle $(1\ 2\ldots m)$ such that $r_{\pi(i)} = \pi_A(r_i)$. Thus π_A is in the cyclic group generated by the cycle $(r_1\ r_2\ldots r_m)$ and we can conclude that G is cyclic. $\qquad\square$

5   Simulation of Cylindrical Circuits by Bounded Depth Circuits

In this section we provide an overview of the proof of Theorem 2 which can be found in the technical report version of this paper [7].

The rough outline is similar to that of the last section. For fixed k we consider the following "circuit value problem" CV_k: Given a width k cylindrical circuit *and* a truth assignment to its input variables, decide if the circuit evaluates to 1. This is then reduced, by an **AC**^{0} reduction, to the word problem of the monoid $\hat{N}_k$ defined next, which will be proved to be solvable. By the result of Barrington and Therien [4] it then follows that CV_k is in **ACC**^{0}.

Consider a width k cylindrical circuit C with k input nodes, all placed in the first layer. We can view this as computing a function mapping $\{0,1\}^k$ to $\{0,1\}^k$ by reading off the values of the nodes in the last layer. We let $\hat{N}_k$ be the monoid of such functions mapping $\{0,1\}^k$ to $\{0,1\}^k$.

This provides the base for the desired **AC**^{0} reduction in the following way: Given an instance of the circuit value problem we substitute constants for the variables according to the truth assignment and then view each layer of the circuit as an element of $\hat{N}_k$ by preceding it with k input nodes. By computing the product of these and evaluating it on the constants given to the first layer, the desired result is obtained.

The monoid $\hat{N}_k$ is shown to be solvable like in the previous section, by proving that all its groups are cyclic. A first step to obtain this is to eliminate constants from the circuits correspond to group elements. Let N_k be the monoid of functions mapping $\{0,1\}^k$ to $\{0,1\}^k$ which are computed by width k cylindrical circuits with k variable input nodes, all placed in the first layer, with constant input nodes disallowed. It is then proved that every group in $\hat{N}_k$ is isomorphic to a group in N_k.

The tool for studying N_k will be an identification of input vectors in $\{0,1\}^k$ with its set of *maximal 1-intervals* as considered in [3], only here we consider *cyclic* intervals. For example is the vector 1010011011 identified with the set of intervals $\{[3,3],[6,7],[9,1]\}$.

Now consider a group G in N_k with identity e, and let $f \in G$. Since $e \circ e = e$ we get that e is the identity mapping on the image of e, $\operatorname{Im} e$. Thus any $f \in G$ is a permutation of $\operatorname{Im} e$, since $f \circ f^{-1} = f^{-1} \circ f = e$ and $e \circ f = f$. Also since $f \circ e = f$ it follows that f is completely described by its restriction to $\operatorname{Im} e$.

The fact that f has an inverse on $\operatorname{Im} e$, is shown to imply that f must preserve the number of intervals in any $x \in \operatorname{Im} e$. The crucial property employed here, is the monotonicity of the gate operations. This furthermore implies that f is completely described by its restriction to the set $\mathcal{I}$ of vectors in $\operatorname{Im} e$ consisting of only a single interval.

Next, using the natural partial order on $\mathcal{I}$ given by lifting the order $0 < 1$ pointwise, one can decompose $\mathcal{I}$ into antichains, onto which $f \in G$ is easy to describe. In fact f is a *cyclic shift* on each of these antichains. Finally by relating these cyclic shifts one can conclude that G is a cyclic group.

6 Conclusion and Open Problems

We have located the class of functions computed by small constant width cylindrical circuits (or nondeterministic branching programs) between $\Pi_2 \circ \mathbf{MOD} \circ \mathbf{AC}^0$ and $\mathbf{ACC}^0$. It would be very interesting to get an exact characterisation of the power of cylindrical circuits and branching programs in terms of bounded depth circuits. It is not known whether $\Pi_2 \circ \mathbf{MOD} \circ \mathbf{AC}^0$ is different from $\mathbf{ACC}^0$ and this seems a difficult problem to resolve, so we cannot hope for an unconditional separation of the power of cylindrical circuits from $\mathbf{ACC}^0$. On the other hand, it seems difficult to generalise the simulation of $\Pi_2 \circ \mathbf{MOD} \circ \mathbf{AC}^0$ by cylindrical branching programs to handle more than one layer of MOD gates and we tend to believe that such a simulation is in general not possible. Thus, one could hope that by better understanding the structure of the monoids we have considered in this paper, it would be possible to prove an upper bound seemingly better than $\mathbf{ACC}^0$, such as for instance $\mathbf{AC}^0 \circ \mathbf{MOD} \circ \mathbf{AC}^0$.

It would also be interesting to separate the power of branching programs from the power of circuits. As circuits can be trivially negated while preserving cylindricality, we immediately have that not only $\Pi_2 \circ \mathbf{MOD} \circ \mathbf{AC}^0$ but also $\Sigma_2 \circ \mathbf{MOD} \circ \mathbf{AC}^0$ can be simulated by small constant width cylindrical circuits. On the other hand, we don't know if $\Sigma_2 \circ \mathbf{MOD} \circ \mathbf{AC}^0$ can be simulated by small constant width cylindrical branching programs. Note that in the upwards planar case, both models capture $\mathbf{AC}^0$ and in the geometrically unrestricted case, both models capture $\mathbf{NC}^1$, so it is not clear if one should a priori conjecture the cylindrical models to have different power. Note that if the models have identical power then they can simulate $\mathbf{AC}^0 \circ \mathbf{MOD} \circ \mathbf{AC}^0$. This follows from the fact that the branching program model is closed under polynomial fan-in AND while the circuit model is closed under negation.

An interesting problems concerns the blowup of width to depth when going from a cylindrical circuit or branching program to an $\mathbf{ACC}^0$ circuit. Our proof does not yield anything better than a doubly exponential blowup. Again, by better understanding the structure of the monoids we have considered, one could hope for a better upper bound.

Acknowledgements. The first two authors are supported by BRICS, Basic Research in Computer Science, a Centre of the Danish National Research Foundation.

References

1. D. A. Barrington. Bounded-width polynomial-size branching programs recognize exactly those languages in $\mathbf{NC}^1$. *J. Comput. System Sci.*, 38(1):150–164, 1989.
2. D. A. M. Barrington, C.-J. Lu, P. B. Miltersen, and S. Skyum. Searching constant width mazes captures the $\mathbf{AC}^0$ hierarchy. In *Proceedings of the 15th Annual Symposium on Theoretical Aspects of Computer Science*, pages 73–83, 1998.

3. D. A. M. Barrington, C.-J. Lu, P. B. Miltersen, and S. Skyum. On monotone planar circuits. In *14th Annual IEEE Conference on Computational Complexity*, pages 24–31. IEEE Computer Society Press, 1999.
4. D. A. M. Barrington and D. Thérien. Finite monoids and the fine structure of $\mathbf{NC}^1$. *Journal of the ACM (JACM)*, 35(4):941–952, 1988.
5. V. Grolmusz and G. Tardos. Lower bounds for (mod$p$ – mod$m$) circuits. *SIAM Journal on Computing*, 29(4):1209–1222, Aug. 2000.
6. K. A. Hansen. Constant width planar computation characterizes $\mathbf{ACC}^0$. Technical Report 25, Electronic Colloquium on Computational Complexity, 2003.
7. K. A. Hansen, P. B. Miltersen, and V. Vinay. Circuits on cylinders. Technical Report 66, Electronic Colloquium on Computational Complexity, 2002.
8. S. Skyum and L. G. Valiant. A complexity theory based on boolean algebra. *Journal of the ACM (JACM)*, 32(2):484–502, 1985.
9. V Vinay. Hierarchies of circuit classes that are closed under complement. In *11th Annual IEEE Conference on Computational Complexity*, pages 108–117. IEEE Computer Society, 1996.

Fast Perfect Phylogeny Haplotype Inference

Peter Damaschke

Chalmers University, Computing Sciences, 41296 Göteborg, Sweden
ptr@cs.chalmers.se

Abstract. We address the problem of reconstructing haplotypes in a population, given a sample of genotypes and assumptions about the underlying population. The problem is of major interest in genetics because haplotypes are more informative than genotypes when it comes to searching for trait genes, but it is difficult to get them directly by sequencing. After showing that simple resolution-based inference can be terribly wrong in some natural types of population, we propose a different combinatorial approach exploiting intersections of sampled genotypes (considered as sets of candidate haplotypes). For populations with perfect phylogeny we obtain an inference algorithm which is both sound and efficient. It yields with high propability the complete set of haplotypes showing up in the sample, for a sample size close to the trivial lower bound. The perfect phylogeny assumption is often justified, but we also believe that the ideas can be further extended to populations obeying relaxed structural assumptions. The ideas are quite different from other existing practical algorithms for the problem.

1 Introduction

Somatic cells of diploid organisms such as higher animals and plants contain two copies of genetic material, in pairs of homologous chromosomes. The material on an arbitrary but fixed part of a single chromosome is called a *haplotype*. Formally we may describe a haplotype as a vector $(a_1, \ldots, a_s)$ where s is the number of sites considered, and a_i is the genetic data at site i. Here the term *site* can refer to a gene, a short subsequence, or even a single nucleotide. The a_i are called *alleles*. The vector of unordered pairs $(\{a_1, b_1\}, \ldots, \{a_n, b_n\})$ resulting from haplotypes $(a_1, \ldots, a_n)$ and $(b_1, \ldots, b_n)$ on homologous chromosomes is called a *genotype*. A site is *homozygous* if $a_i = b_i$, and *heterozygous* (or *ambiguous*) if $a_i \neq b_i$. The terminology in the literature is not completely standardized, in the present paper we use it as introduced above.

Usual sequencing methods yield only genotypes but not the pairs of haplotypes they are built from, the so-called phase information. Haplotyping techniques exist, but they are much more expensive, and it is expected that this relation will stay so for quite many years. On the other hand, haplotype data is often needed for analyzing the background of hereditary dispositions.

For example, a hereditary trait often originates from a single mutation on a chromosome that has been transmitted over generations, and further silent

A. Lingas and B.J. Nilsson (Eds.): FCT 2003, LNCS 2751, pp. 183–194, 2003.
© Springer-Verlag Berlin Heidelberg 2003

mutations (without effect) supervened. This way the trait is associated with a certain subset of haplotypes. If one wants to find the relevant mutation amongst the silent ones, it is useful to recognize haplotypes of affected individuals and to search the corresponding chromosomes only. Genotype information alone is less specific, also for the purpose of prediction of traits. Other applications include questions from population dynamics. Therefore it is important to reconstruct haplotypes from observed genotypes.

A genotype with $k > 0$ ambigous sites can be explained by 2^{k-1} distinct haplotype pairs, and reconstruction is impossible if we consider isolated genotypes only. However if we have a large enough genotype sample from a population and a proper assumption about the structure of this population, we may be able to infer the haplotypes with high confidence. One of them is:

Definition 1. *A population fulfills the random mating assumption (is in Hardy-Weinberg equilibrium) if the haplotypes form pairs at random, according to their frequencies in the population, i.e. the probability to have a specific ordered pair of haplotypes in a randomly chosen genotype is simply the product of their frequencies.*

Although this is not perfectly true in real populations, due to mating preferences and spatial structure, the behaviour of an inference algorithm in such a setting says much about its appropriateness.

We focus attention on the *biallelic* case where each a_i has two possible values which we may denote by Boolean constants 0 and 1. This is not a severe restriction because there exist only two alleles per locus if mutations affect every locus only once, which is typically the case. For notational convenience we write haplotypes as binary strings and genotypes as ternary strings where 0,1, and 2 stand for $\{0, 0\}$, $\{1, 1\}$, and $\{0, 1\}$, respectively.

Definition 2. *For $\beta \subset \{0, 1, 2\}$, the β-set of a genotype or haplotype is the set of all sites whose value is in β. We omit set parantheses in β.*

Sometimes it is convenient to rename the alleles such that some specific haplotype is w.l.o.g. the zero string $00\ldots0$. Note that the 2-sets of genotypes are invariant under this renaming.

One may also think of haplotypes as vertices of the s-dimensional cube of Boolean vectors of length s. Having this picture in mind, we identify a genotype with the subcube c having the generating haplotype pair as one of its diagonals, i.e. with the *set* of haplotypes $a \in c$. This relation holds true iff $a_i = c_i$ for all i in the 0,1-set of c. We will use the notations interchangeably.

Related literature and our contribution. We try to give an overview of various attempts, and we apologize for any omission.

In [2], the following resolution (or subtraction) method has been proposed. Assume that our sample contains a genotype with no or one ambigous site. Then we immediately know one or two haplotypes, respectively, for sure. They are called resolved haplotypes. For any resolved haplotype $a = a_1 \ldots a_n$ and any genotype $c = c_1 \ldots c_n$ such that $a_i \neq 2$ implies $c_i = a_i$, it is possible that c is composed of a and another haplotype b defined by $b_i = a_i$ for $c_i \neq 2$, and $b_i = 1 - a_i$ for $c_i = 2$. We call b the *complement* of a in c. The classical resolution algorithm simply assumes that c is indeed built from a and b, it considers b as a new resolved haplotype, and removes c as a resolved genotype from the sample, and so on, until no further resolution step can be executed.

Objections against this heuristic have been noticed already in [2]: A minor problem is that we may not find a resolved haplotype to start with. A large enough sample will contain some homozygous genotypes w.h.p. (Here and henceforth, w.h.p. means: with high probability.) More seriously, any resolution step may be wrong, i.e. the subcube c containing vertex a may actually be formed by a different haplotype pair. This is called an *anomalous match*. Even worse, further resolution steps starting from a false haplotype b may cause a cascade of such errors. The rush removal of resolved genotypes is yet another source of errors, since the same genotype may well be formed by different haplotype pairs in a population.

Resolution has been further studied in [7,8]. The output depends on the ordering the steps are performed, and the "true" ordering must resolve all genotypes. Unfortunately, the corresponding maximization problem to resolve as many genotypes as possible is Max-SNP hard [7], and moreover, a big number of resolved genotypes does not guarantee that the inferred haplotypes are correct. (There exist some conjectures, heuristic reasoning, and experimental results around this question, but apparently without rigorous theoretical foundation.) More advanced resolution algorithms solve some integer programming problem on a resolution graph constructed from the sample, and they can find good results in experiments [8], but still the reliabilty question remains.

A completlely different approach to haplotype inference is Bayesian statistics under the random mating assumption. We refer to [4,11,12]. Although accuracy has certainly been noticed as an issue, it is not obvious how reliable every single haplotype in output sets of the various algorithms actually is.

In the present paper we address the question of reliable combinatorial haplotype inference methods. For haplotype populations having a perfect phylogenetic tree (definitions are given later) we show that a combinatorial algorithm which is different from resolution is able to infer all haplotypes w.h.p. from a large enough sample, whereas resolution is provably bad.

The perfect phylogeny assumption has first been used for haplotype inference in [9], resulting in an almost linear but very complicated algorithm (via reduction to the graph realization problem). Slower but practical and elegant algorithms have been discovered shortly thereafter independently by [1,3], and they proved useful on real data. The work presented here (including the principal idea to exploit perfect phylogeny structure) was mainly finished before we became aware of [9,1,3]. We propose another elementary algorithm. It happens to be quite different from the algorithms in [1,3] which work with pairs of sites. Our approach is "orthogonal" so to speak, as it works with pairs of genotypes. This can be advantageous for the running time, since only certain pairs of genotypes have to be considered. It should be noticed that the algorithms in [1,3] output a representation of all consistent haplotyping results, whereas our primary goal is to output the haplotypes that can be definitely determined. We also study the size of a random sample that leads to a unique result w.h.p. This does not mean that the method gives a result only in the latter case: It still resolves many haplotypes if fewer genotypes are available, and it is incremental in the sense that new genotype data can be easily incorporated. Due to the different approach and focus, our expected time complexity is not directly comparable to the previous bounds, but under some circumstances it seems to be favourable. (Details follow later.)

We believe that our approach complements the arsenal of haplotype inference methods. It seems that the ideas can be generalized to more complicated populations.

2   Preliminaries

In addition to the notion already introduced, we clarify some more terminology as we use it in the paper.

Definition 3. *The genotype formed by haplotypes a and b (where $a = b$ is allowed) is simply denoted ab. Haplotype b is called the* complement *of a in ab, and vice versa.*

Recall that we sometimes consider genotypes as sets (subcubes) of haplotypes, and note that each haplotype has a unique complement in a genotype.

Definition 4. *A population is a set P of haplotypes, equipped with a frequency of each haplotype in P. Clearly, the frequencies sum up to 1. A sample from P is a multiset G of genotypes (not haplotypes!) ab with $a, b \in P$. (The same genotype may appear several times in G.) An* anomalous match, *with respect to G, is a triple of haplotypes a, b, c such that $a, b \in P$, $ab \in G$, $c \in ab$, but the complement of c in ab is not in P.*

An anomalous match can cause a wrong resolution step, if c is used to resolve ab. (We do not demand $c \in P$ since c may already be result of an earlier false resolution step.)

Since very rare haplotypes are hard to find but, on the other hand, are also of minor significance, we take a parameter n and aim at finding those haplotypes with frequency at least $1/n$. Suppose that n is chosen large enough such that these haplotypes cover the whole P, up to some negligible fraction.

In the following we adopt the random mating assumption and make some technical simplifications for the analysis later on. We emphasize that they are not substantial and do not affect the algorithm itself. Let f_i $(i = 1, 2, \ldots)$ denote the haplotype frequencies. We fix some parameter n and aim at identifying all haplotypes with $f_i \geq 1/n$, where n is chosen large enough such that the $f_i < 1/n$ sum up to a negligible fraction. In the worst case P contains n different haplotypes, all with $f_i = 1/n$. In general we will for simplicity pretend that all f_i are (roughly) integer multiples of $1/n$. Then a haplotype of frequency f_i is considered as a set of $f_i n$ haplotypes which are equal as strings. Henceforth, if we speak of "k haplotypes" or "k genotypes", we do not require that they are pairwise different. We say "identical" and "distinct" when we refer to these copies of haplotypes and genotypes, and "equal" and "different" when we refer to their string values. The probability that a randomly chosen genotype yields a resolved haplotype is $1/n$ in the worst case.

Definition 5. *The* sample graph *of G has vertex set P (consisting of n distinct haplotypes) and edge set G, that is: An edge joins two haplotypes if they produced the genotype corresponding to that edge.*

A sample graph may contain loops (completlely homozygous genotypes) and multiple edges (if the same haplotype pair is sampled several times). Note that the sample graph is of course not "visible", otherwise we would already know P.

Our focus is on asymptotic results, so we consider sums of sufficiently many independent random variables, being sharply concentrated around their expected values, such that we may simply take these expectations for deterministic values.

A well-known result on the coupon collector's problem says that, if we choose one of k objects at random then, after $O(k \log k)$ such trials, we have w.h.p. touched every object at least once (see e.g. [10]). Consequently, if we sample $O(n^2 \log n)$ genotypes then w.h.p. all haplotypes are trivially resolved, because all vertices in the sample graph get loops. The interesting question is what can be accomplished by a smaller sample. Thus, suppose that G has size n^{1+g}, with $g < 1$. Then the sample graph has loops at (expected) n^g distinct vertices and about n^{1+g} further edges between distinct vertices.

3 Populations with Tree Structure

Now we approach the particular contribution of this paper. A natural special type of population has a single founder haplotype and is exposed to random mutations over time. As long as the population is relatively young and the total number of mutations (and hence n) is bounded by some small fraction of $\sqrt{s}$, w.h.p. each of the s sites is affected at most once. (Calculations are simple.) Non-affected sites can be ignored, therefore s henceforth denotes the number of sites where different alleles appear. From the uniqueness of mutations at every site it follows that such a population P forms a phylogenetic tree T that enjoys some strong properties discussed below. We call T a perfect phylogeny [6].

Definition 6. *A population P of s-site haplotypes has a perfect phylogeny T if the following holds:*
(1) T is a tree. The vertices of T are labeled by haplotypes (bit strings) such that:
(1.1) P is a subset of the vertex set of T.
(1.2) Labels of any two vertices joined by an edge in T differ on exactly one site.
(2) Edges of T are labeled by sites, such that:

(2.1) The label of every edge is the site mentioned in (1.2).
(2.2) Each site is the label of at most one edge.
A branch vertex of T is a vertex with degree > 2.

The vertices of T can be seen as the haplotypes that appeared in the history of P. However not every vertex is necessarily in P, since it can have disappeared by extinction. Every edge in T is labeled by the site of the allele that has been changed by the mutation corresponding to that edge. Sometimes we identify vertices and edges of T with their labels, i.e. haplotypes and sites, respectively. Note that T is an undirected tree. (Knowing the root is immaterial for our purpose.) The distance of two vertices in T equals the Hamming distance of their labels.

For every pair of haplotypes a, b let $[a, b] = [b, a]$ denote the unique path (of length 0 if $a = b$) in T connecting a and b. Obviously, edge labels on $[a, b]$ are exactly the members of the 2-set of ab. It follows easily:

Lemma 1. *A haplotype c from T belongs to (the subcube) ab if and only if the vertex labeled c is on $[a, b]$.*

Proof. We have $c \in ab$ iff a, b, c agree at all sites in the 0,1-set of ab. These sites are exactly the labels of edges out of $[a, b]$.   □

Lemma 1 implies that every such triple a, b, c is an anomalous match, unless $c = a$ or $c = b$: If the complement d of c in ab were in P then d is on $[a, b]$, and $[c, d] = [a, b]$, an obvious contradiction. Therefore we have many anomalous matches already in trivial cases: $\Theta(n^3)$ if T is a path. Even in more natural cases such as fat trees, the number of anomalous matches is still in the order of $n^2 \log n$.

In general, suppose that we have n^{2+d} anomalous matches and sampled n^{1+g} random genotypes. Consider any of the h^g haplotypes in P which are resolved right from the beginning. It has the role of c in (expected) n^{1+d} anomalous matches, but it has only $2h^g$ true haplotypes as neighbors in the sample graph. That means that already for $d > g - 1$, almost all resolution results would be false. (In contrast to perfect trees, resolution is a very good method if parts of the genetic material under consideration have a high mutation rate: $O(\log n)$ random sites are enough to destroy all anomalous matches.)

In the next section we address haplotype inference from a genotype sample G, provided that the given population P has a perfect phylogeny. Since resolution is highly misleading then, we follow another natural idea: We utilize intersections of genotypes (considered as subcubes) from sample G.

4   Haplotype Inference in a Perfect Phylogeny

Problem statement: Given an unknown population P of haplotypes and a known sample G of genotypes, as in Definition 4. We assume (or: it is promised) that P has a perfect phylogeny T (unknown, of course). Identify as many as possible haplotypes in P.

We continue analyzing the problem. Note that the intersection of any two paths in T, say $[a, b]$ and $[c, d]$, is either empty or a path, say $[e, f]$. Genotype intersection neatly corresponds to path intersection in T:

Lemma 2. *With the above denotations, the intersection of genotypes ab and cd is the genotype ef.*

Proof. W.l.o.g. let $a - e - f - b$ and $c - e - f - d$ be the ordering of vertices a, b, c, d, e, f (not necessarily distinct) on path $[a, b]$ and $[c, d]$, respectively. Let the label of e be w.l.o.g. the zero string. Let A, B, C, D, F denote the set of edge labels on $[a, e]$, $[b, f]$, $[c, e]$, $[d, f]$, $[e, f]$, respectively. Then the label of a, b, c, d, f has the 1-set $A, B \cup F, C, D \cup F, F$, respectively. Hence ab has the 2-set $A \cup B \cup F$ and the 1-set $\emptyset$. Similarly, cd has the 2-set $C \cup D \cup F$ and the 1-set $\emptyset$. We conclude that $ab \cap cd$ has the 2-set F and the 1-set $\emptyset$. On the other hand, ef has the 2-set F and the 1-set $\emptyset$. Now equality follows.   □

Due to this exact correspondence we sometimes use the notions genotype and path interchangeably if we do not risk confusion.

Definition 7. *For a subset S of vertices in T, the hull $[S]$ of S is the unique smallest subtree of T that includes S.*

Algorithm, phase 1: We reconstruct $[U]$ where U is the set of haplotypes known in the beginning (i.e. genotypes of size 1 and 2), utilizing the algorithm of [5] which runs in $O(ns)$ time. Surely, output $[U]$ is a (correct) subtree of T since this reconstruction problem has a unique solution up to isomorphism.

While the labels of vertices in U are already determined, we have to compute the labels of branch vertices in $[U]$ as well. For any branch vertex d, there exist three vertices $a, b, c \in U$ such that the paths from d to them are pairwise edge-disjoint. By Lemma 1, d belongs to each of ab, ac, bc. Given three binary strings a, b, c of length s, their majority string, also of length s, is simply defined as follows: At each position, the bit in the majority string is the bit appearing there in a, b, c two or three times.

Lemma 3. *With the above denotations, the label of d is the majority string of labels of a, b, c.*

Proof. Consider any bit position i, and w.l.o.g. let 1 be the bit which has majority among a_i, b_i, c_i. W.l.o.g. let be $a_i = b_i = 1$. Since $d \in ab$, we must have $d_i = 1$.
                    □

Algorithm, phase 2: Compute the labels of all branch vertices d in $[U]$ in $O(ns)$ time, using Lemma 3. Note that we can choose some fixed vertex from U as a, and b, c as descendants of two distinct children of d in the tree rooted at a.

Let U' be the union of U and the set of branch vertices in $[U]$. Note that $[U'] = [U]$, and that U' partitions $[U]$ into edge-disjoint paths. Since we have the vertex labels in U', we know the 2-set assigned to each of these paths, but not the internal linear ordering of edge labels. This gives reason to define the following data structure:

Definition 8. *A path-labeled tree consists of:*
- *a tree,*
- *a subset of its vertices called pins,*
- *labels of the pins,*
- *labels of the pin paths,*
where a pin path is a path that connects two pins, without a further pin as internal vertex.

In our case, every pin path label is simply the set of edge labels on that pin path, i.e. we forget the ordering of edge labels, and the set of pins is initially U'. The edge-labeled tree for $[U']$ can be finished in $O(ns)$ time, as we know the labels of pins, including all the branch vertices. Sometimes we abuse notation and identify edges and their labels if the context is clear.

Algorithm, phase 3: For each genotype in G, compute the intersection of its 2-set with $[U]$. Recall that this intersection must be a path in $[U]$ (since the 2-set of every genotype $ab \in G$ with $a, b \in P$ corresponds to path $[a, b]$ in T). In particular, if the 2-set of a genotype is entirely contained in $[U]$, we conclude that the ends of this path are haplotypes in P. All intersections are obviously computable in $O(n^{1+g}s)$ time.

In our path-labeled tree we recover the labels of endvertices of all (at most n^{1+g}) intersection paths $[a, b]$ (where not necessarily $a, b \in P$), as described in the following. Path $[a, b]$ intersects one or more pin paths in $[U]$, and we can recognize these pin paths by nonempty intersection of their labels with the known 2-set of ab. If an end of $[a, b]$, say a, happens to be a pin, then nothing remains to be done with a. Otherwise a is an inner vertex of a pin path with ends denoted by c and d. If $[a, b]$ intersects parts of $[U]$ outside $[c, d]$, let c be that end of $[c, d]$ not included in $[a, b]$. By computing set differences we get the path labels of $[a, d]$ and $[c, a]$. Since we know the label of pin c, and now also the 2-set of ca, we can change exactly those sites of c being in this 2-set and obtain the label of a. (By symmetry we could also start from d.) Due to this refinement of the path-labeled tree, a satisfies all requirements to become a new pin.

A slightly more complicated argument applies if $[a, b]$ is contained in $[c, d]$. Again let a denote the end of $[a, b]$ being closer to c. Since we have the label of c and the 0-,1-, and 2-set of ab, we can split the set of sites in three subsets: the 2-set of ab, and the remaining sites being equal and different, respectively, in c and ab. (Note that their values are 0 or 1.) If we walk the path $[c, d]$ starting in c, the sites in the 2-set and those being equal in c and ab *cannot* be changed before a is reached, whereas the sites being different *must* be changed before a is reached. These conditions uniquely determine the path label of $[c, a]$. Once this path label is established, we recover the label of a as in the previous case.

This refinement of the path-labeled tree is successively done for all genotypes from G. The operations which are merely manipulations along paths in $[U']$ can be implemented in $O(n^{1+g}s)$ time for all genotypes.

We summarize the preliminary results in

Lemma 4. *We can identify, in $O(n^{1+g}s)$ time, all haplotypes $a \in P$ for which there exists another haplotype $b \in P$ such that $ab \in G$, and $a, b \in [U]$.* □

Next we try to identify also haplotypes that do not fulfill the condition in Lemma 4. Let $ab \in G$ be a genotype such that $[a, b]$ intersects $[U]$, in at least one vertex or in some path. The part of $[a, b]$ outside $[U]$ may consist of two paths. Obviously, it is not possible to determine the correct splitting of the 2-set of ab if we solely look at ab. However we shall see that pairwise intersections of genotypes are useful.

Definition 9. *At any moment, the* known part K *of* T *is the subtree represented by our path-labeled tree as described in Definition 8, where each pin is a haplotype from* P *or a branch vertex or both.*

In particular, after the steps leading to Lemma 4 we have $K = [U]$.

Consider $a, b, c, d \in P$ with $ab, cd \in G$, $ab \neq cd$, and $ab \cap cd \neq \emptyset$. W.l.o.g. suppose that $ab \not\subseteq cd$. Due to Lemma 2, these assumptions imply that $[e, f] := [a, b] \cap [c, d] \neq \emptyset$, and that some edge of $[a, b]$ is not in $[e, f]$ but incident to e or f. Let us call this edge an *anchor*. Remember that we can easily compute the 0-, 1- and 2-set of ef from the sampled ab and cd. From the 2-set we get also $K \cap [e, f]$ if this intersection contains at least one edge. By the same method as described in phase 3, using the labels of pins and pin paths, we can also determine the labels of ends of $K \cap [e, f]$ and thus the precise location of $K \cap [e, f]$ in K, and split the path labels of affected pin paths in K accordingly.

With the denotations from the previous paragraph, next suppose that the anchor is also an edge of K. We can recognize if this is true, since we know that $K \cap [a, b]$ is a path in K extending $K \cap [e, f]$, and we know the corresponding 2-sets. In fact, an anchor belongs to K iff the 2-set of $K \cap [a, b]$ properly contains the 2-set of $K \cap [e, f]$.

Definition 10. *With respect to K, we call ab, cd an* anchored pair *of genotypes if they have a nonempty intersection which also intersects K in at least one edge, $[e, f] = [a, b] \cap [c, d]$ is not completely in K, and they have an anchor, i.e. an edge from the set difference, incident to e or f, in K.*

In that case we can conclude that one end of path $[e, f]$ in T is exactly the vertex of K where the anchor is attached to $[e, f]$, since otherwise $[e, f]$ would not be the intersection of $[a, b]$ and $[c, d]$. (This picture of a fixed point where some "rope" ends inspired the naming "anchor".) Finally, if we start at the anchor and trace the edges of K whose labels are in the known 2-set of ef, we can reconstruct the entire path $[e, f]$, thereby adding its last part to K. In particular, e and f and the vertex where $[e, f]$ leaves K become pins in tree K extended by $[e, f] \setminus K$. Thus, if $[e, f]$ is not entirely in K, we have extended the known part of T.

Algorithm, phase 4: Choose an anchored pair of genotypes and extend K. Repeat this step as long as possible. Resolve the genotypes whose paths are completely contained in K, as in Lemma 4.

Rephrasing Definition 10 we see that a pair of genotypes is anchored if their intersection paths with K end in the same vertex, x say, in K, their other ends in K are different, and the part of the intersection of their 2-sets not yet in K is nonempty. Testing any two genotypes from G for nonempty intersection outside K takes $O(s)$ time, and each pair must be tested at most once: If the test fails, the intersection outside K will always be empty, since K only grows. If the test succeeds, the missing part of the intersection is attached to K at x. This gives a naive overall time bound of $O(n^{2(1+g)}s)$. However, the nice thing here is that we need not check all pairs in G in order to find anchored pairs. (The following is simpler than the implementation suggested in an earlier manuscript.)

In a random sample G we can expect that every set of genotypes in G whose intersection paths in K end at the same vertex x is much smaller than n^{1+g}. Since tests can be restricted to paths that end in the same x, this gives already an improvement. Moreover, the remaining 2-sets of genotypes outside K can be maintained in $O(n^{1+g}s)$ time during the course of the algorithm. To find an anchored pair with common end vertex x we may randomly pick such paths, first with mutually distinct other ends, and mark their edges outside K in an array of length $< s$. As long as no intersection is found, the time is within $O(s)$. If the degree of x is smaller than the number of distinct ends, we find a nonempty intersection in $O(s)$ time by the pigeonhole principle. Otherwise, since the sample graph is random, a nonempty intersection involves w.h.p. two paths with distinct ends in K, such that a few extra trials succeed. Thus we conjecture $O(s^2 + n^{1+g}s)$ expected time for all $O(s)$ extension steps, under the probabilistic assumptions made, but the complete analysis could be subtle.

The algorithms in [1,3] both run in guaranteed time $O(n^{1+g}s^2)$ (in our terminology), however recall that they also output a representation of not completely identified haplotypes, and that improved time bounds might be established. It is hard to compare the algorithms directly.

To resume our haplotype inference algorithm for tree populations: First determine the set U of resolved haplotypes (i.e genotypes being homozygous in all positions except at most one), set up the path-labeled tree description of $K = [U]$, and then successively refine and enlarge it by paths from G in K and intersection paths of anchored pairs, as long as possible.

With all the notation from above we can now state the following, still rather technical result:

Lemma 5. *Given a sample G of genotypes from a population P of haplotypes with perfect phylogeny, we can determine, in polynomial time, all haplotypes $v \in P$ that satisfy these two conditions: v belongs to the subtree K of T obtained by successively adding, to the initially known subtree $[U]$, intersection paths of anchored pairs, and v is endpoint of some path from G in the final K.* $\square$

Note that Lemma 5 is a combinatorial statement, saying which haplotypes can at least be inferred from a given sample G. No probabilistic assumptions have been made at this stage. However, if we plug in the random mating assumption, we can expect that singleton intersections and anchors occur frequently enough such that the final subtree K covers the entire population P:

Theorem 1. *Given a population of n haplotypes with perfect phylogeny which form genotypes by random mating, our algorithm reconstructs the population w.h.p., from a random sample of n^{1+g} genotypes, where for any desired confidence level, any $g > 0$ is sufficient for large enough n.*

Proof. (Sketch) In T we may assign to every path from G a random orientation, such that the bundles of roughly n^g paths starting in each vertex of P are pairwise independent random sets. This can only double the sample size estimate, but it simplifies the argument. Recall that initially $K = [U]$ where U is the set of haplotypes known from the beginning. The expected number of elements in U is n^g. A component (maximal subtree) of $T \setminus K$ of size larger than $\tilde{O}(n^{1-g})$ does not exist w.h.p. since it would contain w.h.p. an element from U which is impossible by definition of K.

Now let $v \in P$ be any vertex in any component C of $T \setminus K$. Some pair of paths from G starting in v has an anchor in K that allows to extend K up to v, unless all these paths end in the same component of $T \setminus K$ or at the same vertex in K. Since roughly n^g paths of G start in v and end in random vertices, the probability of this bad event is in the order of $1/n^{qn^g}$ for any single v, and at most n times as large for all v. Thus we will eventually have $K = [P]$ w.h.p., and all haplotypes inside K can be recovered. $\square$

If the haplotype fractions $f_i < 1/n$ sum up to some considerable fraction $r(n)$, the analysis goes through, only at cost of another factor $1/(1 - r(n))^2 = O(1)$ in the sample size.

The tradeoff between error probability and sample size may be further analyzed. Here it was our main concern to show that much fewer than $O(n^2)$ genotypes are sufficient. We may also recognize a larger part of T in the beginning, since one can show that intersections of genotypes with cardinality at most 2 must be vertices of T, on the other hand it costs extra time to find them.

5   Conclusions

Although perfect phylogeny is not only a narrow special case, as discussed in [9, 1,3], some extensions are desirable. Can we still apply the ideas if P has arisen from several founders by mutations, if mutations affected some sites more than once, if several evolutionary paths led to the same haplotype, if mutations are interspersed with a few crossover events, etc.?

If P consists of several perfect phylogenetic trees with pairwise Hamming distance greater than the number of mutations in each tree, the method obviously works with slight modification: Genotypes with 2-set larger than this distance are ignored. Since the others are composed of two haplotypes from the same tree, the trees can be recovered independently. The fraction of "useful" genotypes in a random sample, and thus the blow-up in sample size, is constant, for any constant number of trees. However, this trivial extension is no longer possible if the trees are not so well separated.

Acknowledgments. This work was partially supported by SWEGENE and by The Swedish Research Council (Vetenskapsrådet), project title "Algorithms for searching and inference in genetics", file no. 621-2002-4574. I also thank Olle Nerman (Chalmers, Göteborg) and Andrzej Lingas (Lund) for some inspiring discussions.

References

1. V. Bafna, D. Gusfield, G. Lancia, S. Yooseph: Haplotyping as perfect phylogeny: A direct approach, UC Davis Computer Science Tech. Report CSE-2002-21
2. A. Clark: Inference of haplotypes from PCR-amplified samples of diploid populations, *Mol. Biol. Evol.* 7 (1990), 111–122
3. E. Eskin, E. Halperin, R.M. Karp: Large scale reconstruction of haplotypes from genotype data, *7th Int. Conf. on Research in Computational Molecular Biology RECOMB'2003*, 104–113
4. L. Excoffier, M. Slatkin: Maximum-likelihood estimation of molecular haplotype frequencies in a diploid population, *Amer. Assoc. of Artif. Intell.* 2000
5. D Gusfield: Efficient algorithms for inferring evolutionary trees, *Networks* 21 (1991), 19–28
6. D. Gusfield: *Algorithms on Strings, Trees and Sequences: Computer Science and Computational Biology*, Cambridge Univ. Press 1997
7. D. Gusfield: Inference of haplotypes from preamplified samples of diploid populations, UC Davis, technical report csse-99-6
8. D. Gusfield: A practical algorithm for optimal inference of haplotypes from diploid populations, *8th Int. Conf. on Intell. Systems for Mol. Biology ISMB'2000* (AAAI Press), 183–189
9. D. Gusfield: Haplotyping as perfect phylogeny: Conceptual framework and efficient solutions (extended abstract), *6th Int. Conf. on Research in Computational Molecular Biology RECOMB'2002*, 166–175
10. R. Motwani, P. Raghavan: *Randomized Algorithms*, Cambridge Univ. Press 1995
11. M. Stephens, N.J. Smith, P. Donnelly: A new statistical method for haplotype reconstruction from population data, *Amer. J. Human Genetics* 68 (2001), 978–989
12. J. Zhang, M. Vingron, M.R. Hoehe: On haplotype reconstruction for diploid populations, EURANDOM technical report, 2001

On Exact and Approximation Algorithms for Distinguishing Substring Selection

Jens Gramm*, Jiong Guo**, and Rolf Niedermeier**

Wilhelm-Schickard-Institut für Informatik, Universität Tübingen, Sand 13,
D-72076 Tübingen, Fed. Rep. of Germany
{gramm,guo,niedermr}@informatik.uni-tuebingen.de

Abstract. The NP-complete DISTINGUISHING SUBSTRING SELECTION problem (DSSS for short) asks, given a set of "good" strings and a set of "bad" strings, for a solution string which is, with respect to Hamming metric, "away" from the good strings and "close" to the bad strings. Studying the parameterized complexity of DSSS, we show that DSSS is W[1]-hard with respect to its natural parameters. This, in particular, implies that a recently given polynomial-time approximation scheme (PTAS) by Deng *et al.* cannot be replaced by a so-called *efficient* polynomial-time approximation scheme (EPTAS) unless an unlikely collapse in parameterized complexity theory occurs.
By way of contrast, for a special case of DSSS, we present an exact fixed-parameter algorithm solving the problem efficiently. In this way, we exhibit a sharp border between fixed-parameter tractability and intractability results.

Keywords: Algorithms and complexity, parameterized complexity, approximation algorithms, exact algorithms, computational biology.

1 Introduction

Recently, there has been strong interest in developing polynomial-time approximation schemes (PTAS's) for several string problems motivated by computational molecular biology [6,15,16]. More precisely, all these problems adhere to a scenario where we are looking for a string which is "close" to a given set of strings and, in some cases, which shall also be "far" from another given set of strings (see Lanctot *et al.* [14] for an overview on these kinds of problems and their applications in molecular biology). The underlying distance measure is Hamming metric. The list of problems in this context includes CLOSEST (SUB)STRING [15], CONSENSUS PATTERNS [16], and DISTINGUISHING (SUB)STRING SELECTION [6]. All these problems are NP-complete, hence polynomial-time exact solutions are out of reach and PTAS's might be the best one can hope for. PTAS's, however,

---

* Supported by the Deutsche Forschungsgemeinschaft (DFG), project OPAL (optimal solutions for hard problems in computational biology), NI 369/2.
** Partially supported by the Deutsche Forschungsgemeinschaft (DFG), junior research group PIAF (fixed-parameter algorithms), NI 369/4.

A. Lingas and B.J. Nilsson (Eds.): FCT 2003, LNCS 2751, pp. 195–209, 2003.
© Springer-Verlag Berlin Heidelberg 2003

often carry huge hidden constant factors that make them useless from a practical point of view. This difficulty also occurs with the problems mentioned above. Hence, two natural questions arise.

1. To what extent can the above approximation schemes be made really practical? [1]
2. Are there, besides pure heuristics, theoretically satisfying approaches to solve these problems exactly, perhaps based on a parameterized point of view [2, 10]?

In this paper, we address both these questions, focusing on the DISTINGUISHING SUBSTRING SELECTION problem (DSSS):

> **Input:** Given an alphabet Σ of constant size, two sets of strings over Σ,
> - $S_g = \{s_1, \ldots, s_{k_g}\}$, each string of length at least L (the "good" strings), [2]
> - $S_b = \{s'_1, \ldots, s'_{k_b}\}$, each string of length at least L (the "bad" strings),
>
> and two non-negative integers d_g and d_b.
>
> **Question:** Is there a length-L string s over Σ such that
> - in every $s_i \in S_g$, for *every* length-L substring t_i, $d_H(s, t_i) \geq d_g$ and
> - every $s'_i \in S_b$ has *at least one* length-L substring t'_i with $d_H(s, t'_i) \leq d_b$?

Here, $d_H(s, t_i)$ denotes the Hamming distance between strings s and s_i. Following Deng *et al.* [6], we distinguish DSSS from DISTINGUISHING STRING SELECTION (DSS) in which all good and bad strings have the same length L; note that Lanctot *et al.* [14] did not make this distinction and denoted both problems as DSS.

The above mentioned CLOSEST SUBSTRING is the special case of DSSS where the set of good strings is empty. Furthermore, CLOSEST STRING is the special case of CLOSEST SUBSTRING where all given strings and the goal string have the same length. Since CLOSEST STRING is known to be NP-complete [12,14], the NP-completeness of CLOSEST SUBSTRING and DSSS immediately follows.

All the mentioned problems carry at least two natural input parameters ("distance" and "number of input strings") which often are small in practice when compared to the overall input size. This leads to the important question whether the seemingly inevitable "combinatorial explosion" in exact algorithms for these problems can be restricted to some of the parameters—this is the parameterized

---

[1] As Fellows [10] put it in his recent survey, "it would be interesting to sort out which problems with PTAS's have any hope of practical approximation". Also see the new survey by Downey [7] for a good exposition on this issue.

[2] Deng *et al.* [6] let all good strings be of same length L; we come back to this restriction in Sect. 4. The terminology "good" and "bad" has its motivation in the application [14] of designing genetic markers to distinguish the sequences of harmful germs (to which the markers should bind) from human sequences (to which the markers should not bind).

complexity approach [2,7,8,10]. In [13], it was shown that for CLOSEST STRING this can successfully be done for the "distance" parameter as well as the parameter "number of input strings". However, CLOSEST STRING is the easiest of these problems. As to CLOSEST SUBSTRING, fixed-parameter intractability (in the above sense of restricting combinatorial explosion to parameters) was recently shown with respect to the parameter "number of input strings" [11]. More precisely, a proof of W[1]-hardness (see [8] for details on parameterized complexity theory) was given. It was conjectured that CLOSEST SUBSTRING is also fixed-parameter intractable with respect to the distance parameter, but it is an open question to prove (or disprove) this statement.[3]

Now, in this work, we show that DSSS is fixed-parameter intractable (i.e., W[1]-hard) with respect to all natural parameters as given in the problem definition and, thus, in particular, with respect to the distance parameters. Besides of the interest in its own concerning the impossibility[4] of efficient exact fixed-parameter algorithms, this result also has important consequences concerning approximation algorithms. More precisely, our result implies that no efficient polynomial-time approximation scheme (EPTAS) in the sense of Cesati and Trevisan [5] is available for DSSS. As a consequence, there is strong theoretical support for the claim that the recent PTAS of Deng et al. [6] cannot be made practical. In addition, we indicate an instructive border between fixed-parameter tractability and fixed-parameter intractability for DSSS which lies between alphabets of size two and alphabets of size greater than two. Two proofs in Sect. 4 had to be omitted due to the lack of space.

2   Preliminaries and Previous Work

Parameterized Complexity. Given a graph $G = (V, E)$ with vertex set V, edge set E, and a positive integer k, the NP-complete VERTEX COVER problem is to determine whether there is a subset of vertices $C \subseteq V$ with k or fewer vertices such that each edge in E has at least one of its endpoints in C. VERTEX COVER is *fixed-parameter tractable* with respect to the parameter k. There now are algorithms solving it in less than $O(1.3^k + kn)$ time. The corresponding complexity class is called FPT. By way of contrast, consider the NP-complete CLIQUE problem: Given a graph $G = (V, E)$ and a positive integer k, CLIQUE asks whether there is a subset of vertices $C \subseteq V$ with at least k vertices such that C forms a clique by having all possible edges between the vertices in C. CLIQUE appears to be *fixed-parameter intractable*: It is *not* known whether it can be solved in $f(k) \cdot n^{O(1)}$ time, where f might be an arbitrarily fast growing function only depending on k.

Downey and Fellows developed a completeness program for showing fixed-parameter intractability [8]. We very briefly sketch some integral parts of this theory.

---

[3] In fact, more hardness results for *unbounded* alphabet size are known [11]. Here, we refer to the practically most relevant case of constant alphabet size.

[4] Unless an unlikely collapse in structural parameterized complexity theory occurs [10].

Let $L, L' \subseteq \Sigma^* \times \mathbf{N}$ be two parameterized languages.[5] For example, in the case of CLIQUE, the first component is the input graph and the second component is the positive integer k, that is, the parameter. We say that L *reduces to* L' *by a standard parameterized m-reduction* if there are functions $k \mapsto k'$ and $k \mapsto k''$ from $\mathbf{N}$ to $\mathbf{N}$ and a function $(x, k) \mapsto x'$ from $\Sigma^* \times \mathbf{N}$ to Σ^* such that

1. $(x, k) \mapsto x'$ is computable in time $k''|x|^c$ for some constant c and
2. $(x, k) \in L$ iff $(x', k') \in L'$.

Observe that in the subsequent section we will present a reduction from CLIQUE to DSSS, mapping the CLIQUE parameter k into all *four* parameters of DSSS; i.e., k' in fact is a four-tuple $(k_g, k_b, d_g, d_b) = (1, \binom{k}{2}, k + 3, k - 2)$ (see Sect. 3.1 for details). Notably, most reductions from classical complexity turn out *not* to be parameterized ones. The basic reference degree for fixed-parameter intractability, W[1], can be defined as the class of parameterized languages that are equivalent to the SHORT TURING MACHINE ACCEPTANCE problem (also known as the k-STEP HALTING problem). Here, we want to determine, for an input consisting of a nondeterministic Turing machine M and a string x, whether or not M has a computation path accepting x in at most k steps. This can trivially be solved in $O(n^{k+1})$ time and we would be surprised if this can be much improved. Therefore, this is the parameterized analogue of the TURING MACHINE ACCEPTANCE problem that is the basic generic NP-complete problem in classical complexity theory, and the conjecture that FPT $\neq$ W[1] is very much analogous to the conjecture that P $\neq$ NP. Other problems that are W[1]-*hard* (and also W[1]-*complete*) include CLIQUE and INDEPENDENT SET, where the parameter is the size of the relevant vertex set [8]. W[1]-hardness gives a concrete indication that a parameterized problem with parameter k is unlikely to allow for a solving algorithm with $f(k) \cdot n^{O(1)}$ running time, i.e., restricting the combinatorial explosion to k.

Approximation. In the following, we explain some basic terms of approximation theory, thereby restricting to minimization problems. Given a minimization problem, a solution of the problem is $(1 + \epsilon)$-*approximate* if the cost of the solution is d, the cost of an optimal solution is d_{opt}, and $d/d_{opt} \leq 1 + \epsilon$. A *polynomial-time approximation scheme (PTAS)* is an algorithm that computes, for any given real $\epsilon > 0$, a $(1+\epsilon)$-approximate solution in polynomial time where ϵ is considered to be constant. For more details on approximation algorithms, refer to [4]. Typically, PTAS's have a running time $n^{O(1/\epsilon)}$, often with large constant factors hidden in the exponent which make them infeasible already for moderate approximation ratio. Therefore, Cesati and Trevisan [5] proposed the concept of an *efficient* polynomial-time approximation scheme (EPTAS) where the PTAS is required to have an $f(\epsilon) \cdot n^{O(1)}$ running time where f is an arbitrary function depending only on ϵ and not on n. Notably, most known PTAS's are *not* EPTAS's [7,10].

---

[5] Generally, the second component (representing the parameter) can also be drawn from Σ^*; for most cases, assuming the parameter to be a positive integer (or a tuple of positive integers) is sufficient.

Previous Work. Lanctot *et al.* [14] initiated the research on the algorithmic complexity of distinguishing string selection problems. In particular, besides showing NP-completeness (an independent NP-completeness result was also proven by Frances and Litman [12]), they gave a polynomial-time factor-2-approximation for DSSS. Building on PTAS algorithms for CLOSEST STRING and CLOSEST SUBSTRING [15], Deng *et al.* [6] recently gave a PTAS for DSSS.

There appear to be no nontrivial results on exact or fixed-parameter algorithms for DSSS. Since CLOSEST SUBSTRING is a special case of DSSS, however, the fixed-parameter intractability results for CLOSEST SUBSTRING [11] also apply to DSSS, implying that DSSS is W[1]-hard with respect to the parameter "number of input strings". Finally, the special case DSS of DSSS (where all given input strings have exactly the same length as the goal string) is solvable in $O((k_g + k_b) \cdot L \cdot (\max \{d_b + 1, (d'_g + 1) \cdot (|\Sigma| - 1)\})^{d_b})$ time with $d'_g = L - d_g$ [13], i.e., for constant alphabet size, it is fixed-parameter tractable with respect to the aggregate parameter (d'_g, d_b). In a sense, DSS relates to DSSS as CLOSEST STRING relates to CLOSEST SUBSTRING and, thus, DSS should be regarded as considerably easier and of less practical importance than DSSS.

3   Fixed-Parameter Intractability of DSSS

We show that DSSS is, even for binary alphabet, W[1]-hard with respect to the aggregate parameter (d_g, d_b, k_g, k_b). This also means hardness for every single of these parameters. With [5], this implies that DSSS does not have an EPTAS.

To simplify presentation, in the rest of this section we use the following technical terms. Regarding the good strings, we say that a length-L string s *matches* an $s_i \in S_g$ or, equivalently, s is a *match* for s_i, if $d_H(s, t_i) \geq d_g$ for every length-L substring t_i of s_i. Regarding the bad strings, we say that a length-L string s *matches* an $s'_i \in S_b$ or, equivalently, s is a *match* for s'_i, if there is a length-L substring t'_i of s'_i with $d_H(s, t'_i) \leq d_b$. Both these notions of matching for good as well as for bad strings generalize to sets of strings in the natural way.

Our hardness proof follows a similar structure as the W[1]-hardness proof for CLOSEST SUBSTRING [11]. We give a parameterized reduction from CLIQUE to DSSS. Here, however, the reduction has novel features in two ways. Firstly, from the technical point of view, the reduction becomes much more compact and, thus, more elegant. Secondly, for CLOSEST SUBSTRING with binary alphabet, we could only show W[1]-hardness with respect to the number of input strings. Here, however, we can show W[1]-hardness with respect to, among others, parameters d_g and d_b. This has strong implications: Here, we can conclude that DSSS has no EPTAS, which is an open question for CLOSEST SUBSTRING [11].

3.1   Reduction from Clique to DSSS

A CLIQUE instance is given by an undirected graph $G = (V, E)$, with a set $V = \{v_1, v_2, \ldots, v_n\}$ of n vertices, a set E of m edges, and a positive integer k denoting the desired clique size. We describe how to generate two sets of strings over alphabet $\{0,1\}$, S_g (containing one string s_g of length $L := nk + 5$) and S_b (containing $\binom{k}{2}$ strings, each of length $m \cdot (2nk + 5) + (m - 1)$), such that G has a clique of size k iff there is a length-L string s which is a match for S_g and also for S_b; this means that $d_H(s, s_g) \geq d_g$ with $S_g := \{s_g\}$ and $d_g := k + 3$, and every $s'_b \in S_b$ has a length-L substring t'_b with $d_H(s, t'_b) \leq d_b$ and $d_b := k - 2$. In the following we use "$\circ$" to denote the concatenation of strings.

Good String. $S_g := \{s_g\}$ where $s_g = 0^L$, the all-zero string of length L.

Bad Strings. $S_b := \{s'_{1,2}, \ldots, s'_{1,k}, s'_{2,3}, s'_{2,4}, \ldots, s'_{k-1,k}\}$, where *every* $s'_{i,j}$ has length $m \cdot (2nk + 5) + (m - 1)$ and encodes the whole graph; in the following, we describe how we generate a string $s'_{i,j}$.

We encode a vertex $v_r \in V$, $1 \leq r \leq n$, in a length-n string by setting the rth position of this string to "1" and all other positions to "0", i.e.,

$$\langle \text{vertex}(v_r) \rangle := 0^{r-1} 1 0^{n-r}.$$

In $s'_{i,j}$, we encode an edge $\{v_r, v_s\} \in E$, $1 \leq r < s \leq n$, by a length-(nk) string

$$\langle \text{edge}(i, j, \{v_r, v_s\}) \rangle := \underbrace{0^n \ldots 0^n}_{(i-1)} \circ \langle \text{vertex}(v_r) \rangle \circ \underbrace{0^n \ldots 0^n}_{(j-i-1)} \circ \langle \text{vertex}(v_s) \rangle \circ \underbrace{0^n \ldots 0^n}_{(k-j)}.$$

Furthermore, we define

$$\langle \text{edge\_block}(i, j, \{v_r, v_s\}) \rangle := \langle \text{edge}(i, j, \{v_r, v_s\}) \rangle \circ 01110 \circ \langle \text{edge}(i, j, \{v_r, v_s\}) \rangle.$$

We choose this way of constructing the $\langle \text{edge\_block}(\cdot, \cdot, \cdot) \rangle$ strings for the following reason: Let $\langle \text{edge}(i, j, \{v_r, v_s\}) \rangle[h_1, h_2]$ denote the substring of $\langle \text{edge}(i, j, \{v_r, v_s\}) \rangle$ ranging from position h_1 to position h_2. Then, every length $L = nk + 5$ substring of $\langle \text{edge\_block}(\cdot, \cdot, \cdot) \rangle$ which contains the "01110" substring will have the form

$$\langle \text{edge}(i, j, \{v_r, v_s\}) \rangle[h, nk] \circ 01110 \circ \langle \text{edge}(i, j, \{v_r, v_s\}) \rangle[1, h - 1]$$

for $1 \leq h \leq nk + 1$. This will be important because our goal is that a match for a solution in a bad string contains all information of $\langle \text{edge}(i, j, \{v_r, v_s\}) \rangle$. It is difficult to enforce that a match starts at a particular position but we will show that we are able to enforce that it contains a "111" substring which, by our construction, implies that the match contains all information of $\langle \text{edge}(i, j, \{v_r, v_s\}) \rangle$.

Then, given $E = \{e_1, \ldots, e_m\}$, we set

$$s'_{i,j} := \langle \text{edge\_block}(i, j, e_1) \rangle \circ 0 \circ \langle \text{edge\_block}(i, j, e_2) \rangle \circ \ldots \circ \langle \text{edge\_block}(i, j, e_m) \rangle.$$

Parameter Values. We set $L := nk + 5$ and generate $k_g := 1$ good string, $k_b := \binom{k}{2}$ bad strings, and we set distance parameters $d_g := k + 3$ and $d_b := k - 2$.

Example. Let $G = (V, E)$ with $V := \{v_1, v_2, v_3, v_4\}$ and $E := \{\{v_1, v_3\}, \{v_1, v_4\}, \{v_2, v_3\}, \{v_3, v_4\}\}$ as shown in Fig. 1(a) and let $k = 3$. Fig. 1(b) displays the good string s_g and the $\binom{k}{2} = 3$ bad strings $s'_{1,2}$, $s'_{1,3}$, and $s'_{2,3}$. Additionally, we show the length-$(nk + 5)$, i.e., length-17, string s which is a match for $S_g = \{s_g\}$ and a match for $S_b = \{s'_{1,2}, s'_{1,3}, s'_{2,3}\}$ and, thus, corresponds to the k-clique in G.

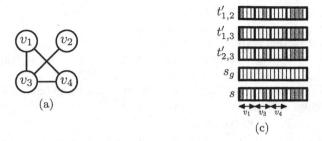

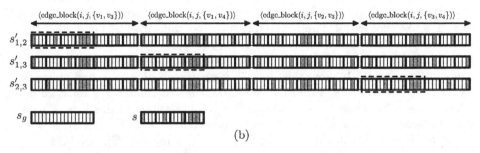

Fig. 1. Example for the reduction from a CLIQUE instance to a DSSS instance with binary alphabet. (a) A CLIQUE instance $G = (V, E)$ with $k = 3$. (b) The produced DSSS instance. We indicate the "1"s of the construction by grey boxes, the "0"s by white boxes. We display the solution s that is found since G has a clique of size $k = 3$; matches of s in $s'_{1,2}$, $s'_{1,3}$, and $s'_{2,3}$ are indicated by dashed boxes. By bold lines we indicate the substrings by which we constructed the bad strings: each $\langle \text{edge\_block}(\cdot, \cdot, e) \rangle$ substring is built from $\langle \text{edge}(\cdot, \cdot, e) \rangle$ for some $e \in E$, consisting of k length-n substrings, followed by "01110", followed again by $\langle \text{edge}(\cdot, \cdot, e) \rangle$. (c) Alignment of the matches $t'_{1,2}$, $t'_{1,3}$, and $t'_{2,3}$ (marked by dashed boxes in (b)) with s_g and s.

3.2 Correctness of the Reduction

We show the two directions of the correctness proof for the above construction by two lemmas.

Lemma 1 *For a graph with a k-clique, the construction in Sect. 3.1 produces an instance of* DSSS *that has a solution, i.e., there is a length-L string s such that $d_H(s, s_g) \geq d_g$ and every $s'_{i,j} \in S_b$ has a length-L substring $t'_{i,j}$ with $d_H(s, t'_{i,j}) \leq d_b$.*

Proof. Let $h_1, h_2, \ldots, h_k$ denote the indices of the clique's vertices, $1 \leq h_1 < h_2 < \cdots < h_k \leq n$. Then, we can find a solution string

$$s := \langle \text{vertex}(v_{h_1}) \rangle \circ \langle \text{vertex}(v_{h_2}) \rangle \circ \cdots \circ \langle \text{vertex}(v_{h_k}) \rangle \circ 01110.$$

For every $s'_{i,j}$, $1 \leq i < j \leq k$, the bad string $s'_{i,j}$ contains a substring $t'_{i,j}$ with $d_H(s, t'_{i,j}) \leq d_b = k - 2$, namely

$$t'_{i,j} := \langle \text{edge}(i, j, \{v_{h_i}, v_{h_j}\}) \rangle \circ 01110.$$

Moreover, we have $d_H(s, s_g) \geq d_g = k + 3$. $\qquad \square$

Lemma 2 *A solution for the* DSSS *instance produced from a graph* G *by the construction in Sect. 3.1 corresponds to a* k-*clique in* G.

Proof. We prove this statement in several steps:

(1) We observe that a solution for the DSSS instance has at least $k + 3$ "1"s since $d_H(s, s_g) \geq d_g = k + 3$ and s_g consists only of "0"s.

(2) We observe that a solution for the DSSS instance has at most $k + 3$ many "1"s: Following the construction, every length-L substring $t'_{i,j}$ of every bad string $s'_{i,j}$, $1 \leq i < j \leq k$, contains at most five "1"s and $d_H(s, t'_{i,j}) \leq k - 2$.

(3) A match $t'_{i,j}$ for s in the bad string $s'_{i,j}$ contains exactly five "1"s: This follows from the observation that *any* length-L substring in a bad string contains *at most* five "1"s together with (1) and (2): Only if $t'_{i,j}$ contains five "1"s and all of them coincide with "1"s in s, we have $d_H(s, t'_{i,j}) \leq (k + 3) - 5 = k - 2$.

(4) All $t'_{i,j}$, $1 \leq i < j \leq k$, and s must contain a "111" substring, located at the same position: To show this, let $t'_{i,j}$ be a match of s in a bad string $s'_{i,j}$ for some $1 \leq i < j \leq k$. From (3), we know that the match $t'_{i,j}$ must contain exactly five "1"s. Thus, since a substring of a bad string contains five "1"s only if it contains a "111" substring, $t'_{i,j}$ must also contain a "111" substring (which separates in $s'_{i,j}$ two substrings $\langle \text{edge}(i, j, e) \rangle$ for some $e \in E$). All "1"s in $t'_{i,j}$ have to coincide with "1"s chosen from the $k - 3$ "1"s in s. In particular, the position of the "111" substring must be the same in the solution and in $t'_{i,j}$ for all $1 \leq i < j \leq k$. This ensures a "synchronization" of the matches.

(5) W.l.o.g., all $t'_{i,j}$, $1 \leq i < j \leq k$, and s all end with the "01110" substring: From (4), we know that all $t'_{i,j}$ contain a "111" substring at the same position. If they do not all end with "01110", we can shift them such that the contained "111" substring is shifted to the appropriate position, as we describe more precisely in the following. Recall that every length-L substring which contains the "111" substring of $\langle \text{edge\_block}(i, j, e) \rangle$ has the form $\langle \text{edge}(i, j, e) \rangle[h, nk] \circ 01110 \circ \langle \text{edge}(i, j, e) \rangle[1, h - 1]$ for $1 \leq h \leq nk$ and $e \in E$. Since all $t'_{i,j}$, $1 \leq i < j \leq k$, contain the "111" substring at the same position, they all have this form for the same h. Then, we can, instead, consider $\langle \text{edge}(i, j, e) \rangle[1, nk] \circ 01110$ and, by a circular shift, move the "111" substring in the solution to the appropriate position. Considering the solution s and the matches $t'_{i,j}$ for all $1 \leq i < j \leq k$ as a character matrix, this is a reordering of columns and, thus, the pairwise Hamming distances do not change.

(6) We divide the first nk positions of the matches and the solution into k "sections", each of length n. In s, each of these sections has the form $\langle \text{vertex}(v) \rangle$ for a vertex $v \in V$ by the following argument: By (5), all matches in bad strings end with "01110" and, by the way we constructed the bad strings, each of their sections either consists only of "0"s or has the form $\langle \text{vertex}(v) \rangle$ for a vertex $v \in V$. If the section encodes a vertex, it contains one "1" which has to coincide with a "1" in s. For the ith section, $1 \leq i \leq k$, the matches in strings $s'_{i,j}$ for $i < j \leq k$ and in strings $s'_{j,i}$ for $1 \leq j < i$, encode a vertex in their ith section. Therefore, every of the k sections in s contains a "1" and, since s (by (1) and (2)) contains $k + 3$ many "1"s and (by (4)) ends with "01110", each of its sections contains exactly one "1". Therefore, every section of s can be read as the encoding $\langle \text{vertex}(v) \rangle$ for a $v \in V$.

Conclusion. Following (6), let v_{h_i}, $1 \leq i \leq k$, be the vertex encoded in the ith length-n section of s. Now, consider some $1 \leq i < j \leq k$. Solution s has a match in $s'_{i,j}$ iff there is an $\langle \text{edge}(i, j, \{v_{h_i}, v_{h_j}\})\rangle \circ 01110$ substring in $s'_{i,j}$ and this holds iff $\{v_{h_i}, v_{h_j}\} \in E$. Since this is true for all $1 \leq i < j \leq k$, all $v_{h_1}, v_{h_2}, \ldots, v_{h_k}$ are pairwisely connected by edges in G and, thus, form a k-clique. $\square$

Lemmas 1 and 2 yield the following theorem.

Theorem 1 DSSS *with binary alphabet is* W[1]*-hard for every combination of the parameters* k_g, k_b, d_g, *and* d_b.[6] $\square$

Theorem 1 means, in particular, that DSSS with binary alphabet is W[1]-hard with respect to every single parameter k_g, k_b, d_g, and d_b. Moreover, it allows us to exploit an important connection between parameterized complexity and the theory of approximation algorithms as follows.

Corollary 1 *There is no EPTAS for* DSSS *unless* W[1] = FPT.

Proof. Cesati and Trevisan [5] have shown that a problem with an EPTAS is fixed-parameter tractable with respect to the parameters that correspond to the objective functions of the EPTAS. In Theorem 1, we have shown W[1]-hardness for DSSS with respect to d_g and d_b. Therefore, we conclude that DSSS cannot have an EPTAS for the objective functions d_g and d_b unless W[1] = FPT. $\square$

4   Fixed-Parameter Tractability for a Special Case

In this section, we give a fixed-parameter algorithm for a *modified version* of DSSS. First of all, we restrict the problem to a binary alphabet $\Sigma = \{0, 1\}$. Then, the problem input consists, similar as in DSSS, of two sets S_g and S_b of binary strings, here with all strings in S_g being of length L. Increasing the number of good strings, we can easily transform an instance of DSSS into one in which all good strings have the same length L by replacing each string $s_i \in S_g$ by a set containing all length-L substrings of s_i. Therefore, in the same way as Deng *et al.* [6] we assume in the following that all strings in S_g have length L. We now consider, instead of the parameter d_g from the DSSS definition, the "dual parameter" $d'_g := L - d_g$ such that we require a solution string s with $d_H(s, s_i) \geq L - d'_g$ for all $s_i \in S_g$. The idea behind is that in some practical cases it might occur that, while d_g is rather large, d'_g is fairly small. Hence, restricting the combinatorial explosion to d'_g might sometimes be more natural than restricting it to d_g. Parameter d'_g is said to be *optimal* if there is an s with $d_H(s, s_i) \geq L - d'_g$ for all $s_i \in S_g$ and if there is no s' with $d_H(s', s_i) \geq L - d'_g + 1$ for all $s_i \in S_g$. The question addressed in this section is to find the minimum integer d_b such that, for the optimal parameter value d'_g, there is a length-L

---

[6] Note that this is the strongest statement possible for these parameters because it means that the combinatorial explosion cannot be restricted to a function $f(k_g, k_b, d_g, d_b)$.

string s with $d_H(s, s_i) \geq L - d'_g$ for every $s_i \in S_g$ and such that every $s'_i \in S_b$ has a length-L substring t'_i with $d_H(s, t'_i) \leq d_b$. Naturally, we also want to compute the length-L solution string s corresponding to the found minimum d_b. We refer to this modified version of DSSS as MDSSS. We can read the set S_g of k_g length-L strings as a $k_g \times L$ character matrix. We call a column in this matrix *dirty* if it contains "0"s as well as "1"s.

In the following, we present an algorithm solving MDSSS. We conclude this section by pointing out the difficulties arising when giving up some of the restrictions concerning MDSSS.

4.1   Fixed-Parameter Algorithm

We present an algorithm that shows the fixed-parameter tractability of MDSSS with respect to the parameter d'_g. There are instances of MDSSS where d'_g is in fact smaller than the parameter d_g. In these cases, solving MDSSS could be a way to circumvent the combinatorial difficulty of computing exact solutions for DSSS; notably, DSSS is not fixed-parameter tractable with respect to d_g (Sect. 3) and we conjecture that it is not fixed-parameter tractable with respect to d'_g. The structure of the algorithm is as follows.

Preprocessing: Process all non-dirty columns of the input set S_g. If there are more than $d'_g \cdot k_g$ dirty columns then reject the input instance. Otherwise, proceed on the thereby reduced set S_g consisting only of dirty columns.

Phase 1: Determine all solutions s such that $d_H(s, s_i) \geq L - d'_g$ for every $s_i \in S_g$ for the optimal d'_g.

Phase 2: For every s found in Phase 1, determine the minimal value of d_b such that every $s'_i \in S_b$ has a length-L substring t'_i with $d_H(s, t'_i) \leq d_b$. Finally, find the minimum value of d_b over all examined choices of s.

Note that, in fact, Phase 1 and Phase 2 are interleaved. Phase 1 of our algorithm extends the ideas behind a bounded search tree algorithm for CLOSEST STRING in [13]. There, however, the focus was on finding *one* solution whereas, here, we require to find *all* solutions for the optimal parameter value. This extension was only mentioned in [13] and it will be described here.

Preprocessing. Reading the set S_g as a $k_g \times L$ character matrix, we set, for an all-"0" (all-"1") column in this matrix, the corresponding character in the solution to "1" ("0"); otherwise, we would not find a solution for an *optimal* d'_g. If the number of remaining dirty columns is larger than $d'_g \cdot k_g$ then we reject the input instance since no solution is possible.

Phase 1. The precondition of this phase is an optimal parameter d'_g. Since, in general, the optimal d'_g is not known in advance, it can be found by looping through $d'_g = 0, 1, 2, \ldots$, each time invoking the procedure described in the following until we meet the optimal d'_g. Notably, for each such d'_g value, we do not have to redo the preprocessing, but only compare the number of dirty columns against $d'_g \cdot k_g$.

Phase 1 is realized as a recursive procedure: We maintain a length-L candidate string s_c which is initialized as $s_c := \text{inv}(s_1)$ for $s_1 \in S_g$, where $\text{inv}(s_1)$ denotes the bitwise complement of s_1. We call a recursive procedure Solve_MDSSS, given in Fig. 2, working as follows.

If s_c is far away from all strings in S_g (i.e., $d_H(s_c, s_i) \geq L - d_g'$ for all $s_i \in S_g$) then s_c already is a solution for Phase 1. We invoke the second phase of the algorithm with the argument s_c. Since it is possible that s_c can be further transformed into another solution, we continue the traversal of the search tree: we select a string $s_i \in S_g$ such that s_c is not allowed to be closer to s_i (i.e., $d_H(s_c, s_i) = L - d_g'$); such an s_i must exist since parameter d_g' is optimal. We try all possible ways to move s_c away from s_i (such that $d_H(s_c, s_i) = L - (d_g' - 1)$), calling the recursive procedure Solve_MDSSS for each of the produced instances.

Otherwise, if s_c is not a solution for Phase 1, we select a string $s_i \in S_g$ such that s_c is too close to s_i (i.e., $d_H(s_c, s_i) < L - d_g'$) and try all possible ways to move s_c away from s_i, calling the recursive procedure for each of the produced instances.

The invocations of the recursive procedure can, thus, be described by a search tree. In the above recursive calls, we omit those calls trying to change a position in s_c which has already been changed before. Therefore, we also omit further invocations of the recursive procedure if the current node of the search tree is already at depth d_g' of the tree; otherwise, s_c would move too close to s_1 (i.e., $d_H(s_c, s_1) < L - d_g'$).

Phase 1 is given more precisely in Fig. 2. It is invoked by Solve_MDSSS($\text{inv}(s_1), d_g'$).

Phase 2. The second phase deals with determining the minimal value of d_b such that there is a string s in the set of the solution strings found in the first phase with $d_H(s, t_i') \leq d_b$ for $1 \leq i \leq k_b$, where t_i' is a length-L substring of s_i'.

For a given solution string s from the first phase and a string $s_i' \in S_b$, we use Abrahamson's algorithm [1] to find the minimum of the number of mismatches between s and every length-L substring of s_i' in $O(|s_i|\sqrt{L \log L})$ time. This minimum is equal to $\min_{t_i'} d_H(s, t_i')$, where t_i' is length-L substring of s_i'. Applying this algorithm to all strings in S_b, we get the value of d_b for s, $\max_{i=1,\ldots,k_b} \min_{t_i'} d_H(s, t_i')$. The minimum value of d_b is then the minimum distance of a solution string from Phase 1 to all bad strings, and s which achieves this minimum distance is the corresponding solution string.

If we are given a fixed d_b and are asked if there is a string s among the solution strings from the first phase which is a match to all strings in S_b, there is a more efficient algorithm by Amir *et al.* [3] for string matching with d_b-mismatches, which takes only $O(|s_i'|\sqrt{d_b \log d_b})$ time to find all length-L substrings in s_i' whose Hamming distance to s is at most d_b.

4.2   Correctness of the Algorithm

Preprocessing. The correctness of the preprocessing follows in a similar way as the correctness of the "problem kernel" for CLOSEST STRING observed by Evans *et al.* [9] (proof omitted).

Recursive procedure Solve_MDSSS($s_c, \Delta d$):
Global variables: Sets S_g and S_b of strings, all strings in S_g of length L, and integer d'_g.
Input: Candidate string s_c and integer Δd, $0 \le \Delta d \le d'_g$.
Output: For optimal d'_g, each length-L string $\hat{s}$ with $d_H(\hat{s}, s_i) \ge L - d'_g$ and $d_H(\hat{s}, s_c) \le \Delta d$.
Remark: The procedure calls, for each computed string $\hat{s}$, Phase 2 of the algorithm.

Method:
(0) **if** ($\Delta d < 0$) **then return**;
(1) **if** ($d_H(s_c, s_i) \le L - (d'_g + \Delta d)$) for some $i \in \{1, \dots, k_g\}$ **then return**;
(2) **if** ($d_H(s_c, s_i) \ge L - d'_g$) for all $i = 1, \dots, k_g$ **then**
  /* s_c already is a solution for Phase 1 */
  call Phase_2(s_c, S_b);
  choose $i \in \{1, \dots, k_g\}$ such that $d_H(s_c, s_i) = L - d'_g$;
  $P := \{ p \mid s_c[p] = s_i[p] \}$;
  for all $p \in P$ **do**
    $s'_c := s_c$;
    $s'_c[p] := \text{inv}(s_c[p])$;
    call Solve_MDSSS($s'_c, \Delta d - 1$);
  end for
else
  /* s_c is not a solution for Phase 1 */
  choose $i \in \{1, \dots, k_g\}$ such that $d_H(s_c, s_i) < L - d'_g$;
  $Q := \{ p \mid s_c[p] = s_i[p] \}$;
  choose any $Q' \subseteq Q$ with $|Q'| = d'_g + 1$;
  for all $q \in Q'$ **do**
    $s'_c := s_c$;
    $s'_c[q] := \text{inv}(s_c[q])$;
    call Solve_MDSSS($s'_c, \Delta d - 1$);
  end for
end if
(3) **return**;

Fig. 2. Recursive procedure realizing Phase 1 of the algorithm for MDSSS.

Lemma 3 *Given an MDSSS instance with the set S_g of k_g good length-L strings, and a positive integer d'_g. If the resulting $k_g \times L$ matrix has more than $k_g \cdot d'_g$ dirty columns then there is no string s with $d_H(s, s_i) \ge L - d'_g$ for all $s_i \in S_g$.* □

Phase 1. From Step (2) in Fig. 2 it is obvious that every string s, which is output of Phase 1 and for which, then, Phase 2 is invoked, satisfies $d_H(s, s_i) \ge L - d'_g$ for all $s_i \in S_g$. The reverse direction, i.e., to show that Phase 1 finds *every* length-L string s with $d_H(s, s_i) \ge d'_g$ for all $s_i \in S_g$, is more involved; the proof is omitted:

Lemma 4 *Given an MDSSS instance, if s is an arbitrary length-L solution string, i.e., $d_H(s, s_i) \geq L - d'_g$ for all $s_i \in S_g$, then s can be found by calling procedure* Solve_MDSSS. □

Phase 2. The second phase is only an application of known algorithms.

4.3   Running Time of the Algorithm

Preprocessing. The preprocessing can easily be done in $O(L \cdot k_g)$ time. Even if the optimal d'_g is not known in advance, we can simply process the non-dirty columns and count the number L_d of dirty ones; therefore, the preprocessing has to be done only once. Then, while looping through $d'_g = 0, 1, 2, \ldots$ in order to find the optimal d'_g, we only have to check, for every value of d'_g in constant time, whether $L_d \leq d'_g \cdot k_g$.

Phase 1. The dependencies of the recursive calls of procedure Solve_MDSSS can be described as a search tree in which an instance of the procedure is the parent node of all its recursive calls. One call of procedure Solve_MDSSS invokes at most $d'_g + 1$ new recursive calls. More precisely, if s_c is a solution then it invokes at most d'_g calls and if s_c is not a solution then it invokes at most $d'_g + 1$ calls. Therefore, every node in the search tree has at most $d'_g + 1$ children. Moreover, Δd is initialized to d'_g and every recursive call decreases Δd by 1. As soon as $\Delta d = 0$, no new recursive calls are invoked. Therefore, the height of the search tree is at most d'_g. Hence, the search tree has a size of $O((d'_g + 1)^{d'_g}) = O((d'_g)^{d'_g})$.

Regarding the running time needed for one call of procedure Solve_MDSSS, note that, after the preprocessing, the instance consists of at most $d'_g \cdot k_g$ columns. Then, a central task in the procedure is to compute the Hamming distance of two strings. To this end, we initially build, in $O(d'_g \cdot k_g^2) = O(L \cdot k_g)$ time, a table containing the distances of s_c to all strings in S_g. Using this table, to determine whether or not s_c is a match for S_g or to find an s_i having at least d'_g positions coinciding with s_c can both be done in $O(k_g)$ time. To identify the positions in which s_c coincides with an $s_i \in S_g$ can be done in $O(d'_g \cdot k_g)$ time. After we change one position in s_c, we only have to inspect one column of the $k_g \times (d'_g \cdot k_g)$ matrix induced by S_g and, therefore, can update the table in $O(k_g)$ time. Summarizing, one call of procedure Solve_MDSSS can be done in $O(d'_g \cdot k_g)$ time.

Together with the $d'_g = 0, 1, 2, \ldots$ loop in order to find the optimal d'_g, Phase 1 can be done in $O((d'_g)^2 \cdot k_g \cdot (d'_g)^{d'_g})$ time.

Phase 2. For every solution string found in Phase 1, the running time of the second phase is $O(N\sqrt{L \log L})$, where N denotes the sum of the length of all strings in S_b [1].

We obtain the following theorem:

Theorem 2 *MDSSS can be solved in $O(L \cdot k_g + ((d'_g)^2 k_g + N\sqrt{L \log L}) \cdot (d'_g)^{d'_g})$ time where $N = \sum_{s'_i \in S_b} |s'_i|$ is the total size of the bad strings.* □

4.4   Extensions of MDSSS

The special requirements imposed on the input of MDSSS seem inevitable in order to obtain the above fixed-parameter tractability result. We discuss the problems arising when relaxing the constraints on the alphabet size and the value of d'_g.

Non-binary alphabet. Already extending the alphabet size in the formulation of MDSSS from two to three makes our approach, described in Sect. 4.1, combinatorially much more difficult such that it does not yield fixed-parameter tractability any more. A reason lies in the preprocessing. When having an all-equal column in the character matrix induced by S_g, for a three-letter alphabet there are two instead of one possible choices for the corresponding position in the solution string. Therefore, to enumerate all solutions s with $d_H(s, s_i) \geq L - d'_g$ for all $s_i \in S_g$, which is essential for our approach, is not fixed-parameter tractable any more; the number of solutions is too large. Let $L' \leq L$ be the number of non-dirty columns and let the alphabet size be three. Then, aside from the dirty columns, we already have $2^{L'}$ assignments of characters to the positions corresponding to non-dirty columns.

Non-optimal d'_g parameter. Also for non-optimal d'_g parameter, the number of solutions s with $d_H(s, s_i) \geq L - d'_g$ for all $s_i \in S_g$ can become too large and it appears to be fixed-parameter intractable with respect to d'_g to enumerate them all. Consider the example where $S_g = \{0^L\}$. Then, there are more than $\binom{L}{d'_g}$ strings s with $d_H(s, 0^L) \geq L - d'_g$. (If the value of d'_g is only a fixed number larger than the optimal one, it could, nevertheless, be possible to enumerate all solution strings of Phase 1.)

5   Conclusion

We have shown that DISTINGUISHING SUBSTRING SELECTION, which has a PTAS, cannot have an EPTAS unless FPT = W[1]. It remains open whether this also holds for the tightly related and similarly important computational biology problems CLOSEST SUBSTRING and CONSENSUS PATTERNS, each of which has a PTAS [15,16] and for each of which it is unknown whether an EPTAS exists. It has been shown that, even for constant size alphabet, CLOSEST SUBSTRING and CONSENSUS PATTERNS are W[1]-hard with respect to the number of input strings [11]; the parameterized complexity with respect to the distance parameter, however, is open for these problems, whereas it has been settled for DSSS in this paper. It would be interesting to further explore the border between fixed-parameter tractability and intractability as initiated in Sect. 4.

References

1. K. Abrahamson. Generalized string matching. *SIAM Journal on Computing*, 16(6):1039–1051, 1987.
2. J. Alber, J. Gramm, and R. Niedermeier. Faster exact solutions for hard problems: a parameterized point of view. *Discrete Mathematics*, 229(1-3):3–27, 2001.
3. A. Amir, M. Lewenstein, and E. Porat. Faster algorithms for string matching with k mismatches. In *Proc. of 11th ACM-SIAM SODA*, pages 794–803, 2000.
4. G. Ausiello, P. Crescenzi, G. Gambosi, V. Kann, A. Marchetti-Spaccamela, and M. Protasi. *Complexity and Approximation – Combinatorial Optimization Problems and their Approximability Properties*. Springer, 1999.
5. M. Cesati and L. Trevisan. On the efficiency of polynomial time approximation schemes. *Information Processing Letters*, 64(4):165–171, 1997.
6. X. Deng, G. Li, Z. Li, B. Ma, and L. Wang. A PTAS for Distinguishing (Sub)string Selection. In *Proc. of 29th ICALP*, number 2380 in LNCS, pages 740–751, 2002. Springer.
7. R. G. Downey. Parameterized complexity for the skeptic (invited paper). In *Proc. of 18th IEEE Conference on Computational Complexity*, July 2003.
8. R. G. Downey and M. R. Fellows. *Parameterized Complexity*. Springer, 1999.
9. P. A. Evans, A. Smith, and H. T. Wareham. The parameterized complexity of p-center approximate substring problems. Technical report TR01-149, Faculty of Computer Science, University of New Brunswick, Canada. 2001.
10. M. R. Fellows. Parameterized complexity: the main ideas and connections to practical computing. In *Experimental Algorithmics*, number 2547 in LNCS, pages 51–77, 2002. Springer.
11. M. R. Fellows, J. Gramm, and R. Niedermeier. On the parameterized intractability of Closest Substring and related problems. In *Proc. of 19th STACS*, number 2285 in LNCS, pages 262–273, 2002. Springer.
12. M. Frances and A. Litman. On covering problems of codes. *Theory of Computing Systems*, 30:113–119, 1997.
13. J. Gramm, R. Niedermeier, and P. Rossmanith. Exact solutions for Closest String and related problems. In *Proc. of 12th ISAAC*, number 2223 in LNCS, pages 441–453, 2001. Springer. Full version to appear in *Algorithmica*.
14. J. K. Lanctot, M. Li, B. Ma, S. Wang, and L. Zhang. Distinguishing string selection problems. In *Proc. of 10th ACM-SIAM SODA*, pages 633–642, 1999.
15. M. Li, B. Ma, and L. Wang. On the Closest String and Substring Problems. *Journal of the ACM*, 49(2):157–171, 2002.
16. M. Li, B. Ma, and L. Wang. Finding similar regions in many sequences, *Journal of Computer and System Sciences*, 65(1):73–96, 2002.

Complexity of Approximating Closest Substring Problems

Patricia A. Evans[1] and Andrew D. Smith[1,2]

[1] University of New Brunswick, P.O. Box 4400, Fredericton N.B., E3B 5A3, Canada
pevans@unb.ca
[2] Ontario Cancer Institute, University Health Network, Suite 703
620 University Avenue, Toronto, Ontario, M5G 2M9 Canada
fax: +1-506-453-3566
asmith@uhnres.utoronto.ca

Abstract. The CLOSEST SUBSTRING problem, where a short string is sought that minimizes the number of mismatches between it and each of a given set of strings, is a minimization problem with a polynomial time approximation scheme [6]. In this paper, both this problem and its maximization complement, where instead the number of matches is maximized, are examined and bounds on their hardness of approximation are proved. Related problems differing only in their objective functions, seeking either to maximize the number of strings covered by the substring or maximize the length of the substring, are also examined and bounds on their approximability proved. For this last problem of length maximization, the approximation bound of 2 is proved to be tight by presenting a 2-approximation algorithm.

Keywords: Approximation algorithms; Hardness of approximation; Closest Substring

1 Introduction

Given a set $\mathcal{F}$ of strings, the CLOSEST SUBSTRING problem seeks to find a string C of a desired length l that minimizes the maximum distance from C to a substring in each member of $\mathcal{F}$. We call such a short string C a *center* for $\mathcal{F}$. The corresponding substrings from each string in $\mathcal{F}$ are the *occurrences* of C. If all strings in $\mathcal{F}$ are the same length n, and the center is also to be of length n, then this special case of the problem is known as CLOSEST STRING. We examine the complexity of approximating three problems related to CLOSEST SUBSTRING with different objective functions. A center is considered to be *optimal* in the context of the problem under discussion, in that it either maximized or minimizes the problem's objective function. This examination of the problems' approximability with respect to their differing objective functions reveals interesting differences between the optimization goals.

In [6], a polynomial time approximation scheme (PTAS) is given for CLOSEST SUBSTRING that has a performance ratio of $1 + \frac{1}{2r-1} + \epsilon$, for any $1 \leq r \leq m$ where $m = |\mathcal{F}|$, and $\epsilon > 0$.

A. Lingas and B.J. Nilsson (Eds.): FCT 2003, LNCS 2751, pp. 210–221, 2003.
© Springer-Verlag Berlin Heidelberg 2003

While CLOSEST SUBSTRING minimizes the number of mismatches, MAX CLOS-
EST SUBSTRING maximizes the number of matches. We show that the MAX CLOS-
EST SUBSTRING problem cannot be approximated in polynomial time with ratio
better than $(\log m)/4$, unless P=NP. As the maximization complement of the
CLOSEST SUBSTRING problem, its reduction can also be applied to CLOSEST SUB-
STRING. This application produces a similarly complementary result indicating
the necessity of the $\frac{1}{O(m)}$ term in the PTAS [6]. While the hard ratio for CLOSEST
SUBSTRING disappears asymptotically when m approaches infinity (as is to be
expected given the PTAS [6]), it indicates a connection between the objective
function and the number of strings given as input. This result supports the position that the term $\frac{1}{O(m)}$ in the PTAS performance ratio cannot be significantly
improved by a polynomial time algorithm.

In [8], Sagot presents an exponential exact algorithm for the decision problem
version of CLOSEST SUBSTRING, also known as COMMON APPROXIMATE SUB-
STRING. Sagot also extends the problem to quorums, finding strings that are
approximately present in at least a specified number of the input strings. This
quorum size can be maximized as an alternate objective function, producing the
MAXIMUM COVERAGE APPROXIMATE SUBSTRING problem. A restricted version
of this problem was examined in [7], and erroneously claimed to be as hard to
approximate as clique. We give a reduction from the MAXIMUM COVERAGE version of SET COVER, showing that the problem is hard to approximate within
$e/(e-1) - \epsilon$ (where e is the base of the natural logarithm) for any $\epsilon > 0$.

The LONGEST COMMON APPROXIMATE SUBSTRING problem seeks to maximize the length of a center string that is within some specified distance d from
every occurrence. We give a 2-approximation algorithm for this problem and
show that 2 is optimal unless P=NP.

2   Preliminary Definitions

Definition 1. *Let x be an instance of optimization problem Π with optimal
solution $opt(x)$. Let A be an algorithm solving Π, and $A(x)$ the solution value
produced by A for x. The* performance ratio *of A with respect to x is*

$$\max \left\{ \frac{A(x)}{opt(x)}, \frac{opt(x)}{A(x)} \right\} .$$

*A is a ρ-approximation algorithm if and only if A always returns a solution with
performance ratio less than or equal to ρ.*

Definition 2. *Let Π and Π' be two minimization problems. A* gap-preserving
reduction *(GP-reduction, $\leq_{GP}$) from Π to Π' with parameters $(c, \rho),(c', \rho')$ is a
polynomial-time algorithm f. For each instance I of Π, f produces an instance
$I' = f(I)$ of Π'. The optima of I and I', say $opt(I)$ and $opt(I')$ respectively,
satisfy the following properties:*

$$opt(I) \leq c \Rightarrow opt(I') \leq c' \ ,$$

$$opt(I) > c\rho \Rightarrow opt(I') > c'\rho',$$

where (c, ρ) and (c', ρ') are functions of $|I|$ and $|I'|$ respectively, and $\rho, \rho' > 1$.

Observe that the above definition of gap preserving reduction specifically refers to minimization problems, but can easily be adapted for maximization problems. Although it is implied by the name, GP-reductions do not require the size of the gap to be preserved, only that some gap remains [1].

We now formally specify the problems treated in this paper. All of these can be seen as variations on the CLOSEST SUBSTRING problem. Note that $d_H(x, y)$ represents the number of mismatches, or *Hamming distance*, between two strings x and y of equal length $|x| = |y|$.

MAX CLOSEST SUBSTRING

Instance: A set $\mathcal{F} = \{S_1, \ldots, S_m\}$ of strings over alphabet Σ such that $\max_{1 \leq i \leq m} |S_i| = n$, integer l, $(1 \leq l \leq n)$.

Question: Maximize $\min_i(l - d_H(\mathcal{C}, s_i))$, such that $\mathcal{C} \in \Sigma^l$ and s_i is a substring of S_i, $(1 \leq i \leq m)$.

MAXIMUM COVERAGE APPROXIMATE SUBSTRING

Instance: A set $\mathcal{F} = \{S_1, \ldots, S_m\}$ of strings over alphabet Σ such that $\max_{1 \leq i \leq m} |S_i| = n$, integers d and l, $(1 \leq d < l \leq n)$.

Question: Maximize $|\mathcal{F}'|$, $\mathcal{F}' \subseteq \mathcal{F}$, such that for some $\mathcal{C} \in \Sigma^l$ and for all $S_i \in \mathcal{F}'$, there exists a substring s_i of S_i such that $d_H(\mathcal{C}, s_i) \leq d$.

LONGEST COMMON APPROXIMATE SUBSTRING

Instance: A set $\mathcal{F} = \{S_1, \ldots, S_m\}$ of strings over alphabet Σ such that $\max_{1 \leq i \leq m} |S_i| = n$, integer d, $(1 \leq d < n)$.

Question: Maximize $l = |\mathcal{C}|$, $\mathcal{C} \in \Sigma^*$, such that $d_H(\mathcal{C}, s_i) \leq d$ and s_i is a substring of S_i, $(1 \leq i \leq m)$.

Throughout this paper, when discussing different problems the values of d, l and m may refer to either the optimal values of objective functions or the values specified as part of the input. These symbols are used in accordance with their use in the formal statement of whatever problem is being discussed.

3   Max Closest Substring

3.1   Hardness of Approximating Max Closest Substring

In this section we use a gap preserving reduction from SET COVER to show inapproximability for MAX CLOSEST SUBSTRING. Lund and Yannakakis [2], with a reduction from LABEL COVER to SET COVER, showed that SET COVER could not be approximated in polynomial time with performance ratio better than

$(\log |\mathcal{B}|)/4$ (where $\mathcal{B}$ is the base set) unless $\mathrm{NP} = \mathrm{DTIME}(2^{\mathrm{poly}(\log n)})$. A result of Raz and Safra [3] indirectly strengthened the conjecture; SET COVER is now known to be NP-hard to approximate with ratio better than $(\log |\mathcal{B}|)/4$.

SET COVER

Instance: A set $\mathcal{B}$ of elements to be covered and a collection of sets $\mathcal{L}$ such that $\mathcal{L}_i \subseteq \mathcal{B}, (1 \leq i \leq |\mathcal{L}|)$.

Question: Minimize $|R|$, $R \subseteq \mathcal{L}$, such that $\cup_{j=1}^{|R|} R_j = \mathcal{B}$.

Let $I = \langle \mathcal{B}, \mathcal{L} \rangle$ be an instance of SET COVER. The reduction constructs, in polynomial time, a corresponding instance $I' = \langle \mathcal{F}, l \rangle$ of MAX CLOSEST SUBSTRING. For all $\rho > 1$, there exists a $\rho' > 1$ such that a solution for I with a ratio of ρ can be obtained in polynomial time from a solution to I' with ratio ρ'.

The Alphabet. The strings of $\mathcal{F}$ are composed of characters from the alphabet $\Sigma = \Sigma_1 \cup \Sigma_2$. The characters of Σ_1 are referred to as *set characters*, and identify sets in $\mathcal{L}$. The characters of Σ_2 are referred to as *element characters* and are in one-to-one correspondence with elements of the base set $\mathcal{B}$.

$$\Sigma_1 = \{p_i : 1 \leq i \leq |\mathcal{L}|\} \, ,$$

$$\Sigma_2 = \{u_i : 1 \leq i \leq |\mathcal{B}|\} \, .$$

Substring Gadgets. The strings of $\mathcal{F}$ are made up of two types of substring gadgets. We use the function f, defined below, to ensure that the substring gadgets are sufficiently large. The gadgets are defined as follows:

Subset Selectors: $\langle set(i) \rangle = p_i^{f(|\mathcal{B}|)}$

Separators: $\langle separator(j) \rangle = u_j^{f(|\mathcal{B}|)}$

The Reduction. The string set $\mathcal{F}$ contains $|\mathcal{B}|$ strings, corresponding to the elements of $\mathcal{B}$. For each $j \in \mathcal{B}$, let $L_j \subseteq \mathcal{L}$ be the subfamily of sets containing the element j. With product notation referring to concatenation, define the string

$$S_j = \prod_{q \in L_j} \langle set(q) \rangle \langle separator(j) \rangle \, .$$

The function $f : \mathbb{N} \mapsto \mathbb{N}$ must be defined. It is necessary for f to have the property that for all positive integers $x < |\mathcal{B}|$,

$$\left\lfloor \frac{f(|\mathcal{B}|)}{x} \right\rfloor > \left\lfloor \frac{f(|\mathcal{B}|)}{x+1} \right\rfloor \, .$$

It is straightforward to check that $f(y) = y^2$ has this property. The maximum length of any member of $\mathcal{F}$ is $n = 2|\mathcal{L}||\mathcal{B}|^2$, the size of $\mathcal{F}$ is $m = |\mathcal{B}|$, the length of the center is $l = f(|\mathcal{B}|) = |\mathcal{B}|^2$ and the alphabet size is $|\Sigma| = |\mathcal{L}| + |\mathcal{B}|$. We call any partition of $\mathcal{F}$ whose equivalence relation is the property of having an exact common substring a *substring induced partition*. For any two occurrences s, s' of a center, we call s and s' disjoint if for all $1 \leq q \leq |s|$, $s[q] \neq s'[q]$. Observe that the maximum distance to an optimal center, for any set of disjoint occurrences, increases with the size of the set.

Lemma 1. *Let F be a set of occurrences of an optimal center C such that $|F| = k$. If for each pair $s, s' \in F$, $d_H(s, s') = l$, then for every $s \in F$, $l - d_H(C, s) \geq \lfloor l/k \rfloor$. Also, there is at least one $s \in F$ such that $l - d_H(C, s) = \lfloor l/k \rfloor$.*

Proof. There are l total positions and for any position p, there is a unique $s \in F$ such that $s[p] = C[p]$. If some $s \in F$ had $l - d_H(C, s) < \lfloor l/k \rfloor$, then the center C would not be optimal, as a better center can be constructed by taking position symbols evenly from the k occurrences. If all $s \in F$ have $l - d_h(C, s) > \lfloor l/k \rfloor$, then the total number of matches exceeds l, some pair of matches would have the same position, and thus some pair $s, s' \in F$ have $d_H(s, s') < l$. $\square$

The significance of our definition for f is apparent from the above proof. It is essential that, under the premise of Lemma 1, values of k (the number of distinct occurrences of a center) can be distinguished based on the maximum distance from any occurrence to the optimal center.

Lemma 2. SET COVER $\leq_{GP}$ MAX CLOSEST SUBSTRING.

Proof. Suppose the optimal cover R for $\langle \mathcal{B}, \mathcal{L} \rangle$ has size less than or equal to c. Construct string C of length $|\mathcal{B}|^2$ as follows. To the positions in C, assign in equal amounts the set characters representing members of R. Then C is a center for $\mathcal{F}$ with maximum similarity $\lfloor |\mathcal{B}|^2/c \rfloor$.

Suppose $|R| > c$. Let $\mathcal{F}'$ be the largest subset of $\mathcal{F}$ having a substring induced c-partition. By the reduction, since $|R| > c$, $\mathcal{F}' \neq \mathcal{F}$. Let S be any string in $\mathcal{F} \backslash \mathcal{F}'$. By Lemma 1, any optimal center for $\mathcal{F}'$ must have minimum similarity $\lfloor |\mathcal{B}|^2/c \rfloor$, and therefore has at least $\lfloor |\mathcal{B}|^2/c \rfloor$ characters from a substring of every string in $\mathcal{F}'$. But the occurrence in S is disjoint from the occurrences in $\mathcal{F}'$, forcing the optimal center to match an equal number of positions in more than c disjoint occurrences. Hence, also by Lemma 1, the optimal center matches no more than $\lfloor |\mathcal{B}|^2/(c+1) \rfloor < \lfloor |\mathcal{B}|^2/c \rfloor$ characters in some occurrence. The gap preserving property of the reduction follows since $\lfloor |\mathcal{B}|^2/c \rfloor$ is a decreasing function of c. $\square$

Theorem 1. MAX CLOSEST SUBSTRING *is not approximable within* $(\log m)/4$ *in polynomial time unless* P=NP.

Proof. The theorem follows from the fact that the NP-hard ratio for MAX CLOSEST SUBSTRING remains identical to that of the source problem SET COVER. $\square$

As MAX CLOSEST SUBSTRING is the complementary maximization version of CLOSEST SUBSTRING, and there is a bijection between feasible solutions to the complementary problems that preserves the order of solution quality, this reduction also applies to CLOSEST SUBSTRING. The form of the hard performance ratio for CLOSEST SUBSTRING provides evidence that the two separate sources of error, $1/O(m)$ and ϵ, are necessary in the PTAS of [6].

Theorem 2. CLOSEST SUBSTRING *cannot be approximated with performance ratio* $1 + \frac{1}{\omega(m)}$ *in polynomial time unless* P=NP.

Proof. Since the NP-hard ratio for SET COVER is $\rho = (1/4)\log|\mathcal{B}|$, the NP-hard ratio obtained for CLOSEST SUBSTRING in the above reduction is

$$\rho' = \frac{c\rho-1}{c\rho-\rho}$$

$$= 1 + \left(\frac{\rho-1}{\rho}\right) \cdot \left(\frac{1}{c-1}\right)$$

$$\geq 1 + \frac{1}{O(m)} \; .$$

$\square$

3.2   An Approximation Algorithm for Max Closest Substring

The preceding subsection showed that MAX CLOSEST SUBSTRING cannot be approximated within $(\log m)/4$. Here, we show that this bound is within a factor of $4 \cdot |\Sigma|$ of being tight, by presenting an approximation algorithm that achieves a bound of $|\Sigma|\log m$ for MAX CLOSEST SUBSTRING.

Due to the complementary relationship between MAX CLOSEST SUBSTRING and CLOSEST SUBSTRING, we start by presenting a greedy algorithm for CLOSEST STRING. The greedy nature of the algorithm is due to the fact that it commits to a local improvement at each iteration. The algorithm also uses a lazy strategy that bases each decision on information obtained by examining a restricted portion of the input. This is the most naive form of local search; the algorithm is not expected to perform well. The idea of the algorithm is to read the input strings column by column, and for each column i, assign a character to $\mathcal{C}[i]$ before looking at any column j such that $j > i$. Algorithm 1 describes this procedure, named GREEDYANDLAZY, in pseudocode.

Algorithm 1: Pseudocode for the GREEDYANDLAZY algorithm.
Input: A set of strings $\mathcal{F} = \{S_1, \ldots, S_m\}$.
Output: An $m(1 - |\Sigma|^{-1})$-approximate center $\mathcal{C}$ for $\mathcal{F}$.
GREEDYANDLAZY($\mathcal{F}$)

```
1.        C ← λ
2.        for i ← 1 to n
3.            Let Fᵢ = {S₁[1..i], ..., Sₘ[1..i]}
4.            for each α ∈ Σ
5.                Xᵅ ← {S ∈ Fᵢ : d_H(Cα, S) = max_{S'∈Fᵢ} d_H(Cα, S')}
6.            Let α = S₁[i]
7.            for each β ∈ Σ
8.                if |Xᵝ| < |Xᵅ|
9.                    α = β
10.           C ← Cα
11.       output(C)
```

Lemma 3. *The greedy and lazy algorithm for* CLOSEST STRING *produces a center string with radius within a factor of* $m(1 - \frac{1}{|\Sigma|})$ *of the optimal radius.*

Proof. Consider the number of iterations required to guarantee that each $S \in \mathcal{F}$ matches $\mathcal{C}$ in at least one position. Let J_i be the set of strings that do not match any position of $\mathcal{C}$ after the ith iteration, then

$$J_{i+1} \leq \left(\frac{|\Sigma| - 1}{|\Sigma|} \right) J_i \leq \exp(-1/|\Sigma|) J_i \ .$$

This is because the algorithm always selects the column majority character of those strings in J_i. Let x be the number of iterations required before all members of $\mathcal{F}$ match $\mathcal{C}$ in at least one position. A bound on the value of x is given by the following inequality:

$$\frac{1}{m} > \exp\left(-\frac{x}{|\Sigma|} \right) \ .$$

Hence, for any strictly positive ϵ, after $x = |\Sigma| \ln m + \epsilon$ iterations, each member of $\mathcal{F}$ matches $\mathcal{C}$ in at least one position. After the final iteration, the total distance from $\mathcal{C}$ to any member of $\mathcal{F}$ is at most $n - n/(|\Sigma| \ln m)$. The optimal distance is at least n/m, otherwise some positions are identical in $\mathcal{F}$ (and thus should not be considered). Therefore the performance ratio of GREEDYANDLAZY is

$$\frac{n - n/(|\Sigma| \ln m)}{n/m} \leq m \left(1 - \frac{1}{|\Sigma|} \right) \ .$$

$\square$

The running time of GREEDYANDLAZY, for m sequences of length n, is $O(|\Sigma| m n^2)$.

Now consider applying GREEDYANDLAZY to the MAX CLOSEST SUBSTRING problem by selecting an arbitrary set of substrings of length l to reduce the problem to a MAX CLOSEST STRING problem. The number of matches between any string in $\mathcal{F}$ and the constructed center will be at least $\Omega(l/(|\Sigma| \log m))$.

Corollary 1. GREEDYANDLAZY *is a* $O(|\Sigma| \log m)$-*approximation algorithm for* MAX CLOSEST SUBSTRING.

Since MAX CLOSEST SUBSTRING is hard to approximate with ratio better than $(\log m)/4$, this approximation algorithm is within $4 \cdot |\Sigma|$ of optimal.

4   Maximum Coverage Approximate Substring

The incorrect reduction given in [7] claimed an NP-hard ratio of $O(n^\epsilon)$, $\epsilon = \frac{1}{4}$, for MAXIMUM COVERAGE APPROXIMATE SUBSTRING when $l = n$ and $|\Sigma| = 2$. Its error resulted from applying Theorem 5 of [5], proven only for alphabet size at least three, to binary strings. Hardness of approximation for the general problem is shown here by a reduction from MAXIMUM COVERAGE.

MAXIMUM COVERAGE

> *Instance:* A set $\mathcal{B}$ of elements to be covered and a collection of sets $\mathcal{L}$ such that $\mathcal{L}_i \subseteq \mathcal{B}, (1 \leq i \leq |\mathcal{L}|)$, a positive integer k.
>
> *Question:* Maximize $|B|$, $B \subseteq \mathcal{B}$, such that $B = \cup_{j=1}^k \mathcal{L}_j$, where $\mathcal{L}_j \in \mathcal{L}$.

Given an instance $\langle \mathcal{B}, L, k \rangle$ of MAXIMUM COVERAGE, we construct an instance $\langle \mathcal{F}, l, d \rangle$ of MAXIMUM COVERAGE APPROXIMATE SUBSTRING where $m = |\mathcal{B}|$, $l = k$, $d = k - 1$ and $n \leq k|\mathcal{L}|$. The construction of $\mathcal{F}$ is similar to the construction used when reducing from SET COVER to CLOSEST SUBSTRING in Section 3; unnecessary parts are removed.

The Alphabet. The strings of $\mathcal{F}$ are composed of characters from the alphabet Σ. The characters of Σ correspond to the sets $\mathcal{L}_i \in \mathcal{L}$ that can be part of a cover, so $\Sigma = \{x_i : 1 \leq i \leq |\mathcal{L}|\}$.

The Reduction. The string set $\mathcal{F} = \{S_1, \ldots, S_{|\mathcal{B}|}\}$ will contain strings corresponding to the elements of $\mathcal{B}$. To construct these strings for each $j \in \mathcal{B}$, let $L_j \subseteq \mathcal{L}$ be the subfamily of sets containing the element j. For each $j \in \mathcal{B}$, define

$$S_j = \prod_{x_i \in L_j} x_i^k .$$

Set $d = k - 1$ and $l = k$. We seek to maximize the number of strings in $\mathcal{F}$ containing occurrences of some center $\mathcal{C}$.

Lemma 4. MAXIMUM COVERAGE $\leq_{GP}$ MAXIMUM COVERAGE APPROXIMATE SUBSTRING.

Proof. Suppose $\langle \mathcal{L}, \mathcal{B}, k \rangle$ is an instance of MAXIMUM COVERAGE with a solution set $R \subset \mathcal{L}$, such that $|R| = k$ and R covers $b \leq |\mathcal{B}|$ elements. Then there is a center $\mathcal{C}$ for $\mathcal{F}$ of length $l = k$ that has distance at most $d = k - 1$ from a substring of b strings in $\mathcal{F}$. Let the k positions in $\mathcal{C}$ be assigned characters representing the k sets in the cover, *i.e.* for each $x_i \in R$, there is a position p such that $\mathcal{C}[p] = x_i$. All b members of $\mathcal{F}$ corresponding to those covered elements in $\mathcal{B}$ contain a substring matching at least one character in $\mathcal{C}$, and mismatch at most $k - 1$ characters. Suppose one cannot obtain a k cover with ratio better than ρ. Then one cannot obtain a center for $\mathcal{F}$ that occurs in more than b/ρ strings of $\mathcal{F}$, so the hard ratio is $\rho' = \frac{b}{b/\rho} = \rho$. $\square$

Theorem 3. MAXIMUM COVERAGE APPROXIMATE SUBSTRING *cannot be approximated with performance ratio* $e/(e-1) - \epsilon$, *for any* $\epsilon > 0$, *unless* P=NP.

Proof. It was shown in [4] that the NP-hard ratio for MAXIMUM COVERAGE is $e/(e-1) - \epsilon$. This result combined with Lemma 4 proves the theorem. $\square$

Note that this reduction shows hardness for the general version of the problem, and leaves open the restricted case of $l = n$ with $|\Sigma| = 2$. No approximation algorithms with nontrivial ratios are known.

5   Longest Common Approximate Substring

The LONGEST COMMON APPROXIMATE SUBSTRING problem seeks to maximize the length of a center that is within a given distance from each string in the problem instance. That a feasible solution always exists can be seen by considering the case of a single character, since the problem is defined with $d > 0$. This problem is useful in finding seeds of high similarity for sequence comparisons.

Here we show that a simple algorithm always produces a valid center that is at least half the optimal length. A valid center is any string that has distance at most d from at least one substring of each string in $\mathcal{F}$. The algorithm simply evaluates each substring of members of $\mathcal{F}$ and tests them as centers. The following procedure EXTEND accomplishes this with a time complexity of $\Theta(m^2 n^3)$.

Algorithm 2: Pseudocode for the EXTEND algorithm.
Input: A set of strings $\mathcal{F} = \{S_1, \ldots, S_m\}$ and an integer d.
Output: A valid center $\mathcal{C}$ for $\mathcal{F}$.
EXTEND($\mathcal{F}, d$)
1.   Let $\mathcal{C}$ be λ, the empty string
2.   **for each** string S_i in $\mathcal{F}$
3.   **for each** substring c of S_i
4.   **if** c is a valid center for $\mathcal{F}$ and $|c| > |\mathcal{C}|$ **then** let $\mathcal{C}$ be c
5.   **output**($\mathcal{C}$)

Theorem 4. EXTEND *is a 2-approximation algorithm for* LONGEST COMMON APPROXIMATE SUBSTRING.

Proof. Let $\mathcal{C}$ be the optimal center for $\mathcal{F}$. For each $S_i \in \mathcal{F}$, let s_i be the occurrence of $\mathcal{C}$ from S_i; observe that $|s_i| = |\mathcal{C}|$. Define $s_{i,1}$ as the substring of s_i consisting of the first $|\mathcal{C}|/2$ positions of s_i, and $s_{i,2}$ as the substring consisting of the remaining positions. Similarly, define $\mathcal{C}_1$ and $\mathcal{C}_2$ as the first and last half of $\mathcal{C}$. For $x \in \{1, 2\}$, let c_x be equal to the string $s_{i,x}$ that satisfies

$$d_H(s_{i,x}, \mathcal{C}_x) \leq \min_{s_{j,x}, j \neq i} d_H(s_{j,x}, \mathcal{C}_x) .$$

Define c such that

$$c = \begin{cases} c_1 & \text{if } d_H(c_1, \mathcal{C}_1) \leq d_H(c_2, \mathcal{C}_2), \\ c_2 & \text{otherwise.} \end{cases}$$

Note that $d_H(c, \mathcal{C}_x) \leq d/2$, for some $x \in \{1, 2\}$. Suppose, for contradiction, that c is not a valid center. Assume, without loss of generality, that $c = s_{i,1}$ for some i. Then there is some $s_{i,1}$ such that $d_H(c, s_{i,1}) > d$. Since $d_H(c, \mathcal{C}_1) = d/2 - y$ for some $1 \leq y \leq d/2$, by the triangle inequality $d_H(s_{i,1}, \mathcal{C}_1) \geq d/2 + y + 1$. This implies that $d_H(s_{i,2}, \mathcal{C}_2) \leq d/2 - y - 1 < d_H(c, \mathcal{C}_1)$, contradicting the definition of c. Hence c is a valid center. Since c is a substring of one of the input strings, it will be found by EXTEND. It is half the length of the optimal length center C, so a center will be found that is at least half the length of the longest center. $\square$

The performance ratio of 2 is optimal unless P=NP. We use a transformation from the VERTEX COVER decision problem that introduces a gap in the objective function.

VERTEX COVER

Instance: A graph $G = (V, E)$ and a positive integer k.
Question: Does G have a vertex cover of size at most k, *i.e.*, a set of vertices $V' \subseteq V$, $|V'| \le k$, such that for each edge $(u, v) \in E$, at least one of u and v belongs to V'?

Suppose for some graph G, we seek to determine if G contains a vertex cover of size k. We construct an instance of LONGEST COMMON APPROXIMATE SUBSTRING with $|E|$ strings corresponding to the edges of G. The intuition behind the reduction is that an occurrence of the center in each string corresponds to the occurrence of a cover vertex in the corresponding edge. Before giving values of n and d, we describe the gadgets used in the reduction.

The Alphabet. The string alphabet is $\Sigma = \Sigma_1 \cup \Sigma_2 \cup \{A\}$. We refer to these as vertex characters (Σ_1), unique characters (Σ_2), and the alignment character (A), where $\Sigma_1 = \{v_i : 1 \le i \le |V|\}$ and $\Sigma_2 = \{u_{ij} : (i, j) \in E\}$.

Substring Gadgets. We next describe the two "high level" component substrings used in the construction. The function f is any arbitrarily large polynomial function of $|G|$.

Vertex Selectors: $\langle vertex(x, i, j, z)\rangle = A^{f(k)} u_{ij}^{(z-1)} v_x u_{ij}^{(k-z)} A^{f(k)}$

Separators: $\langle separator(i, j)\rangle = u_{ij}^{3f(k)}$

The Reduction. We construct $\mathcal{F}$ as follows. For any edge $(i, j) \in E$:

$$S_{ij} = \prod_{1 \le z \le k} \langle vertex(i, i, j, z)\rangle \langle separator(i, j)\rangle \langle vertex(j, i, j, z)\rangle \langle separator(i, j)\rangle$$

The length of each string is then $n = k(10f(k) + 2k)$. The threshold distance is $d = k - 1$.

Theorem 5. LONGEST COMMON APPROXIMATE SUBSTRING *cannot be approximated in polynomial time with performance ratio better than* $2 - \epsilon$, *for any* $\epsilon > 0$, *unless* P=NP.

Proof. For any set of strings $\mathcal{F}$ so constructed, there is an exact common substring of length $f(k)$ corresponding to the $f(k)$ repeats of the alignment character A. Suppose there is a size k cover for the source instance of VERTEX COVER. Construct a center C for $\mathcal{F}$ as follows. Assign the alignment character A to the first $f(k)$ positions in C. To positions $f(k)+1$ through $f(k)+k$, assign the characters corresponding to the vertices in the vertex cover. These may be assigned in any order. Finally, assign the alignment character A to the remaining $f(k)$ positions of C. Each string in $\mathcal{F}$ contains a substring that matches $2f(k)+1$ positions in C, so C is a valid center.

If there is no k cover for the source instance of VERTEX COVER, then for any length $f(k)+k$ string there will be some $S \in \mathcal{F}$ that mismatches k positions. As f can be any arbitrarily large polynomial function of k, the NP-hard performance ratio is

$$\frac{2f(k)+k}{f(k)+k} \geq 2 - \epsilon \ ,$$

for any constant $\epsilon > 0$.

To show hardness for $2 - \epsilon$, where ϵ is not a constant (it can be a function of l), consider that we can manipulate the hard ratio into the form

$$2 - \frac{k}{f(k)+k} \ .$$

Since l is the optimal length and $l = 2f(k)+k$, substitute $f(k) = l/2 - k/2$ in the performance ratio:

$$2 - \frac{k}{l/2 - k/2 + k} = 2 - \frac{2k}{l+k} \ .$$

Suppose we select $l = k^c$ during the reduction, where c is any arbitrarily large constant. Then we have shown a hard performance ratio of

$$2 - \frac{2l^{1/c}}{l + l^{1/c}} \geq 2 - \frac{2l^{1/c}}{l} = 2 - \frac{2}{l^{(c-1)/c}} = 2\left(1 - \frac{1}{l^{(c-1)/c}}\right) \ .$$

$\square$

6   Conclusion

These results show that, unless P=NP, the MAX CLOSEST SUBSTRING, MAXIMUM COVERAGE APPROXIMATE SUBSTRING, and LONGEST COMMON APPROXIMATE SUBSTRING problems all have limitations on their approximability.

The relationships between the different objective functions produce an interesting interplay between the approximability of minimizing d with l fixed, maximizing l with d fixed, and maximizing their difference $l - d$. While this last variant, the MAX CLOSEST SUBSTRING problem, has a hard performance ratio directly related to the number of strings m, the two variants that fix one parameter and attempt to maximize the difference by optimizing the other parameter have lower ratios of approximability. It is NP-hard to approximate MAX CLOSEST SUBSTRING with a performance ratio better than $(\log m)/4$, and we have provided a $(|\Sigma|\log m)$-approximation. For LONGEST COMMON APPROXI-MATE SUBSTRING, with d fixed, the length can be approximately maximized with a ratio of 2, and it is NP-hard to approximate for any smaller ratio. The best ratio of approximation is for CLOSEST SUBSTRING, where l is fixed and d is minimized; the PTAS of [6] achieves a ratio of $(1 + \frac{1}{2r-1} + \epsilon)$, for any $1 \leq r \leq m$, and we have now shown that unless P=NP it cannot be approximated closer than $1 + \frac{1}{O(m)}$.

For the quorum variant of CLOSEST SUBSTRING, where the number of strings covered is instead the objective function to be maximized, then it is NP-hard to obtain a performance ratio better than $e/(e-1)$. The restricted variant with $l = n$ and $|\Sigma| = 2$ once thought to be proven hard by [7] is still open, without either hardness or a nontrivial approximation algorithm.

Our reductions use alphabets whose size will increase. The complexity of variants of these new problems where the alphabet size is treated as a constant is open, except as they relate to known results for constant alphabets [6,7].

References

1. Sanjeev Arora. *Probabilistic checking of proofs and the hardness of approximation problems.* PhD thesis, UC Berkeley, 1994.
2. Carsten Lund and Mihalis Yannakakis. On the hardness of approximating mini-mization problems. *Journal of the ACM*, 41(5), 1994.
3. Ran Raz and Shmuel Safra. A sub-constant error-probability low-degree test, and a sub-constant error-probability PCP characterization of NP. In *Proceedings of the Annual ACM Symposium on Theory of Computing*, 475–484, 1997.
4. Uriel Feige. A threshold of $\log n$ for approximating set cover. *Journal of the ACM*, 45(4):634–652, 1998.
5. J. K. Lanctot, M. Li, B. Ma, S. Wang, and L. Zhang. Distinguishing string selec-tion problems. In *Proceedings of the Annual ACM-SIAM Symposium on Discrete Algorithms*, 633–642. ACM Press, 1999.
6. Ming Li, Bin Ma, and Lusheng Wang. On the closest string and substring problems. *Journal of the ACM*, 49(2):157–171, 2002.
7. Bin Ma. A polynomial time approximation scheme for the closest substring prob-lem. In *Combinatorial Pattern Matching (CPM 2000), Lecture Notes in Computer Science 1848*, 99–107. Springer, 2000.
8. Marie-France Sagot. Spelling approximate repeated or common motifs using a suffix tree. In *LATIN'98, Lecture Notes in Computer Science 1380*, 374–390. Springer, 1998.

On Lawson's Oriented Walk in Random Delaunay Triangulations*

Binhai Zhu

Department of Computer Science
Montana State University
Bozeman, MT 59717-3880 USA
bhz@cs.montana.edu

Abstract. In this paper we study the performance of Lawson's Oriented Walk, a 25-year old randomized point location algorithm without any preprocessing and extra storage, in 2-dimensional Delaunay triangulations. Given n pseudo-random points drawn from a convex set C with unit area and their Delaunay triangulation $\mathcal{D}$, we prove that the algorithm locates a query point q in $\mathcal{D}$ in expected $O(\sqrt{n \log n})$ time. We also present an improved version of this algorithm, Lawson's Oriented Walk with Sampling, which takes expected $O(n^{1/3})$ time. Our technique is elementary and the proof is in fact to relate Lawson's Oriented Walk with Walkthrough, another well-known point location algorithm without preprocessing. Finally, we present empirical results to compare these two algorithms with their siblings, Walkthrough and Jump&Walk.

Keywords: Random Delaunay triangulation, point location, average-case analysis.

1 Introduction

Point location is one of the classical problems in computational geometry, GIS, graphics and solid modeling. In general, point location deals with the following problem: given a set of disjoint geometric objects, determine the object containing a query point. The theoretical problem is well studied in the computational geometry literature and several theoretically optimal algorithms have been proposed since early 1980s; see e.g., Snoeyink's recent survey [Sn97]. In the last couple of years, optimal or close to optimal solutions (sometimes even in the average-case) are proposed with simpler data structures [ACMR00,AMM00, AMM01a,AMM01b,GOR97]. All these (theoretically) faster algorithms require preprocessing to obtain fast query bounds.

However, it should be noted that in practice point location is mainly used as a subroutine for computing and updating large scale triangulations, like in mesh generation. Therefore, extra preprocessing and building additional data structure

* The research is partially supported by NSF CARGO grant DMS-0138065 and a MONTS grant.

A. Lingas and B.J. Nilsson (Eds.): FCT 2003, LNCS 2751, pp. 222–233, 2003.
© Springer-Verlag Berlin Heidelberg 2003

is hard, if not impossible, to perform in practice. We need practical point location solutions which performs no or very little preprocessing in practice; moreover, as Delaunay triangulation is used predominantly in areas like mesh generation, finite-element analysis (FEA) and GIS we in fact need efficient practical point location algorithms in Delaunay triangulations.

Practical point location in Delaunay triangulations only receives massive attention from computational geometers very recently [DMZ98,De98,MSZ99, DLM99]. All these works are somehow based on an idea due to Green and Sibson to use the "walkthrough" method to perform point location in Delaunay triangulation, a common data structure used in these areas. In particular the Jump&Walk method of [DMZ98,MSZ99] uses random sampling to select a good starting point to walk toward the destination while others mix the "walkthrough" idea with some extra simple tree-like data structure to make the algorithm more general [De98,DLM99] (e.g., deal with arbitrary-distributed data [De98] or handle extremely large input while bounding the query time [DLM99]). Some of these algorithms, e.g., Jump&Walk, has been used in important software packages [Sh96,TG+96,BDTY00]. Theoretically, for pseudo-uniformly distributed points in a convex set C, in 2D Jump&Walk is known to have a running time of $O(n^{1/3})$ when the query point is slightly away from the boundary of C [DMZ98]. Similar result holds in 3D [MSZ99]. (We remark that similar "walk" ideas have also been used in ray shooting [AF97,HS95].)

Lawson's Oriented Walk, another randomized point location algorithm without preprocessing, was proposed in 1977 [La77]. It is known that, unlike the Walkthrough method, it could run into loops in arbitrary triangulations. But in Delaunay triangulations it always terminates [Ed90,DFNP91]. Almost no theoretical analysis was ever done on its performance and this question was raised again recently [DPT01]. In this paper, we focus on proving the expected performance of Lawson's Oriented Walk algorithm in a random Delaunay triangulation (i.e., Delaunay triangulation of n random points). (We remark that given these random data, when enough preprocessing, i.e., $\Theta(n)$ expected time and space, is performed we can answer point location queries in expected $O(1)$ time [AEI+85].)

Delaunay Triangulations. For completeness, we briefly mention the following definitions. Further details can be found in some standard textbooks like [PS85]. The *convex hull* of a finite point set X is the smallest convex set containing X, denoted as $CH(X)$. The convex hull of a set of $k+1$ affinely independent points in $\mathcal{R}^d$, for $0 \le k \le d$, is called a k-simplex; i.e., a vertex, an edge, a triangle, or a tetrahedron, etc. If $k = d$, we also say the simplex is *full dimensional*. A *triangulation* $\mathcal{T}$ of X is a subdivision of the convex hull of X consisting of simplices with the following two properties: (1) for every simplex in $\mathcal{T}$, all its faces are also simplices in $\mathcal{T}$; (2) the intersection of any two simplices in $\mathcal{T}$ is either empty or a face of both, in which case it is again a simplex in $\mathcal{T}$. A *Delaunay triangulation* $\mathcal{D}$ of X is a triangulation in which the circumsphere of every full-dimensional simplex is empty, i.e., contains no points of X in its interior.

Point Location by Walking. The basic idea is straightforward; it goes back to early work on constructing Delaunay triangulations in 2D and 3D [GS78, Bo81]. Given a Delaunay triangulation $\mathcal{D}$ of a set X of n points in $\mathcal{R}^d$, and a query point q; in order to locate the (full-dimensional) simplex in $\mathcal{D}$ containing q, start at some arbitrary simplex in $\mathcal{D}$ and then "walk" from the center of that simplex to neighboring simplex "in the general direction" of the target point q. Figure 1 shows an example for the straight Walkthrough method walking from an edge e to q. Other simple variations of this kind of "walk" are possible, e.g., the Orthogonal Walk [DPT01]. The underlying assumption for "walk" is that the $\mathcal{D}$ is given by an internal representation allowing constant-cost access between neighboring simplices (for example, in 2D, a linked list of triangles suffices as long as each triangle store its corresponding local information, i.e., the coordinates of its three vertices and pointers to its three edges and three neighboring triangles). The list of other suitable data structures includes the 2D quad-edge data structure [GS85], the edge-facet structure in 3D [DL89], its specialization and compactification to the domain of 3D triangulations [Mu93], or its generalization to d dimensions [Br93], etc.

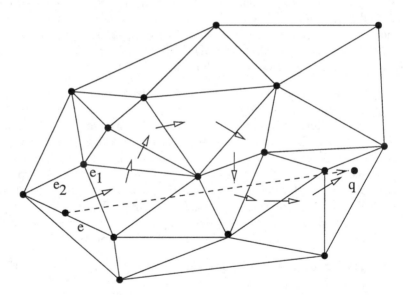

Fig. 1. An example for the walkthrough method and Lawson's Oriented Walk.

Lawson's Oriented Walk.   Given the Delaunay triangulation $\mathcal{D}$ of these n points $\{X_1, X_2, \ldots, X_n\}$, and a query point q, Lawson's Oriented Walk algorithm locates the simplex of $\mathcal{D}$ containing q, if such a simplex exists, as follows (Figure 1).

(1) Select an edge $e = \overline{Y_1 Y_2}$ at random from $\mathcal{D}$.
(2) Determine the triangle t adjacent to e such that t and q are on the same side of the line containing e. Let the other two edges of t be e_1, e_2.

(3) Determine $e_i, i = 1, 2$, such that the halfplane passing through e_i and not containing t, h_i, contains q. If both e_i's have this property, randomly pick up one. If neither e_i's have this property, return t as the triangle containing q.

(4) Update $e \leftarrow e_i$ and repeat step (2)-(4).

The advantage of Lawson's Oriented Walk is that it handles geometric degeneracy better in practice compared with the Walkthrough method (in which some edges of $\mathcal{D}$ might be collinear with the walking segment). In the following, we focus on proving the expected performance of Lawson's Oriented Walk algorithm under the assumption that the Delaunay triangulation $\mathcal{D}$ of n points $X_1, ..., X_n$ are pseudo-uniformly distributed in a compact convex set C.

2   Theoretical Analysis

We start by recalling some fundamental definitions. Let C be a compact convex set of $\mathcal{R}^2$ and let α and β be two reals such that $0 < \alpha < \beta$. We say that a probability measure P is an (α, β)-*measure* over C if $P[C] = 1$ and if we have $\alpha \lambda(S) \leq P[S] \leq \beta \lambda(S)$ for every measurable subset S of C, where λ is the usual Lebesgue measure. An $\mathcal{R}^2$-valued random variable X is called an (α, β)-random variable over C if its probability law $\mathcal{L}(X)$ is an (α, β)-measure over C. A particular and important example of an (α, β)-measure P is when P is a probability measure with density $f(x)$ such that $\alpha \leq f(x) \leq \beta$ for all $x \in C$. This probabilistic model was slightly more general than the uniform distribution and we will loosely call it *pseudo-uniform* or *pseudo-random*.

Throughout this section, c_i's are constants related to the local geometry (but not related to n). The idea of our analysis on Lawson's Oriented Walk in random Delaunay triangulations is as follows. When $e = \overline{Y_1 Y_2}$ is selected, we consider two situations. In case 1, the segment $\overline{pq}$, where p is any point on e, is $O(\sqrt{\frac{\log n}{n}})$ distance away from the boundary of C, ∂C. In case 2, the segment $\overline{pq}$ could be very close to ∂C (but this event has a very small probability). In both cases, we argue that the number of triangles visited by Lawson's Oriented Walk is proportional to the number of triangles crossed by the segment $\overline{pq}$. To estimate the number of triangles of $\mathcal{D}$ crossed by a line segment $\overline{pq}$ when $\overline{pq}$ is $O(\sqrt{\frac{\log n}{n}})$ distance away from ∂C, we need the following lemma of [BD98] which is reorganized as follows.

Lemma 1. *Let C be a compact convex set with unit area in $\mathcal{R}^2$ and let $X_1, \ldots, X_n$ be n points drawn independently in C from an (α, β)-measure. Let $\mathcal{D}$ be the Delaunay triangulation of $X_1, \ldots, X_n$. If $\mathcal{L}$ is a fixed line segment of length $|\mathcal{L}|$ in C and is $O(\sqrt{\frac{\log n}{n}})$ distance away from the boundary of C and if $\mathcal{L}$ is independent of X, then the expected number of triangles or edges of the Delaunay triangulation $\mathcal{D}$ crossed by $\mathcal{L}$ is bounded by*

$$c_3 + c_4 |\mathcal{L}| \sqrt{n} .$$

We now prove the following lemma.

Lemma 2. *Let* $\mathbf{E}[T_1(e, q)]$, *where* $e = \overline{Y_1 Y_2}$ *is a random edge picked by Lawson's Oriented Walk and the query point* q *is independent of* $X_1, \ldots, X_n$ *and both* e, q *are* $O(\sqrt{\frac{\log n}{n}})$ *distance away from* ∂C, *be the expected number of triangles crossed by (or, visited by the walkthrough method along) a straight segment* $\overline{pq}$, *where* $p \in e$ *is any point of* e. *We have* $\mathbf{E}[T_1(e, q)] \leq c_5 + c_6 \mathbf{E}|pq|\sqrt{n}$.

Proof. Let $\mathcal{D}_e$ be the Delaunay triangulation for data points $\{X_1, \ldots, X_n\} - \{Y_1, Y_2\}$. Then $\mathcal{L} = \overline{pq}$, the line segment connecting p and q, is independent of the data points $\{X_1, \ldots, X_n\} - \{Y_1, Y_2\}$. By Lemma 1, $\overline{pq}$ crosses an expected number of $c_3 + c_4 \mathbf{E}|pq|\sqrt{n-2}$ edges in $\mathcal{D}_e$.

Let $T_1(e, q)$ denote the number of triangles in $\mathcal{D}$ crossed by $\overline{pq}, p \in e$. Clearly $\mathbf{E}T_1(e, q)$ is bounded by the number of triangles in $\mathcal{D}$ crossed by $\overline{pq}$ which is in turn bounded by the number of triangles of $\mathcal{D}_e$ crossed by $\overline{pq}$ plus the sum of the degrees of Y_1, Y_2 in the Delaunay triangulation $\mathcal{D}_e$. To see this, note that $\mathcal{L}$ either crosses a triangle without one of Y_1 and Y_2 as a vertex (in which case the triangle is identical in $\mathcal{D}$ and $\mathcal{D}_e$) or with one of Y_1 and Y_2 as a vertex. The total number of the latter kind of triangles does not exceed S. The expected value of S is, by symmetry, 2 times the expected degree of Y_1, which is at most 6 by Euler's formula. Therefore, we have

$$\mathbf{E}T_1(e, q) \leq 6 \times 2 + c_3 + c_4 \cdot \mathbf{E}|pq|\sqrt{n-2}$$
$$\leq 12 + c_3 + c_4 \mathbf{E}|pq|\sqrt{n}$$
$$= c_5 + c_6 \mathbf{E}|pq|\sqrt{n}, c_5 > 12 .$$

This concludes the proof of Lemma 2. $\square$

Lemma 2 has a very interesting implication which will be useful in the proof of Theorem 1. We simply list it as a corollary.

Corollary 1. *Let* e, q, c_5, c_6 *be as in Lemma 2. If* $c_5 + c_6 \mathbf{E}|p'q|\sqrt{n}$, *for some* $p' \in e$, *is greater than a value, then so is* $c_5 + c_6 \mathbf{E}|pq|\sqrt{n}$, *for every* $p \in e$.

Now we are ready to prove the following theorem regarding the expected performance of Lawson's Oriented Walk in a random Delaunay triangulation.

Theorem 1. *Let* C *be a compact convex set with unit area in* $\mathcal{R}^2$, *and let* $X_1, \ldots, X_n$ *be* n *points drawn independently in* C *from an* (α, β)-*measure. If the query point* q *is independent of* $X_1, \ldots, X_n$ *and is* $O(\sqrt{\frac{\log n}{n}})$ *distance away from* ∂C, *then the expected number of triangles visited by Lawson' Oriented Walk is bounded by*

$$c_1 + c_2 \sqrt{n \log n} .$$

Proof of Theorem 1. Let B be the event that e is $O(\sqrt{\frac{\log n}{n}})$ distance away from the boundary of C, i.e., $B = \{e \text{ is } O(\sqrt{\frac{\log n}{n}}) \text{ distance away from}$

∂C}. Clearly, $P[B] \geq 1 - \beta \cdot O(\sqrt{\frac{\log n}{n}})$ and $P[\overline{B}] \leq \beta \cdot O(\sqrt{\frac{\log n}{n}})$ following the property of (α, β) measure.

Let $\mathbf{E}[T(e,q)]$, $e = \overline{Y_1 Y_2}$, be the expected number of triangles of $\mathcal{D}$ visited by Lawson's Oriented Walk. We first consider $\mathbf{E}[T(e,q)|B]$. Let t be the triangle incident to e such that t and q are on the same side of the line through e. Let $t = \triangle Y_1 Y_2 Y_3$. We have two cases: (a) Y_3 is inside $\triangle q Y_1 Y_2$; and (b) Y_3 is outside of $\triangle q Y_1 Y_2$. We prove by induction that $\mathbf{E}[T(e,q)|B] \leq c_7 + c_8 \cdot \mathbf{E}|pq|\sqrt{n}$, for any point $p \in e$; moreover, $c_7 = c_5$ and $c_8 = c_6$ suffices.

Notice that in case (a), the algorithm needs to pick up e_1 or e_2 randomly. Without loss of generality, assume that algorithm picks e_1. We have

$$\mathbf{E}[T(e,q)|B] = 1 + \mathbf{E}[T(e_1, q)|B].$$

In this case the distance from any point on e_1 to q is always smaller than the distance from some point on e to q. We extend $\overline{qY_3}$ which intersects e at Y_4 and we have $\overline{qY_4} = \overline{qY_3} + \overline{Y_3 Y_4}$ (Figure 2 (a)). We prove by induction that in this case $\mathbf{E}[T(e,q)|B] \leq c_7 + c_8 \cdot \mathbf{E}|pq|\sqrt{n}$ for any $p \in e$. (The induction is on the number of edges visited by the algorithm, in reverse order.) The basis is straightforward: if q is inside a triangle incident to e and p is any point on e, then $\mathbf{E}[T(e,q)|B] = 1$ and following Lemma 2, $c_7 + c_8 \cdot \mathbf{E}|pq|\sqrt{n}$ is less than on equal to $c_7 + c_8 \cdot O(\sqrt{\frac{\log n}{n}})\sqrt{n}$. (This is due to the fact that $|pq|$ is less than the maximal edge length of the triangle containing q, following [BEY91,MSZ99], the expected maximal edge length in $\mathcal{D}$ is $O(\sqrt{\frac{\log n}{n}})$ when the edge is $O(\sqrt{\frac{\log n}{n}})$ distance away from the boundary of C.) Clearly, $1 \leq c_7 + c_8 \cdot O(\sqrt{\frac{\log n}{n}})\sqrt{n} = c_7 + c_8 O(\sqrt{\log n})$ (if we set $c_7 = c_5 > 12$). Let the inductive hypothesis be $\mathbf{E}[T(e_1, q)|B] \leq c_7 + c_8 \cdot \mathbf{E}|qY'|\sqrt{n}$, for any $Y' \in e_1$. Consequently, $\mathbf{E}[T(e_1, q)|B] \leq c_7 + c_8 \cdot \mathbf{E}|qY_3|\sqrt{n}$, as $Y_3 \in e_1$. We have

$$\mathbf{E}[T(e,q)|B] = 1 + \mathbf{E}[T(e_1, q)|B]$$
$$\leq 1 + c_7 + c_8 \mathbf{E}|qY_3|\sqrt{n}$$
$$= c_7 + c_8 \mathbf{E}(|qY_3| + |Y_3 Y|)\sqrt{n} + (1 - c_8\sqrt{n}\mathbf{E}|YY_3|),$$

which is bounded by $c_7 + c_8 \cdot \mathbf{E}|qY|\sqrt{n}, Y \in \overline{Y_1 Y_2}$, if we set $1 - c_8\mathbf{E}|YY_3|\sqrt{n} < 0$, i.e., $c_8 \geq \frac{1}{\mathbf{E}|YY_3|\sqrt{n}}$. Following [BEY91,MSZ99], $\mathbf{E}|YY_3| \leq \sqrt{\frac{c_9 \log n}{n}}$. So in this case we just need to set $c_8 = \max\{c_6, \frac{1}{\sqrt{c_9 \log n}}\}$, which is c_6 when n is sufficiently large. To finish our inductive proof for case (a) using Corollary 3, we can simple set $c_7 = c_5$. In other words, $\mathbf{E}[T(e,q)|B] \leq c_7 + c_8 \cdot \mathbf{E}|pq|\sqrt{n}$, for any point $p \in e$; moreover, $c_7 = c_5$ and $c_8 = c_6$.

Notice that in case (b), the algorithm can only pick up one of e_1 and e_2. Without loss of generality, assume that algorithm picks e_1. Let the intersection of $\overline{qY_1}$ and e_1 be Y (Figure 2 (b)). In this case we still have $\mathbf{E}[T(e,q)|B] = 1 + \mathbf{E}[T(e_1, q)|B]$.

In this case, we can again prove by induction that $\mathbf{E}[T(e,q)|B]$ is bounded by $\mathbf{E}[T(e,q)|B] \leq c_7 + c_8 \cdot \mathbf{E}|pq|\sqrt{n}$, for any $p \in e$. We consider the line segment

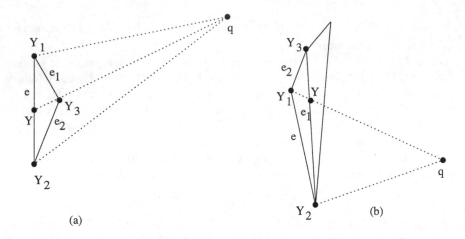

Fig. 2. Illustration for the proof of Theorem 1.

$\overline{qY_1} = \overline{qY} + \overline{YY_1}$. From the inductive hypothesis we further have

$$\mathbf{E}[T(e_1, q)|B] \leq c_7 + c_8 \cdot \mathbf{E}|qY|\sqrt{n}.$$

Therefore, in this case we also have

$$\begin{aligned}
\mathbf{E}[T(e, q)|B] &= 1 + \mathbf{E}[T(e_1, q)|B] \\
&\leq 1 + c_7 + c_8 \cdot \mathbf{E}|qY|\sqrt{n} \\
&= c_7 + c_8 \cdot \mathbf{E}|qY_1|\sqrt{n} + (1 - c_8\mathbf{E}|YY_1|\sqrt{n}).
\end{aligned}$$

To make $\mathbf{E}[T(e, q)|B] \leq c_7 + c_8 \cdot \mathbf{E}|qY_1|\sqrt{n}$, we just need to set $c_8 \geq \frac{1}{\mathbf{E}|YY_1|\sqrt{n}}$. Again, following [BEY91,MSZ99], $\mathbf{E}|YY_1| \leq \sqrt{\frac{c_6 \log n}{n}}$, in this case we also need to set $c_8 = \max\{c_6, \frac{1}{\sqrt{c_9 \log n}}\} = c_6$. Similarly, we can set $c_7 = c_5$ and finish the inductive proof for case (b).

By definition, we have

$$\mathbf{E}[T(e, q)] = \mathbf{E}[T(e, q)|B] \cdot P[B] + \mathbf{E}[T(e, q)|\overline{B}] \cdot P[\overline{B}].$$

To conclude the proof, we note that $\mathbf{E}|pq|$ is of length $\Theta(1)$ in both cases. To see this, let p be any point on $\overline{Y_1Y_2}$. and note that $|pq|^2\pi$ is the probability contents of the circle at q of radius $|pq|$, and is therefore distributed as an i.i.d. (independently identically distributed) uniform $[0, c_{10}]$ random variables, which we call Z. Clearly, $\mathbf{E}\{Z\} = c_{10}/2$. Following the Cauchy-Schwarz inequality, $\mathbf{E}|pq| \leq \sqrt{\mathbf{E}|pq|^2} = \sqrt{\mathbf{E}(Z/\pi)} = \sqrt{\frac{c_{10}}{2\pi}}$.

Also, note that $\mathbf{E}[T(e, q)|\overline{B}]$ is bounded by the size of $\mathcal{D}$, i.e., $\mathbf{E}[T(e, q)|\overline{B}] = O(n)$. A final calculation shows that

$$\mathbf{E}[T(e, q)] \leq c_1 + c_2\sqrt{n \log n} .$$

$\square$

3 Lawson's Oriented Walk with Sampling

We notice that it is very easy to generalize Lawson's Oriented Walk by starting at a 'closer' edge e using random sampling, as done in [DMZ98]. The algorithm is presented as follow.

(1) Select m edges at random and without replacement from $\mathcal{D}$. Let $e = \overline{Y_1 Y_2}$ be the closest one from q.

(2) Determine the triangle t adjacent to e such that t and q are on the same side of the line containing e. Let the other two edges of t be e_1, e_2.

(3) Determine $e_i, i = 1, 2$, such that the halfplane passing through e_i and not containing t, h_i, contains q. If both e_i's have this property, randomly pick up one. If neither e_i's have this property, return t as the triangle containing q.

(4) Update $e \leftarrow e_i$ and repeat step (2)-(4).

In Step (1), the distance between a sample edge and q can be measured as the distance between the midpoint of the sample edge and q. The following theorem can be obtained in very much the way as in [DMZ98]. We hence omit the proof.

Theorem 2. *Let C be a compact convex set with unit area in $\mathcal{R}^2$, and let $X_1, \ldots, X_n$ be n points drawn independently in C from an (α, β)-measure. If the query point q is independent of $X_1, \ldots, X_n$ and is $O(\sqrt{\frac{\log n}{n}})$ distance away from ∂C, then the expected time of Lawson' Oriented Walk with Sampling is bounded by*

$$c_{11} m + c_{12} \sqrt{n/m} .$$

If $m = \Theta(n^{1/3})$, then the running time is optimized to $O(n^{1/3})$, provided that q is $O(\sqrt{\frac{\log n}{n^{1/3}}})$ distance away from ∂C.

4 Empirical Results

In this section, we present some empirical results to compare the following algorithms: Green and Sibson's Walkthrough method (Walk), Lawson's Oriented Walk (Lawson), Jump and Walk (J&W) and Lawson's Oriented Walk with Sampling (L&S). All the data points and query points are within a unit square Q bounded by (0,0) and (1,1). (Throughout this section, we define an axis-parallel square by giving the coordinates of its lower-left and upper-right corner points.) We mainly consider two classes of data: random (uniformly generated) points in Q and three clusters of random points in Q. The latter case does not satisfy the conditions of the theorems we have proved in this paper, but it covers practical situation when data points could be clustered.

The 3-cluster contains three cluster squares defined by lower-left and upper-right corner points: (0.40,0.10) and (0.63,0.33); (0.70,0.67) and (0.93,0.90); and, (0.10,0.67) and (0.33,0.90). Each cluster square has an area of 0.0529 (or 5.29%

Fig. 3. 200 random data points in Q and 200 random data points within the three-cluster.

of the area of Q). In Figure 3 we show two examples for random data and 3-cluster data when there are 200 data points. In both situations, we include the four corner points of Q as data points.

Our empirical results are summarized in Table 1 and Table 2. For each n, we record the average cost (i.e., # of triangles visited) over 10000 queries. The actual cost is also related to the actual implementation, especially the geometric primitives used. For Jump&Walk and Lawson's Oriented Walk with Sampling, we use either $s_1 = \lfloor n^{1/3} \rfloor$ or $s_2 = \lceil n^{1/3} \rceil$ sample edges, depending on whether $|n - s_1^3|$ or $|n - s_2^3|$ is smaller.

Table 1. Comparison of Walk, Jump&Walk, Lawson's Oriented Walk and Lawson's Oriented Walk with Sampling when the data points are random.

n	10000	15000	20000	25000	30000	35000	40000	45000	50000
$Walk$	110	130	155	182	197	211	227	235	257
$Lawson$	127	140	173	193	209	244	243	258	265
$J\&W$	24	28	31	33	35	38	39	40	42
$L\&S$	25	29	33	35	37	41	42	43	45

From Table 1, we can see that when the data points are randomly generated Lawson's Oriented Walk usually visits an extra (small) constant number of triangles compared with Green and Sibson's walkthrough method. This conforms with the proof of Theorem 1 (in which we set $c_8 = c_6$, i.e., the number of triangles visited by the two algorithms is bounded by the same function). For Jump & Walk and Lawson's Oriented Walk with Sampling, the difference is even smaller.

Table 2. Comparison of Walk, Jump&Walk, Lawson's Oriented Walk and Lawson's Oriented Walk with Sampling when the data points are clustered.

n	10000	15000	20000	25000	30000	35000	40000	45000	50000
$Walk$	87	114	137	148	156	170	184	184	187
$Lawson$	103	132	151	156	175	189	207	225	237
$J\&W$	27	33	34	36	37	40	41	44	45
$L\&S$	29	33	36	38	39	41	44	46	47

From Table 2, we can see that when the data points are clustered similar fact can be observed: Lawson's Oriented Walk usually visits an extra constant number of triangles compared with Green and Sibson's walkthrough method and the difference between Jump & Walk and Lawson's Oriented Walk with Sampling is very small. One interesting observation is that the costs for walkthrough and Lawson's Oriented Walk algorithms when data are clustered are lower than the corresponding costs when data are random. The reason is probably the following: As the three clusters have a total area of 15.87% of Q, most parts of the Delaunay triangulation in Q are 'sparse'. Since the 10000 query points are randomly generated, we can say that most of the time these algorithms traverse those 'sparse' regions.

5   Closing Remarks

We remark that similar results for Theorem 1 and Theorem 2 hold for $d = 3$, with a polylog factor and extra boundary conditions inherit from [MSZ99]. It is an interesting question whether we can generalize these results into any fixed dimension, possibly with no extra polylog factor.

The theoretical results in this paper implies that within random Delaunay triangulations Lawson's Oriented Walk performs in very much the same way as the Walkthrough method. Empirical results show the Walkthrough performs slightly better. Still, if we know in advance that degeneracies could appear in the data then Lawson's Oriented Walk might be a better choice. It seems that when the input data points are random then such degeneracies do not occur.

Acknowledgement. The author would like to thank Sunil Arya for communicating his research results.

References

[AEI+85]   T. Asano, M. Edahiro, H. Imai, M. Iri, and K. Murota. Practical use of bucketing techniques in computational geometry. In G. T. Toussaint, editor, *Computational Geometry*, pages 153–195. North-Holland, Amsterdam, Netherlands, 1985.

[AF97]   B. Aronov and S. Fortune. Average-case ray shooting and minimum weight triangulations. In *Proceedings of the 13th Symposium on Computational Geometry*, pages 203–212, 1997.

[ACMR00]   S. Arya, S.W. Cheng, D. Mount and H. Ramesh. Efficient expected-case algorithms for planar point location. In *Proceedings of the 7th Scand. Workshop on Algorithm Theory*, pages 353–366, 2000.

[AMM00]   S. Arya, T. Malamatos and D. Mount. Nearly optimal expected-case planar point location. In *Proceedings of the 41th IEEE Symp on Foundation of Computer Science*, pages 208–218, 2000.

[AMM01a]   S. Arya, T. Malamatos and D. Mount. A simple entropy-based algorithm for planar point location. In *Proceedings of the 12th ACM/SIAM Symp on Discrete Algorithms*, pages 262–268, Jan, 2001.

[AMM01b] S. Arya, T. Malamatos and D. Mount. Entropy-preserving cuttings and space-efficient planar point location. In *Proceedings of the 12th ACM/SIAM Symp on Discrete Algorithms*, pages 256–261, Jan, 2001.

[BD98] P. Bose and L. Devroye. Intersections with random geometric objects. *Comp. Geom. Theory and Appl.*, 10:139–154, 1998.

[BDTY00] J. Boissonnat, O. Devillers, M. Teillaud and M. Yvinc. Triangulations in CGAL triangulation. *Proc. 16th Symp. On Computational Geometry*, pages 11–18, 2000.

[BEY91] M. Bern, D. Eppstein, and F. Yao. The expected extremes in a Delaunay triangulation. *International Journal of Computational Geometry & Applications*, 1:79–91, 1991.

[Bo81] A. Bowyer. Computing Dirichlet tessellations. *The Computer Journal*, 24:162–166, 1981.

[Br93] E. Brisson. Representing geometric structures in d dimensions: Topology and Order. *Discrete & Computational Geometry*, 9(4):387–426, 1993.

[De98] O. Devillers. Improved incremental randomized Delaunay triangulation. In *Proceedings of the 14th Symposium on Computational Geometry*, pages 106–115, 1998.

[DFNP91] L. De Floriani, B. Falcidieno, G. Nagy and C. Pienovi. On sorting triangles in a Delaunay tessellation. *Algorithmica*, 6: 522–532, 1991.

[DLM99] L. Devroye, C. Lemaire and J-M. Moreau. Fast Delaunay point location with search structures. In *Proceedings of the 11th Canadian Conf on Computational Geometry*, pages 136–141, 1999.

[DMZ98] L. Devroye, E. P. Mücke, and B. Zhu. A note on point location in Delaunay triangulations of random points. *Algorithmica*, Special Issue on Average Case Analysis of Algorithms, 22(4):477–482, 1998.

[DL89] D. P. Dobkin and M. J. Laszlo. Primitives for the manipulation of three-dimensional subdivisions. *Algorithmica*, 4(1):3–32, 1989.

[DPT01] O. Devillers, S. Pion, and M. Teillaud. Walking in a triangulation. In *Proceedings of 17th ACM Symposium on Computational Geometry (SCG'01)*, pages 106–114, 2001.

[Ed90] H. Edelsbrunner. An acyclicity theorem for cell complexes in d dimensions. *Combinatorica*, 10(3):251–280, 1990.

[GOR97] M. T. Goodrich, M. Orletsky, and K. Ramaiyer. Methods for achieving fast query times in point location data structures. In *Proceedings of Eighth Annual ACM-SIAM Symposium on Discrete Algorithms (SODA '97)*, pages 757–766, 1997.

[GS78] P. J. Green and R. Sibson. Computing Dirichlet tessellations in the plane. *The Computer Journal*, 21:168–173, 1978.

[GS85] L. J. Guibas and J. Stolfi. Primitives for the manipulation of general subdivisions and the computation of Voronoi diagrams. *ACM Transactions on Graphics*, 4(2):74–123, 1985.

[HS95] J. Hershberger and S. Suri. A pedestrian approach to ray shootings: shoot a ray, take a walk. *J. Algorithms*, 18:403–431, 1995.

[La77] C. L. Lawson. Software for C^1 surface interpolation. In J.R. Rice, editor, *Mathematical Software III*, pages 161–194. Academic Press, NY, 1977.

[Mu93] E. P. Mücke. Shapes and Implementations in Three-Dimensional Geometry. Ph.D. thesis. Technical Report UIUCDCS-R-93-1836. Department of Computer Science, University of Illinois at Urbana-Champaign, Urbana, Illinois, 1993.

[MSZ99]   E. P. Mücke, I. Saias and B. Zhu. Fast randomized point location with-
           out preprocessing in two and three-dimensional Delaunay triangulations.
           Comp. Geom. Theory and Appl., Special Issue for SoCG'96, 12(1/2):63–
           83, 1999.

[PS85]   F. P. Preparata and M.I. Shamos. *Computational Geometry: An Intro-
           duction.* Springer-Verlag, 1985.

[Sh96]   J. R. Shewchuk. Triangle: Engineering a 2D quality mesh generator and
           Delaunay triangulator. In *Proceedings of the First ACM Workshop on
           Applied Computational Geometry*, pages 124–133, 1996.

[Sn97]   J. Snoeyink. Point location. In J. E. Goodman and J. O'Rourke, editors,
           Handbook of Discrete and Computational Geometry, pages 559–574. CRC
           Press, Boca Raton, 1997.

[TG⁺96]   H. Trease, D. George, C. Gable, J. Fowler, E. Linnbur, A. Kuprat and
           A. Khamayseh. *The X3D Grid Generation System.* In *Proceedings of the
           5th International Conference on Numerical Grid Generation in Computa-
           tional Field Simulations*, 239–244, 1996.

Competitive Exploration of Rectilinear Polygons

Mikael Hammar[1], Bengt J. Nilsson[2], and Mia Persson[2]

[1] Department of Computer Science, Salerno University, Baronissi (SA) - 84081, Italy.
hammar@dia.unisa.it
[2] Technology and Society, Malmö University College, S-205 06 Malmö, Sweden.
{Bengt.Nilsson,Mia.Persson}@ts.mah.se

Abstract. Exploring a polygon with a robot, when the robot does not have a map of its surroundings can be viewed as an online problem. Typical for online problems is that you must make decisions based on past events without complete information about the future. In our case the robot does not have complete information about the environment. Competitive analysis can be used to measure the performance of methods solving online problems. The competitive ratio of such a method is the ratio between the method's performance and the performance of the best method having full knowledge of the future. We are interested in obtaining good upper bounds on the competitive ratio of exploring polygons and prove a 3/2-competitive strategy for exploring a simple rectilinear polygon in the L_1 metric.

1 Introduction

Exploring an environment is an important and well studied problem in robotics. In many realistic situations the robot does not possess complete knowledge about its environment, e.g., it may not have a map of its surroundings [1,2,4,6,7,8,9].

The search of the robot can be viewed as an *online* problem since the robot's decisions about the search are based only on the part of its environment that it has seen so far. We use the framework of *competitive analysis* to measure the performance of an online search strategy S. The *competitive ratio* of S is defined as the maximum of the ratio of the distance traveled by a robot using S to the optimal distance of the search.

We are interested in obtaining good upper bounds for the competitive ratio of exploring a rectilinear polygon. The search is modeled by a path or closed tour followed by a point sized robot inside the polygon, given a starting point for the search. The only information that the robot has about the surrounding polygon is the part of the polygon that it has seen so far. Deng *et al.* [4] show a deterministic strategy having competitive ratio two for this problem if distance is measured according to the L_1-metric. Hammar *et al.* [5] prove a strategy with competitive ratio 5/3 and Kleinberg [7] proves a lower bound of 5/4 for the competitive ratio of any deterministic strategy. We will show a deterministic strategy obtaining a competitive ratio of 3/2 for searching a rectilinear polygon in the L_1-metric.

A. Lingas and B.J. Nilsson (Eds.): FCT 2003, LNCS 2751, pp. 234–245, 2003.
© Springer-Verlag Berlin Heidelberg 2003

The paper is organized as follows. In the next section we present some definitions and preliminary results. In Section 3 we give an overview of the strategy by Deng *et al.* [4]. Section 4 contains an improved strategy giving a competitive ratio of 3/2.

2   Preliminaries

We will henceforth always measure distance according to the L_1 metric, i.e., the distance between two points p and q is defined by

$$||p, q|| = |p_x - q_x| + |p_y - q_y|,$$

where p_x and q_x are the x-coordinates of p and q and p_y and q_y are the y-coordinates. We define the x-distance between p and q to be $||p, q||_x = |p_x - q_x|$ and the y-distance to be $||p, q||_y = |p_y - q_y|$.

If C is a polygonal curve, then the length of C, denoted $length(C)$, is defined the sum of the distances between consecutive pairs of segment end points in C.

Let $\mathbf{P}$ be a simple rectilinear polygon. Two points in $\mathbf{P}$ are said to *see* each other, or be *visible* to each other, if the line segment connecting the points lies in $\mathbf{P}$. Let p be a point somewhere inside $\mathbf{P}$. A *watchman route* through p is defined to be a closed curve C that passes through p such that every point in $\mathbf{P}$ is seen by some point on C. The shortest watchman route through p is denoted by SWR_p. It can be shown that the shortest watchman route in a simple polygon is a closed polygonal curve [3].

Since we are only interested in the L_1 length of a polygonal curve we can assume that the curve is *rectilinear*, that is, the segments of the curve are all axis parallel. Note that the shortest rectilinear watchman route through a point p is not necessarily unique.

For a point p in $\mathbf{P}$ we define four *quadrants* with respect to p. Those are the regions obtained by cutting $\mathbf{P}$ along the two maximal axis parallel line segments that pass through p. The four quadrants are denoted $\mathbf{Q}_1(p)$, $\mathbf{Q}_2(p)$, $\mathbf{Q}_3(p)$, and $\mathbf{Q}_4(p)$ in anti-clockwise order from the top right quadrant to the bottom right quadrant. We let $\mathbf{Q}_{i,j}(p)$ denote the union of $\mathbf{Q}_i(p)$ and $\mathbf{Q}_j(p)$.

Consider a reflex vertex of $\mathbf{P}$. The two edges of $\mathbf{P}$ connecting at the reflex vertex can each be extended inside $\mathbf{P}$ until the extensions reach a boundary point. The segments thus constructed are called *extensions* and to each extension a direction is associated. The direction is the same as that of the collinear polygon edge as we follow the boundary of $\mathbf{P}$ in clockwise order. We use the four compass directions *north*, *west*, *south*, and *east* to denote the direction of an extension.

Lemma 1. (CHIN, NTAFOS [3]) *A closed curve is a watchman route for* $\mathbf{P}$ *if and only if the curve has at least one point to the right of every extension of* $\mathbf{P}$.

Our objective is thus to present a competitive online strategy that enables a robot to follow a closed curve from the start point s in $\mathbf{P}$ and back to s with the curve being a watchman route for $\mathbf{P}$.

An extension e splits $\mathbf{P}$ into two sets $\mathbf{P}_l$ and $\mathbf{P}_r$ with $\mathbf{P}_l$ to the left of e and $\mathbf{P}_r$ to the right. We say a point p is *to the left of* e if p belongs to $\mathbf{P}_l$. *To the right* is defined analogously.

As a further definition we say that an extension e is *a left extension* with respect to a point p, if p lies to the left of e, and an extension e *dominates* another extension e', if all points of $\mathbf{P}$ to the right of e are also to the right of e'. By Lemma 1 we are only interested in the extensions that are left extensions with respect to the starting point s since the other ones already have a point (the point s) to the right of them. So without loss of clarity when we mention extensions we will always mean extensions that are left extensions with respect to s.

3   An Overview of GO

Consider a rectilinear polygon $\mathbf{P}$ that is not a priori known to the robot. Let s be the robot's initial position inside $\mathbf{P}$. For the starting position s of the robot we associate a point f^0 on the boundary of $\mathbf{P}$ that is visible from s and call f^0 the *principal projection point* of s. For instance, we can choose f^0 to be the first point on the boundary that is hit by an upward ray starting at s. Let f be the end point of the boundary that the robot sees as we scan the boundary of $\mathbf{P}$ in clockwise order; see Figure 1(a). The point f is called the current *frontier*.

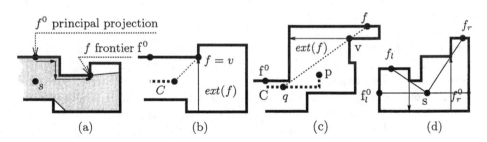

Fig. 1. Illustrating definitions.

Let C be a polygonal curve starting at s. Formally a *frontier* f of C is a vertex of the visibility polygon, $\mathbf{VP}(C)$ of C adjacent to an edge e of $\mathbf{VP}(C)$ that is not an edge of $\mathbf{P}$. Extend e until it hits a point q on C and let v be the vertex of $\mathbf{P}$ that is first encountered as we move along the line segment $[q, f]$ from q to f. We denote the left extension with respect to s associated to the vertex v by $ext(f)$; see Figures 1(b) and (c).

Deng *et al.* [4] introduce an online strategy called *greedy-online*, GO for short, to explore a simple rectilinear polygon $\mathbf{P}$ in the L_1 metric. If the starting point s lies on the boundary of $\mathbf{P}$, their strategy, we call it BGO, goes as follows: from the starting point scan the boundary clockwise and establish the first frontier f.

Move to the closest point on $ext(f)$ and establish the next frontier. Continue in this fashion until all of **P** has been seen and move back to the starting point.

Deng *et al.* show that a robot using strategy BGO to explore a rectilinear polygon follows a tour with shortest length, i.e., BGO has competitive ratio one. They also present a similar strategy, called IGO, for the case when the starting point s lies in the interior of **P**. For IGO they show a competitive ratio of two, i.e., IGO specifies a tour that is at most twice as long as the shortest watchman route through s.

IGO shoots a ray upwards to establish a principal projection point f^0 and then scans the boundary clockwise to obtain the frontier. Next, it proceeds exactly as BGO, moving to the closest point on the extension of the frontier, updating the frontier, and repeating the process until all of the polygon has been seen.

It is clear that BGO could just as well scan the boundary anti-clockwise instead of clockwise when establishing the frontiers and still have the same competitive ratio. Hence, BGO can be seen as two strategies, one scanning clockwise and the other anti-clockwise. We can therefore parameterize the two strategies so that BGO(p, *orient*) is the strategy beginning at some point p on the boundary and scanning with orientation *orient* where *orient* is either clockwise cw or anti-clockwise aw.

Similarly for IGO, we can not only choose to scan clockwise or anti-clockwise for the frontier but also choose to shoot the ray giving the first principal projection point in any of the four compass directions north, west, south, or east. Thus IGO in fact becomes eight different strategies that we can parameterize as IGO(p, *dir*, *orient*) and the parameter *dir* can be one of *north*, *south*, *west*, or *east*.

We further define partial versions of GO starting at boundary and interior points. Strategies PBGO(p, *orient*, *region*) and PIGO(p, *dir*, *orient*, *region*) apply GO until either the robot has explored all of *region* or the robot leaves the region *region*. The strategies return as result the position of the robot when it leaves *region* or when *region* has been explored. Note that PBGO(p, *orient*, **P**) and PIGO(p, *dir*, *orient*, **P**) are the same strategies as BGO(p, *orient*) and IGO(p, *dir*, *orient*) respectively except that they do not move back to p when all of **P** has been seen.

4   The Strategy CGO

We present a new strategy *competitive-greedy-online*(CGO) that explores two quadrants simultaneosly without using up too much distance. We assume that s lies in the interior of **P** since otherwise we can use BGO and achieve an optimal route. The strategy uses two frontier points simultaneously to improve the competitive ratio. However, to initiate the exploration, the strategy begins by performing a scan of the polygon boundary to decide in which direction to start the exploration. This is to minimize the loss inflicted upon us by our choice of initial direction.

The initial scan works as follows: construct the visibility polygon $\mathbf{VP}(s)$ of the initial point s. Consider the set of edges in $\mathbf{VP}(s)$ not coinciding with the boundary of $\mathbf{P}$. The end points of these edges define a set of frontier points each having an associated left extension. Let e denote the left extension that is furthest from s (distance being measured orthogonally to the extension), breaking ties arbitrarily. Let l be the infinite line containing e. We rotate the view point of s so that $\mathbf{Q}_3(s)$ and $\mathbf{Q}_4(s)$ intersect l whereas $\mathbf{Q}_1(s)$ and $\mathbf{Q}_2(s)$ do not. Hence, e is a horizontal extension lying below s. The initial direction of exploration is upwards through $\mathbf{Q}_1(s)$ and $\mathbf{Q}_2(s)$. The two frontier points used by the strategy are obtained as follows: the *left frontier* f_l is established by shooting a ray towards the left for the left principal projection point f_l^0 and then scan the boundary in clockwise direction for f_l; see Figure 1(d). The *right frontier* f_r is established by shooting a ray towards the right for the right principal projection point f_r^0 and then scan the boundary in anti-clockwise direction for f_r; see Figure 1(d). To each frontier point we associate a left extension $ext(f_l)$ and a right extension $ext(f_r)$ with respect to s.

The strategy CGO, presented in pseudo code below makes use of three different substrategies: CGO-0, CGO-1, and CGO-2, that each takes care of specific cases that can occur.

Our strategy ensures that whenever it performs one of the substrategies this is the last time that the outermost while-loop is executed. Hence, the loop is repeated only when the strategy does not enter any of the specified substrategies. The loop will lead the strategy to follow a straight line and we will maintain the invariant during the while-loop that all of the region $\mathbf{Q}_{3,4}(p) \cap \mathbf{Q}_{1,2}(s)$ has been seen.

We distinguish four classes of extensions. $\mathcal{A}$ is the class of extensions e whose defining edge is above e, $\mathcal{B}$ is the class of extensions e whose defining edge is below e. Similarly, $\mathcal{L}$ is the class of extensions e whose defining edge is to the left of e, and $\mathcal{R}$ is the class of extensions e whose defining edge is to the right of e. For conciseness, we use $\mathcal{C}_1\mathcal{C}_2$ as a shorthand for the Cartesian product $\mathcal{C}_1 \times \mathcal{C}_2$ of the two classes $\mathcal{C}_1$ and $\mathcal{C}_2$.

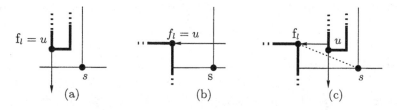

Fig. 2. Illustrating the key point u.

We define two key vertices u and v together with their extensions $ext(u)$ and $ext(v)$ that are useful to establish the correct substrategy to enter. The vertex u lies in $\mathbf{Q}_2(s)$ and v in $\mathbf{Q}_1(s)$. If $ext(f_l) \in \mathcal{A} \cup \mathcal{B}$, then u is the vertex issuing

$ext(f_l)$ and $ext(u) = ext(f_l)$. If $ext(f_l) \in \mathcal{L}$ and $ext(f_l)$ crosses the vertical line through s, then u is the vertex issuing $ext(f_l)$ and again $ext(u) = ext(f_l)$. If $ext(f_l) \in \mathcal{L}$ does not cross the vertical line through s, then u is the leftmost vertex of the bottommost edge visible from the robot, on the boundary going from f_l clockwise until we leave $\mathbf{Q}_2(s)$. The extension $ext(u)$ is the left extension issued by u, and hence, $ext(u) \in \mathcal{A}$; see Figures 2(a), (b), and (c). The vertex v is defined symmetrically in $\mathbf{Q}_1(s)$ with respect to f_r.

Each of the substrategies is presented in sequence and for each of them we claim that if CGO executes the substrategy, then the competitive ratio of CGO is bounded by $3/2$. Let FR_s be the closed route followed by strategy CGO starting at an interior point s. Let $FR_s(p, q, orient)$ denote the subpath of FR_s followed in direction $orient$ from point p to point q, where $orient$ can either be cw (clockwise) or aw (anti-clockwise). Similarly, we define the subpath $SWR_s(p, q, orient)$ of SWR_s. We denote by $SP(p, q)$ a shortest rectilinear path from p to q inside $\mathbf{P}$.

Strategy   CGO

1   Establish the exploration direction by performing the initial scan of the polygon boundary

2   Establish the left and right principal projection points f_l^0 and f_r^0 for $\mathbf{Q}_2(s)$ and $\mathbf{Q}_1(s)$ respectively

3   **while** $\mathbf{Q}_1(s) \cup \mathbf{Q}_2(s)$ is not completely seen **do**

3.1   Obtain the left and right frontiers, f_l and f_r

3.2   **if** f_l lies in $\mathbf{Q}_2(s)$ and f_r lies in $\mathbf{Q}_1(s)$ **then**

3.2.1   Update vertices u and v as described in the text

3.2.2   **if** $(ext(u), ext(v)) \in \mathcal{LR}$ **or** $\big((ext(u), ext(v)) \in \mathcal{AR} \cup \mathcal{LA}$ **and** $ext(u)$ crosses $ext(v)\big)$ **then**

3.2.2.1   Go to the closest horizontal extension

          elseif $(ext(u), ext(v)) \in \mathcal{BR} \cup \mathcal{LB}$ **or** $\big((ext(u), ext(v)) \in \mathcal{AR} \cup \mathcal{LA}$ **and** $ext(u)$ does not cross $ext(v)\big)$ **then**

3.2.2.2   Apply substrategy CGO-1

          elseif $(ext(u), ext(v)) \in \mathcal{AA} \cup \mathcal{AB} \cup \mathcal{BA} \cup \mathcal{BB}$ **then**

3.2.2.3   Apply substrategy CGO-2

              endif

        else

3.2.3   Apply substrategy CGO-0

          endif

     endwhile

4   **if** $\mathbf{P}$ is not completely visible **then**

4.1   Apply substrategy CGO-0

     endif

End   CGO

We claim the following two simple lemmas without proof.

Lemma 2. *If t is a point on some tour SWR_s, then*

$$length(SWR_t) \leq length(SWR_s).$$

Lemma 3. *Let S be a set of points that are enclosed by some tour SWR_s, and let $S_1 = S \cap \mathbf{Q}_{1,2}(s)$, $S_2 = S \cap \mathbf{Q}_{2,3}(s)$, $S_3 = S \cap \mathbf{Q}_{3,4}(s)$, and $S_4 = S \cap \mathbf{Q}_{1,4}(s)$. Then*

$$length(SWR_s) \geq 2 \max_{p \in S_1}\{||s,p||_y\} + 2 \max_{p \in S_2}\{||s,p||_x\} +$$
$$+ 2 \max_{p \in S_3}\{||s,p||_y\} + 2 \max_{p \in S_4}\{||s,p||_x\}.$$

The structure of the following proofs are very similar to each other. In each case we will establish a point t that we can ensure is passed by SWR_s and that either lies on the boundary of $\mathbf{P}$ or can be viewed as to lie on the boundary of $\mathbf{P}$. We then consider the tour SWR_t and compare its length to the length of FR_s. By Lemma 2 we know that $length(SWR_t) \leq length(SWR_s)$, hence the difference in length between FR_s and SWR_t is an upper bound on the loss produced by CGO.

We start by presenting CGO-0, that does the following: Let p be the current robot position. If $\mathbf{Q}_1(p)$ is completely seen from p then we run PIGO(p, *north*, *aw*, $\mathbf{P}$) and move back to the starting point s, otherwise $\mathbf{Q}_2(p)$ is completely seen from p and we run PIGO(p, *north*, *cw*, $\mathbf{P}$) and move back to the starting point s.

Lemma 4. *If CGO-0 is applied, then $length(FR_s) = length(SWR_s)$.*

Proof. Assume that CGO-0 realizes that when FR_s reaches the point p, then $\mathbf{Q}_1(p)$ is completely seen from p. The other case, that $\mathbf{Q}_2(p)$ is completely seen from p is symmetric.

Since the path $FR_s(s, p, orient)$ that the strategy has followed when it reaches point p is a straight line, the point p is the currently topmost point of the path. Hence, we can add a vertical spike issued by the boundary point immediately above p, giving a new polygon $\mathbf{P}'$ having p on the boundary and furthermore with the same shortest watchman route through p as $\mathbf{P}$. This means that performing strategy IGO(p, *north*, *orient*) in $\mathbf{P}$ yields the same result as performing BGO(p, *orient*) in $\mathbf{P}'$, p being a boundary point in $\mathbf{P}'$, and *orient* being either *cw* or *aw*. The tour followed is therefore a shortest watchman route through the point p in both $\mathbf{P}'$ and $\mathbf{P}$.

Also the point p lies on an extension with respect to s, by the way p is defined, and it is the closest point to s such that all of $\mathbf{Q}_1(s)$ has been seen by the path $FR_s(s, p, orient) = SP(s, p)$. Hence, there is a route SWR_s that contains p and by Lemma 2 $length(SWR_p) \leq length(SWR_s)$. The tour followed equals $FR_s = SP(s, p) \cup SWR_p(p, s, aw)$, and we have that $length(FR_s) = length(SWR_p) \leq length(SWR_s)$, and since FR_s cannot be strictly shorter than SWR_s the equality holds which concludes the proof.

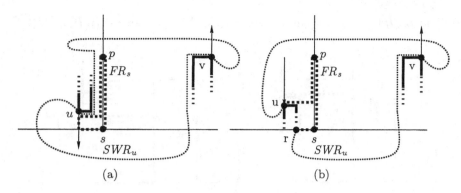

Fig. 3. Illustrating the cases in Lemma 5 when $||s,p||_y + ||s,u||_x \leq ||s,v||_x$.

Next we present CGO-1. Let u and v be vertices as defined earlier. The strategy does the following: if $(ext(u), ext(v)) \in \mathcal{LA} \cup \mathcal{LB}$, we mirror the polygon $\mathbf{P}$ at the vertical line through s and swap the names of u and v. Hence, $(ext(u), ext(v)) \in \mathcal{AR} \cup \mathcal{BR}$. We continue moving upwards updating f_r and v until either all of $\mathbf{Q}_1(s)$ has been seen or $ext(v)$ no longer crosses the vertical line through s.

If all of $\mathbf{Q}_1(s)$ has been seen then we explore the remaining part of $\mathbf{P}$ using PIGO$(p, east, aw, \mathbf{P})$, where p is the current robot position.

If $ext(v)$ no longer crosses the vertical line through s then we either need to continue the exploration by moving to the right or return to u and explore the remaining part of the polygon from there.

If $||s,p||_y + ||s,u||_x \leq ||s,v||_x$ we choose to return to u. If $ext(u) \in \mathcal{A}$ we run PBGO$(u, aw, \mathbf{P})$ and if $ext(u) \in \mathcal{B}$ we use PBGO$(u, cw, \mathbf{P})$; see Figure 3. Otherwise, $||s,p||_y + ||s,u||_x > ||s,v||_x$ and in this case we move to the closest point v' on $ext(v)$. By definition, the extension of v is either in $\mathcal{A}$ or $\mathcal{B}$ in this case.

If $ext(v) \in \mathcal{B}$ then $v = v'$ and we choose to run PBGO$(v, aw, \mathbf{P})$.

Otherwise, $ext(v) \in \mathcal{A}$. If $\mathbf{Q}_1(v')$ is seen from v' then the entire quadrant has been explored and we run PIGO$(v', east, aw, \mathbf{P})$ to explore the remainder of the polygon. If $\mathbf{Q}_1(v')$ is not seen from v' then there are still things hidden from the robot in $\mathbf{Q}_1(v)$. We explore the rest of the quadrant using PBGO$(v', north, aw, \mathbf{Q}_1(v))$ reaching a point q where a second decision needs to be made.

If v is seen from the starting point and $||s,q||_x \leq ||s,v||$, we go back to v and run PBGO$(v, aw, \mathbf{P})$, otherwise we run PIGO$(q, east, cw, \mathbf{P})$ from the interior point q; see Figure 5.

If v is not seen from the starting point s then we go back to v and run PBGO$(v, aw, \mathbf{P})$.

To finish the substrategy CGO-1 our last step is to return to the starting point s.

Lemma 5. *If* CGO-1 *is applied, then* $length(FR_s) \leq \frac{3}{2} length(SWR_s)$.

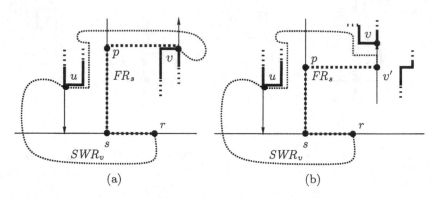

Fig. 4. Illustrating the proof of Lemma 5 when $||s,p||_y + ||s,u||_x > ||s,v||_x$.

Proof. We handle each case separately. Assume for the first case that when FR_s reaches the point p, then $\mathbf{Q}_1(p)$ is completely visible. Hence, we have the same situation as in the proof of Lemma 4 and using the same proof technique it follows that $length(FR_s) = length(SWR_s)$.

Assume for the second case that CGO-1 decides to go back to u, i.e., that $||s,p||_y + ||s,u||_x \leq ||s,v||_x$; see Figures 3(a) and (b). The tour followed equals one of

$$FR_s = \begin{cases} SP(s,p) \cup SP(p,u) \cup SWR_u \cup SP(u,s) \\ SP(s,p) \cup SP(p,u) \cup SWR_u(u,r,cw) \cup SP(r,s) \end{cases}$$

where r is the last intersection point of FR_s with the horizontal line through s. Using that $||s,p||_y + ||s,u||_x \leq ||s,v||_x$ it follows that the length of FR_s in both cases is bounded by

$$length(FR_s) = length(SWR_u) + 2||s,p||_y + 2||s,u||_x \leq length(SWR_s) +$$

$$+ ||s,p||_y + ||s,u||_x + ||s,v||_x \leq \frac{3}{2} length(SWR_s).$$

The inequalities follow from the assumption together with Lemmas 2 and 3.

Assume for the third case that CGO-1 goes to the right, i.e., that $||s,p||_y + ||s,u||_x > ||s,v||_x$. We begin by handling the different subcases that are independent of whether s sees v; see Figures 4(a) and (b). The tour followed equals one of

$$FR_s = \begin{cases} SP(s,v) \cup SWR_v(v,r,aw) \cup SP(r,s) \\ SP(s,v') \cup SWR_{v'}(v',r,aw) \cup SP(r,s) \end{cases}$$

Since $||s,v||_x = ||s,v'||_x$ the length of FR_s is in both subcases bounded by

$$length(FR_s) \leq length(SWR_s) + 2||s,v||_x < length(SWR_s) +$$

$$+ ||s,p||_y + ||s,u||_x + ||s,v||_x \leq \frac{3}{2} length(SWR_s),$$

The inequalities follow from Lemmas 2 and 3.

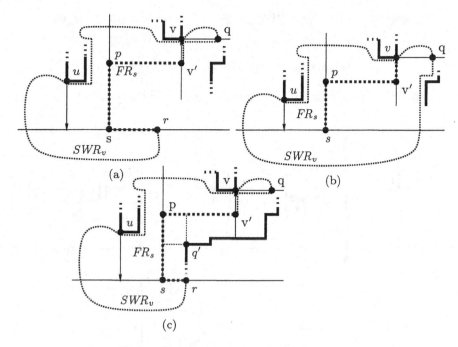

Fig. 5. Illustrating the proof of Lemma 5.

Assume now that CGO-1 goes to the right, i.e., that $||s,p||_y + ||s,u||_x > ||s,v||_x$ and that v is indeed seen from s; see Figures 5(a) and (b). The tour followed in this case is one of

$$FR_s = \begin{cases} SP(s,v) \cup SWR_v(v,q,cw) \cup SP(q,v) \cup SWR_v(v,r,aw) \cup SP(r,s) & (*) \\ SP(s,v) \cup SWR_v \cup SP(v,s) \end{cases}$$

where q is the resulting location after exploring $Q_1(v)$. Here we use that v is seen from s, and hence, that the initial scan guarantees that there is a point t of SWR_s in $Q_{3,4}(s)$ such that $||s,t||_y \geq ||s,v||_x$, thus FR_s is bounded by

$$\begin{aligned} length(FR_s) = length(SWR_v) + 2\min\{||s,v||, ||s,q||_x\} & \leq length(SWR_s) + \\ + ||s,v||_y + ||s,v||_x + ||s,q||_x & < length(SWR_s) + \\ + ||s,v||_y + ||s,t||_y + ||s,q||_x + ||s,u||_x & \leq \frac{3}{2} length(SWR_s). \end{aligned}$$

On the other hand, when v is not seen from s, the tour follows the path marked with (*) above; see Figure 5(c). Thus, the polygon boundary obscures the view from s to v, and hence, there is a point q' on the boundary such that the shortest path from s to v' contains q'. The path our strategy follows between s

and v' is a shortest path and we can therefore assume that it also passed through q'. We use that $||s, q'||_x \leq ||s, v||_x \leq ||s, q||_x$ to get the bound.

$$
\begin{aligned}
length(FR_s) &= length(SWR_v) + 2||s, q'||_x &\leq& \ length(SWR_s) + \\
&+ ||s, v||_x + ||s, q||_x &<& \ length(SWR_s) + \\
&+ ||s, v||_y + ||s, u||_x + ||s, q||_x &\leq& \ \frac{3}{2} length(SWR_s).
\end{aligned}
$$

The inequalities above follow from Lemmas 2 and 3 and this concludes the proof.

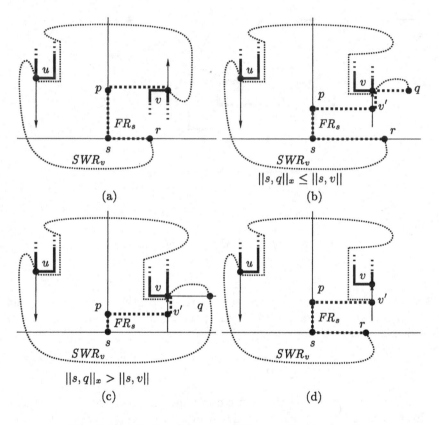

(a)

$||s, q||_x \leq ||s, v||$

(b)

$||s, q||_x > ||s, v||$

(c)

(d)

Fig. 6. Illustrating the cases in the proof of Lemma 6.

We continue the analysis by first showing the substrategy CGO-2 and then claiming its competitive ratio. The strategy does the following: if $||s, u||_x \leq ||s, v||_x$ then we mirror **P** at the vertical line through s also swapping the names of u and v. This means that v is closer to the current point p with respect to x-distance than u. Next, go to v', the closest point on $ext(v)$. If $ext(v) \in \mathcal{B}$, run PBGO$(v, aw, \mathbf{P})$ since $v = v'$. If $ext(v) \in \mathcal{A}$ and $\mathbf{Q}_1(v)$ is seen from v' then we run PIGO$(v', east, aw, \mathbf{P})$. If $ext(v) \in \mathcal{A}$ but $\mathbf{Q}_1(v)$ is not completely seen from v' then we explore $\mathbf{Q}_1(v)$ using PBGO$(v', north, cw, \mathbf{Q}_1(v'))$.

Once $\mathbf{Q}_1(v)$ is explored we have reached a point q and we make a second decision. If $||s,q||_x \leq ||s,v||$, go back to v and run PBGO$(v, aw, \mathbf{P})$, otherwise run PIGO$(q, east, cw, \mathbf{P})$. Finally go back to s.

We claim the following lemma without proof. The proof idea is the same as that of Lemma 5.

Lemma 6. *If* CGO-2 *is applied, then* $length(FR_s) \leq \frac{3}{2} length(SWR_s)$.

We have the following theorem.

Theorem 1. *CGO is 3/2-competitive.*

5   Conclusions

We have presented a 3/2-competitive strategy to explore a rectilinear simple polygon in the L_1 metric.

An obvious open problem is to reduce the gap between the lower bound of 5/4 and our upper bound of 3/2 for deterministic strategies. It would also be interesting to look at variants of this problem, e.g., what if we are only interested in finding a shortest path and not a closed tour that sees all of the polygon; see Deng *et al.* [4].

References

1. M. BETKE, R.L. RIVEST, M. SINGH. Piecemeal Learning of an Unknown Environment. *Machine Learning*, 18(2–3):231–254, 1995.
2. K-F. CHAN, T.W. LAM. An on-line algorithm for navigating in an unknown environment. *International Journal of Computational Geometry & Applications*, 3:227–244, 1993.
3. W. CHIN, S. NTAFOS. Optimum Watchman Routes. *Information Processing Letters*, 28:39–44, 1988.
4. X. DENG, T. KAMEDA, C.H. PAPADIMITRIOU. How to Learn an Unknown Environment I: The Rectilinear Case. *Journal of the ACM*, 45(2):215–245, 1998.
5. M. HAMMAR, B.J. NILSSON, S. SCHUIERER. Improved Exploration of Rectilinear Polygons. *Nordic Journal of Computing*, 9(1):32–53, 2002.
6. F. HOFFMANN, C. ICKING, R. KLEIN, K. KRIEGEL. The Polygon Exploration Problem. *SIAM Journal on Computing*, 31(2):577–600, 2001.
7. J.M. KLEINBERG. On-line search in a simple polygon. In *Proc. of 5th ACM-SIAM Symp. on Discrete Algorithms*, pages 8–15, 1994.
8. A. MEI, Y. IGARASHI. An Efficient Strategy for Robot Navigation in Unknown Environment. *Inform. Process. Lett.*, 52:51–56, 1994.
9. C.H. PAPADIMITRIOU, M. YANNAKAKIS. Shortest Paths Without a Map. *Theoret. Comput. Sci.*, 84(1):127–150, 1991.

An Improved Approximation Algorithm for Computing Geometric Shortest Paths*

Lyudmil Aleksandrov[1], Anil Maheshwari[2], and Jörg-Rüdiger Sack[2]

[1] Bulgarian Academy of Sciences, CICT,
Acad. G. Bonchev Str. Bl. 25-A, 1113 Sofia, Bulgaria
[2] School of Computer Science, Carleton University,
Ottawa, Ontario K1S5B6, Canada

Abstract. Consider a polyhedral surface consisting of n triangular faces where each face has an associated positive weight. The cost of travel through each face is the Euclidean distance traveled multiplied by the weight of the face. We present an approximation algorithm for computing a path such that the ratio of the cost of the computed path with respect to the cost of a shortest path is bounded by $(1 + \varepsilon)$, for a given $0 < \varepsilon < 1$. The algorithm is based on a novel way of discretizing the polyhedral surface. We employ a generic greedy approach for solving shortest path problems in geometric graphs produced by such discretization. We improve upon existing approximation algorithms for computing shortest paths on polyhedral surfaces [1,4,5,10,12,15].

1 Introduction

Shortest path problems are among the fundamental problems studied in computational geometry and graph algorithms. These problems arise naturally in application areas such as motion planning, navigation and geographical information systems. Aside from the importance of shortest paths problems in their own right, often they appear in the solutions of other problems. Existing algorithms for many shortest path problems, are quite complex in design and implementation or have very large time and space complexities. Hence they are unappealing to practitioners and pose a challenge to theoreticians. This coupled with the fact that geographic and spatial models are approximations of reality and high-quality paths are favored over optimal paths that are "hard" to compute, approximation algorithms are suitable and necessary.

In this paper we present algorithms for computing approximate shortest paths on (weighted) polyhedral surfaces. Our solutions employ the paradigm of partitioning a continuous geometric search space into a discrete combinatorial search space. Discretization methods are natural, theoretically interesting, and enable implementation. They transform geometric shortest path problems into combinatorial shortest path problems in graphs. Shortest path problems in graphs are well studied and general solutions with implementations are readily

---

* Research supported in part by NSERC

A. Lingas and B.J. Nilsson (Eds.): FCT 2003, LNCS 2751, pp. 246–257, 2003.
© Springer-Verlag Berlin Heidelberg 2003

available. We consider surfaces that are polyhedral 2-manifolds, whereas most of the previous algorithms were designed to handle particular geometric instances, such as convex polyhedra, or possibly non-convex hole-free polyhedra, etc. Also, we allow arbitrary (positive) weights to be assigned to the faces of the domain thus generalizing from uniform and obstacle avoidance scenarios. While shortest paths graph algorithms are available and applicable to the graphs generated here, the geometric structure of shortest path problems can be exploited for the design of more efficient algorithms.

Brief Literature Review: Shortest path problems can be categorized by various factors which include the dimensionality of the space, the type and the number of objects or obstacles, and the distance measure used. We discuss those contributions which relate directly to this paper. The following table summarizes the results for shortest path problems on polyhedral surfaces. We need a few preliminaries in order to comprehend the table. Let P be a polyhedral surface in 3-dimensional Euclidean space consisting of n triangular faces. A path π' is an $1+\epsilon$ approximation of a shortest path π between two vertices of P if $||\pi'|| \leq (1+\epsilon)||\pi||$, where $||\pi||$ denotes the length of π and $\epsilon > 0$. A natural generalization of the Euclidean shortest path problems are shortest path problems set in weighted surfaces. In this problem a triangulated polyhedral surface is given consisting of n faces, where each face has a positive weight representing the cost of traveling through that face. The cost of a path is defined to be the sum of Euclidean lengths multiplied by the face weights of the sub-paths within each face traversed. (Results on weighted shortest paths involve geometric parameters and they have been omitted for the sake of clarity.)

Surface	Cost Metric	Approx. Ratio	Time Complexity	Reference
Convex	Euclidean	Exact	$O(n^3 \log n)$	[14]
Non-convex	Euclidean	Exact	$O(n^2 \log n)$	[11]
Non-convex	Euclidean	Exact	$O(n^2)$	[7]
Non-convex	Euclidean	Exact	$O(n \log^2 n)$	[9]
Convex	Euclidean	2	$O(n)$	[8]
Convex	Euclidean	$1 + \epsilon$	$O(n \log \frac{1}{\epsilon} + 1/\epsilon^3)$	[3]
Convex	Euclidean	$1 + \epsilon$	$O(n/\sqrt{\epsilon} + 1/\epsilon^4)$	[2]
Non-convex	Euclidean	$7(1 + \epsilon)$	$O(n^{5/3} \log^{5/3} n)$	[1]
Non-convex	Euclidean	$15(1 + \epsilon)$	$O(n^{8/5} \log^{8/5} n)$	[1]
Non-convex	Weighted	$(1 + \epsilon)$	$O(n^8 \log \frac{n}{\epsilon})$	[12]
Non-convex	Weighted	Additive	$O(n^3 \log n)$	[10]
Non-convex	Weighted	$(1 + \epsilon)$	$O(\frac{n}{\epsilon^2} \log n \log \frac{1}{\epsilon})$	[4]
Non-convex	Weighted	$(1 + \epsilon)$	$O(\frac{n}{\epsilon} \log \frac{1}{\epsilon}(\frac{1}{\sqrt{\epsilon}} + \log n))$	[5]
Non-convex	Weighted	$(1 + \epsilon)$	$O(\frac{n}{\epsilon} \log \frac{n}{\epsilon} \log \frac{1}{\epsilon})$	[15]
Non-convex	Weighted	$(1 + \epsilon)$	$O(\frac{n}{\sqrt{\epsilon}} \log \frac{n}{\epsilon} \log \frac{1}{\epsilon})$	**This paper**

From practical point of view the "exact" algorithms are unappealing, since they are fairly complex, numerically unstable and may require exponential number of bits to perform the computation associated to "unfolding" of faces. These drawbacks have motivated researchers to look into practical approximation algorithms. Approximation algorithms of [8,2,10,15,5,4] have been implemented.

New Results - Overview and Significance: Results of this paper are

1. We provide a new discretization of polyhedral surfaces. For a given approximation parameter $\varepsilon \in (0,1)$, the size of the discretization for a polyhedral surface consisting of n triangular faces is $O(\frac{n}{\sqrt{\varepsilon}} \log \frac{1}{\varepsilon})$. We precisely evaluate the constants hidden in the big-O notation. (Section 2)

2. We define approximation graphs with nodes corresponding to the Steiner points of the discretization. We show that the distance between any pair of nodes in the approximation graph is within a factor of $(1+\varepsilon)$ times the cost of a shortest path in the corresponding surface. (Section 3)

3. We describe a greedy approach for solving the single source shortest path (SSSP) problem in the approximation graph and obtain an $O(\frac{n}{\sqrt{\varepsilon}} \log \frac{n}{\varepsilon} \log \frac{1}{\varepsilon})$ time $(1 + \varepsilon)$-approximation algorithm for SSSP problem on a polyhedral surface. (Section 4)

Our scheme places Steiner points, for the first time, in the interior of the faces and not on the face boundaries. While this is somewhat counter-intuitive, we can show that the desired approximation properties can still be proven, but now using a much sparser mesh. (In addition this leads to algorithmic simplifications by avoiding the construction of "cones" used in [5].) The size of the discretization is smaller than those previously established and the improvement is by a factor of $\sqrt{\varepsilon}$. A greedy approach for computing SSSP in the approximation graph has been proposed in [15]. Edges in our approximation graphs do not correspond to line segments as required in their algorithm, as well as their approach does not seem to generalize to 3-dimensions. We propose an alternative greedy algorithm, which is applicable here as well as generalizes to 3-dimensions.

Geographical information systems are an immediate application domain for shortest path problems on polyhedral surfaces and terrains. In such applications, the number of faces, n, may be huge (several million). Storage and time complexities are functions on n and constants are critical. In terms of computational complexity our algorithm improves upon previous approximation algorithms for solving shortest path problems on polyhedral surfaces [1,4,5,10,12,15]. The running time of our algorithm improves upon the most recent algorithm of [15] by a factor of $\sqrt{\varepsilon}$. Ignoring the geometric parameters, the original algorithm of [12] has been improved by about $1/n^7$. The algorithm of [12] uses $O(n^4)$ space. This was improved substantially in [5,15]. The discretization presented here improves further on the storage requirement by reducing the number of Steiner points by $\sqrt{\varepsilon}$ over [5,15].

The practicality of discretization for solving geodesic shortest path problems has been demonstrated in [10,15,16]. From a theoretical viewpoint the discretization scheme proposed here is more complex and requires very careful analysis, its implementation would however be similar to our previous ε-schemes [4,5]. These have been implemented and experimentally verified in [16]. More precisely, the algorithm presented here does not require any complex data structures (just linked lists, binary search trees, and priority queues). Existing software libraries for computing shortest paths in graphs (Dijkstra's algorithm) can be used. We provide explicit calculation of key constants often hidden through the use of the $big - O$-notation. The constant in the estimate on the total number of Steiner points (Lemma 1) is $12\Gamma \log L$, where Γ is the average of the reciprocals of the sinuses of the angles of the faces of P. For example, if no face of P has angles smaller than $10°$, then $\Gamma \leq 5$. Moreover the simplicity of our algorithm, coupled with the fact that we obtain theoretically guaranteed approximation factors, should make it a very promising candidate for the application domain. It is important to note that the edges and Steiner points of the discretization can be produced on-the-fly. When Dijkstra's algorithm requests edges incident to the current vertex all incident edges (connecting Steiner points) are generated.

2   Preliminaries and Discretization

Let P be a triangulated polyhedral surface in the 3-dimensional Euclidean space. P can be any polyhedral 2-manifold. We do not assume any additional geometrical or topological properties such as convexity, being a terrain, or absence of holes, etc. Assume that P consists of n triangular faces denoted by $t_1, \ldots, t_n$. Positive weights $w_1, \ldots, w_n$ are associated with triangles $t_1, \ldots, t_n$ representing the cost of traveling inside them. The cost of traveling along an edge is the minimum of the weights of the triangles incident to that edge. Edges are assumed to be part of the triangle, from which they inherit their weight. Any continuous (rectifiable) curve lying in P is called a path. The cost of a path π is defined by $\|\pi\| = \sum_{i=1}^{n} w_i |\pi_i|$, where $|\pi_i|$ denotes the Euclidean length of the intersection of π with triangle t_i, i.e., $\pi_i = \pi \cap t_i$. Given two distinct points u and v in P a minimum cost path $\pi(u, v)$ joining u and v is called a *geodesic path*. Without loss of generality we may assume that u and v lie on a boundary of a face. In this setting, it is well known that geodesic paths are simple (non self-intersecting) and consist of a sequence of segments, whose endpoints are on the edges of P. The intersection of a geodesic path with the interior of faces or edges is a set of disjoint segments. More precisely, each segment on a geodesic path is of one of the following two types: 1) *face-crossing* – a segment which crosses a face joining two points on its boundary; 2) *edge-using* – a sub-segment of an edge. We define *linear paths* to be simple paths consisting of face-crossing and edge-using segments exclusively. Thus, any geodesic path is a linear path. A linear path $\pi(u, v)$ is represented as a sequence of its segments $\{s_1, \ldots, s_l\}$ or equivalently as a sequence of points $a_0, \ldots, a_{l+1}$ lying on the edges, that are endpoints of these segments, i.e., $s_i = (a_{i-1}, a_i)$, $u = a_0$, and $v = a_{l+1}$. Points a_i that are not vertices of P are called *bending points* of the path. Geodesic paths satisfy Snell's law of refraction at each of their bending points (see [12] for details).

In the following we introduce a function $d(x)$ defined as the minimum Euclidean distance from a point $x \in P$ to the edges around x. The distance $d(x)$ is a lower bound on the length of a face-crossing segment incident to x and plays essential role in our constructions.

Definition 1. *Given a point $x \in P$ let $E(x)$ be the set of edges of triangles incident to x minus the edges incident to x. The distance $d(x)$ is defined as the minimum Euclidean distance from x to the edges in $E(x)$.*

Throughout the paper ε is a real number in $(0, 1)$. Next, we define a set of points on P, called Steiner points, that together with vertices of P constitute an $(1+\varepsilon)$-approximation mesh for the set of linear paths on P. That is, we define a graph G_ε whose set of nodes consists of the vertices of P and the Steiner points. The edges of G_ε correspond to local shortest paths between their endpoints and have cost equal to the cost of their corresponding path. Then we show how the graph G_ε can be used to approximate geodesic paths between vertices of P. Using Definition 1 above, for each vertex v of P we define a weighted radius

$$r(v) = \frac{w_{\min}(v)}{7w_{\max}(v)} d(v), \tag{1}$$

where $w_{\max}(v)$ and $w_{\min}(v)$ are the maximum and the minimum weights of the faces incident to v. By using the weighted radius $r(v)$ for each face incident to v we define a "small" isosceles triangle with two sides of length $\varepsilon r(v)$ incident to v. These triangles around v form a star shaped polygon $S(v)$, which we call a *vertex-vicinity* of v.

In all previous approximation schemes Steiner points have been placed on the edges of P. Here we place Steiner points inside faces of P. In this way we reduce the total number of Steiner points by a factor of $\sqrt{\varepsilon}$. We will need to show that the $(1+\varepsilon)$-approximation property of the resulting mesh is preserved. Let triangle t be a face of P. Steiner points inside t are placed along the three bisectors of t as follows. Let v be a vertex of t and ℓ be the bisector of the angle α of t at v. We define a set of Steiner points $p_1, \ldots, p_k$ on ℓ by

$$|p_{i-1}p_i| = \sin(\alpha/2)\sqrt{\varepsilon/2}|vp_{i-1}|, \quad i = 1, \ldots, k, \tag{2}$$

where p_0 is the point on ℓ and on the boundary of the vertex vicinity $S(v)$ (Figure 1). The next lemma establishes estimates on the number of Steiner points inserted on a particular bisector and on their total number.

Lemma 1. *(a) The number of Steiner points inserted in a bisector ℓ of an angle α at a vertex v is bounded by $C(\ell)\frac{1}{\sqrt{\varepsilon}}\log_2\frac{2}{\varepsilon}$, where the constant $C(\ell) < \frac{4}{\sin\alpha}\log_2\frac{|\ell|}{r(v)\cos(\alpha/2)}$. (b) The total number of Steiner points on P is less than*

$$C(P)\frac{n}{\sqrt{\varepsilon}}\log_2\frac{2}{\varepsilon}, \tag{3}$$

where $C(P) < 12\Gamma \log L$ and L is the maximum of the ratios $|\ell(v)|/r(v)\cos(\alpha/2)$ and Γ is the average of the reciprocals of the sinuses of angles on P, i.e. $\Gamma = \frac{1}{3n}\sum_{i=1}^{3n}\frac{1}{\sin\alpha_i}$.

Proof: We estimate the number of Steiner points on a bisector ℓ of an angle α at a vertex v. From (2) it follows, that $|vp_i| = \lambda^i \varepsilon r(v) \cos(\alpha/2)$, where $\lambda = (1 + \sqrt{\varepsilon/2} \sin(\alpha/2))$. Therefore the number of the Steiner points on ℓ is

$$k \leq \log_\lambda \frac{|\ell|}{\varepsilon r(v) \cos(\alpha/2)} = \frac{\ln \frac{|\ell|}{2r(v)\cos(\alpha/2)} + \ln \frac{2}{\varepsilon}}{\ln(1 + \sqrt{\varepsilon/2}\sin(\alpha/2))} \leq \frac{4\log_2 \frac{|\ell|}{r(v)\cos(\alpha/2)}}{\sin\alpha\sqrt{\varepsilon}} \log_2 \frac{2}{\varepsilon}.$$

This proves (a). Estimate (b) is obtained by summing up (a) over all bisectors on P. □

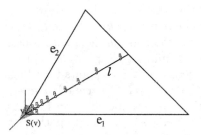
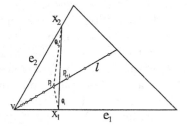

Fig. 1. (Left) Steiner points inserted in a bisector ℓ are shown. (Right) Proof of Lemma 2 is illustrated: Sinuses of angles $\angle p_i x_1 p_{i+1}$ and $\angle p_i x_2 p_{i+1} \leq \sqrt{\varepsilon/2}$, implying $|x_1 p_i| + |p_i x_2| \leq (1 + \varepsilon/2)|x_1 x_2|$.

The set of Steiner points partitions bisectors into intervals, that we call *Steiner intervals*. The following lemma establishes two important properties of Steiner intervals (Figure 1).

Lemma 2. *(a) Let ℓ be the bisector of the angle formed by edges e_1 and e_2 of P. If (p_i, p_{i+1}) is a Steiner interval on ℓ and x is a point on e_1 or e_2, then*

$$\sin(\angle p_i x p_{i+1}) \leq \sqrt{\varepsilon/2}. \tag{4}$$

(b) Let x_1 and x_2 be points on e_1 and e_2 and outside the vertex vicinity of the vertex incident to e_1 and e_2. If p is the Steiner point closest to the intersection between the segment (x_1, x_2) and ℓ, then

$$|x_1 p| + |p x_2| \leq (1 + \varepsilon/2)|x_1 x_2|. \tag{5}$$

Proof: The statement (a) follows easily from the definition of Steiner points. Here we prove (b). Let us denote by θ, θ_1, and θ_2 the angles of the triangle $p x_1 x_2$ at p, x_1 and x_2 respectively. From (a) and $\varepsilon \leq 1$ it follows that $\theta \geq \pi/2$ and we have $|x_1 p| + |p x_2| = (1 + \frac{2\sin(\theta_1/2)\sin(\theta_2/2)}{\sin(\theta/2)})|x_1 x_2| = (1 + \frac{\sin(\theta_1)\sin(\theta_2)}{2\sin(\theta/2)\cos(\theta_1/2)\cos(\theta_2/2)})|x_1 x_2| \leq (1 + \frac{\varepsilon}{4\sin^2(\theta/2)})|x_1 x_2| \leq (1 + \varepsilon/2)|x_1 x_2|.$ □

3   Discrete Paths

Next, we define a graph $G_\varepsilon = (V(G_\varepsilon), E(G_\varepsilon))$. The set of nodes $V(G_\varepsilon)$ consists of the set of vertices of P and the set of Steiner points. The set of edges $E(G_\varepsilon)$ is defined as follows. A node that is a vertex of P is connected to all Steiner points on bisectors in the faces incident to this vertex. The cost of these edges equals the cost of the shortest path between its endpoints restricted to lie inside the triangle containing them. These shortest paths consist either of a single segment joining the vertex and the corresponding Steiner point or of two segments the first of which follows one of the edges incident to the vertex. The rest of the edges of G_ε join pairs of Steiner points lying on neighboring bisectors as follows. Let e be an edge of P. In general, there are four bisectors incident to e. We define graph edges between pairs of nodes (Steiner points) on these four bisectors. We refer to all these edges as edges of G_ε crossing the edge e of P. Let (p,q) be an edge between Steiner points p and q crossing e. The cost of (p,q) is defined as the cost of the shortest path between p and q restricted to lie inside the quadrilateral formed by the two triangles around e, that is $\|pq\| = \min_{x,y \in e} \|px\| + \|xy\| + \|yq\|$. (Note that we do not need edges in G_ε between pairs of Steiner points for which the local shortest paths do not intersect e.) Paths in G_ε are called *discrete paths*. The cost of a discrete path π is the sum of the costs of its edges and is denoted by $\|\pi\|$. Note that if we replace each of the edges in a discrete path with the corresponding segments (at most three) forming the shortest path used to compute its cost we obtain a path on P of the same cost.

Theorem 1. *Let $\tilde{\pi}(v_0, v)$ be a linear path joining two different vertices v_0 and v on P. There exists a discrete path $\pi(v_0, v)$, such that $\|\pi\| \leq (1 + \varepsilon)\|\tilde{\pi}\|$.*

Proof: First, we discuss the structure of linear paths. Following from the definition, a linear path $\tilde{\pi}(v_0, v)$ consists of face-crossing and edge-using segments and is determined by the sequence of their endpoints, called bending points, which are located on the edges of P. Following the path from v_0 and on, let a_0 be the last bending point on $\tilde{\pi}$ that is inside the vertex vicinity $S(v_0)$. Next, let b_1 be the first bending point after a_0 that is in a vertex vicinity, say $S(v_1)$, and let a_1 be the last bending point in $S(v_1)$. Continuing in this way, we define a sequence of vertices of $v_0, v_1, \ldots, v_l = v$ and a sequence of bending points $a_0, b_1, a_1, \ldots, a_{l-1}, b_l$ on $\tilde{\pi}$, such that for $i = 0, \ldots, l$, points b_i, a_i are in $S(v_i)$ (we assume $b_0 = v_0$, $a_l = v$). Furthermore, portions of $\tilde{\pi}$ between a_i and b_i do not intersect vertex vicinities. Thereby, the path $\tilde{\pi}$ is partitioned into portions

$$\tilde{\pi}(v_0, a_0), \tilde{\pi}(a_0, b_1), \tilde{\pi}(b_1, a_1), \ldots, \tilde{\pi}(b_l, v). \qquad (6)$$

Portions $\tilde{\pi}(a_i, b_{i+1})$ for $i = 0, \ldots, l-1$ are called *between vertex vicinities* portions and portions $\tilde{\pi}(b_i, a_i)$ for $i = 0, \ldots, l$ ($b_0 = v_0$), are called *vertex vicinities* portions. Consider a between vertex vicinity portion $\tilde{\pi}(a_i, b_{i+1})$ for some $0 \leq i < l$. We define $\tilde{\pi}'(v_i, v_{i+1})$ to be the linear path from v_i to v_{i+1} along the sequence of inner bending points of $\tilde{\pi}(a_i, b_{i+1})$. Using triangle inequality and the definition of vertex vicinities (1) we obtain

$$\|\tilde{\pi}'(v_i, v_{i+1})\| \leq \|\tilde{\pi}(a_i, b_{i+1})\| + \|v_i a_i\| + \|b_{i+1} v_{i+1}\| \leq$$
$$\|\tilde{\pi}(a_i, b_{i+1})\| + \frac{\varepsilon}{7}(w_{\min}(v_i)d(v_i) + w_{\min}(v_{i+1})d(v_{i+1})). \quad (7)$$

Changing all between vertex vicinities portions in this way we obtain a linear path $\tilde{\pi}'(v_0, v) = \{\tilde{\pi}'(v_0, v_1), \tilde{\pi}'(v_1, v_2), \ldots, \tilde{\pi}'(v_{l-1}, v)\}$, consisting of between vertex vicinities portions only.

Next, we approximate each of these portions by a discrete path. Consider a portion $\tilde{\pi}'_i = \tilde{\pi}'(v_i, v_{i+1})$ for some $0 \leq i < l$ and let $s_j = (x_{j-1}, x_j)$, $j = 1, \ldots, \nu$ be the segments forming this portion $(x_0 = v_i, x_\nu = v_{i+1})$. Segments s_j are face-crossing and edge-using segments. Indeed, there are no consecutive edge-using segments. Let s_j be a face-crossing segment. Then s_j intersects the bisector ℓ_j of the angle formed by the edges of P containing the end-points of s_j. We define p_j to be the closest Steiner point to the intersection between s_j and ℓ_j. Now we replace each of the face-crossing segments s_j of $\tilde{\pi}'_i$ by two segments path x_{j-1}, p_j, x_j and denote the obtained path by $\tilde{\pi}''_i$. From (5) it follows that $\|\tilde{\pi}''_i\| \leq (1 + \varepsilon/2)\|\tilde{\pi}'_i\|$. The sequence of bending points of $\tilde{\pi}''_i$ contains as a subsequence the Steiner points $p_{j_1}, \ldots, p_{j_{\nu_1}}$, $(\nu_1 \leq \nu)$ corresponding to the face-crossing segments of $\tilde{\pi}'_i$. Note that pairs (v_i, p_{i_1}) and $(p_{i_{\nu_1}}, v_{i+1})$ are adjacent in G_ε. Furthermore, between any two consecutive Steiner points $p_{j_\mu}, p_{j_{\mu+1}}$ there is at most one edge-using segment and, according our definition of the graph G_ε, they are connected in G_ε. The cost of each edge $(p_{j_\mu}, p_{j_{\mu+1}})$ is at most the cost of the portions of $\tilde{\pi}''_i$ from p_{j_μ} to $p_{j_{\mu+1}}$. Therefore, the sequence of nodes $\{v_i, p_{j_1}, \ldots, p_{j_{\nu_1}}, v_{i+1}\}$ defines a discrete path $\pi(v_i, v_{i+1})$ such that

$$\|\pi(v_i, v_{i+1})\| \leq \|\tilde{\pi}''_i\| \leq (1 + \varepsilon/2)\|\tilde{\pi}'(v_i, v_{i+1})\|. \tag{8}$$

We combine discrete paths $\pi(v_0, v_1), \ldots, \pi(v_{l-1}, v)$ and obtain a discrete path $\pi(v_0, v)$ from v_0 to v. We complete the proof by estimating the cost of this path. We denote $w_{\min}(v_i)d(v_i) + w_{\min}(v_{i+1})d(v_{i+1})$ by κ_i and use (8), (7) obtaining

$$\|\pi(v_0, v)\| = \sum_{i=0}^{l-1} \|\pi((v_i, v_{i+1})\| \leq (1 + \varepsilon/2) \sum_{i=0}^{l-1} \|\tilde{\pi}'(v_i, v_{i+1})\| \leq$$

$$(1 + \varepsilon/2) \sum_{i=0}^{l-1} (\|\tilde{\pi}(a_i, b_{i+1})\| + \varepsilon\kappa_i/7) \leq (1 + \varepsilon/2)\|\tilde{\pi}(v_0, v)\| + (3\varepsilon/14) \sum_{i=0}^{l-1} \kappa_i. \tag{9}$$

It remains to estimate the sum $\sum_{i=0}^{l-1} \kappa_i$ appearing above. From the definitions of $d(\cdot)$, (6), and (1) it follows that $\kappa_i \leq 2\|\tilde{\pi}(a_i, b_{i+1})\| + \|v_i a_i\| + \|b_{i+1} v_{i+1}\| \leq 2\|\tilde{\pi}(a_i, b_{i+1})\| + \kappa_i/7$. Thus $\kappa_i \leq (7/3)\|\tilde{\pi}(a_i, b_{i+1})\|$ and substituting this in (9) we obtain the desired estimate $\|\pi(v, v_0)\| \leq (1 + \varepsilon)\|\tilde{\pi}(v_0, v)\|$. $\qquad\square$

4   Algorithms

In this section we discuss algorithms for solving the Single Source Shortest Paths (SSSP) problem in approximation graphs G_ε. Straightforwardly, one can apply Dijkstra's algorithm. When implemented using Fibonacci heaps it would solve SSSP problem in $O(|E_\varepsilon| + |V_\varepsilon| \log |V_\varepsilon|)$ time. By Lemma 1, $|V_\varepsilon| = O(\frac{n}{\sqrt{\varepsilon}} \log \frac{1}{\varepsilon})$ and by the definition of edges $|E_\varepsilon| = O(\frac{n}{\varepsilon} \log^2 \frac{1}{\varepsilon})$. Thus it follows that the SSSP problem can be solved by Dijkstra's algorithm in $O(\frac{n}{\varepsilon} \log \frac{n}{\varepsilon} \log \frac{1}{\varepsilon})$ time. Already this time matches the best previously known bound [15]. In the remainder of this section we show how geometric properties of our model can be used to obtain a more efficient algorithm for SSSP in the corresponding approximation graph. More precisely, we present an algorithm that runs in $O(|V_\varepsilon| \log |V_\varepsilon|) = O(\frac{n}{\sqrt{\varepsilon}} \log \frac{n}{\varepsilon} \log \frac{1}{\varepsilon})$ time.

First, we discuss the general structure of our algorithm. Let $G(V, E)$ be a directed graph with positive costs (lengths) assigned to its edges and s be a fixed vertex of G. The SSSP problem is to find shortest paths from s to any other vertex of G. The standard greedy approach for solving the SSSP problem works as follows: a subset of vertices S to which the shortest path has already been found and a set of edges $E(S)$ connecting S with $S^a \subset V \setminus S$ is maintained. The set S^a consists of vertices not in S but adjacent to S. In each iteration an *optimal* edge $e(S) = (u, v)$ in $E(S)$ is selected. Its target v is added to S and $E(S)$ is updated correspondingly. An edge $e = e(S)$ is optimal for S if it minimizes the value $\delta(u) + c(e)$, where $\delta(u)$ is the distance from s to u and $c(e)$ is the cost of e. The correctness of this approach follows from the fact that when $e = (u, v)$ is optimal the distance $\delta(v)$ is equal to $\delta(u) + c(e)$.

Different strategies for maintaining information about $E(S)$ and finding an optimal edge $e(S)$ in each iteration result in different algorithms for computing SSSP. For example, Dijkstra's algorithm maintains only a subset $Q(S)$ of $E(S)$, which however always contains an optimal edge. Namely, for each vertex v in S^a Dijkstra's algorithm keeps in $Q(S)$ one edge only – the one that ends the shortest path to v using vertices in S only. Alternatively, one may maintain a subset $Q(S)$ of $E(S)$ containing one edge per vertex $u \in S$. The target vertex of this edge is called *representative* of u and is denoted by $\rho(u)$. The vertex u itself is called *predecessor* of its representative. The representative $\rho(u)$ is defined to be the target of the minimum cost edge in the *propagation set* $I(u)$ of u, where $I(u) \subset E(S)$ consists of all edges (u, v) such that $\delta(u) + c(u, v) \le \delta(u') + c(u', v)$ for any other vertex $u' \in S$ (ties are broken arbitrarily). The union of propagation sets forms a subset $Q(S)$ of $E(S)$, that always contains an optimal edge. Propagation sets $I(u)$ for u form a partition of $Q(S)$, which we call a *Propagation Diagram* and denote by $\mathcal{I}(S)$. Similar scheme has been used by [15].

A possible implementation of this alternative strategy is to maintain the set of representatives $R \subset S^a$ organized in a priority queue, where a key of a vertex $\rho(u)$ in R is defined to be $\delta(u) + c(u, \rho(u))$. Observe that the edge corresponding to the minimum in R is an optimal edge for S. In each iteration the minimum key node v in R is selected and the following three steps are implemented:

Step 1. The vertex v is moved from R into S. Then the propagation set $I(v)$ is computed and the propagation diagram $\mathcal{I}(S)$ is updated accordingly.

Step 2. The representative $\rho(v)$ of v and a new representative $\rho(u)$ for the predecessor u of v are computed.

Step 3. The new representatives $\rho(u)$ and $\rho(v)$ are either inserted into R together with their corresponding keys, or (if they are already in R) their keys are updated and the decrease key operation is executed in R if necessary.

Clearly, this leads to a correct algorithm for solving the SSSP problem in G. The total time for the priority queue operations if R is implemented with Fibonacci heaps is $O(|V| \log |V|)$. Therefore the efficiency of this strategy depends on the maintenance of the propagation diagrams, the complexity of the propagation sets and efficient updates of the new representatives.

Our approach is as follows. We partition the set of edges $E(S)$ into groups, so that the propagation sets and the corresponding propagation diagrams when restricted to a fixed group become simple and allow efficient updates. Then for each vertex u in S we will keep multiple representatives in R, one for each group, where edges incident to u participate. As a result a vertex in S^a will eventually have multiple predecessors. As we describe below, the number of groups where u can participate will be $O(1)$. We will be able to compute new representatives in $O(1)$ time and update propagation diagrams in logarithmic time in our approximation graphs G_ε. Next, we present some details and state the complexity of the resulting algorithm.

The edges of the approximation graph G_ε were defined to join pairs of nodes (Steiner points) lying on neighboring bisectors, where two bisectors are neighbors if the angles they split share an edge of P. Since the polyhedral surface P is triangulated a fixed bisector may have at most six neighbors. We can partition the set of edges of G_ε into groups $E(\ell, \ell_1)$ corresponding to pairs of neighboring bisectors ℓ and ℓ_1. For a node u on a bisector ℓ we maintain one representative $\rho(u, \ell_1)$ per each bisector ℓ_1 neighboring ℓ. The representative $\rho(u, \ell_1)$ is defined to be the target of the minimum cost edge in the propagation set $I(u; \ell, \ell_1)$, consisting of the edges (u, v) in $E(\ell, \ell_1)$, such that $\delta(u) + c(u, v) \leq \delta(u') + c(u', v)$ for any node $u' \in \ell \cap S$. A node on ℓ with a non-empty propagation set on ℓ_1 will be called *active* for $E(\ell, \ell_1)$.

Consider now an iteration of our greedy algorithm. Let v be the node produced by Extract_min operation in the priority queue R comprising of representatives. Denote the set of predecessors of v by $R^{-1}(v)$. Our task is to compute new representatives for v and for each of the predecessors $u \in R^{-1}(v)$. Consider first the case when v is a vertex of the polyhedral surface P. We assume that the edges incident to a vertex v have been sorted with respect to their cost and when a new representative for v is required we simply report the target of the smallest cost edge joining v with S^a. Thereby the new representative for a node that is a vertex of P can be computed in constant time. The total number of edges incident to vertices of P is $O(\frac{n}{\sqrt{\varepsilon}} \log \frac{1}{\varepsilon})$ and their sorting in a preprocessing step takes $O(\frac{n}{\sqrt{\varepsilon}} \log^2 \frac{1}{\varepsilon})$ time. Consider now the case when v is a node on a bisector say ℓ. An efficient computation of representatives in this case is based on the following two lemmas.

Lemma 3. *The propagation set $I(v; \ell, \ell_1)$ for an active node v is characterized by an interval (x_1, x_2) on ℓ_1, i.e., it consists of all edges in $E(\ell, \ell_1)$ whose targets belong to (x_1, x_2). Furthermore, the function $dist(v, x)$, measuring the cost of the shortest path from v to x restricted to lie in the union of the two triangles containing ℓ and ℓ_1, is convex in (x_1, x_2).*

Lemma 4. *Let* $v_1, \ldots, v_k$ *be the active vertices for* $E(\ell, \ell_1)$. *The propagation diagram* $\mathcal{I}(\ell, \ell_1) = \mathcal{I}(v_1, \ldots, v_k)$ *is characterized by* k *intervals. Updating the diagram* $\mathcal{I}(v_1, \ldots, v_k)$ *to the propagation diagram* $\mathcal{I}(v_1, \ldots, v_k, v)$, *where* v *is a new active node in* ℓ *takes* $O(\log k)$ *time.*

Thus to compute a new representative of v on a neighboring bisector ℓ_1 we update the propagation diagram $\mathcal{I}(\ell, \ell_1)$. Then we consider the interval characterizing the propagation set $I(v; \ell, \ell_1)$ and select the minimum cost edge whose target is in that interval and in S^a. Assume that nodes on ℓ_1 currently in S^a are maintained in a doubly linked list with their positions on ℓ_1. Using the convexity of the function $dist(v, x)$ this selection can be done in time logarithmic on the number of these nodes, which is $O(\log \frac{1}{\varepsilon})$. There are at most six new representatives of v corresponding to bisectors around ℓ to be computed. Thus the total time for updates related to v is $O(\log \frac{1}{\varepsilon})$. The update of the representative for a node $u \in R^{-1}(v)$ on ℓ takes constant time since no change in the propagation set $I(u; \cdot, \ell)$ occurred and the new representative of u is a neighbor to the current one in the list of nodes in S^a on ℓ. The set of predecessors $R^{-1}(v)$ contains at most six vertices and thus their representatives are updated in constant time. So computing representatives in an iteration takes $O(\log \frac{1}{\varepsilon})$ time and in total $O(|V_\varepsilon| \log \frac{1}{\varepsilon})$. The following theorem summarizes the result of this section.

Theorem 2. *The SSSP problem in the approximation graph* G_ε *for a polyhedral surface* P *can be solved in* $O(\frac{n}{\sqrt{\varepsilon}} \log \frac{n}{\varepsilon} \log \frac{1}{\varepsilon})$ *time.*

In the following theorem we summarize the main result of this paper. Starting from a vertex v_0 our algorithm solves SSSP problem in the graph G_ε and construct shortest paths tree rooted at v_0. According to Theorem 1 output distances from v_0 to other vertices of P are within a factor of $1 + \varepsilon$ from the cost of the shortest paths. Using the definition of the edges of G_ε an approximate shortest path between pair of vertices can be output in time proportional to the number of segments in this path. The approximate shortest paths tree rooted at v_0 and containing all Steiner points and vertices of P can be output in $O(|V_\varepsilon|)$ time. Thus we have established the following theorem.

Theorem 3. *Let* P *be a weighted polyhedral surface with* n *triangular faces and* $\varepsilon \in (0, 1)$. *Shortest paths from a vertex* v_0 *to all other vertices of* P *can be approximated within a factor of* $(1 + \varepsilon)$ *in* $O(\frac{n}{\sqrt{\varepsilon}} \log \frac{n}{\varepsilon} \log \frac{1}{\varepsilon})$ *time.*

Extensions: We briefly comment on how our approach can be applied to approximate shortest paths in weighted polyhedral domains and formulate the corresponding result. In 3-dimensional space most shortest path problems are difficult. Given a set of pairwise disjoint polyhedra in 3D and two points s and t, the *Euclidean 3-D Shortest Path Problem* is to compute a shortest path between s and t that avoids the interiors of polyhedra seen as obstacles. Canny and Reif have shown that this problem is NP-hard [6] (even for the case of axis parallel triangles in 3D). Papadimitriou [13] gave the first fully polynomial $(1 + \epsilon)$-approximation algorithm for the 3D problem. There are numerous other results on this problem, but due to the space constraints we omit their discussion and refer the reader to the most recent work [5] for a literature review.

Let P be a tetrahedralized polyhedral domain in the 3-dimensional Euclidean space, consisting of n tetrahedra. Assume that positive weights are assigned to the tetrahedra of P and that the cost of traveling inside a tetrahedron t is equal to the Euclidean distance traveled multiplied by the weight of t. Using the approach of this paper we are able to approximate shortest paths in P within $(1+\varepsilon)$ factor as follows: Discretization in this case is done by inserting Steiner points in the bisectors of the dihedral angles of the tetrahedra of P. The total number of Steiner points in this case is $O(\frac{n}{\varepsilon^2}\log\frac{1}{\varepsilon})$. The construction of Steiner points and the proof of the approximation properties of the resulting graph G_ε in this case involves more elaborate analysis because of the presence of *edge vicinities* – small spindle like regions around edges – in addition to vertex vicinities. Nevertheless, an analogue to Theorem 1 holds. SSSP in the graph G_ε can be computed by following a greedy approach like that in Section 4.

References

1. K.R. Varadarajan and P.K. Agarwal, "Approximating Shortest Paths on Nonconvex Polyhedron", SIAM Jl. Comput. 30(4): 1321–1340 (2000).
2. P.K. Agarwal, S. Har-Peled, and M.Karia, "Computing approximate shortest paths on convex polytopes", Algorithmica 33:227–242, 2002.
3. P.K. Agarwal et al., "Approximating Shortest Paths on a Convex Polytope in Three Dimensions", Jl. ACM 44:567–584, 1997.
4. L. Aleksandrov, M. Lanthier, A. Maheshwari, J.-R. Sack, "An ε-approximation algorithm for weighted shortest paths", SWAT, LNCS 1432:11–22, 1998.
5. L. Aleksandrov, A. Maheshwari, and J.-R. Sack, "Approximation Algorithms for Geometric Shortest Path Problems", *32nd STOC*, 2000, pp. 286–295.
6. J. Canny and J. H. Reif, "New Lower Bound Techniques for Robot Motion Planning Problems", *28th FOCS*, 1987, pp. 49–60.
7. J. Chen and Y. Han, "Shortest Paths on a Polyhedron", *6th SoACM-CG*, 1990, pp. 360–369. Appeared in "Internat. J. Comput. Geom. Appl.", 6: 127–144, 1996.
8. J. Hershberger and S. Suri, "Practical Methods for Approximating Shortest Paths on a Convex Polytope in $\Re^3$", *6SODA*, 1995, pp. 447–456.
9. S. Kapoor, "Efficient Computation of Geodesic Shortest Paths", *31st STOC*, 1999.
10. M. Lanthier, A. Maheshwari and J.-R. Sack, "Approximating Weighted Shortest Paths on Polyhedral Surfaces", Algorithmica 30(4): 527–562 (2001).
11. J.S.B. Mitchell, D.M. Mount and C.H. Papadimitriou, "The Discrete Geodesic Problem", SIAM Jl. Computing, 16:647–668, August 1987.
12. J.S.B. Mitchell and C.H. Papadimitriou, "The Weighted Region Problem: Finding Shortest Paths Through a Weighted Planar Subdivision", JACM, 38:18–73, 1991.
13. C.H. Papadimitriou, "An Algorithm for Shortest Path Motion in Three Dimensions", *IPL*, **20**, 1985, pp. 259–263.
14. M. Sharir and A. Schorr, "On Shortest Paths in Polyhedral Spaces", *SIAM J. of Comp.*, **15**, 1986, pp. 193–215.
15. Z. Sun and J. Reif, "BUSHWACK: An approximation algorithm for minimal paths through pseudo-Euclidean spaces", 12th ISAAC, LNCS 2223:160–171, 2001.
16. M. Ziegelmann, Constrained Shortest Paths and Related Problems Ph.D. thesis, Universität des Saarlandes (Max-Planck Institut für Informatik), 2001.

Adaptive and Compact Discretization for Weighted Region Optimal Path Finding

Zheng Sun and John H. Reif

Department of Computer Science, Duke University, Durham, NC 27708, USA
{sunz,reif}@cs.duke.edu

Abstract. This paper presents several results on the *weighted region optimal path problem*. An often-used approach to approximately solve this problem is to apply a discrete search algorithm to a graph $\mathcal{G}_\epsilon$ generated by a discretization of the problem; this graph guarantees to contain an ϵ-approximation of an optimal path between given source and destination points. We first provide a discretization scheme such that the size of $\mathcal{G}_\epsilon$ does not depend on the ratio between the maximum and minimum unit weights. This leads to the first ϵ-approximation algorithm whose complexity is not dependent on the unit weight ratio. We also introduce an empirical method, called *adaptive discretization method*, that improves the performance of the approximation algorithms by placing discretization points densely only in areas that may contain optimal paths. BUSHWHACK is a discrete search algorithm used for finding optimal paths in $\mathcal{G}_\epsilon$. We added two heuristics to BUSHWHACK to improve its performance and scalability.

1 Introduction

In the past two decades the geometric optimal path problems have been extensively studied (see [1] for a review). These problems have a wide range of applications in robotics and geographical information systems.

In this paper we study the path planning problem for a point robot in a $2\mathcal{D}$ space consisting of n triangular regions, each of which is associated with a distinct unit weight. Such a space can be used to model an area consisting of different geographical features, such as deserts, forests, grasslands, and lakes, in which the traveling costs for the robot are different. The goal is to find between given source and destination points s and t an *optimal path* (a path with the minimum weighted length).

Unlike the un-weighted $2\mathcal{D}$ optimal path problem, which can be solved in $O(n \log n)$ time, this problem is believed to be very difficult. Much of the effort has been focused on *ϵ-approximation algorithms* that can guarantee to find ϵ-*good approximate optimal paths* (see [2,3,4,5,6]). For any two points s and t in the space, we say that a path p connecting s and t is an ϵ-good approximate optimal path if $\|p\| < (1 + \epsilon)\|p_{opt}(s,t)\|$, where $p_{opt}(s,t)$ represents an optimal path from s to t and $\|\cdot\|$ represents the weighted length, or the *cost*, of a path. Equivalently, we say that p is $p_{opt}(s,t)$'s ϵ-approximation.

A. Lingas and B.J. Nilsson (Eds.): FCT 2003, LNCS 2751, pp. 258–270, 2003.
© Springer-Verlag Berlin Heidelberg 2003

Before we give a review of previous works, we first define some notations. We let V be the set of vertices of all regions, and let E be the set of all boundary edges. We use w_r to denote the unit weight of any region r. For a boundary edge e separating two regions r_1 and r_2, the unit weight w_e of e is defined to be $\min\{w_{r_1}, w_{r_2}\}$. We define *unit weight ratio* μ to be $\frac{w_{max}}{w_{min}}$, where w_{max} (w_{min}, respectively) is the maximum (minimum, respectively) unit weight among all regions. We use $|p|$ to denote the Euclidean length of path p, and use $p_1 + p_2$ to denote the concatenation of two paths p_1 and p_2.

The first ϵ-approximation algorithm on this problem was given by Mitchell and Papadimitriou [2]. Their algorithm uses "Snell's Law" and "continuous Dijkstra method" to give an optimal-path map for any given source point s. The time complexity of their algorithm is $O(n^8 \log \frac{n\mu}{\epsilon})$. In practice, however, the time complexity is expected to be much lower. Later Mata and Mitchell [3] presented another ϵ-approximation algorithm based on constructing a "pathnet graph" of size $O(nk)$, where $\epsilon = O(\frac{\mu}{k})$. The time complexity, in terms of ϵ and n, is $O(\frac{n^3\mu}{\epsilon})$.

Some of the existing algorithms construct from the original continuous space a weighted graph $\mathcal{G}_\epsilon(V', E')$ by placing discretization points, called *Steiner points*, on boundary edges. The node set V' of $\mathcal{G}_\epsilon$ contains all Steiner points as well as vertices of the regions. The edge set E' of $\mathcal{G}_\epsilon$ contains every edge $\overline{v_1v_2}$ such that v_1 and v_2 are on the border of the same region. The weight of edge $\overline{v_1v_2}$ is determined by the weighted length of segment $\overline{v_1v_2}$ in the original weighted space. $\mathcal{G}_\epsilon$ guarantees to contain an ϵ-good approximate optimal path between s and t, and therefore the task of finding an ϵ-good approximate optimal path is reduced to computing a shortest path in $\mathcal{G}_\epsilon$, which we call *optimal discrete path*, using a discrete search algorithm such as Dijkstra's algorithm or BUSHWHACK [5,7].

In the remainder of this paper, we will mainly discuss techniques for approximation algorithms using this approach. Since an optimal discrete path from s to t in $\mathcal{G}_\epsilon$ is used as an ϵ-approximation for the *real* optimal path, the phrases "optimal discrete path" and "ϵ-good approximate optimal path" are used interchangeably, and are both denoted by $p'_{opt}(s, t)$.

Aleksandrov *et al.* [4,6] proposed two discretization schemes that place $O(\frac{1}{\epsilon} \log \frac{1}{\epsilon} \log \mu)$ Steiner points on each boundary edge to construct $\mathcal{G}_\epsilon$ for a given ϵ. Combining the discretization scheme of [6] with a "pruned" Dijkstra's algorithm, they provided an ϵ-approximation algorithm that runs in roughly $O(\frac{n}{\epsilon}(\frac{1}{\sqrt{\epsilon}} + \log n) \log \frac{1}{\epsilon} \log \mu)$ time.

It is important to note, however, that the discretization size (and therefore the time complexity) for these approximation algorithms also depends on various geometric parameters, such as the smallest angle between two adjacent boundary edges, maximum integer coordinate of vertices, etc. These parameters are omitted here since they are irrelevant to our discussion.

In this paper we present the following results on finding ϵ-good approximate optimal paths in weighted regions:

Compact Discretization Scheme. The complexity of each of the approximation algorithms we have mentioned above depends more or less on μ, either linearly ([3]) or logarithmically ([2,4,6]). This dependency is caused by the corresponding discretization scheme used. In particular, the discretization scheme of Aleksandrov *et al.* [6] places $O(\frac{1}{\epsilon} \log \frac{1}{\epsilon} \log \mu)$ Steiner points on each boundary edge. Here again we omit the other geometric parameters.

The main obstacle for removing the dependency on μ from the size of $\mathcal{G}_\epsilon$ is that otherwise it is difficult to prove that for each optimal path p_{opt} there exists in $\mathcal{G}_\epsilon$ a discrete path that is an ϵ-approximation of p_{opt}. One traditional proof technique used in proving the existence of such a discrete path is to decompose p_{opt} into k subpaths $p_1, p_2, \cdots, p_k$ and then construct a discrete path $p' = p'_1 + p'_2 + \cdots + p'_k$ such that $\|p'_i\| \leq (1 + \epsilon)\|p_i\|$ for each i. Ideally, we could choose p'_i such that p_i and p'_i lie in the same region, and therefore the discretization just needs to make sure that $|p'_i| \leq (1 + \epsilon)|p_i|$. However, due to the discrete nature of $\mathcal{G}_\epsilon$, it is not always possible to find such p'_i for each p_i. For example, as shown in Figure 1.a, p_{opt} could cross a series of boundary edges near a vertex v. The point where it crosses each boundary edge e is between v and the closest Steiner point from v on e. In that case, p'_i could travel in regions different from where p_i lies in, and therefore to bound $\|p'_i\|$ with respect to $\|p_i\|$, the discretization scheme has to take into consideration variance of unit weights.

By modifying the above proof technique, we provide in Section 2 an improvement on the discretization scheme of Aleksandrov *et al.* [6]. The number of Steiner points inserted by this new discretization scheme is $O(\frac{1}{\epsilon} \log \frac{1}{\epsilon})$, with the dependency on other geometric parameters unchanged. Combining BUSH-WHACK with this discretization scheme, we can have the first ϵ-approximation algorithm whose time complexity is not dependent on μ.

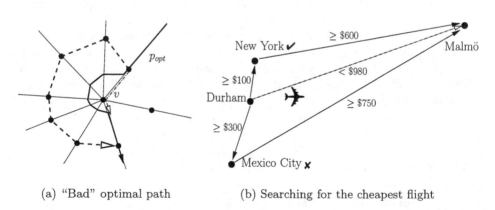

(a) "Bad" optimal path   (b) Searching for the cheapest flight

Fig. 1.

Adaptive Discretization Method. The traditional approximation algorithms construct from the original space a graph $\mathcal{G}_\epsilon$ and compute with a discrete search algorithm an optimal discrete path in $\mathcal{G}_\epsilon$ in a one-step manner. We call this method the *fixed discretization method*. For single query problem, this method is rather inefficient in that, although the goal is to find an ϵ-good approximate optimal path $p'_{opt}(s,t)$ from s to t, it actually computes an ϵ-good approximate optimal path from s to any point v in $\mathcal{G}_\epsilon$, as long as the cost of such a path is less than that of $p'_{opt}(s,t)$. Much of the effort is unnecessary as most of these points would not help to find an ϵ-good approximate optimal path from s to t.

We use flight ticket booking as an example. When trying to find the cheapest flight from Durham to Malmö with one stop (supposing no direct flight is available), a travel agent does not need to consider Mexico City as a candidate for the connecting airport if she knows the following: a) there is *always* a route from Durham to Malmö with one stop that costs less than \$980; b) any direct flight from Durham to Mexico City costs no less than \$300; and c) any direct flight from Mexico City to Malmö costs no less than \$750. Therefore, she does not need to find out the exact prices of the direct flights from Durham to Mexico City and from Mexico City to Malmö, saving two queries to the ticketing database.

Analogously, we do not need to compute $p'_{opt}(s,v)$ and $p'_{opt}(v,t)$ for a point $v \in \mathcal{G}_\epsilon$ if we know in advance that v does not connect any optimal discrete path between s and t. However, while the travel agent can rely on knowledge she previously gained, the approximation algorithms using the fixed discretization method have no prior knowledge to draw upon. In Section 3 we discuss a multiple-stage discretization method that we call *adaptive discretization method*. It starts with a coarse discretization $\mathcal{G}' = \mathcal{G}_{\epsilon_1}$ for some $\epsilon_1 > \epsilon$ and adaptively refines $\mathcal{G}'$ until it guarantees to contain an ϵ-good approximate optimal path from s to t. Approximate optimal path information acquired in each stage is used to identify the areas where no optimal path from s to t will pass through and therefore no further Steiner point needs to be inserted in the next stage.

Heuristics for BUSHWHACK. The BUSHWHACK algorithm is an alternative algorithm for computing optimal discrete paths in $\mathcal{G}_\epsilon$. It uses a number of complex data structures to keep track of all potential optimal paths. When m, the number of Steiner points placed on each boundary edge, is small, the efficiency gained by accessing only a subgraph of $\mathcal{G}_\epsilon$ is outweighed by the cost of establishing and maintaining these data structures. Another weakness of BUSHWHACK is that its performance improvement diminishes when the number of regions in the space is large. These weaknesses affect the practicability of BUSHWHACK as in most cases the desired quality of approximation does not require too many Steiner points for each boundary edge, while in the given $2\mathcal{D}$ space there can be arbitrary number of regions. In Section 4 we introduce two cost-saving heuristics for the original BUSHWHACK algorithm to overcome the weaknesses mentioned above.

2   Compact Discretization Scheme

In this section we provide an improvement on the discretization scheme of Aleksandrov *et al.* [6] by removing the dependency of the size of $\mathcal{G}_\epsilon$ on the unit weight ratio μ.

For any point v, we let $E(v)$ be the set of boundary edges incident to v and let $d(v)$ be the minimum distance between v and boundary edges in $E \backslash E(v)$. For each edge $e \in E$, we let $d(e) = \sup\{d(v) \mid v \in e\}$ and let v_e be the point on e so that $d(v_e) = d(e)$. For each vertex v of a region, the *radius* $r'(v)$ of v is defined to be $\frac{d(v)}{5}$, and the *weighted radius* $r(v)$ of v is defined to be $\frac{w_{min}(v)}{w_{max}(v)} \cdot r'(v)$, where $w_{min}(v)$ and $w_{max}(v)$ are the minimum and maximum unit weights among all regions incident to v, respectively.

According to the discretization scheme of Aleksandrov *et al.* [6], for each boundary edge $e = \overline{v_1 v_2}$, the Steiner points on e are chosen as the following. Each vertex v_i has a "vertex-vicinity" $S(v_i)$ of radius $r_\epsilon(v_i) = \epsilon r(v_i)$ and the Steiner points $v_{i,1}, v_{i,2}, \cdots, v_{i,k_i}$ are placed on the segment of e outside the vertex-vicinities so that $|\overline{v_i v_{i,1}}| = r_\epsilon(v_i)$, $|\overline{v_{i,j} v_{i,j+1}}| = \epsilon d(v_{i,j})$ and $\overline{v_{i,k_i} v_i} + \epsilon d(v_{i,k_i}) \geq |\overline{v_i v_e}|$. The number of Steiner points placed on e can be bounded by $C(e) \cdot \frac{1}{\epsilon} \log \frac{1}{\epsilon}$, where $C(e) = O(|e|/d(e) \cdot \log(|e|/\sqrt{r(v_1)r(v_2)})) = O(|e|/d(e) \cdot (\log(|e|/\sqrt{r'(v_1)r'(v_2)}) + \log \mu))$. This discretization can guarantee a 3ϵ-good approximate optimal path.

Observe that, for this discretization scheme, on each boundary edge e Steiner points are placed more densely in the portion of e closer to the two endpoints, with the exception that no Steiner point is placed inside the vertex-vicinities. Therefore, the larger the vertex vicinities are, the less Steiner points the discretization needs to use. In the following we show that the radius $r_\epsilon(v)$ of the vertex-vicinity of v can be increased to $\epsilon r'(v)$ while still guaranteeing the same error bound. Here we assume that $\epsilon \leq \frac{1}{2}$.

A piecewise linear path p is said to be a *normalized path* if it does not cross region boundaries inside vertex vicinities other than at the vertices. That is, for each bending point u of p, if u is located on boundary edge $e = \overline{v_1 v_2}$, then either u is one of the endpoints of e, or $|\overline{v_i u}| \geq r_\epsilon(v_i)$ for $i = 1, 2$. For example, the path shown in Figure 2 is not a normalized path, as it passes through u_1 and u_2, both of which are inside the vertex vicinity of v. We first state the following lemma:

Lemma 1. *For any path p from s to t, there is a normalized path $\hat{p}$ from s to t such that $\|\hat{p}\| = (1 + \frac{\epsilon}{2}) \cdot \|p\|$.*

Proof. In the following, for a path p and two points $u_1, u_2 \in p$, we use $p[u_1, u_2]$ to denote the subpath of p between u_1 and u_2.

Refer to Figure 2. Suppose path p passes through the vertex vicinity $S(v)$ of v, as shown in Figure 2. We use u_1 (u_2, respectively) to denote the first (last, respectively) bending point of p inside $S(v)$, and use u_1'' (u_2'') to denote the first (last, respectively) bending point of p on the border of the union of all regions incident to v. By the definition of $d(v)$, we have $|p[u_1'', u_1]| + |\overline{u_1 v}| \geq d(v)$ and $|p[u_2, u_2'']| + |\overline{v u_2}| \geq d(v)$. Therefore, $|\overline{u_1 v}|/|p[u_1'', u_1]| \leq \frac{\epsilon \cdot d(v)/5}{d(v) - \epsilon \cdot d(v)/5} = \frac{\epsilon}{5 - \epsilon} \leq \frac{\epsilon}{4}$, as $|\overline{u_1 v}| \leq \frac{\epsilon d(v)}{5}$. Similarly, we can prove that $|\overline{v u_2}|/|p[u_2, u_2'']| \leq \frac{\epsilon}{4}$.

We let r_1 be the region with the minimum unit weight among all regions crossed by subpath $p[u_1'', u_1]$, and u_1' be the point where $p[u_1'', u_1]$ enters region r_1 for the first time. Similarly, we let r_2 be the region with the minimum unit weight among all regions crossed by subpath $p[u_2, u_2'']$, and let u_2' be the point where $p[u_2, u_2'']$ leaves region r_1 for the last time.

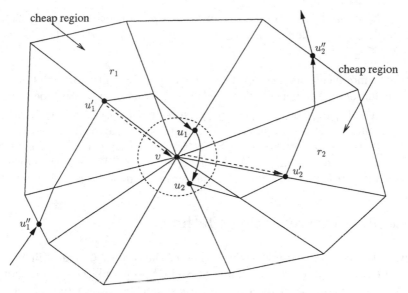

Fig. 2. Path passing through vicinity of a vertex

Consider replacing subpath $p[u_1'', u_2'']$ by this normalized subpath: $\hat{p}[u_1'', u_2''] = p[u_1'', u_1'] + \overline{u_1'v} + \overline{vu_2'} + p[u_2', u_2'']$. We have the following inequality:

$$\|\hat{p}[u_1'', u_2'']\| - \|p[u_1'', u_2'']\|$$
$$= w_{r_1} \cdot |\overline{u_1'v}| + w_{r_2} \cdot |\overline{vu_2'}| - \|p[u_1', u_1]\| - \|p[u_1, u_2]\| - \|p[u_2, u_2']\|$$
$$\leq (w_{r_1} \cdot |\overline{u_1'v}| - \|p[u_1', u_1]\|) + (w_{r_2} \cdot |\overline{vu_2'}| - \|p[u_2, u_2']\|)$$
$$\leq w_{r_1}(|\overline{u_1'v}| - |p[u_1', u_1]|) + w_{r_2}(|\overline{vu_2'}| - |p[u_2, u_2']|)$$
$$\leq w_{r_1} \cdot |\overline{u_1v}| + w_{r_2} \cdot |\overline{vu_2}| \leq w_{r_1} \cdot \frac{\epsilon \cdot |p[u_1'', u_1]|}{4} + w_{r_2} \cdot \frac{\epsilon \cdot |p[u_2, u_2'']|}{4}$$
$$\leq \frac{\epsilon}{4} \cdot (\|p[u_1'', u_1]\| + \|p[u_2, u_2'']\|) \leq \frac{\epsilon}{4} \cdot \|p[u_1'', u_2'']\|$$

Therefore, $\|\hat{p}[u_1'', u_2'']\| \leq (1+\frac{\epsilon}{4})\|p[u_1'', u_2'']\|$. Suppose p passes through k vertex vicinities, $S(v_1), S(v_2), \cdots, S(v_k)$. For each v_i, we replace the subpath p_i of p that passes through $S(v_i)$ by a normalized subpath $\hat{p}_i$ as we described above. Let $\hat{p}$ be the resulting normalized path. Note that the sum of the weighted lengths of $p_1, p_2, \cdots, p_k$ is less than twice of the weighted length of p, we have $\|\hat{p}\| \leq \|p\| + \frac{\epsilon}{4}\sum_{i=1}^{k}\|p_i\| \leq (1 + \frac{\epsilon}{2})\|p\|$. □

We call a segment of a boundary edge bounded by two adjacent Steiner points a *Steiner segment*. Each segment $\overline{u_1u_2}$ of a normalized path $\hat{p}$ is significantly long as compared to the Steiner segment on which u_1 or u_2 lies. Therefore, it is easy to find a discrete path in $\mathcal{G}_\epsilon$ that is an ϵ-approximation of $\hat{p}$. With Lemma 1, we can prove the claimed error bound for this modified discretization:

Theorem 1. *The discretization constructed with $r_\epsilon(v) = \epsilon r'(v)$ contains a 3ϵ-good approximation for an optimal path p_{opt} from s to t, for any two vertices s and t.*

Proof. We first construct a normalized path $\hat{p}$ such that $\|\hat{p}\| \leq (1 + \frac{\epsilon}{2})\|p_{opt}\|$. Then we can use a proof similar to the one provided in [6] to show that, for any normalized path $\hat{p}$, there is a discrete path p' so that $\|p'\| \leq (1 + 2\epsilon)\|\hat{p}\|$. Therefore, $\|p'\| \leq (1 + 2\epsilon)(1 + \frac{\epsilon}{2})\|p_{opt}\| = (1 + \frac{5}{2}\epsilon + \epsilon^2)\|p_{opt}\| \leq (1 + 3\epsilon)\|p_{opt}\|$, assuming $\epsilon \leq \frac{1}{2}$. □

With the modification on the radius of each vertex vicinity, for each boundary edge e the number of Steiner points placed on e is reduced to $C'(e) \cdot \frac{1}{\epsilon} \log \frac{1}{\epsilon}$, where $C'(e) = O(|e|/d(e) \log(|e|/\sqrt{r'(v_1)r'(v_2)}))$. Note that $C'(e)$ is independent of μ.

The significance of this compact discretization scheme is that, combining it with either Dijkstra's algorithm or BUSHWHACK, we can get an approximation algorithm whose time complexity does not depend on μ. To our best knowledge, all previous ϵ-approximation algorithms have time complexities dependent on μ.

3   Adaptive Discretization Method

Even with the compact discretization scheme, the size of $\mathcal{G}_\epsilon$ can still be very large even for a modest ϵ, as the number of Steiner points placed on each boundary edge is also determined by a number of geometric parameters. Therefore, computing an ϵ-good approximate optimal path by directly applying a discrete search algorithm to $\mathcal{G}_\epsilon$ may be very costly. In particular, a discrete search algorithm such as Dijkstra's algorithm will compute an optimal discrete path from s to every point $v \in \mathcal{G}_\epsilon$ that is closer to s than t is, meaning that it has to search through a large space with the same (small) error tolerance ϵ.

Here we further elaborate the flight ticket booking example. With the knowledge accumulated through past experiences, the travel agent may know, for any intermediate airport A, a *lower bound* $L_{D,A}$ of the cost of a direct flight from Durham to A as well as a *lower bound* $L_{A,M}$ of the cost of a direct flight from A to Malmö. Further, she also knows an *upper bound*, say, $980, of the cost of the *cheapest* flight (with one stop) from Durham to Malmö. In that case, the travel agent would only consider airport A as a possible stop between Durham and Malmö if $L_{D,A} + L_{A,M} < 980$. For example, it at least worths the effort to check the database to find out the exact cost of the flight from Durham to Malmö via New York, as shown in Figure 1.b.

The A* algorithm partially addresses this issue as it would first explore points that are estimated using a heuristic function to be closer to the destination point t. However, if the unit weights of the regions vary significantly, it is difficult for a heuristic function to provide a close estimation of the weighted distance between any point and t. As a result, the A* algorithm may still have to search through many points in $\mathcal{G}_\epsilon$ unnecessarily.

Here we introduce a multi-stage approximation algorithm that uses an adaptive discretization method. For each i, $1 \leq i \leq d$, this method computes an ϵ_i-good approximate path from s to t in a subgraph $\mathcal{G}'_{\epsilon_i}$ of $\mathcal{G}_{\epsilon_i}$, where $\epsilon_1 > \epsilon_2 > \cdots > \epsilon_{d-1} > \epsilon_d = \epsilon$. In each stage, with the approximate optimal path information acquired through the previous stage, the algorithm can identify for each boundary edge the portion of the edge where more Steiner points need to placed to guarantee an approximate optimal path with a reduced error bound.

For the rest portion of the boundary edge, no further Steiner point needs to be placed.

We say that a path p' *neighbors* an optimal path p_{opt} if, for any Steiner segment that p_{opt} crosses, p' passes through one of the two Steiner points that bound the Steiner segment. Our method requires that the discretization scheme satisfy the following property (which is the case for the discretization schemes of [4,6] and the one described in Section 2):

Property 1. For any two vertices v_1 and v_2 in the original (continuous) space and any optimal path p_{opt} from v_1 and v_2, there is a discrete path from v_1 to v_2 in the discretization with a cost no more than $(1 + \epsilon) \cdot \|p_{opt}(v_1, v_2)\|$ that neighbors p_{opt}.

For any two points $v_1, v_2 \in \mathcal{G}'_{\epsilon_i}$, we denote the optimal discrete path found from v_1 to v_2 in the i-th stage by $p'_{\epsilon_i}(v_1, v_2)$. We say that a point $v \in \mathcal{G}'_{\epsilon_i}$ is a *searched point* if an optimal discrete path $p'_{\epsilon_i}(s, v)$ from s to v in $\mathcal{G}'_{\epsilon_i}$ is determined. For each searched point v, we also compute an optimal discrete path $p'_{\epsilon_i}(v, t)$ from v to t. We say that a point v is a *useful point* if either $\|p'_{\epsilon_i}(s, v)\| + \|p'_{\epsilon_i}(v, t)\| \leq (1 + \epsilon_i) \cdot \|p'_{\epsilon_i}(s, t)\|$ or v is a vertex; we say that a Steiner segment is a *useful segment* if at least one of its endpoints is useful. An optimal path p_{opt} will not pass through a useless segment, and therefore in the next stage the algorithm can avoid putting more Steiner points in this segment.

```
1.   i ← 1
2.   construct a discretization G'_{ε_i} = G_{ε_i}.
3.   repeat
4.       compute p'_{ε_i}(s, t) in G'_{ε_i}.
5.       if i = d then  return p'_{ε_i}(s, t).
6.       continue to compute p'_{ε_i}(s, v) for each point v in G'_{ε_i} until ‖p'_{ε_i}(s, v)‖ grows
             beyond (1 + ε_i) · ‖p'_{ε_i}(s, t)‖.
7.       apply Dijkstra's algorithm in a reversed way, and compute p'_{ε_i}(v, t) for
             any searched point v.
8.       G'_{ε_{i+1}} ← ∅
9.       for each useful point v ∈ G'_{ε_i}
10.          add v into G'_{ε_{i+1}}
11.      for each point v ∈ G_{ε_{i+1}}
12.          if v is located inside a useful Steiner segment of G'_{ε_i} then
13.              add v into G'_{ε_{i+1}}
14.      i ← i + 1
```

Algorithm 1: Adaptive

Each stage contains a forward search and a backward search. These two searches can be performed simultaneously using Dijkstra's two-tree algorithm [8].

To prove the correctness of our multiple-stage approximation algorithm, it suffices to show the following theorem:

Theorem 2. *For any optimal path $p_{opt}(s, t)$, in each $\mathcal{G}'_{\epsilon_i}$ there is a discrete path $p'(s, t)$ with a cost no more than $(1 + \epsilon_i) \cdot \|p_{opt}(s, t)\|$ that neighbors $p_{opt}(s, t)$.*

Proof. We prove by induction.

Basic Step: When $i = 1$, $\mathcal{G}'_{\epsilon_i} = \mathcal{G}_{\epsilon_1}$, and therefore the proposition is true, according to Property 1.

Inductive Step: We assume that, for any optimal path $p_{opt}(s, t)$, $\mathcal{G}'_{\epsilon_i}$ contains a discrete path $p'(s, t)$ neighboring $p_{opt}(s, t)$ such that $\|p'(s, t)\| \leq (1 + \epsilon_i) \cdot \|p_{opt}(s, t)\|$. We first show that $p_{opt}(s, t)$ will not pass through any useless Steiner segment $\overline{u_1 u_2}$ in $\mathcal{G}'_{\epsilon_i}$. Suppose otherwise that $p_{opt}(s, t)$ passes through a point between u_1 and u_2. According to the induction hypothesis, we can construct a discrete path $p'(s, t)$ from s to t with a cost no more than $(1 + \epsilon_i) \cdot \|p_{opt}(s, t)\|$ that neighbors $p_{opt}(s, t)$. This implies that $p'(s, t)$ passes through either u_1 or u_2. W.L.O.G. we assume that $p'(s, t)$ passes through u_1. Because $\|p_{opt}(s, t)\| \leq \|p'_{\epsilon_i}(s, t)\|$, the cost of $p'(s, t)$ is no more than $(1 + \epsilon_i) \cdot \|p'_{\epsilon_i}(s, t)\|$. This is a contradiction to the fact that $\|p'_{\epsilon_i}(s, u_1)\| + \|p'_{\epsilon_i}(u_1, t)\| > (1 + \epsilon_i) \cdot \|p'_{\epsilon_i}(s, t)\|$, as $p'(s, t)$ cannot be better than the concatenation of $p'_{\epsilon_i}(s, u_1)$ and $p'_{\epsilon_i}(u_1, t)$.

Since any optimal path from s to t will not pass through a useless Steiner segment, $\mathcal{G}'_{\epsilon_{i+1}}$, which includes all the Steiner points of $\mathcal{G}_{\epsilon_{i+1}}$ except those inside useless Steiner segments, contains every discrete path in $\mathcal{G}_{\epsilon_{i+1}}$ that neighbors one of the optimal paths from s to t. This finishes the proof.   □

The adaptive discretization method has both pros and cons when compared against the fixed discretization method. It has to run a discrete search algorithm on d different graphs, and each time it involves both forward and backward searches. However, in the earlier stages it explores approximate optimal paths with high error tolerance, while in later stages, as it gradually reduces the error tolerance, it only searches approximate optimal paths in a small subspace (that is, the useful segments of the boundary edges) instead of the entire original space (all boundary edges). Our experimental results show that, when the desired error tolerance ϵ is small, the adaptive discretization method performs more efficiently than the fixed discretization.

This discretization method can also be applied to other geometric optimal path problems, such as the time-optimum movement planning problem in regions with flows [9], the anisotropic optimal path problem [10,11], and the $3\mathcal{D}$ Euclidean shortest path problem [12,13].

4   Heuristics for BUSHWHACK

The BUSHWHACK algorithm was originally designed for the weighted region optimal path problem [5] and was later generalized to a class of *piecewise pseudo-Euclidean optimal path problems* [7]. BUSHWHACK, just like Dijkstra's algorithm, is used to compute optimal discrete paths in a graph $\mathcal{G}_\epsilon$ generated by a discretization scheme. Unlike Dijkstra's algorithm, which applies to any arbitrary weighted graph, BUSHWHACK is adept at finding optimal discrete paths in graphs derived from geometric spaces with certain properties, one of which being the following:

Property 2. Two optimal discrete paths that originate from a same source point cannot intersect in the interior of any region.

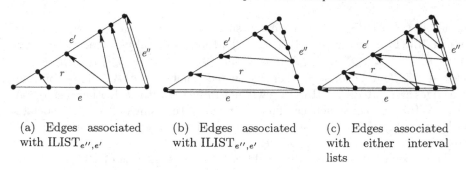

(a) Edges associated with ILIST$_{e'',e'}$

(b) Edges associated with ILIST$_{e'',e'}$

(c) Edges associated with either interval lists

Fig. 3. Intersecting Edges Associated with Two Interval Lists

One implication of Property 2 is that, if two edges $\overline{v_1 v_2}$ and $\overline{u_1 u_2}$ of $\mathcal{G}_\epsilon$ intersect inside region r, they cannot both be *useful*. An edge is said to be useful if it contributes to optimal discrete paths that originate from s. To exploit this property, BUSHWHACK maintains a list ILIST$_{e,e'}$ of *intervals* for each pair of boundary edges e and e' such that e and e' are on the border of the same region r. A point v is said to be *discovered* if an optimal discrete path $p'_{opt}(s, v)$ has been determined. ILIST$_{e,e'}$ contains for each discovered point $v \in e$ an interval $I_{v,e,e'}$ defined as the following: $I_{v,e,e'} = \{v^* \in e' | w_r \cdot |\overline{vv^*}| + \|p'_{opt}(s, v)\| \leq w_r \cdot |\overline{v'v^*}| + \|p'_{opt}(s, v')\| \| \forall\ v' \in \text{PLIST}_e \}$. Here PLIST$_e$ is the list of all discovered points on e. We say that edge $\overline{vv^*}$ is *associated* with interval list ILIST$_{e,e'}$ if $v \in e$ and $v^* \in I_{v,e,e'}$.

It is easy to see that any edge $\overline{vv^*}$ that crosses region r is useful only if it is associated with an interval list inside r. If m is the number of Steiner points placed on each boundary edge, the total number of edges associated with interval lists inside a region r is $\Theta(m)$. Dijkstra's algorithm, on the other hand, has to consider all $\Theta(m^2)$ edges inside r. By avoid accessing most of the useless edges, BUSHWHACK takes only $O(nm \log nm)$ time to compute an optimal discrete path from s to t, as compared to $O(nm^2 + nm \log nm)$ time for Dijkstra's algorithm.

In this section we introduce BUSHWHACK$^+$, a variation of BUSHWHACK. On the basis of the original BUSHWHACK algorithm, BUSHWHACK$^+$ uses several cost-saving heuristics. The necessity of the first heuristic is rather obvious. Let r be a triangular region with boundary edges e, e' and e''. There are six interval lists for each triangular region r, one for each ordered pair of boundary edges of r. Although the edges associated with the same interval list do not intersect with each other, two edges associated with different interval lists may still intersect inside r. Therefore, BUSHWHACK may still use some intersecting edges to construct candidate optimal paths. Figure 3.a and 3.b show the edges associated with ILIST$_{e,e'}$ and ILIST$_{e'',e'}$, respectively. Figure 3.c shows that these two sets of edges intersect with each other, meaning that some of them must be useless.

To address this issue, BUSHWHACK$^+$ merges ILIST$_{e,e'}$ and ILIST$_{e'',e'}$ into a single list ILIST$_{r,e'}$. Any point $v^* \in e'$ is included in one and only one interval in this list. (In BUSHWHACK, every such point is included in two intervals, one in $\text{ILIST}_{e,e'}$ and one in $\text{ILIST}_{e'',e'}$.) More specifically, for any discovered point $v \in e \cup e''$, $v^* \in I_{v,r,e'}$ if and only if $w_r \cdot |\overline{vv^*}| + \|p'_{opt}(s,v)\| \leq w_r \cdot |\overline{v'v^*}| + \|p'_{opt}(s,v')\|$ for any other discovered point $v' \in e \cup e''$. Therefore, any two edges associated with $\text{ILIST}_{r,e'}$ will not intersect with each other inside r. As BUSHWHACK$^+$ constructs candidate optimal paths using only edges associated with interval lists, it would avoid using both of two intersecting edges $\overline{v_1 v_1^*}$ and $\overline{v_2 v_2^*}$ if $v_1, v_2 \in e \cup e''$ and $v_1^*, v_2^* \in e'$.

The second heuristic is rather subtle. It reduces the size of QLIST, the list of candidate optimal paths. Possible operations on this list include inserting a new candidate optimal path and deleting the minimum cost path in the list. On average, each such operation costs $O(\log(nm))$ time. As each iteration of the algorithm will invoke one or more such operations, it is very important to contain the size of QLIST.

In the original BUSHWHACK, for any point $v \in e$, QLIST may contain six or more candidate optimal paths from s to v. Among these paths, four of them are propagated through edges associated with interval lists, while the remaining ones are extended to v from left and right along the edge e. This is a serious disadvantage against Dijkstra-based approximation algorithm, which keeps only one path from s to v in the Fibonacci heap for each Steiner point v. When n is relatively large, the performance gain of BUSHWHACK by accessing only a small subgraph of $\mathcal{G}_\epsilon$ will be totally offset by the time wasted on a larger path list.

If multiple candidate optimal paths for v are inserted into QLIST, BUSH-WHACK keeps each of them until it is time to extract that path from QLIST, even though it can be immediately decided that all of those paths except one cannot be optimal (by comparing the costs of those paths). This is because BUSHWHACK would generate new candidate optimal paths using these paths in different ways. A (non-optimal) path may lead to the generation of a true optimal discrete path and therefore it cannot be simply discarded. What BUSH-WHACK does is to keep the path in QLIST until this path becomes the minimum cost path. At that time, it will be extracted from QLIST and a new candidate optimal path generated from the old path will be inserted into QLIST.

BUSHWHACK$^+$, however, uses a slightly different propagation scheme to avoid keeping multiple paths with the same ending point. Let $p(s,v')$ be a candidate optimal path from s to v' that has just been inserted into QLIST. If there is already another candidate optimal path $p'(s,v')$ in QLIST, instead of keeping both of them in QLSIT, BUSHWHACK$^+$ will take the more costly one, say $p'(s,v')$, and immediately extract it from QLIST. This extracted path will be processed as if it had been extracted in the normal situation (in which it would have been the minimum cost path in the list). This is, in essence, a "propagation-in-advance" strategy that is somewhat contradictory to "lazy" propagation scheme of BUSHWHACK. It may cause accessing edges unnecessarily. It is a trade-off between reducing the path list size and reducing the number of edges accessed.

5 Preliminary Experimental Results

In order to provide a performance comparison, we implemented using Java the following three algorithms: 1) BUSHWHACK$^+$; 2) *pure Dijkstra's algorithm*, which searches every incident edge of a Steiner point in $\mathcal{G}_\epsilon$; 3) two-stage adaptive discretization method, which uses pure Dijkstra's algorithm for each stage and chooses $\epsilon_1 = \frac{\epsilon}{2}$. All the timed results were acquired from a Sun Blade-1000 workstation with 4GB memory.

For our experiments we chose triangulations converted from terrain maps in grid data format. More specifically, we used the DEM (Digital Elevation Model) file of Kaweah River basin. It is a 1424x1163 grid with 30m between two neighboring grid points. We randomly took twenty 60x45 patches and converted them to TINs by connecting two grid points diagonally for each grid cell. Therefore, in each example there are 5192 triangular faces. For each triangular face r, we assign to r a unit weight w_r that is equal to $1 + 10\tan\alpha_r$, where α_r is the angle between r and the horizontal plane.

Table 1. Statistics of running time (in seconds) and number of visited edges per region

Algorithm	BUSHWHACK$^+$	pure Dijkstra	adaptive discretization
$\frac{1}{\epsilon} = 3$	156.9 / 2371	243.0 / 16558	281.3 / 10877
$\frac{1}{\epsilon} = 5$	290.7 / 4603	711.0 / 55797	570.2 / 24041
$\frac{1}{\epsilon} = 7$	440.6 / 7098	1506.0 / 124086	1054.7 / 40827
$\frac{1}{\epsilon} = 9$	631.9 / 9795	2672.5 / 224987	1528.9 / 60495

For each TIN, we ran the three algorithms five times, each time choosing randomly generated source and destination points. For each algorithm, we took the average of the running times of all experiments. We repeated the experiments with $\frac{1}{\epsilon} = 3, 5, 7$ and 9. From Table 1, it is easy to see that, when $\frac{1}{\epsilon}$ grows, the running times of the BUSHWHACK$^+$ algorithm and adaptive discretization method are growing much slower than that of the pure Dijkstra's algorithm. We also list the average number of visited edges per region for each algorithm and each ϵ value. It occurs to us that, the number of visited edges per region and the running time are closely correlated.

6 Conclusion

In this paper we provided several improvements on the approximation algorithms for the weighted region optimal path problem: 1) a compact discretization scheme that removes the dependency on the unit weight ratio; 2) an adaptive discretization that selectively put Steiner points with high density on boundary edges; and 3) a revised BUSHWHACK algorithm with two cost-saving heuristics.

Acknowledgement. This work is supported by NSF ITR Grant EIA-0086015, DARPA/AFSOR Contract F30602-01-2-0561, NSF EIA-0218376, and NSF EIA-0218359.

References

1. Mitchell, J.S.B.: Geometric shortest paths and network optimization. In Sack, J.R., Urrutia, J., eds.: Handbook of Computational Geometry. Elsevier Science Publishers B.V. North-Holland, Amsterdam (2000) 633–701
2. Mitchell, J.S.B., Papadimitriou, C.H.: The weighted region problem: Finding shortest paths through a weighted planar subdivision. Journal of the ACM **38** (1991) 18–73
3. Mata, C., Mitchell, J.: A new algorithm for computing shortest paths in weighted planar subdivisions. In: Proceedings of the 13th Annual ACM Symposium on Computational Geometry. (1997) 264–273
4. Aleksandrov, L., Lanthier, M., Maheshwari, A., Sack, J.R.: An ϵ-approximation algorithm for weighted shortest paths on polyhedral surfaces. In: Proceedings of the 6th Scandinavian Workshop on Algorithm Theory. Volume 1432 of Lecture Notes in Computer Science. (1998) 11–22
5. Reif, J.H., Sun, Z.: An efficient approximation algorithm for weighted region shortest path problem. In: Proceedings of the 4th Workshop on Algorithmic Foundations of Robotics. (2000) 191–203
6. Aleksandrov, L., Maheshwari, A., Sack, J.R.: Approximation algorithms for geometric shortest path problems. In: Proceedings of the 32nd Annual ACM Symposium on Theory of Computing. (2000) 286–295
7. Sun, Z., Reif, J.H.: BUSHWHACK: An approximation algorithm for minimal paths through pseudo-Euclidean spaces. In: Proceedings of the 12th Annual International Symposium on Algorithms and Computation. Volume 2223 of Lecture Notes in Computer Science. (2001) 160–171
8. Helgason, R.V., Kennington, J., Stewart, B.: The one-to-one shortest-path problem: An empirical analysis with the two-tree dijkstra algorithm. Computational Optimization and Applications **1** (1993) 47–75
9. Reif, J.H., Sun, Z.: Movement planning in the presence of flows. In: Proceedings of the 7th International Workshop on Algorithms and Data Structures. Volume 2125 of Lecture Notes in Computer Science. (2001) 450–461
10. Lanthier, M., Maheshwari, A., Sack, J.R.: Shortest anisotropic paths on terrains. In: Proceedings of the 26th International Colloquium on Automata, Languages and Programming. Volume 1644 of Lecture Notes in Computer Science. (1999) 524–533
11. Sun, Z., Reif, J.H.: On energy-minimizing paths on terrains for a mobile robot. In: Proceedings of the 2003 IEEE International Conference on Robotics and Automation. (2003) To appear.
12. Papadimitriou, C.H.: An algorithm for shortest-path motion in three dimensions. Information Processing Letters **20** (1985) 259–263
13. Choi, J., Sellen, J., Yap, C.K.: Approximate Euclidean shortest path in 3-space. In: Proceedings of the 10th Annual ACM Symposium on Computational Geometry. (1994) 41–48

On Boundaries of Highly Visible Spaces and Applications

John H. Reif and Zheng Sun

Department of Computer Science, Duke University, Durham, NC 27708, USA
{reif,sunz}@cs.duke.edu

Abstract. The purpose of this paper is to investigate the properties of a certain class of highly visible spaces. For a given geometric space S containing *obstacles* specified by disjoint subsets of S, the *free space* $\mathcal{F}$ is defined to be the portion of S not occupied by these obstacles. The space is said to be *highly visible* if at each point in $\mathcal{F}$ a viewer can see at least an ϵ fraction of the entire $\mathcal{F}$. This assumption has been used for robotic motion planning in the analysis of random sampling of points in the robot's configuration space, as well as the upper bound of the minimum number of guards needed for art gallery problems. However, there is no prior result on the implication of this assumption to the geometry of the space under study. For the two-dimensional case, with the additional assumptions that S is bounded within a rectangle of constant aspect ratio and that the volume ratio between $\mathcal{F}$ and S is a constant, we show by "charging" each obstacle boundary by a certain portion of S that the total length of all obstacle boundaries in S is $O(\sqrt{n\mu(\mathcal{F})/\epsilon})$, if S contains polygonal obstacles with a total of n boundary edges; or $O(\sqrt{n\mu(\mathcal{F})/\epsilon})$, if S contains n convex obstacles that are piecewise smooth. In both cases, $\mu(\mathcal{F})$ is the volume of $\mathcal{F}$. For the polygonal case, this bound is tight as we can construct a space whose boundary size is $\Theta(\sqrt{n\mu(\mathcal{F})/\epsilon})$. These results can be partially extended to three dimensions. We show that these results can be applied to the analysis of certain probabilistic roadmap planners, as well as a variation of the art gallery problem.

1 Introduction

Computational geometry is now a mature field with a multiplicity of well-defined foundational problems associated with, for many cases, efficient algorithms as well as well-established applications over a broad range of areas including computer vision, robotic motion planning and rendering. However, as compared to some other fields, the field of computational geometry has not yet explored as much the methodology of looking at reasonable sub-cases of inputs that appear in practice for practical problems. For example, in matrix computation, there is a well-established set of specialized matrices, such as sparse matrices, structured matrices, and banded matrices, for which there are especially efficient algorithms.

One assumption that has been used in a number of previous works in computational geometry is the assumption that, for a given geometric space S with

A. Lingas and B.J. Nilsson (Eds.): FCT 2003, LNCS 2751, pp. 271–283, 2003.
© Springer-Verlag Berlin Heidelberg 2003

a specified set of obstacles, a viewer can see at every point of the free space $\mathcal{F}$ an ϵ fraction of the entire volume of $\mathcal{F}$. Here *obstacles* are defined to be compact subsets of $\mathcal{S}$, while the *free space* $\mathcal{F}$ is defined to be the portion of $\mathcal{S}$ not occupied by the obstacles. In this paper we will call this assumption ϵ-visibility (though please note that some of the prior authors called it instead ϵ-goodness).

1.1   Probabilistic Roadmap Planners

The ϵ-visibility assumption, in particular, has been used in the analysis of randomized placements of points in the robot's configuration space for probabilistic roadmap (PRM) planners [1,2]. A classic PRM planner [3,4] randomly picks in the free space of the robot's configuration space a set of points, called *milestones*. With these milestones, it constructs a *roadmap* by connecting each pair of milestones between which a collision-free path can be computed using a simple local planner. For any given initial and goal configurations s and t, the planner first finds two milestones s' and t' such that a simple collision-free path can be found connecting s (t, respectively) with s' (t', respectively) and then searches the roadmap for a path connecting s' and t'. The PRM planners have proved to be very effective in practice, capable of solving robotic motion planning problems with many degrees of freedom. They also find applications in other areas such as computer animation, computational biology, etc.

The performance of a PRM planner depends on two key features of the roadmaps it constructs, *visibility* and *connectivity*. Firstly, for any given (initial or goal) configuration v, there should exist in the roadmap a milestone v' such that a local planner can find a path connecting v and v'. Since in practice most PRM planners use local planners that connect configurations by straight line segments, this implies that the milestones collectively need to see the entire (or at least a significant portion of) free space. Secondly, the roadmap should capture the connectivity of the free space it represents. Any two milestones in the same connected component of the free space should also be connected via the roadmap, or otherwise the planner would give "false negative" answers to some queries.

The earlier PRM planners pick milestones with a uniform distribution in the free space. The success of these planners motivated Kavraki *et al.*[1] to establish a theoretical foundation for the effectiveness of this sampling method. They showed that, for an ϵ-visible configuration space, $O(\frac{1}{\epsilon} \log \frac{1}{\epsilon})$ milestones uniformly sampled in the free space are needed to *adequately* cover the free space with a *high* probability.

1.2   Art Gallery Problems

The ϵ-visibility assumption has also been used in bounding the number of guards needed for art gallery problems [5,6,7,8]. Potentially, this assumption might also allow for much more efficient algorithms in this case. The assumption appears to be reasonable in large number of practical cases as long as the considered area is within a closed area (such as a room).

The original art gallery problem was first proposed by V. Klee, who described the problem as the following: how many guards are necessary, and how many guards are sufficient, to guard the paintings and works of art in an art gallery with n walls? Later, Chvátal [9] showed that $\lfloor \frac{n}{3} \rfloor$ guards are always sufficient and occasionally necessary to guard a simple polygon with n edges. Since then, there have been numerous variations of the art gallery problem, including, but not limited to, vertex guard problem, edge guard problem, fortress and prison yard problems, etc. (See [10] for a comprehensive review of various art gallery problems.)

Although for the worst case the number of guards needed is $\Theta(n)$ for polygonal galleries with n edges, intuitively, one would expect that galleries that are ϵ-visible should require much fewer guards. By translating the result of Kavraki et al.[1] into the context of art gallery problems, a uniformly random placement of $O(\frac{1}{\epsilon} \log \frac{1}{\epsilon})$ guards is very likely to guard an adequate portion of the gallery. Kavraki et al.[1] also conjectured that in d-dimensional space any ϵ-visible polygonal gallery with h holes can be guarded by at most $f_d(h, \frac{1}{\epsilon})$ guards, for some polynomial function f_d. Following some ideas of an earlier work by Kalai and Matoušek [5], Valtr [6] confirmed the 2D version of the conjecture by showing that $f_2(h, \frac{1}{\epsilon}) = (2 + o(1))\frac{1}{\epsilon} \log \frac{1}{\epsilon} \log(h + 2)$. However, Valtr [7] disapproved the 3D version of the conjecture by constructing for any integer k a $\frac{5}{9}$-visible art gallery that cannot be guarded by k guards. Kirkpatrick [8] later showed that $64\frac{1}{\epsilon} \log \log \frac{1}{\epsilon}$ vertex guards are needed to guard all vertices of a simply connected polygon P that has the property that each vertex of P can see at least ϵ fraction of the other vertices of P. He also gave a similar result for boundary guards.

It has been proved that, for various art galleries problems, finding the *minimum* number of guards is difficult. Lee and Lin [11] proved that the minimum vertex guard problem for polygons is NP-hard. Schuchardt and Hecker[12] further showed that even for orthogonal polygons, whose edges are parallel to either the x-axis or the y-axis, the minimum vertex and point guard problems are NP-hard. Ghosh [13] presented an $O(n^5 \log n)$ algorithm that can compute a vertex guard set whose size is at most $O(\log n)$ times the minimum number of guards needed.

However, with the assumption of ϵ-visibility, one can use a simple and efficient randomized approximation algorithm based on the result of Kavraki et al.[1] for the original art gallery problem. Moreover, this approximation algorithm does not require the assumption that the space is polygonal.

1.3   Our Result

Intuitively, for an ϵ-visible space, the total size of all obstacle boundaries cannot be arbitrarily large; an excessive size of obstacle boundaries would inevitably cause a point in $\mathcal{F}$ to lose ϵ-visibility by blocking a significant portion of its view. Our main result of this paper is an upper bound of the boundary size of ϵ-visible spaces in two and (in some special cases) three dimensions. The upper bound of the boundary size not only is a fundamental property for the geometric spaces of this type, but also may have implications to other applications that use this assumption.

We show that, for an ϵ-visible $2D$ space, the total length of all obstacle boundaries is $O(\sqrt{n\mu(\mathcal{F})/\epsilon})$, if the space contains polygonal obstacles with a total of n boundary edges; or $O(\sqrt{n\mu(\mathcal{F})}/\epsilon)$, if the space contains n convex obstacles that are piecewise smooth. In both cases, $\mu(\mathcal{F})$ is the area of $\mathcal{F}$. For the case of polygonal obstacles, this bound is tight as one can construct an ϵ-visible space containing obstacle boundaries with a total length of $\Theta(\sqrt{n\mu(\mathcal{F})/\epsilon})$.

Our result can be used to bound the number of guards needed for the following variation of the original art gallery problem: given a space with a specified set of obstacles, how to put points on boundaries of obstacles so that these points see the entire (or a significant portion of) space. We call this problem *boundary art gallery problem*. This problem can find applications in practical situations where the physical constraints would only allow points to be placed on obstacle boundaries. For example, one might need to install lights on the walls to enlighten a closed space consisting of rooms and corridors.

If this result can be extended to higher dimensions, we can also apply it to bounding the number of randomly sampled boundary points needed to adequately cover the free space. Although it is difficult to uniformly sample points on the boundary of a space without an explicit algebraic description, there exist PRM planners [14,15] that place milestones "pseudo-uniformly" on the boundary of the free space using various techniques. These planners have proved to be more effective in capturing the connectivity of the configuration space with the presence of narrow passages.

2   Bounding Boundary Size for $2D$ and $3D$ ϵ-Visible Spaces

In this section we prove an upper bound of the boundary size of $2D$ ϵ-visible spaces. We also show that this result can be partially extended to $3D$ ϵ-visible spaces.

2.1   Preliminaries

Suppose $\mathcal{S}$ is the $2D$ space bounded inside a rectangle $\mathcal{R}$. We let $\mathcal{B}$ denote the union of all obstacles in $\mathcal{S}$, and let $\partial\mathcal{B}$ denote the boundaries of all obstacles. For each point $v \in \mathcal{F}$, we let $\mathcal{V}_v = \{v' | \text{ line segment } \overline{vv'} \subset \mathcal{F}\}$. That is, $\mathcal{V}_v$ is the set of all free space points that can be seen from v.

We assume that the *aspect ratio* of $\mathcal{R}$, defined to be the ratio between the lengths of the shorter and longer sides of $\mathcal{R}$, is no less than λ, where $0 < \lambda < 1$. We also assume that $\mu(\mathcal{F}) \geq \rho \cdot \mu(\mathcal{S})$, for some constant $\rho > 0$. In the full version of the paper, we will give examples where the boundary size cannot be bounded if λ and ρ are not bounded by constants.

A segment of the boundary (which we call *sub-boundary*) of an obstacle is said to be *smooth* if the curvature is continuous along the curve defining the boundary. The boundary of an obstacle is said to be *piecewise smooth* if it consists of a finite number of smooth sub-boundaries. In this section we assume that the boundaries of all obstacles inside $\mathcal{R}$ are piecewise smooth.

For a smooth sub-boundary c, the *turning angle*, denoted by $\mathcal{A}(c)$, is defined to be the integral of the curvature along c. For a piecewise sub-boundary c, the turning angle is defined to be the sum of the turning angles of all smooth sub-boundaries of c, plus the sum of the instantaneous angular changes at the joint points. Observe that the turning angle of the boundary of an obstacle is 2π if the obstacle is convex, or greater than 2π if it is non-convex. In some sense, the turning angle of the boundary of an obstacle reflects the geometric complexity of the obstacle.

For each sub-boundary c, we use $|c|$ to denote the length of c, and use $c[u_1, u_2]$ to denote the part of c between points u_1 and u_2 on c. For any point $v \notin c$, we let u_1 and u_2 be the two points on c such that c is lying between the two rays $\overrightarrow{vu_1}$ and $\overrightarrow{vu_2}$. We call u_1 and u_2 *bounding points* of c by v. We define the *viewing angle of c from v* to be $\angle u_1 v u_2$.

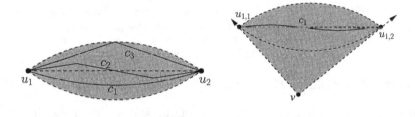

(a) Various ϵ-flat sub-boundaries bounded between two arcs

(b) Blocked visibility near ϵ-flat sub-boundary

Fig. 1. Lines and curves are not drawn proportionally.

For each obstacle, we decompose its boundary into minimum number of ϵ-flat sub-boundaries. A sub-boundary c is said to be ϵ-*flat* if $\mathcal{A}(c) \leq \pi - \theta_\epsilon$, where $\theta_\epsilon = \frac{\lambda \rho}{16(1+\lambda^2)} \cdot \epsilon$. Let u_1 and u_2 be the two endpoints of c. Observe that c is bounded between two minor arcs each with chord $\overline{u_1 u_2}$ and angle $2\theta_\epsilon$, as shown in Figure 1.a. Therefore, the width of c, defined by $|\overline{u_1 u_2}|$, is no less than $|c| \cdot \cos \frac{\theta_\epsilon}{2}$, while the height of c, defined by the maximum distance between any point $u \in c$ and line segment $\overline{u_1 u_2}$, is no more than $\frac{|c|}{2} \cdot \sin \frac{\theta_\epsilon}{2}$.

Since ϵ-flat sub-boundaries are "relatively" flat, any point $v \in \mathcal{F}$ "sandwiched" between two ϵ-flat sub-boundaries will have a limited visibility, as we show in the follow lemma:

Lemma 1. *If $v \in \mathcal{F}$ is a point between two ϵ-flat sub-boundaries c_1 and c_2 and the total viewing angle of c_1 and c_2 from v is more than $2\pi - 6\theta_\epsilon$, then v is not ϵ-visible.*

Proof Abstract. For each $i = 1, 2$, let $u_{i,1}$ and $u_{i,2}$ be the two endpoints of c_i. $\mathcal{V}_v$ is the union of the following three regions: I) the region bounded by sub-boundary c_1, $\overline{vu_{1,1}}$ and $\overline{vu_{1,2}}$; II) the region bounded by sub-boundary c_2, $\overline{vu_{2,1}}$

and $\overline{vu_{2,2}}$; and III) the region not inside either $\angle u_{1,1}vu_{1,2}$ or $\angle u_{2,1}vu_{2,2}$. Since the total viewing angle of v blocked by c_1 and c_2 is more than $2\pi - 6\theta_\epsilon$, and $u_{1,1}vu_{1,2} \leq \pi + \theta_\epsilon$ and $\angle u_{2,1}vu_{2,2} \leq \pi + \theta_\epsilon$, we have $\angle u_{1,1}vu_{1,2} > \pi - 7\theta_\epsilon$ and $\angle u_{2,1}vu_{2,2} > \pi - 7\theta_\epsilon$. Since c_1 is ϵ-flat, the volume of Region I is bounded by the union of $\triangle u_{i,1}vu_{i,2}$ and the arc with chord $|c_1|$ and angle $2\theta_\epsilon$, as shown in Figure 1.b. Since $|c_1| \cdot \cos(\theta_\epsilon/2) \leq |\overline{u_{1,1}u_{1,2}}| \leq L_{\mathcal{R}} \leq \sqrt{\frac{\lambda^2+1}{\lambda\rho}\mu(\mathcal{F})}$, where $L_{\mathcal{R}}$ is the length of the diagonal of $\mathcal{R}$, the volume of Region I is bounded by $O(\epsilon\mu(\mathcal{F}))$. Region III is the union of two (possibly merged) cones with a total angle of $6\theta_\epsilon$, and therefore the volume of Region III is also $O(\epsilon\mu(\mathcal{F}))$. Hence, the region visible from v has a total volume of $O(\epsilon\mu(\mathcal{F}))$. (In the full version of the paper we will show that the volume is actually less than $\epsilon\mu(\mathcal{F})$.) Therefore, v is not ϵ-visible.   □

In the rest of this section we will prove the following theorem:

Theorem 1. *If the boundaries of all obstacles can be divided into n ϵ-flat sub-boundaries, the total length of all obstacle boundaries is bounded by $O(\sqrt{\frac{n\mu(\mathcal{F})}{\epsilon}})$.*

However, to prove Theorem 1 we need two lemmas, which we will prove in the next subsection. In Subsection 2.3 we will show the proof of this theorem as well as its corollaries.

2.2   Forbidden Neighborhoods of ϵ-Flat Sub-boundaries

For each ϵ-flat sub-boundary c with endpoints u_1 and u_2, we divide it into 15 equal-length segments, and let u_1' and u_2' be the two endpoints of the middle segment. The ϵ-*neighborhood* of c, denoted by $\mathcal{N}_\epsilon(c)$, is defined to be the union of points from each of which the viewing angle of $c[u_1', u_2']$ is greater than $\pi - \theta_\epsilon$, as show in Figure 2.a. It is easy to see that, for any $v \in \mathcal{N}_\epsilon(c)$, the distance between v and line segment $\overline{u_1'u_2'}$ is no more than $\frac{|c[u_1', u_2']|}{2} \cdot \tan\theta_\epsilon = \frac{|c|}{30} \cdot \tan\theta_\epsilon$. The distance between v and line segment $\overline{u_1u_2}$ is no more than the sum of the distance between u and $\overline{u_1'u_2'}$ and the maximum distance between $\overline{u_1'u_2'}$ and $\overline{u_1u_2}$, which is $\frac{|c|}{30} \cdot \tan\theta_\epsilon + \frac{|c|}{2} \cdot \sin\frac{\theta_\epsilon}{2}$.

These neighborhoods are "forbidden" in the sense that they do not overlap with each other if the corresponding sub-boundaries are roughly the same length, as we will show in Lemma 2. By "charging" a certain portion of $\mathcal{S}$ to each ϵ-flat sub-boundary, we show that the total length of all ϵ-flat sub-boundaries, that is, the length of $\partial\mathcal{B}$, can be upper-bounded.

Lemma 2. *The ϵ-neighborhoods of two sub-boundaries c_1 and c_2 do not overlap if $\frac{|c_1|}{2} \leq |c_2| \leq 2|c_1|$.*

Proof. Suppose for the sake of contradiction $v \in \mathcal{S}$ is a point inside $\mathcal{N}_\epsilon(c_1) \cap \mathcal{N}_\epsilon(c_2)$, where the length ratio between c_1 and c_2 is between $\frac{1}{2}$ and 2. For each $i = 1, 2$, we let $u_{i,1}$ and $u_{i,2}$ be the two endpoints of c_i, and let $u_{i,1}'$ and $u_{i,2}'$ be the endpoints of the portion of c_i incident to the ϵ-neighborhood of c_i. Let v_i be

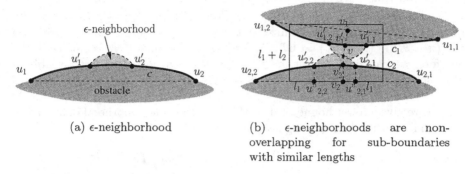

(a) ϵ-neighborhood

(b) ϵ-neighborhoods are non-overlapping for sub-boundaries with similar lengths

Fig. 2. Lines and curves are not drawn proportionally.

the projection of v on line segment $\overline{u_{i,1}u_{i,2}}$, and let v_i' be the intersection of c_i and the straight line that passes both v_i and v.

The intuition here is as the following: since c_1 and c_2 are "relatively" flat, non-intersecting, and about the same length, for $\mathcal{N}_\epsilon(c_1)$ and $\mathcal{N}_\epsilon(c_2)$ to overlap, $\overline{u_{1,1}u_{1,2}}$ and $\overline{u_{2,1}u_{2,2}}$ have to be "almost" parallel and also close to each other. That way, we can find in the free space between c_1 and c_2 a point that can only see less than $\epsilon\mu(\mathcal{F})$ of the free space as its visibility is mostly "blocked" by c_1 and c_2, leading to a contradiction to the assumption that $\mathcal{S}$ is ϵ-visible.

There are a number of cases corresponding to different geometric arrangements of the points, line segments and curves (sub-boundaries). In the following we assume that $\overline{u_{1,1}u_{1,2}}$ and $\overline{u_{2,1}u_{2,2}}$ do not intersect, v lies between $\overline{u_{1,1}u_{1,2}}$ and $\overline{u_{2,1}u_{2,2}}$, and v_1' (v_2', respectively) lies between v and v_1 (v_2, respectively), as shown in Figure 2.b. The other cases can be analyzed in an analogous manner.

Since line segments $\overline{u_{1,1}u_{1,2}}$ and $\overline{u_{2,1}u_{2,2}}$ do not intersect, either both $\overline{v_1u_{2,1}}$ and $\overline{v_1u_{2,2}}$ lie between $\overline{u_{1,1}u_{1,2}}$ and $\overline{u_{2,1}u_{2,2}}$, or both $\overline{v_2u_{1,1}}$ and $\overline{v_2u_{1,2}}$ lie between $\overline{u_{1,1}u_{1,2}}$ and $\overline{u_{2,1}u_{2,2}}$. Without loss of generality we assume that it is the former case. Let $l_1 = |\overline{vv_1}|$ and $l_2 = |\overline{vv_2}|$. Let $u_{2,1}''$ ($u_{2,2}''$, respectively) be the projection of $u_{2,1}'$ ($u_{2,2}'$, respectively) on $\overline{u_{2,1}u_{2,2}}$. Observe that v_1' lies inside the small rectangle of width $|\overline{u_{2,1}''u_{2,2}''}| + 2l_1$ and height $l_1 + l_2$ (the solid rectangle in Figure 2.b). Since $|\overline{u_{2,2}u_{2,2}''}| = |\overline{u_{2,2}u_{2,1}}| - |\overline{u_{2,2}''u_{2,1}}| > |\overline{u_{2,2}u_{2,1}}| - |c[u_{2,2}', u_{2,1}]|$, we have

$$\tan\angle v_1'u_{2,1}u_{2,2} \le \frac{l_1 + l_2}{|\overline{u_{2,2}u_{2,1}}| - |c[u_{2,2}', u_{2,1}]| - l_1}$$

$$\le \frac{(\frac{1}{30} \cdot \tan\theta_\epsilon + \frac{1}{2} \cdot \sin\frac{\theta_\epsilon}{2}) \cdot (|c_1| + |c_2|)}{|c_2| \cdot \cos\frac{\theta_\epsilon}{2} - \frac{8|c_2|}{15} - (\frac{1}{30} \cdot \tan\theta_\epsilon + \frac{1}{2} \cdot \sin\frac{\theta_\epsilon}{2}) \cdot |c_1|}.$$

Applying $|c_1| \le 2|c_1|$ and $\theta_\epsilon < \frac{1}{12}$, we now have

$$\tan\angle v_1'u_{2,1}u_{2,2} \le \frac{\theta_\epsilon \cdot (\frac{1}{30\cos\theta_\epsilon} + \frac{1}{4}) \cdot 3|c_2|}{(\cos\frac{\theta_\epsilon}{2} - \frac{8}{15} - (\frac{1}{15} \cdot \tan\theta_\epsilon + \sin\frac{\theta_\epsilon}{2})) \cdot |c_2|}$$

$$\le \frac{5\theta_\epsilon}{2} \le \frac{5}{2}\tan\theta_\epsilon \le \tan\frac{5\theta_\epsilon}{2}.$$

It follows that $\angle v_1' u_{2,1} u_{2,2} \leq \frac{5\theta_\epsilon}{2}$. Similarly, we can show that $\angle v_1' u_{2,2} u_{2,1} \leq \frac{5\theta_\epsilon}{2}$, and therefore $\angle u_{2,1} v_1' u_{2,2} \geq \pi - 5\theta_\epsilon$. Since v_1' is on c_1, $\angle u_{1,1} v_1' u_{1,2} \geq \pi - \theta_\epsilon$. Therefore, the viewing angle from v_1' not blocked by c_1 and c_2 is no more than $2\pi - (\pi - \theta_\epsilon) - (\pi - 5\theta_\epsilon) = 6\theta_\epsilon$. According to Lemma 1 v_1' is not ϵ-visible. Therefore, we can find a point $v_1^* \in \mathcal{F}$ close to v_1' who is also not ϵ-visible, a contradiction to the assumption that $\mathcal{S}$ is ϵ-visible.   □

Next we give a lower bound of the volume of the ϵ-neighborhood of any ϵ-flat sub-boundary with the following lemma:

Lemma 3. *For any ϵ-flat sub-boundary c, the volume of $\mathcal{N}_\epsilon(c)$ is $\Omega(\theta_\epsilon \cdot |c|^2)$.*

Proof. We will show that, the ϵ-neighborhood of c has a volume no less than $\mu_0 = \frac{\theta_\epsilon |c[u_1', u_2']|^2}{18\kappa_1}$, for some constant $\kappa_1 > 1$. (We will explain later how this constant κ_1 is chosen.)

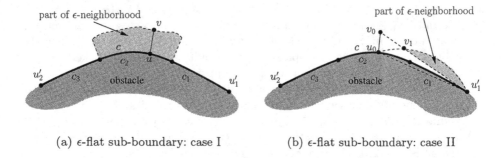

(a) ϵ-flat sub-boundary: case I   (b) ϵ-flat sub-boundary: case II

Fig. 3. In the figures we only show the portion of sub-boundary c between u_1' and u_2'

We divide $c[u_1', u_2']$ into three equal-length segments, c_1, c_2, and c_3. For any point u on $c[u_1', u_2']$, we say that $v \in \mathcal{F}$ is the *lookout point* of u if line segment $\overline{vu}$ is normal to $c[u_1', u_2']$ and the viewing angle of $c[u_1', u_2']$ from v is $\pi - \theta_\epsilon$. We call the length of $\overline{uv}$ the *lookout distance* of $c[u_1', u_2']$ at u.

We first consider Case I, where for each point $u \in c_2$ the length of the lookout distance of c at u is at least $l = \frac{\theta_\epsilon |c[u_1', u_2']|}{3\kappa_1}$, as shown in Figure 3.a. In this case, the volume of the ϵ-neighborhood of c outside c_2 is at least $|c_2| \cdot l - \frac{l^2 \cdot \theta_\epsilon}{2} = \frac{|c[u_1', u_2']|^2 \cdot \theta_\epsilon}{9\kappa_1} \cdot (1 - \frac{\theta_\epsilon^2}{2\kappa_1}) \geq \frac{|c[u_1', u_2']|^2 \cdot \theta_\epsilon}{18\kappa_1} = \mu_0$, and therefore the volume of the ϵ-neighborhood of c is no less than μ_0.

Now we consider Case II, where there exists a point $u_0 \in c_2$ such that the lookout distance at u_0 is less than l, as shown in Figure 3.b. Let v_0 be the lookout point of u_0. Since $\mathcal{A}(c[u_1', u_2']) \leq \mathcal{A}(c) \leq \theta_\epsilon$, v_0 will see at least one of the two endpoints of $c[u_1', u_2']$, or otherwise the viewing angle of v_0 is less than $\pi - \theta_\epsilon$. Without loss of generality we let u_1' be an endpoint of $c[u_1', u_2']$ that is visible from v_0. $c[u_0, u_1']$, the part of c between u_0 and u_1', lies below line segments $\overline{v_0 u_1'}$. Since $u_0 \in c_2$, we have $|c[u_0, u_1']| \geq |c_1| = \frac{|c[u_1', u_2']|}{3}$.

Since curve $c[u_0, u_1']$ is also ϵ-flat, we have $|\overline{u_0 u_1'}| \geq |c[u_0, u_1']| \cdot \cos \frac{\theta_\epsilon}{2} > \frac{|c[u_1', u_2']|}{6}$. We use $\overline{u_0 u_1'}$ as the chord to draw a minor arc of angle $2\theta_\epsilon$ outside $\overline{u_0 u_1'}$. The radius of this arc is $r_0 = \frac{|\overline{u_0 u_1'}|}{2 \sin \theta_\epsilon} \geq \frac{|c[u_1', u_2']|}{12\theta_\epsilon}$. Let v_1 be the point where arc $\widehat{u_0 u_1'}$ intersects $\overline{v_0 u_1'}$. We claim that any point v' inside the closed region bounded by arc $\widehat{u_0 u_1'}$ and chord $\overline{u_1' v_1}$ belongs to the ϵ-neighborhood of c. First of all, v' is outside $c[u_0, u_1']$, as $c[u_0, u_1']$ lies below $\overline{v_0 u_1'}$. Secondly, the viewing angle of $c[u_1', u_2']$ from v' should be no less than the viewing angle of $c[u_0, u_1']$ from v', which is at least $\pi - \theta_\epsilon$.

Now we consider the volume of the region bounded by $\widehat{u_0 u_1'}$ and $\overline{u_1' v_1}$. This is actually an arc $\widehat{u_1' v_1}$ with angle $\theta_0 = 2\theta_\epsilon - 2\angle u_0 u_1' v_0$ and radius r_0. Since $\angle u_0 u_1' v_0 < \frac{|\overline{u_0 v_0}|}{|\overline{u_0 u_1'}|} < \frac{l}{|c[u_1', u_2']|/6} = \frac{2\theta_\epsilon}{\kappa_1}$. As long as we choose κ_1 large enough, we can have $\angle u_0 u_1' v_0 < \frac{\theta_\epsilon}{2}$ and therefore $\theta_0 > \theta_\epsilon$. The volume of arc $\widehat{u_1' v_1}$, therefore, is $\frac{r_0^2}{2}(\theta_0 - \sin \theta_0) \geq \frac{r_0^2 \cdot \theta_0^3}{14} \geq \frac{|c[u_1', u_2']|^2 \theta_\epsilon}{14 \cdot 12^2}$. Once again, if we choose κ_1 large enough, we can have $\mu(\widehat{u_1' v_1}) > \frac{\theta_\epsilon |c[u_1', u_2']|^2}{18\kappa_1} = \mu_0$, and therefore the volume of the ϵ-neighborhood of c is greater than μ_0.

Since $|c[u_1', u_2']| = \frac{|c|}{15}$, we have $\mu(\mathcal{N}_\epsilon(c)) = \Omega(\theta_c \cdot |c|^2)$. □

2.3 Putting It Together

With the lemmas established in the last subsection, we are ready to prove Theorem 1:

Proof of Theorem 1. Let L_{max} be the maximum length of all ϵ-flat sub-boundaries inside $\mathcal{R}$. We divide all ϵ-flat sub-boundaries into subsets $S_1, S_2, \cdots, S_k$. For each i, S_i contains the boundaries edges whose lengths are between $\frac{L_{max}}{2^i}$ and $\frac{L_{max}}{2^{i-1}}$,

We let $c_{i,1}, c_{i,2}, \cdots, c_{i,n_i}$ be the n_i sub-boundaries in S_i. By Lemma 2, $\mathcal{N}_\epsilon(c_{i,j}) \cap \mathcal{N}_\epsilon(c_{i,j'}) = \emptyset$, for any j and j', $1 \leq j, j' \leq n_i$. By Lemma 3, there exists a constant $K > 0$ such that $\mu(\mathcal{N}_\epsilon(c_{i,j})) \geq K \cdot \theta_\epsilon \cdot |c_{i,j}|^2$ for all i and j. Therefore, we have

$$\frac{\mu(\mathcal{F})}{\rho} \geq \mu(\mathcal{S}) \geq \mu(\bigcup_{j=1}^{n_i} \mathcal{N}_\epsilon(c_{i,j})) = \sum_{j=1}^{n_i} \mu(\mathcal{N}_\epsilon(c_{i,j}))$$

$$= \sum_{j=1}^{n_i} K \cdot \theta_\epsilon \cdot |c_{i,j}|^2 \geq n_i \cdot K \cdot \theta_\epsilon \cdot \frac{L_{max}^2}{4^i}.$$

Hence we have $n_i \leq \frac{4^i \cdot \mu(\mathcal{F})}{K \cdot \theta_\epsilon \cdot L_{max}^2 \cdot \rho}$. Let $K' = \frac{\mu(\mathcal{F})}{K \cdot \theta_\epsilon \cdot L_{max}^2 \cdot \rho}$. Now we are to give an upper bound of $|\partial \mathcal{B}|$, which is defined to be $\sum_{i=1}^{k} \sum_{j=1}^{n_i} |c_{i,j}|$, the sum of all ϵ-flat sub-boundaries. Since $|c_{i,j}| \leq \frac{L_{max}}{2^{i-1}}$, we have $|\partial \mathcal{B}| \leq L_{max} \cdot \sum_{i=1}^{k} n_i \cdot 2^{-i+1}$. Observe that $\sum_{i=1}^{k} n_i = n$, $\sum_{i=1}^{k} n_i \cdot 2^{-i+1}$ is maximized when $n_i = K' \cdot 4^i$ for

$i < \log_4 \frac{3n}{K'}$ and $n_i = 0$ for $i \geq \log_4 \frac{3n}{K'}$. Therefore, we have

$$\sum_{i=1}^{k} n_i \cdot 2^{-i+1} \leq \sum_{i=1}^{\log_4 \frac{3n}{K'}-1} K' \cdot 4^i \cdot 2^{-i+1} = 2K' \sum_{i=1}^{\log_4 \frac{3n}{K'}-1} 2^i$$

$$< 2K' \cdot 2^{\log_4 \frac{3n}{K'}} = \sqrt{12n \cdot K'} = \sqrt{\frac{12n \cdot \mu(\mathcal{F})}{K \cdot \theta_\epsilon \cdot L_{max}^2 \cdot \rho}}.$$

Therefore, $|\partial B|$ is no more than $\sqrt{\frac{12n \cdot \mu(\mathcal{F})}{K \cdot \theta_\epsilon \cdot \rho}}$. Recall that K and ρ are constants and that $\theta_\epsilon = \Theta(\epsilon)$, we have $|\partial B| = O(\sqrt{\frac{n\mu(\mathcal{F})}{\epsilon}})$.   □

If all the obstacles inside $\mathcal{S}$ are polygons, each boundary edge is an ϵ-flat sub-boundary, and therefore we have the following corollary:

Corollary 1. *If $\mathcal{S}$ contains polygonal obstacles with a total of n edges, $|\partial B|$ is* $O(\sqrt{\frac{n\mu(\mathcal{F})}{\epsilon}})$.

If all obstacles inside $\mathcal{S}$ are convex, the boundary of each obstacle can be decomposed into $\frac{2\pi}{\theta_\epsilon}$ ϵ-flat sub-boundaries, and therefore we have:

Corollary 2. *If $\mathcal{S}$ contains n convex obstacles that are piecewise smooth, $|\partial B|$ is* $O(\frac{1}{\epsilon}\sqrt{n\mu(\mathcal{F})})$.

In some sense, the upper bound stated in Corollary 1 is tight, as one can construct an ϵ-visible space inside a square consisting of $n = \frac{1}{\epsilon}$ rectangular free space "cells," each with length $\sqrt{\mu(\mathcal{F})}$ and width $\epsilon \cdot \sqrt{\mu(\mathcal{F})}$. The total length of obstacle boundaries is $\Theta(\frac{1}{\epsilon}\sqrt{\mu(\mathcal{F})}) = \Theta(\sqrt{\frac{n\mu(\mathcal{F})}{\epsilon}})$.

Nonetheless, we still conjecture that the best bound should be the following:

Conjecture 1. $|\partial B|$ is $O(\frac{1}{\epsilon}\sqrt{\mu(\mathcal{F})})$.

2.4   Extension to Three Dimensions

In this subsection we show how to generalize our proof of Theorem 1 to $3\mathcal{D}$ spaces. For simplicity, we assume that the boundary (surface) of each obstacle is smooth, meaning that the curvature is continuous everywhere on the surface.

To replicate the proofs of Lemmas 1, 2, and 3 for the $3\mathcal{D}$ case, we first need to define the ϵ-flat surface patch, the $3\mathcal{D}$ counterpart of ϵ-flat sub-boundary. A surface patch s is said to be ϵ-flat if, for any point $u \in s$ and any plane p that contains the line $l_{s,u}$, the curve $c = p \cap s$ is ϵ-flat. Here $l_{s,u}$ is the line that passes through u and is normal to s. Moreover, we also need the surface patch to be "relatively round." More specifically, we require that for each ϵ-flat surface patch s there exists a "center" v_s such that, $\max\{|\overline{v_s v}| | v \in \partial s\} / \min\{|\overline{v_s v}| | v \in \partial s\}$ is bounded by a constant. Here ∂s is the closed curve that defines the boundary of s. We call $R_{s,v_s} = \min\{|\overline{v_s v}| | v \in \partial s\}$ the *minimum radius* of s at center v_s.

We define the ϵ-neighborhood $\mathcal{N}_\epsilon(s)$ for an ϵ-flat surface patch similarly to the case of ϵ-flat sub-boundary. We choose a small "sub-patch" s' of s at the center of s so that the distance between v_s and every point on the boundary of s' is $k_1 \cdot R_{s,v_s}$, for some constant $k_1 < 1$. For any point v outside the obstacle that s is bounding, $v \in \mathcal{N}_\epsilon(s)$ if and only if there exist two points $u_1, u_2 \in s'$ such that $\angle u_1 v u_2 > \pi - k_2 \epsilon$ for some constant $k_2 > 0$.

We use a sequence of planes each containing l_{v,s_v} to "sweep" through the volume of $\mathcal{N}_\epsilon(s)$. Each such plane p contains a "slice" of $\mathcal{N}_\epsilon(s)$ with an area of no less than $\Theta(\epsilon \cdot R_{s,v_s}^2)$, following the same argument of the proof of Lemma 3. Therefore, the total volume of $\mathcal{N}_\epsilon(s)$ is $\Theta(\epsilon \cdot R_{s,v_s}^3) = \Theta(\epsilon \cdot \mu(s)^{\frac{3}{2}})$. We leave the details of the proof as well as the proofs of the $3\mathcal{D}$ versions of the other lemmas to the full version of the paper, and only state the result as the following:

Theorem 2. *If $\mathcal{S}$ contains convex obstacles bounded by a total of n ϵ-flat surface patches, $|\partial\mathcal{B}|$ is $O((\frac{n\mu(\mathcal{F})^2}{\epsilon^2})^{1/3})$.*

3   Applications and Open Problems

It is easy to see that in a $2\mathcal{D}$ ϵ-visible space $\partial\mathcal{B}_v = \Omega(\epsilon\sqrt{\mu(\mathcal{F})})$ for any $v \in \mathcal{F}$. Therefore, we can arrive at a lower bound of the fraction of all obstacle boundaries that each free space point can see for various cases by using Corollaries 1 and 2. In particular, if Conjecture 1 holds, each free space point can see at least $\Omega(\epsilon^2)$ fraction of all obstacle boundaries. Then, using the same proof technique as [1][1], we can show that $O(\frac{1}{\epsilon^2}\log\frac{1}{\epsilon})$ randomly sampled boundary points can view a significant portion of $\mathcal{F}$ with a high probability. These results can be applied to the boundary art gallery problem to provide an upper bound of the number of boundary guards needed to adequately guard the space.

It occurs to us that, although one can construct an example where there exists a free space point that can only see obstacle boundaries of size $\Theta(\epsilon\sqrt{\mu(\mathcal{F})})$, the total volume of such points could be upper-bounded. In particular, we have the following conjecture:

Conjecture 2. Every point in $\mathcal{F}$, except for a small subset of volume $O(\sqrt{\epsilon}\mu(\mathcal{F}))$, can see obstacle boundaries of size $\Omega(\sqrt{\epsilon\mu(\mathcal{F})})$.

If we can prove both Conjecture 1 and Conjecture 2, we can reduce the number of boundary points needed to adequately cover the space with high probability to $O(\frac{1}{\epsilon^{3/2}}\log\frac{1}{\epsilon})$.

So far our results are limited to $2\mathcal{D}$ ϵ-visible spaces and some special cases of $3\mathcal{D}$ ϵ-visible spaces. If we can extend these results to higher dimensions, we will be able to provide a theoretical foundation for analyzing the effectiveness of the PRM planners [14,15] that (randomly) pick milestones close to boundaries of obstacles. These planners have shown to be more efficient than the earlier

---

[1] The difference is that, in our proof, every point v in the free space sees at least ϵ^2 fraction of obstacle boundaries, and therefore the probability that k points uniformly sampled on obstacle boundaries cannot see v is $(1 - \epsilon^2)^k$.

PRM planners based on uniform sampling in the free space by better capturing narrow passages in the configuration space; that is, the roadmaps they construct have better connectivity. However, there has been no prior theoretical result on the visibility of the roadmaps constructed using the sampled boundary points. With upper bound results analogous to the ones for $2\mathcal{D}$ and $3\mathcal{D}$ cases, we will be able to prove an upper bound of the number of milestones uniformly sampled on obstacle boundaries needed to adequately cover free space $\mathcal{F}$ with a high probability, an result similar to the one provided by Kavraki [1] for uniform sampling method.

4 Conclusion

In this paper we provided some preliminary results as well as several conjectures on the upper bound of the boundary size of ϵ-visible spaces in $2\mathcal{D}$ and $3\mathcal{D}$ spaces. These results can be used to bound the number of guards needed for the boundary art gallery problem. Potentially, they can also be applied to the analysis of a certain class of PRM planners that sample points close to obstacle boundaries.

Acknowledgement. This work is supported by NSF ITR Grant EIA-0086015, DARPA/AFSOR Contract F30602-01-2-0561, NSF EIA-0218376, and NSF EIA-0218359.

References

1. Kavraki, L.E., Latombe, J.C., Motwani, R., Raghavan, P.: Randomized query processing in robot motion planning. In: Proceedings of the 27th Annual ACM Symposium on Theory of Computing. (1995) 353–362
2. Hsu, D., Kavraki, L., Latombe, J.C., Motwari, R., Sorkin, S.: On finding narrow passages with probabilistic roadmap planners. In: Proceedings of the 3rd Workshop on Algorithmic Foundations of Robotics. (1998)
3. Kavraki, L., Latombe, J.C.: Randomized preprocessing of configuration space for fast path planning. In: Proceedings of the 1994 International Conference on Robotics and Automation. (1994) 2138–2145
4. Overmars, M.H., Švestka, P.: A probabilistic learning approach to motion planning. In: Proceedings of the 1st Workshop on Algorithmic Foundations of Robotics. (1994) 19–37
5. Kalai, G., Matoušek, J.: Guarding galleries where every point sees a large area. Israel Journal of Mathematics **101** (1997) 125–139
6. Valtr, P.: Guarding galleries where no point sees a small area. Israel Journal of Mathematics **104** (1998) 1–16
7. Valtr, P.: On galleries with no bad points. Discrete & Computational Geometry **21** (1999) 193–200
8. Kirkpatrick, D.: Guarding galleries with no nooks. In: Proceedings of the 12th Canadian Conference on Computational Geometry. (2000) 43–46

9. Chvátal, V.: A combinatorial theorem in plane geometry. Journal of Combinatorial Theory **Series B 18** (1975) 39–41
10. Urrutia, J.: Art gallery and illumination problems. In Sack, J.R., Urrutia, J., eds.: Handbook of Computational Geometry. Elsevier Science Publishers B.V. North-Holland, Amsterdam (2000) 973–1026
11. Lee, D.T., Lin, A.K.: Computational complexity of art gallery problems. IEEE Transactions on Information Theory **32** (1986) 276–282
12. Schuchardt, D., Hecker, H.: Two NP-hard art-gallery problems for ortho-polygons. Mathematical Logic Quarterly **41** (1995) 261–267
13. Ghosh, S.K.: Approximation algorithms for art gallery problems. In: Proceedings of Canadian Information Processing Society Congress. (1987)
14. Amato, N.M., Bayazit, O.B., Dale, L.K., Jones, C., Vallejo, D.: OBPRM: An obstacle-based PRM for 3d workspaces. In: Proceedings of the 3rd Workshop on Algorithmic Foundations of Robotics. (1998) 155–168
15. Boor, V., Overmars, M.H., Stappen, A.F.: The Gaussian sampling strategy for probabilistic roadmap planners. In: Proceedings of the 1999 IEEE International Conference on Robotics and Automation. (1999) 1018–1023

Membrane Computing

Gheorghe Păun

Institute of Mathematics of the Romanian Academy
PO Box 1-764, 70700 Bucureşti, Romania, and
Research Group on Mathematical Linguistics
Rovira i Virgili University
Pl. Imperial Tárraco 1, 43005 Tarragona, Spain
gpaun@imar.ro, gp@astor.urv.es

Abstract. This is a brief overview of membrane computing, at about five years since this area of natural computing has been initiated. One informally introduces the basic ideas and the basic classes of membrane systems (P systems), some directions of research already well developed (mentioning only some central results or types of results along these directions), as well as several research topics which seem to be of interest.

1 Foreword

Membrane computing is a branch of natural computing which abstracts distributed parallel computing models from the structure and functioning of the living cell. The devices investigated in this framework, called membrane systems or P systems, are both able of Turing universal computations and, in certain cases where an enhanced parallelism is provided, able to solve intractable problems in a polynomial time (by trading space for time). The domain is well developed at the mathematical level, still waiting for implementations of a practical computational interest, but several applications in modelling various biological (but also related to ecology, artificial life, abstract chemistry, even to linguistics) phenomena have been reported.

At less than five years since the paper [6] was circulated on Internet, the bibliography of the domain is pretty large and continuously growing, hence the present survey will only mention the main directions of research and their central results, as well as some topics for further investigation. The goal is to let the reader to have an idea about what membrane computing is dealing with, rather than to provide a formal presentation of membrane systems of various types or a list of precise results. Also, we do not give complete references. The domain is fastly evolving – in particular, several results are repeatedly improved – hence we suggest to the interested reader to consult the web page http://psystems.disco.unimib.it for up-dated details and references. Of a special interest can be the collective volumes available in the web page, those devoted to the series of Workshops on Membrane Computing (held in Curtea de Argeş, Romania, in 2000, 2001, and 2002, and in Tarragona, Spain, in 2003),

A. Lingas and B.J. Nilsson (Eds.): FCT 2003, LNCS 2751, pp. 284–295, 2003.
© Springer-Verlag Berlin Heidelberg 2003

as well as the proceedings of the Brainstorming Week on Membrane Computing, held in Tarragona, in February 2003. For a comprehensive introduction to membrane computing one can also use the monograph [7].

2   The Basic Class of P Systems

The fundamental ingredients of a membrane system are the (1) *membrane structure* and the sets of (2) *evolution rules* which process (3) *multisets* of (4) *objects* placed in the compartments of the membrane structure.

A membrane structure is a hierarchically arranged set of membranes (understood as three dimensional vesicles), as suggested in Figure 1. We distinguish the external membrane (corresponding to the plasma membrane and usually called the *skin* membrane) and several internal membranes (corresponding to the membranes present in a cell, around the nucleus, in Golgi apparatus, vesicles, etc); a membrane without any other membrane inside it is said to be *elementary*. Each membrane determines a compartment, also called *region*, the space delimited from above by it and from below by the membranes placed directly inside, if any exists. The correspondence membrane-region is one-to-one, that is why we sometimes use interchangeably these terms; also, we identify by the same label a membrane and its associated region. (Mathematically, a membrane structure is represented by the unordered tree which describes it, or by a sequence of matching labelled parentheses.)

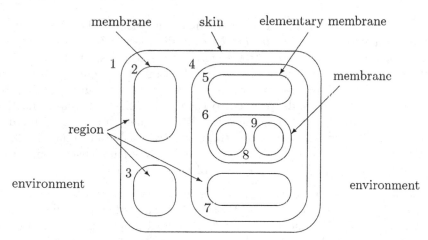

Fig. 1. A membrane structure

In the basic variant of P systems, each region contains a multiset of symbol-objects, which correspond to the chemicals swimming in a solution in a cell compartment; these chemicals are considered here as unstructured, that is why we describe them by symbols from a given alphabet.

The objects evolve by means of evolution rules, which are also localized, associated with the regions of the membrane structure. The rules correspond to the chemical reactions possible in the compartments of a cell. The typical form of such a rule is $aad \rightarrow (a, here)(b, out)(b, in)$, with the following meaning: two copies of object a and one copy of object d react and the reaction produces one copy of a and two copies of b; the new copy of a remains in the same region (indication $here$), one of the copies of b exits the compartment (indication out) and the other enters one of the directly inner membranes (indication in). We say that the objects a, b, b are *communicated* as indicated by the commands associated with them in the right hand member of the rule. When an object exits a compartment, it will go to the surrounding compartment; in the case of the skin membrane this is the environment, hence the object is "lost", it never comes back into the system. If no inner membrane exists (that is, the rule is associated with an elementary membrane), then the indication in cannot be followed, and the rule cannot be applied.

The communication of objects through membranes reminds of the fact that the biological membranes contain various (protein) channels through which the molecules can pass (in a passive way, due to concentration difference, or in an active way, with a consumption of energy), in a rather selective manner.

A rule as above, with several objects in its left hand member, is said to be *cooperative*; a particular case is that of *catalytic* rules, of the form $ca \rightarrow cu$, where a is an object and c is a catalyst, always appearing only in such rules, never changing. A rule of the form $a \rightarrow u$, where a is an object, is called *non-cooperative*.

The rules associated with a compartment are applied to the objects from that compartment, in a *maximally parallel way*: all objects which can evolve by means of local rules should do it (we assign objects to rules, until no further assignment is possible). The used objects are "consumed", the newly produced objects are placed in the compartments of the membrane structure according to the communication commands assigned to them. The rules to be used and the objects to evolve are chosen in a nondeterministic manner. In turn, all compartments of the system evolve at the same time, synchronously (a common clock is assumed for all membranes). Thus, we have two levels of parallelism, one at the level of compartments and one at the level of the whole "cell".

A membrane structure and the multisets of objects from its compartments identify a *configuration* of a P system. By a nondeterministic maximally parallel use of rules as suggested above we pass to another configuration; such a step is called a *transition*. A sequence of transitions constitutes a *computation*. A computation is successful if it halts, it reaches a configuration where no rule can be applied to the existing objects. With a halting computation we can associate a *result* in various ways. The simplest possibility is to count the objects present in the halting configuration in a specified elementary membrane; this is called *internal output*. We can also count the objects which leave the system during the computation, and this is called *external output*. In both cases the result is a number. If we distinguish among different objects, then we can have as the result a vector of natural numbers. The objects which leave the system can also be arranged in a sequence according to the moments when they exit the skin membrane, and in this case the result is a string.

This last possibility is worth emphasizing, because of the qualitative difference between the data structure used inside the system (multisets of objects, hence numbers) and the data structure of the result, which is a string, it contains a positional information, a syntax. A string can also be obtained by following the *trace* of a distinguished object (a "traveller") through membranes.

Because of the nondeterminism of the application of rules, starting from an initial configuration, we can get several successful computations, hence several results. Thus, a P system *computes* (one also uses to say *generates*) a set of numbers, or a set of vectors of numbers, or a language.

We stress the fact that the data structure used in this basic type of P systems is the multiset (of symbols), hence membrane computing can be considered as a biologically inspired algorithmic framework for processing multisets (in a distributed, parallel, nondeterministic manner). Moreover, the main type of evolution rules are rewriting-like rules. Thus, membrane computing has natural connections with many areas of (theoretical) computer science: formal languages (L systems, commutative languages, formal power series, grammar systems, regulated rewriting), automata theory, DNA (more general: molecular) computing, the chemical abstract machine, the Gamma language, Petri nets, complexity theory, etc.

3   Further Ingredients

With motivations coming from biology (trying to have systems as adequate as possible to the cell structure and functioning), from computer science (looking for computationally powerful and/or efficient models), or from mathematics (minimalistic models, even if they are not realistic, are more elegant, challenging, appealing), many types of P systems were introduced and investigated. The number of features considered in this framework is very large.

For instance, we can add a partial order relation to each set of rules, interpreted as a *priority* relation among rules (this corresponds to the fact that certain reactions are more likely to appear – are more active – than others), and in this way the nondeterminism is decreased.

The rules can also have other effects than changing the multisets of objects, namely, they can control the membrane permeability (this corresponds to the fact that the protein channels from cell membranes can sometimes be closed, e.g., when an undesirable substance should be kept isolated, and they are reopen when the "poison" vanishes). If a membrane is non-permeable, then no rule which asks for passing an object through it can be used. In this way, the processes taking place in a membrane system can be controlled ("programmed"). In particular, membranes can be *dissolved* (all objects and membranes from a dissolved membrane are left free in the surrounding compartment – the skin membrane is never dissolved, because this destroys the "computer"; the rules of the dissolved membrane are removed, they are supposed to be specific to the reaction conditions from the former compartment, hence they cannot be applied in the upper compartment, which has its own rules), *created*, and *divided* (like in biology, when a membrane is divided, its content is replicated in the newly obtained membranes).

Furthermore, the rules can be used in a conditional manner, depending on the contents of the region where they are applied. The conditions can be of a permitting context type (a rule is applied only if certain associated objects are present) or of a forbidding context type (a rule is applied only if certain associated objects are not present). This also reminds of biological facts, the promoters and the inhibitors which regulate many biochemical reactions.

Several other ingredients can be considered but we do not enter here into details.

4   Processing Structured Objects

The case of symbol-objects corresponds to a level of approaching ("zooming") the cell where we distinguish the internal compartmentalization and the chemicals from compartments, but not the structure of these chemicals. However, most of the molecules present in a cell have a complex structure, and this observation makes necessary to consider structured objects also in P systems. A particular case of interest is that where the chemicals can be described by strings (this is the case with DNA, RNA, etc).

String-objects were considered in membrane systems from the very beginning. There are two possibilities: to work with sets of strings (hence languages, in the usual sense) or with multisets of strings, where we count the different copies of the same string. In both cases we need evolution rules based on string processing operations, while the second case makes necessary the use of operations which increase and decrease the number of (copies of) strings. Among the operations used in this framework, the basic ones were rewriting and splicing (well-known in DNA computing: two strings are cut at specific sites and the fragments are recombined), but also less popular operations were used, such as rewriting with replication, splitting, conditional concatenation, etc.

The next step is to consider trees or arbitrary graphs as objects, with corresponding operations, then two-dimensional arrays, or even more complex pictures. The bibliography from the mentioned web page contains titles which refer to all these possibilities.

A common feature of the membrane systems which work with strings or with more complex objects is the fact that the halting condition can be avoided when defining the successful computations and their result: a number is not "completely computed" until the computation is finished, it can grow at any further step, but a string sent out of the system at any time remains unchanged, irrespective whether or not the computation continues. Thus, if we compute/generate languages, then the powerful "programming technique" of the halting condition can be ignored (this is also biologically motivated, as, in general, the biological processes aim to last as much as possible, not to reach a "dead state").

5   Universality

From a computability point of view, it is quite interesting that many types of P systems (this means, many combinations of ingredients as those described in the previous sections), of rather restricted forms, are computationally universal. In the case when numbers are computed, this means that these systems can compute all Turing computable sets of natural numbers. When the result of a computation is a string or a set of strings, we get characterizations of the family of recursively enumerable languages. This is true even for systems with simple rules (catalytic), with a very reduced number of membranes (most of the universality results recalled in [7] refer to systems with less that five membranes).

The proof techniques frequently used in such universality results are based on the universality of matrix grammars with appearance checking (in certain normal forms) or on the universality of register machines – and this is rather interesting, as both these machineries are "old stuff" in computer science, being well investigated already three to four decades ago (in both cases, improvements of old results were necessary, motivated by the applications to membrane computing; for instance, new normal forms for matrix grammars, sharper than those known from the literature were recently proved).

The abundance of universality results obtained in membrane computing, on the one hand, shows that "the cell is a powerful computer", on the other hand, asks for an "explanation" of this phenomenon. Roughly speaking, the explanation lies in the fact that Turing computability is based on the possibility to use an arbitrarily large work space, and this means to really *use* it, that is, to control all this space, to send messages at an arbitrary distance (in general, this can be reformulated as *context-sensitivity*); besides context-sensitivity, essential is the possibility of *erasing*. Membrane systems possess erasing by definition (sending objects to the environment or to a "garbage collector" membrane can mean erasing), while the synchronized use of rules (the maximal parallelism) together with the compartmentalization and the halting condition provide "sufficient" context-sensitivity. Thus, the universality is expected, the only challenge is to get it by using systems with a small number of membranes, using as restricted features as possible.

For instance, by using catalytic rules also having associated a priority relation it is rather easy to get the universality; not so easy is to replace the priority with the possibility to control the membrane permeability, but this can be done. However, it is surprising to get the universality by using catalytic rules only and no other ingredient. An additional problem concerns the number of catalysts. The initial proof (by P. Sosik) of the universality of catalytic P systems used eight catalysts, then the number was decreased to six, then to five (R. Freund and P. Sosik), it was shown that one catalyst does not suffice (O.H. Ibarra et al), but the question which is the optimal result from this point of view remains open. Similar "races" for the best result can be found in the case of the number of membranes for various other types of P systems (just one example: for a while, matrix grammars without appearance checking were simulated by rewriting string-object P systems with four membranes, but recently the result was improved to three – M. Madhu – without knowing whether this is an optimal result).

6   Computing by Communication Only

The chemicals do not pass always alone through membranes, but a *coupled transport* is often met, where two solutes pass together through a protein channel, either in the same direction or in the opposite directions. In the first case the process is called *symport*, in the latter case it is called *antiport*. For completeness, *uniport* names the case when a single molecule passes through a membrane.

The idea of a coupled transport can be captured in membrane computing terms in a rather easy way: for the symport case, consider rules of the form (ab, in) or (ab, out), while for the antiport case write $(a, out; b, in)$, with the obvious meaning. Mathematically, we can generalize this idea and consider rules which move arbitrarily many objects through a membrane.

The use such rules suggests a very interesting question (research topic): can we compute only by communication, only by transferring objects through membranes? This question leads to considering systems which contain only symport/antiport rules, which only change the places of objects, but not their "names" (no object is created or destroyed). One starts with (finite) multisets of objects placed in the regions of the system, and with certain objects available in the environment in arbitrarily many copies (the environment is an inexhaustible provider of "raw materials", otherwise we can only deal with the finite number of objects given at the beginning; note that by symport and/or antiport rules associated with the skin membrane we can bring objects from the environment into the system); the symport/antiport rules associated with the membranes are used in the standard nondeterministic maximally parallel manner – and in this way we get a computation.

Note that such systems have several interesting properties, besides the fact that they compute by communication only: the rules are directly inspired from biology, the environment takes part to the process, nothing is created, nothing is destroyed, hence the conservation law is observed – and all these features are rather close to reality.

Surprising at the first sight, but expected in view of the context-sensitivity and erasing possibilities available in symport/antiport P systems, these systems are again universal, even when using a small number of membranes, symport rules and/or antiport rules of small "weights" (the *weight* of a rule is the number of objects it involves).

7   P Automata

Up to now we have discussed only P systems which behave like a grammar: one starts from an initial configuration and one evolves according to the given evolution rules, collecting some results, numbers or strings, in a specified membrane or in the environment. Also an automata-like behavior is possible, especially in the case of systems using only symport/antiport rules. For instance, we can say that a string is accepted by a P system if it consists of symbols brought into the system during a halting computation (we can imagine that a tape is present in the environment, the symbols of which are taken by symport or by antiport rules and introduced into the system; if the computation halts, then the contents of the tape is accepted).

This is a simple and natural definition, considered by R. Freund and M. Oswald. More automata ingredients were considered by E. Csuhaj-Varju and G. Vaszil (the contents of regions are considered states, which control the computation, while only symport rules of the form (x, in) are used, hence the communication is done in a one-way manner; further features are considered, but we omit them here), and by K. Krithivasan, M. Mutyam, and S.V. Varma (special objects are used, playing the role of states, which raises interesting questions concerning the minimisation of P automata both from the point of view of the number of membranes and of states).

The next step is to consider not only an input but also an output of a P system, and this step was also done, by considering P transducers (G. Ciobanu, Gh. Păun, and Gh. Ştefănescu).

As expected, also in the case of P automata (and P transducers) we get the universality: the recursively enumerable languages (the Turing translations, respectively) are characterized in all circumstances mentioned above, always with systems of a reduced size.

8   Computational Efficiency

The computational power is only one criterion for assessing the quality of a new computing machinery; from a practical point of view at least equally important is the *efficiency* of the new device. The P systems display a high degree of parallelism. Moreover, at the mathematical level, rules of the form $a \to aa$ are allowed and by iterating such rules we can produce an exponential number of objects in a linear time. The parallelism and the possibility to produce an exponential working space are standard ways to speed-up computations. In the general framework of P systems with symbol-objects (and without membrane division or membrane creation) these ingredients do not suffice in order to solve computationally hard problems (e.g., **NP**-complete problems) in a polynomial time: in [11] it is proved that any deterministic P system can be simulated by a deterministic Turing machine with a linear slowdown.

However, pleasantly enough, if additional features are considered, either able to provide an enhanced parallelism (for instance, by membrane division, which may produce exponentially many membranes in a linear time), or to better structurate the multisets of objects (by membrane creation), then **NP**-complete problems can be solved in a polynomial (often, linear) time. The procedure is as follows (it has some specific features, slightly different from the standard computational complexity requirements). Given a decision problem, we construct in polynomial time a family of P systems (each one of a polynomial size) which will solve the instance of the problem in the following sense. In a well specified time, bounded by a given function, the system corresponding to the instances of a given size of the problem will sent to its environment a special object *yes* if and only if the instance of the problem introduced into the initial configuration of the system has a positive answer. During the computation, the system can grow exponentially (as the number of objects and/or the number of membranes) and can work in a nondeterministic manner; important is that it *always* halt. Standard problems for illustrating this approach are SAT (satisfiability of propositional formulas in the conjunctive normal form) and HPP (the existence of an Hamiltonian path in a directed graph), but many other problems were also considered. Details can be found in [7] and [9].

There is an interesting point here: we have said that the family of P systems solving a given problem is constructed in polynomial time, but this does not necessarily mean that the construction is *uniform*: it may not start from n but from the nth instance of the problem. Because the construction (done by a Turing machine) takes a polynomial time, it is *honest*, it cannot hide the solution of the problem in the system itself which solves the problem. This "semi-uniformity" (we may call it fairness/honestity) is usual in molecular computing. However, if we insist on having uniform constructions in the classic sense of complexity theory, then this can also be obtained in many cases. A series of results in this direction were obtained by the Sevilla membrane computing group (M.J. Perez-Jimenez, A. Romero-Jimenez, F. Sancho-Caparrini, etc).

Recently, a surprising result was reported by P. Sosik: P systems with membrane division can also solve in polynomial time problems known to be PSPACE-complete. P. Sosik has shown this for QBF (satisfiability of quantified propositional formulas). The family of P systems used in the proof is constructed in the semi-uniform manner mentioned above and the systems use the division operation not only for elementary membranes but also for arbitrary membranes. It is an open problem whether or not the result can be improved from these two points of view.

All previous remarks refer to P systems with symbol-objects. Polynomial (often linear) solutions to **NP**-complete problems can be obtained also in the framework of string-objects, for instance, when string replication is used for obtaining an exponential work space.

9   Resent Research Topics

The two types of attractive results mentioned in the previous sections – computational universality and computational efficiency – as well as the versatility of the P systems explain the very rapid development of the membrane computing area. Besides the topics discussed above, many others were investigated (normal forms in what concerns the shape of the membrane structure, the number and the type of used rules, decidability problems, links with Eilenberg X machines, parallel rewriting of string-objects, ways to avoid the communication deadlock in this case, associating energy to objects or to reactions, and so on and so forth), but we do not enter here into details. Instead, we just briefly mention some topics which were considered in the last time, some of them promising to open new research vistas in membrane computing.

A P system is a computing model, but at the same time it is a model of a cell, whatever reductionistic it is in a given form, hence one can consider its evolution, its "life" as the main topic of investigation and not a number/string produced at the end of a halting computation. This leads to interpreting P systems as dynamic systems, possibly evolving forever, and this viewpoint raises specific questions, different from the computer science ones. Such an approach (P systems as dynamic systems) was started by V. Manca and F. Bernardini, and promises to be of interest for biological applications (see also the next section).

At a theoretical level, a fruitful recent idea is to associate with a P system (with string-objects) not only one language, as usual for grammar or automata-like devices, but a family of languages. This reminds the "old" idea of grammar forms, but also the forbidding-enforcing systems [3]. Actually, M. Cavaliere and N. Jonoska have started from such a possible bridge between forbidding-enforcing systems and membrane systems, considering P systems with a way to define the new populations of strings in terms of forbidding-enforcing conditions. A different idea in defining a family of languages as "generated" by a P system was followed by A. Alhazov.

Returning to the abundance of universality results, which somehow end the research interest for the respective classes of P systems (the equivalence with Turing machines directly implies conclusions regarding decidability, complexity, closure properties, etc), a related question of interest is to investigate the sub-universal classes of P systems. For instance, several universality results refer to systems with arbitrary catalytic rules (of the form $ca \to cu$), used together with non-catalytic rules; also, a given number of membranes is necessary (although in many cases one does not know the sharp borderline between universality and sub-universality from this point of view). What about the power and the properties of P systems which are not universal? Some problems are shown to be decidable for them; which is the complexity of these problems? Which are the closure properties of the associated families of numbers or of languages? Topics of this type were addressed from time to time, but recently O.H. Ibarra and his group started a systematic study, considering both new (restricted) classes of P systems and new problems (e.g., the reachability of a configuration and the complexity of deciding it).

Rather promising seems to be the use of P systems for handling two-dimensional objects. There are several papers in this area, dealing with graphs, arrays, other types of pictures (R. Freund, M. Oswald, K. Krithivasan and her group, R. Ceterchi, R. Gramatovici, N. Jonoska, K.G. Subramanian, etc). Especially interesting is the following idea (suggested several times in membrane computing papers and now followed by R. Ceterchi and her colleagues in Tarragona): instead of using a membrane structure as a support of a computation whose "main" subject are the objects present in the regions of the membrane structure, let us take the tree which describes the membrane structure as the subject of the computation, and use the contents of the regions as auxiliary tools in the computation.

A very important direction of research – important especially from the point of view of applications in biology and related areas – is to bring to membrane computing some approximate reasoning tools, some non-crisp mathematics, in the probability theory, fuzzy sets, or rough sets sense – or in a mixture of all these. Randomized P algorithms, which solve hard problems in polynomial time, using a polynomial space, with a controlled probability, were already proposed by A. Obtulowicz, who has also started a systematic study of the possibility to model the uncertainty in membrane computing.

It is highly probable that all these topics will be much investigated in the near future, with a special emphasis on complexity matters and on issues related to applications, to the adequacy of membrane computing to the biological reality.

10   Implementations and Applications

Some branches of natural computing, such as neural computing and evolutionary computing, starts from biology and try to improve the way we use the existing electronic computers, while DNA computing has the ambition to find a new support for computations, a new hardware. For membrane computing is not yet clear in which direction we have to look for implementations. Anyway, it seems too early to try to implement computations at the level of a cell, whatever attractive this seems to be.

However, there are several attempts to implement (actually, to simulate) P systems on the usual computers. Of course, the biochemically inspired nice features of P systems (in special, the nondeterminism and the parallelism) are lost, as they can be only simulated on the deterministic usual computers, but the obtained simulators still can be useful for certain practical purposes (not to mention their didactical usefulness). At this moment, there are reported at least one dozen of programs for implementing P systems of various types – see references in the web page, where some programs are available, too.

On the other hand, several applications of membrane computing were reported in the literature, in general, of the following type: one takes a piece of reality, most frequently from cell biology, but also from artificial life, abstract chemistry, biology of eco-systems, one constructs a P system modelling this piece of reality, then one writes a program which simulates this P system and one runs experiments, carefully arranging the system parameters (especially, the form of rules and their probabilities to be applied); statistics about the populations of objects in various compartments of the system are obtained, sometimes suggesting interesting conclusions. Typical examples can be found in [1] (including an approach to the famous Brusselator model, with conclusions which fit with the known ones, obtained by using continuous mathematics – by Y. Suzuki et al, an investigation of photosynthesis – by T. Nishida, signaling patways and T cell activation – by G. Ciobanu and his collaborators). Several other (preliminary) applications of P systems to cryptography, linguistics, distributed computing can be found in the volumes [1,8], while [2] contains a promising application in writing algorithms for sorting.

The turning of the domain towards applications in biology is rather natural: P systems are (discrete, algorithmic, well investigated) models of the cell and the cell biologists miss efficient global models of the cell, in spite of the fact that modelling and simulating the living cell is a very important task (as it was stated in several places, this is one of the main challenges of bioinformatics for this beginning of millennium).

11   Final Remarks

At the end of this brief and informal excursion to membrane computing, we stress the fact that our goal was only to give a general impression about this fastly growing research area, hence we strongly suggest to the interested reader to access the web page mentioned in the first section of the paper for any additional information. The page contains the full current bibliography, many downloadable papers, the addresses of people who have contributed to membrane computing, lists of open problems, calls for participation to related meetings, some software for simulating P systems, etc.

References

1. C.S. Calude, Gh. Păun, G. Rozenberg, A. Salomaa, eds., *Multiset Processing. Mathematical, Computer Science, and Molecular Computing Points of View, Lecture Notes in Computer Science*, 2235, Springer, Berlin, 2001.
2. M. Cavaliere, C. Martin-Vide, Gh. Păun, eds., *Proceedings of the Brainstorming Week on Membrane Computing; Tarragona, February 2003*, Technical Report 26/03, Rovira i Virgili University, Tarragona, 2003.
3. A. Ehrenfeucht, G. Rozenberg, Forbidding-enforcing systems, *Theoretical Computer Science*, 292 (2003), 611–638.
4. O.H. Ibarra, On the computational complexity of membrane computing systems, submitted, 2003.
5. K. Krithivasan, S.V. Varma, On minimising finite state P automata, submitted, 2003.
6. Gh. Păun, Computing with membranes, *Journal of Computer and System Sciences*, 61, 1 (2000), 108–143.
7. Gh. Păun, *Computing with Membranes: An Introduction*, Springer, Berlin, 2002.
8. Gh. Păun, G. Rozenberg, A. Salomaa, C. Zandron, eds., *Membrane Computing. International Workshop, WMC-CdeA 2002, Curtea de Argeş, Romania, Revised Papers, Lecture Notes in Computer Science*, 2597, Springer, Berlin, 2003.
9. M. Perez-Jimenez, A. Romero-Jimenez, F. Sancho-Caparrini, *Teoría de la Complejidad en Modelos de Computatión Celular con Membranas*, Editorial Kronos, Sevilla, 2002.
10. P. Sosik, The computational power of cell division in P systems: Beating down parallel computers?, *Natural Computing*, 2003 (in press).
11. C. Zandron, *A Model for Molecular Computing: Membrane Systems*, PhD Thesis, Universitá degli Studi di Milano, 2001.

Classical Simulation Complexity of Quantum Machines*

Farid Ablayev and Aida Gainutdinova

Dept. of Theoretical Cybernetics,
Kazan State University
420008 Kazan, Russia
{ablayev,aida}@ksu.ru

Abstract. We present a classical probabilistic simulation technique of quantum Turing machines. As a corollary of this technique we obtain several results on relationship among classical and quantum complexity classes such as: $PrQP = PP$, $BQP \subseteq PP$ and $PrQSPACE(S(n)) = PrPSPACE(S(n))$.

1 Introduction

Investigations of different aspects of quantum computations in last decade became a very intensively growing area of mathematics, computer science, physics and technology. A good source of information on quantum computations is Nielsen's and Chuang's book [8].

Notice that in quantum mechanic and quantum computations traditionally used "right-left" presentation of computational process. That is, current general state of quantum system is presented as column-vector $|\psi\rangle$ which is multiplied by unitary transition matrix U to obtain next general state $|\psi'\rangle = U|\psi\rangle$.

In this paper we use "left-right" presentation of quantum computational process (as it is used to use for presentation of classical deterministic and stochastic computational processes). That is, current *general state* of quantum system is presented as row-vector $\langle\psi|$ (elements of $\langle\psi|$ are complex conjugates of elements of $|\psi\rangle$) which is multiplied by unitary transition matrix $W = U^\dagger$ to obtain next *general state* $\langle\psi'| = \langle\psi|W$.

In the paper we consider probabilistic and quantum complexity classes. Here $BQSpace(S(n))$ and $PrQSpace(S(n))$ stand for complexity classes determined by $O(S(n))$ space bounded quantum Turing machines that recognize languages with bounded and unbounded error respectively. $PrSpace(S(n))$ stands for complexity class determined by $O(S(n))$ space bounded classical probabilistic Turing machines that recognize languages with unbounded error. $BQTime(T(n))$ and $PrQTime(T(n))$ stand for complexity classes determined by $O(T(n))$ time bounded quantum Turing machines that recognize languages with bounded and

---

* Supported by the Russia Fund for Basic Research under the grant 03-01-00769

A. Lingas and B.J. Nilsson (Eds.): FCT 2003, LNCS 2751, pp. 296–302, 2003.
© Springer-Verlag Berlin Heidelberg 2003

unbounded error respectively. $PrTime(T(n))$ stands for complexity class determined by $O(T(n))$ time bounded classical probabilistic Turing machines that recognize languages with unbounded error. We assume $T(n) \geq n$ and $S(n) \geq \log n$ are fully time and space constructible respectively. For most of the paper, we will refer to the polynomial-time case, where $T(n) = S(n) = n^{O(1)}$.

Classical simulations of quantum computational models use different techniques, see for example [3,9,10,6,7]. In our paper we view a computation process of classical one-tape probabilistic Turing machines (PTM) and quantum Turing machines (QTM) as a linear process. That is, a computation on PTM for particular input u is the Markov process, in which a vector of probabilities distribution of configurations at a given step is multiplied by a fixed stochastic transition matrix M to obtain the vector of probabilities distribution of configurations at the next step. A computation on QTM is a unitary-linear process similar to the Markov process. A quantum computation step corresponds to multiplying a general state (the vector of amplitudes distribution of all possible configurations) at the current step by fixed complex unitary transition matrix to obtain a general state at the next step. We refer to the paper [6] for more information.

In the paper we present classical Simulation Theorem 2 (simulation technique of quantum computation process) which states that having unitary-linear process we can construct equivalent (in the sense of language presentation) Markov process. This simulation technique allows to gather together different complexity results on classical simulation of quantum computations. As a corollary of the Theorem 2 we have the following relations among complexity classes.

Theorem 1.
$PrQTime(T(n)) = PrTime(T(n))$.
In particular $PrQP = PP$.
$BQTime(T(n)) \subseteq PrTime(T(n))$ [1],
In particular $BQP \subseteq PP$
$BQSpace(S(n)) \subseteq PrSpace(S(n))$ [10] and
$PrQSpace(S(n)) = PrSpace(S(n))$ [10]

Proof (Sketch): Quantum simulation technique of classical probabilistic Turing machines is well known, see for example [5,8]. This technique establishes inclusions $PrSpace(S(n)) \subseteq PrQSpace(S(n))$ and $PrTime(T(n)) \subseteq PrQTime(T(n))$. The Simulation Theorem 2 and observation (section 4) prove inclusions:
$BQTime(T(n)) \subseteq PrTime(T(n))$,
$PrQTime(T(n)) \subseteq PrTime(T(n))$,
$BQSpace(S(n)) \subseteq PrSpace(S(n))$,
$PrQSpace(S(n)) \subseteq PrSpace(S(n))$.   ∎

2   Classical Simulation of Quantum Turing Machines

We consider a two-tape Turing machine (probabilistic and quantum) with read-only input tape and read-write tape. We call Turing machine $\mathcal{M}$ $t(n)$-time,

$s(n)$-space machine if every computation of $\mathcal{M}$ on input of length n halts in at most $t(n)$ steps and uses at most $s(n)$ cells on the read-write tape during a computation. We assume $t(n) \geq n$ and $s(n) \geq \log n$ are fully time and space constructible respectively. We will always have $s(n) \leq t(n) \leq 2^{O(s(n))}$. By a configuration C of Turing machine we mean the content of its read-write tape, tape pointers, and current state of the machine.

Definition 1 *A probabilistic Turing machine (PTM) $\mathcal{P}$ consists of a finite set Q of states, a finite input alphabet Σ, a finite tape alphabet Γ, and a transition function*

$$\delta : Q \times \Sigma \times \Gamma \times Q \times \Gamma \times \{L, R\} \times \{L, R\} \to [0, 1]$$

where $\delta(q, \sigma, \gamma, q', \gamma', d_1, d_2)$ gives the probability with which the machine in state q reading σ and γ will enter state q', write γ', and move in direction d_1 and d_2 on read and read-write tapes respectively.

Definition 2 *A quantum Turing machine QTM $\mathcal{Q}$ consists of a finite set Q of states, a finite input alphabet Σ, a finite tape alphabet Γ, and a transition function*

$$\delta : Q \times \Sigma \times \Gamma \times Q \times \Gamma \times \{L, R\} \times \{L, R\} \to \mathbf{C}$$

where $\mathbf{C}$ is the set of complex numbers, $\delta(q, \sigma, \gamma, q', \gamma', d_1, d_2)$ gives the amplitude with which the machine in state q reading σ and γ will enter state q', write γ', and move in direction d_1 and d_2 on read and read-write tapes respectively.

Vector-Matrix Machine. From now we will view Turing machine computation as a linear process described in [6]. Below we present formal description of probabilistic and quantum machine in matrix form. For fairness we should only allow efficiently computable matrix entries, where we can compute i-th bit in time polynomial in i.

First we define a general d-dimensional, t-time "vector-matrix machine" $(d, t) - VMM$ that feeds our needs for linear presentation of computation procedure of probabilistic and quantum machines. Fix an input u.

$$VMM(u) = \langle \langle a(0)|, T, F \rangle$$

where $\langle a(0)| = (a_1, \ldots, a_d)$ is an initial row-vector for an input u, T is a $d \times d$ transition matrix, $F \subseteq \{1, \ldots, d\}$ is an accepting set of states.

$VMM(u)$ proceeds in t steps as follows: in each step i a current vector $\langle a(i)|$ is multiplied by $d \times d$ matrix T to obtain the next vector $\langle a(i+1)|$ that is, $\langle a(i+1)| = \langle a(i)|T$. From the resulting vector $\langle a(t)|$ we determine numbers $Pr^1_{accept}(u)$ and $Pr^2_{accept}(u)$ as follows:

1. $Pr^1_{accept}(u) = \sum_{i \in F} |a_i(t)|$;
2. $Pr^2_{accept}(u) = \sum_{i \in F} |a_i(t)|^2$.

These numbers will express probability of u acceptance for probabilistic and quantum machines respectively. We call $VMM(u)$ that uses $Pr^1(VMM(u))$ $(Pr^2(VMM(u)))$ for probability acceptance *Type I VMM(u) (Type II VMM(u))*.

Linear Presentation of Probabilistic Machine. Let $\mathcal{P}$ be a $t(n)$-time, $s(n)$-space PTM. Computation on an input u of length n by $\mathcal{P}$ can be presented by a finite Markov chain with $d(n) = 2^{O(s(n))}$ states (states of this Markov chain correspond to configurations of PTM) and $d(n) \times d(n)$ stochastic matrix M. Notice that for polynomial-time computation, given configurations C_i, C_j and input u one can in polynomial-time compute probability $M(i, j)$ of transition from C_i to C_j, even though the whole transition matrix M is too big to write down in polynomial-time. Formally computation on input u, $|u| = n$, can be described by stochastic machine $SM(u)$

$$SM(u) = \langle\langle p(0)|, M, F\rangle$$

where SM is Type I $(d(n), t(n)) - VMM$ with the following restrictions: $\langle p(0)| = (p_1, \ldots, p_{d(n)})$ is stochastic row-vector of initial probabilities distribution of configurations. That is, $p_i = 1$ and $p_j = 0$ for $j \neq i$, where C_i is the initial configuration of $\mathcal{P}$ for the input u. M is the stochastic matrix defined above. $F \subseteq \{1, \ldots, d(n)\}$ is a set of indexes of accepting configurations of $\mathcal{P}$.

Linear Presentation of Quantum Machine. Consider $t(n)$-time, $s(n)$-space QTM $\mathcal{Q}$. Computation on an input u of length n by $\mathcal{Q}$ can be presented by the following restricted quantum system (unitary-linear process) with $d(n)$ $(d(n) = 2^{O(s(n))})$ basis states corresponding to configurations of QTM and $d(n) \times d(n)$ complex valued unitary matrix W. Notice that for polynomial-time computation, given configurations C_i, C_j and input u one can in polynomial-time compute amplitude $W(i, j)$ of transition from C_i to C_j, as for PTM. Formally computation on input u, $|u| = n$, can be described by linear machine $LM(u)$

$$LM(u) = \langle\langle \mu(0)|, W, F\rangle$$

where $LM(u)$ is Type II $(d(n), t(n)) - VMM$ with the following restrictions: $\langle \mu(0)| = (z_1, \ldots, z_{d(n)})$ is the initial general state (complex row-vector of initial amplitudes distribution of configurations). Namely, $z_j = 0$ for $j \neq i$ and $z_i = 1$ where C_i is the initial configuration of $\mathcal{Q}$ for the input u. W is the unitary matrix defined above. $F \subseteq \{1, \ldots, d(n)\}$ is a set of indexes of accepting configurations of $\mathcal{Q}$.

Language Acceptance Criteria. We use standard unbounded error and bounded error acceptance criteria. For a language L, for an $n \geq 1$ denote $L_n = L \cap \Sigma^n$. We say that language L_n is unbounded error recognized by Type I (Type II) $(d(n), t(n)) - VMM$ if for arbitrary input $u \in \Sigma^n$ there exists Type I (Type II) $(d(n), t(n)) - VMM(u)$ such that it is holds that $Pr(VMM(u)) > 1/2$ for $u \in L_n$ and $Pr(VMM(u)) < 1/2$ for $u \notin L_n$. Similarly we say that language L_n is $(d(n), t(n)) - VMM$ bounded error recognized by Type I (Type II) $(d(n), t(n)) - VMM$ if for $\epsilon \in (0, 1/2)$, arbitrary $u \in \Sigma^n$ there exists Type I (Type II) $(d(n), t(n)) - VMM(u)$ such that it is holds that $Pr(VMM(u)) \geq 1/2 + \epsilon$ for $u \in L_n$ and $Pr(VMM(u)) \leq 1/2 - \epsilon$ for $u \notin L_n$. We say that $VMM(u)$ process its input u with threshold $1/2$.

Let $\mathcal{M}$ be a classic probabilistic $\mathcal{P}$ or quantum $\mathcal{Q}$ Turing machine. We say that $\mathcal{M}$ unbounded (bounded) error recognizes language $L \subseteq \Sigma^*$ if for all $n \geq 1$ corresponding $(d(n), t(n)) - VMM$ unbounded (bounded) error recognizes language L_n.

Theorem 2 (Simulation Theorem). *Let language L_n be unbounded error (bounded error) recognized by quantum machine $(d(n), t(n)) - LM$. Then there exists stochastic machine $(d'(n), t'(n)) - SM$ that unbounded error recognizes L_n with $d'(n) \leq 4d^2(n) + 3$, and $t'(n) = t(n)$.*

We present the sketch of the proof of Theorem 2 in the next section.

3   Proof of Simulation Theorem

For the proof let us fix arbitrary input u, $|u| = n$, and let $d = d(n)$ and $t = t(n)$. We call $VMM(u)$ complex-valued (real-valued) if VMM has complex-valued (real-valued) entries for initial vector and transition matrix.

Lemma 1. *Let $LM(u)$ be complex-valued $(d, t) - LM(u)$. Then there exists real-valued $(2d, t) - LM'(u)$ such that $Pr(LM(u)) = Pr(LM'(u))$.*

Proof: The proof uses the real-valued simulation of complex-valued matrix multiplication (which is now folklore) and is omitted.   ∎

Next Lemma states complexity relation among machines of Type I and Type II (among "linear" and "non linear" extracting a result of computation).

Lemma 2. *Let $LM(u)$ be real-valued $(d, t) - LM(u)$. Then there exists real-valued Type I $(d^2, t) - VMM(u)$ such that $Pr(VMM(u)) = Pr(LM(u))$.*

Proof: Let $LM(u) = \langle\langle\mu(0)|, W, F\rangle$. We construct $VMM(u) = \langle\langle\tau(0)|, T, F'\rangle$ as follows. The initial general state $\langle\tau(0)| = \langle\mu(0) \otimes \mu(0)|$ — is d^2-dimension vector, $T = W \otimes W$ is $d^2 \times d^2$ matrix. Accepting set $F' \subseteq \{1, \ldots, d^2(n)\}$ of states is defined in according to $F \subseteq \{1, \ldots, d\}$ as follows $F' = \{j : j = (i-1)d+i, \ i \in F\}$.

We denote $|i\rangle$ – d-dimensional unit column-vector with value 1 at i and 0 elsewhere. Using the fact that for real valued vectors c, b it is holds that $\langle c|b\rangle^2 = \langle c \otimes c|b \otimes b\rangle$ we have that $T^t = (W \otimes W)^t = W^t \otimes W^t$ and

$$Pr(VMM(u)) = \sum_{j \in F'} \langle\tau(0)|T^t|j\rangle = \sum_{i \in F} \langle\mu(0) \otimes \mu(0)|W^t \otimes W^t|i \otimes i\rangle$$

$$= \sum_{i \in F} \langle\mu(0)|W^t|i\rangle^2 = Pr(LM(u)).$$

∎

Lemma 3. *Let $(d, t) - VMM(u)$ be real-valued Type I machine with k, $k \leq d$, accepting states. Then there exists real-valued Type I $(d, t) - VMM'(u)$ with unique accepting state such that $Pr(VMM(u)) = Pr(VMM'(u))$.*

Proof: The proof uses standard technique from Linear Automata Theory (see for example the book [4]) and is omitted. ∎

Next lemma presents classical probabilistic simulation complexity of linear machines.

Lemma 4. *Let $VMM(u)$ be real-valued Type I $(d,t) - VMM(u)$. Then there exists stochastic machine $(d+2,t) - SM(u)$ such that*

$$Pr(SM(u)) = c^t Pr(VMM(u)) + 1/(d+2)$$

where constant $c \in (0,1]$ depends on $VMM(u)$.

Proof: Let $VMM(u) = \langle\langle\tau(0)|, T, F\rangle$. In according to Lemma 3 we consider $VMM(u)$ with unique accepting state. We construct $SM(u) = \langle\langle p(0)|, M, F'\rangle$ as follows. For $d \times d$ matrix T we define $(d+2) \times (d+2)$ matrix

$$A = \begin{pmatrix} 0 & 0\ldots0 & 0 \\ b & T & \vdots \\ \beta & q & 0 \end{pmatrix},$$

such that sum of elements of each row and each column of A is zero (we are free to select elements of column b, row q and number β).

Matrix A has the property: sum of elements of each row and each column of A is zero. k-th power A^k of A preserves this property.

Now let R be stochastic $(d+2) \times (d+2)$ matrix who's (i,j)-entry is $1/(d+2)$. Select positive constant $c \le 1$ such that matrix M, defined as

$$M = cA + R$$

is stochastic matrix. Further by induction on k we have that k-th power M^k of M is also stochastic matrix and has the same structure. That is,

$$M^k = c^k A^k + R.$$

By selecting suitable initial probabilities distribution $\langle p(0)|$ and accepting state we can pick up from M^t entry we need (entry that gives u accepting probability). From the construction of stochastic machine $((d+2),t)$-$SM(u)$ we have that $Pr(SM(u)) = c^t Pr(VMM(u)) + 1/(d+2)$. ∎

Lemma 4 says that having Type I $(d,t) - VMM(u)$ that process its input u with threshold $1/2$ one can construct stochastic machine $(d+2,t) - SM(u)$ that process u with threshold $\lambda = c^t 1/2 + 1/(d+2)$.

Lemma 5. *Let (d,t)-$SM(u)$ be stochastic machine that process its input u with threshold $\lambda \in [0,1)$. Then for arbitrary $\lambda' \in (\lambda,1)$ there exists $(d+1,t)$-$SM'(u)$ that process u with threshold λ'.*

Proof: The proof uses standard technique from Probabilistic Automata Theory (see for example the book [4]) and is omitted. ∎

4   Observation

For machines presented in vector-matrix form Theorem 2 states complexity characteristics of classical simulation of quantum machines. Vector-matrix technique keep the dimension of classical machine close to dimension of quantum machine, and amazingly we have that the simulation time does not increase. But from Lemma 4 we have that the stochastic simulation of linear machine is not completely free of charge — we lose ϵ-isolation of threshold (bounded error acceptance property) of the machine.

Notice that we present our classical simulation technique of quantum computation process (Simulation Theorem) in a form of vector-matrix machine VMM and omit a description how to come back to the uniform Turing machine. Obviously we have that in the case of Turing machines we will have slowdown of such simulations but this slowdown keeps simulations in polynomial time restriction. Remind that threshold changing technique for Turing machine models is well known (it was used for proving $NP \subseteq PP$ inclusion, see for example [2]).

Acknowledgments. We are grateful to referees for helpful remarks and on mentioning that the technique of the paper [1] also works for proving the first statement $PrQTime(T(n)) = PrTime(T(n))$ of Theorem 1.

References

1. L. Adleman, J. Demarrais, M. Huang, Quantum computability, SIAM J. on Computing. 26(5), (1997), 1524–1540.
2. J. Balcázar, J. Díaz and J. Gabarró, Structural Complexity I, An EATCS series, Springer-Verlag, 1995.
3. E. Bernstein and U. Vazirany, Quantum complexity theory, SIAM J. Comput, Vol. 26, No. 5, (1997), 1411–1473.
4. R. Bukharaev. The Foundation of Theory of Probabilistic Automata. Moscow, Nauka, 1985. (In Russian).
5. J. Gruska. Quantum computing. The McGraw-Hill Publishing Company. 1999.
6. L. Fortnow. One complexity theorist's view of quantum computing. Theoretical Computer Science, 292(3), (2003), 597–610.
7. C. Moore, J. Crutchfield. Quantum Automata and Quantum Grammars. Theoretical Computer Science 237, (2000), 275–306.
8. M. Nielsen and I. Chuang. Quantum Computation and Quantum Information. Cambridge University Press. 2000.
9. D. Simon, On the power of quantum computation, SIAM J. Comput, Vol. 26, No. 5, (1997), 1474–1483.
10. J. Watrous. Space-bounded quantum complexity. Journal of Computer and System Sciences, 59(2), (1999), 281–326.

Using Depth to Capture Average-Case Complexity

Luís Antunes[1*], Lance Fortnow[2], and N.V. Vinodchandran[3**]

[1] DCC-FC & LIACC-University of Porto
R.Campo Alegre, 823, 4150-180 Porto, Portugal
lfa@ncc.up.pt
[2] NEC Laboratories America
4 Independence way, Princeton, NJ 08540
fortnow@nec-labs.com
[3] Department of Computer Science and Engineering
University of Nebraska
vinod@cse.unl.edu

Abstract. We give the first characterization of Turing machines that run in polynomial-time on average. We show that a Turing machine M runs in average polynomial-time if for all inputs x the Turing machine uses time exponential in the computational depth of x, where the computational depth is a measure of the amount of "useful" information in x.

1 Introduction

In theoretical computer science we analyze most algorithms based on their worst-case performance. Many algorithms with bad worse-case performance nevertheless perform well in practice. The instances that require a large running-time rarely occur. Levin [Lev86] developed a theory of average-case complexity to capture this issue. Levin gives a clean definition of *Average Polynomial Time* for a given language L and a distribution μ. Some languages may remain hard in the worst case but can be solved in Average Polynomial Time for all reasonable distributions. We give a crisp formulation of such languages using computational depth as developed by Antunes, Fortnow and van Melkebeek [AFvM01].

Define $depth_t(x)$ as the difference of $K^t(x)$ and $K(x)$ where $K(x)$ is the usual Kolmogorov complexity and $K^t(x)$ is the version where the running times are bounded by time t. The $depth_t$ function [AFvM01] measures in some sense the "useful information" of a string.

We have two main results that hold for every language L.

1. If (L, μ) is in Average Polynomial Time for all P-samplable distributions μ then there exists a Turing machine M computing L and a polynomial p such that for all x, the running time of $M(x)$ is bounded by $2^{O(depth_p(x)+\log |x|)}$.

* Research done during an academic internship at NEC. This author is partially supported by funds granted to LIACC through the Programa de Financiamento Plurianual, Fundação para a Ciência e Tecnologia and Programa POSI.
** Research done while a post doctoral scientist at NEC Research Institute, Princeton.

A. Lingas and B.J. Nilsson (Eds.): FCT 2003, LNCS 2751, pp. 303–310, 2003.
© Springer-Verlag Berlin Heidelberg 2003

2. If there exists a Turing machine M and a polynomial p such that M computes L and for all inputs x, the running time of $M(x)$ is bounded by $2^{O(depth_p(x)+\log|x|)}$, then (L, μ) is in Average Polynomial Time for all P-computable distributions.

We do not get an exact characterization from these results. The first result requires P-samplable distributions and the second holds only for the smaller class of P-computable distributions. However, we can get an exact characterization by considering the time-bounded universal distribution $\mathbf{m}^t$. We show that the following are equivalent for every language L and every polynomial p:

- $(L, \mathbf{m}^p)$ is in Average Polynomial Time.
- There is some Turing machine M computing L such that for all inputs x the running time of M is bounded by $2^{O(depth_p(x)+\log|x|)}$.

Since the polynomial-time bounded universal distribution is dominated by a P-samplable distribution and dominates all P-computable distributions (see [LV97]) our main results follow from this characterization.

We prove our results for arbitrary time bounds t and as we take t towards infinity we recover Li and Vitányi's [LV92] result that under (non-time-bounded) universal distribution, the average-case complexity and the worst-case complexity coincide. Our theorems could be viewed as a time-bounded version of Li and Vitányi's result. This directly addresses the issue raised by Miltersen [Mil93] of relating a time-bounded version of Li and Vitányi with Levin's average-case complexity.

2   Preliminaries

We use binary alphabet $\Sigma = \{0, 1\}$ for encoding strings. Our computation model will be *prefix free* Turing machines: Turing machines with a one-way input tape (the input head can only read from left to right), a one-way output tape and a two-way work tape. The function log denote $\log_2$. All explicit resource bounds we use in this paper are time-constructible.

2.1   Kolmogorov Complexity and Computational Depth

We give essential definitions and basic result in Kolmogorov complexity for our needs and refer the reader to the textbook by Li and Vitányi [LV97] for more details. We are interested in self-delimiting Kolmogorov complexity (denoted by $K(.)$).

Definition 1. *Let U be a fixed prefix free universal Turing machine. Then for any string $x \in \{0, 1\}^*$, the Kolmogorov complexity of x is, $K(x) = \min_p\{|p| : U(p) = x\}$.*
For any time constructible t, the t-time-bounded Kolmogorov complexity of x is, $K^t(x) = \min\{|p| : U(p) = x \text{ in at most } t(|x|) \text{ steps}\}$.

Kolmogorov complexity of a string is a rigorous measure of the amount of information contained in it. A string with high Kolmogorov complexity contains lots of information. A random string has high Kolmogorov complexity and hence very informative. However, intuitively, the very fact that it is random restricts its utility in computational complexity theory. How can we measure the nonrandom information in a string?

Antunes, Fortnow and van Melkebeek [AFvM01] propose a notion of *Computational Depth* as a measure of nonrandom information in a string. Intuitively strings of high depth are low Kolmogorov complexity strings (and hence nonrandom), but a resource bounded machine cannot identify this fact. Indeed, Bennett's logical depth [Ben88] can be viewed as such a measure, but its definition is rather technical. Antunes, Fortnow and van Melkebeek suggest that the difference between two Kolmogorov complexity measures captures the intuitive notion of nonrandom information. Based on this intuition and with simplicity in mind, in this work we use the following depth measure.

Definition 2 (Antunes-Fortnow-van Melkebeek). *Let t be a constructible time bound. For any string $x \in \{0,1\}^*$,*

$$depth_t(x) = K^t(x) - K(x).$$

Average Case Complexity

We give definitions from average case complexity theory necessary for our purposes [Lev86]. For more details readers can refer to the survey by Jie Wang [Wan97]. In average case complexity theory, a computational problem is a pair (L, μ) where $L \subseteq \Sigma^*$ and μ is a probability distribution. The probability distribution is a function from Σ^* to the real interval $[0, 1]$ such that $\sum_{x \in \Sigma^*} \mu(x) \leq 1$. For probability distribution μ, the *distribution function*, denoted by μ^* is given by $\mu^*(x) = \sum_{y \leq x} \mu(x)$. The notion of *polynomial on average* is central to the theory of average case completeness.

Definition 3. *Let μ be a probability distribution function on $\{0,1\}^*$. A function $f : \Sigma^+ \to \mathbf{N}$ is polynomial on μ-average if there exists an $\epsilon > 0$ such that $\sum_x \frac{f(x)^\epsilon}{|x|} \mu(x) < \infty$.*

From the definition it follows that any polynomial is polynomial on μ-average for any μ. It is easy to show that if functions f and g are polynomial on μ-average, then the functions $f.g$, $f + g$, and f^k for some constant k are also polynomial on μ-average.

Definition 4. *Let μ be a probability distribution and $L \subseteq \Sigma^*$. Then the pair (L, μ) is in Average Polynomial time (denoted as Avg-P) if there is a Turing machine accepting L whose running time is polynomial on μ-average.*

We need the notion of *domination* for comparing distributions. The next definition formalizes this notion.

Definition 5. *Let μ and ν be two distributions on Σ^*. Then μ dominates ν if there is a constant c such that for all $x \in \Sigma^*$, $\mu(x) \geq \frac{1}{|x|^c}\nu(x)$. We also say ν is dominated by μ.*

Proposition 1. *If a function f is polynomial on μ-average, then for all distributions ν dominated by μ, f is also polynomial on ν-average.*

Average case analysis is, in general, sensitive to the choice of distribution, if we allow arbitrary distributions then average case complexity classes take the form of traditional worst-case complexity classes [LV92]. So it is important to restrict attention to distributions which are in some sense *simple*. Usually simple distributions are identified with the polynomial-time computable or polynomial-time samplable distributions.

Definition 6. *Let t be a time constructible function. A probability distribution function μ on $\{0,1\}^*$ is said to be t-time computable, if there is a deterministic Turing machine that on every input x and a positive integer k, runs in time $t(|x|+k)$, and outputs a fraction y such that $|\mu^*(x) - y| \leq 2^{-k}$.*

The most controversial definition in the average case complexity theory is the association of the class of *simple* distributions with P-computable, which may seem too restricting. Ben-David *et al.* in [BCGL92] introduced a wider family of natural distributions, P-samplable, consisting of distributions that can be sampled by randomized algorithms, working in time polynomial in the length of the sample generated.

Definition 7. *A probability distribution μ on $\{0,1\}^*$ is said to be P-samplable, if there is a probabilistic Turing machine M which on input 0^k produces a string x such that $|Pr(M(0^k) = x) - \mu(x)| \leq 2^{-k}$ and M runs in time $poly(|x|+k)$.*

Every P-computable distribution is also P-samplable, however the converse is unlikely.

Theorem 1 ([BCGL92]). *If one-way functions exists, then there is a P-samplable probability distribution μ which is not dominated by any polynomial-time computable probability distribution ν.*

Universal Distributions

The Kolmogorov complexity function $K(.)$ naturally defines a probability distribution on Σ^*: for any string x assign a probability of $2^{-K(x)}$. Kraft's inequality implies that this indeed is a probability distribution. This distribution is called the *universal distribution* and is denoted by **m**. Universal distribution has many equivalent formulations and has many nice properties. Refer to the textbook by Li and Vitányi [LV97] for an in-depth study on **m**. The main drawback of **m** is that it is not computable. In this paper we consider the resource-bounded version of the universal distribution.

Definition 8. *The t-time bounded universal distribution, $\mathbf{m}^t$ is given by $\mathbf{m}^t(x)$ $= 2^{-K^t(x)}$.*

One important property of $\mathbf{m}^t$ is that it dominates certain computable distributions.

Theorem 2 ([LV97]). $\mathbf{m}^t$ *dominates any t/n-time computable distribution.*

Proof. (Sketch) Let μ be a t/n-time computable distribution and let μ^* denote the distribution of μ. We will show that for any $x \in \Sigma^n$, $K^t(x) \leq -\log(\mu(x)) + C_\mu$ for a constant C_μ which depends on μ. Let $B_i = \{x \in \Sigma^n | 2^{-(i+1)} \leq \mu(x) < 2^{-i}\}$. Since for any x in B_i, $\mu(x) \geq 2^{-(i+1)}$, we have that $|B_i| \leq 2^i$. Consider the real interval $[0, 1]$. Divide it into intervals of size 2^{-i}. Since $\mu(x) \geq 2^{-i}$, we have for any $j, 0 \leq j \leq 2^i$, the jth interval $[j2^{-i}, (j + 1)2^{-i}]$ will have at most one $x \in B_i$ such that $\mu(x) \in [j2^{-i}, (j + 1)2^{-i}]$. Since μ is t/n-computable, for any $x \in B_i$, given j, we can do a binary search to output the unique x satisfying $\mu(x) \in [j2^{-i}, (j + 1)2^{-i}]$. This involves computing μ^* correct up to $2^{-(i+1)}$. So the total running time of the process will be bounded by $O((t/n)n)$. Hence we have the theorem.

Note that $\mathbf{m}^t$ approaches $\mathbf{m}$ as $t \to \infty$. In the proof of Theorem2, $\mathbf{m}^t$ very strongly dominates t/n-time computable distributions, in the sense that $\mathbf{m}^t(x) \geq \frac{1}{2^{C_\mu}}\mu(x)$. The definition of domination that we follow only needs $\mathbf{m}^t$ to dominate μ within a polynomial.

It is then natural to ask if there exists a polynomial-time computable distribution dominating $\mathbf{m}^t$. Schuler [Sch99] showed that if such a distribution exists then no polynomially secure pseudo-random generators exists. Pseudo-random generators are efficiently computable functions which stretches a seed into a long string so that for a random input the output looks random for a resource-bounded machine.

Theorem 3 ([Sch99]). *If there exists a polynomial time computable distribution that dominates $\mathbf{m}^t$ then pseudo-random generators do not exist.*

While, it is unlikely that there are polynomial-time computable distributions dominating universal distributions, we show that there are P-samplable distributions dominating the time-bounded universal distributions.

Lemma 1. *For any polynomial t, there is a P-samplable distribution μ which dominates $\mathbf{m}^t$.*

Proof. (Sketch) We will define a samplable distribution μ_t by prescribing a sampling algorithm for μ_t as follows. Let U be the universal machine.

Sample $n \in \mathbf{N}$ with probability $\frac{1}{n^2}$
Sample $1 \leq j \leq n$ with probability $1/n$
Sample uniformly $y \in \Sigma^j$
Run $U(y)$ for t steps. If U stops and outputs a string $x \in \Sigma^n$, output x.

For any string x of length n, $K^t(x) \leq n$. Hence it is clear that the probability that x is at least $\frac{1}{n^3}2^{-K^t(x)}$.

3 Computational Depth and Average Polynomial Time

We state our main theorem which relates computational depth to average polynomial time.

Theorem 4. *Let T be a constructible time bound. Then for any time constructible t, the following statements are equivalent.*

1. $T(x) \in 2^{O(depth_t(x) + \log |x|)}$.
2. T *is polynomial on* $\mathbf{m}^t$-*average.*

In [LV92], Li and Vitányi showed that when the inputs to any algorithm are distributed according to the universal distribution, the algorithm's average case complexity is of the same order of magnitude as its worst case complexity. Rephrasing this connection in the setting of average polynomial time we can make the following statement.

Theorem 5 (Li-Vitányi). *Let T be a constructible time bound. The following statements are equivalent*

1. $T(x)$ *is bounded by a polynomial in* $|x|$.
2. T *is polynomial on* $\mathbf{m}$-*average.*

As $t \to \infty$, K^t approaches K. So $depth_t$ approaches 0 and $\mathbf{m}^t$ approaches $\mathbf{m}$. Hence our main theorem can be seen as a generalization of Li and Vitányi's theorem.

We can apply the implication $(1 \Rightarrow 2)$ of the main theorem in the following way. Let M be a Turing machine and let $L(M)$ denote the language accepted by M. Let T_M denote its running time. If $T_M(x) \in 2^{O(depth_t(x) + \log |x|)}$ then $(L(M), \mu)$ is in Avg-P for any μ which is computable in time t/n. The following corollary follows from our main theorem and the universality of $\mathbf{m}^t$ (Theorem 2).

Corollary 1. *Let M be a deterministic Turing machine whose running time is bounded by $2^{O(depth_t(x) + \log |x|)}$, for some polynomial t. Then for any t/n-computable distribution μ, the pair $(L(M), \mu)$ is in Avg-P.*

Hence a sufficient condition for a language L (accepted by M) to be in Avg-P with respect to all polynomial-time computable distributions is that the running time of M is bounded by exponential in $depth_t$, for all polynomials t. An obvious question that arises is whether this condition is necessary. We have already partially answered this question (Lemma 1) by exhibiting an efficiently *samplable* distribution μ_t that dominates $\mathbf{m}^t$. Hence if $(L(M), \mu_t)$ is in Avg-P then $(L(M), \mathbf{m}^t)$ is also in Avg-P. From the implication $(2 \Rightarrow 1)$ of the main theorem, we have that $T_M(x) \in 2^{O(depth_t(x) + \log |x|)}$.

From Lemma 1, we get that if a machine runs in time polynomial on average for all P-samplable distributions then it runs in time exponential in its depth.

Corollary 2. *Let M be a machine which runs in time T_M. Suppose for all distributions μ in P-samplable, T_M is polynomial on μ-average, then $T_M(x) \in 2^{O(depth_t(x) + \log |x|)}$, for some polynomial t.*

We now prove our main theorem.

Proof. (*Theorem 4*) (1 $\Rightarrow$ 2). We will show that the statement 1 implies that $T(x)$ is polynomial on $\mathbf{m}^t$-average. Let $T(x) \in 2^{O(depth_t(x)+\log|x|)}$. Because of the closure properties of functions which are polynomial on average, it is enough to show that the function $T'(x) = 2^{depth_t(x)}$ is polynomial on $\mathbf{m}^t$-average. This essentially follows from the definitions and Kraft's inequality. The details are as follows. Consider the sum

$$\sum_{x \in \Sigma^*} \frac{T'(x)}{|x|} \mathbf{m}^t(x) = \sum_{x \in \Sigma^*} \frac{2^{depth_t(x)}}{|x|} 2^{-K^t(x)}$$

$$= \sum_{x \in \Sigma^*} \frac{2^{K^t(x)-K(x)}}{|x|} 2^{-K^t(x)}$$

$$\leq \sum_{x \in \Sigma^*} \frac{2^{-K(x)}}{|x|} < \sum_{x \in \Sigma^*} 2^{-K(x)} < 1$$

The last inequality is the Kraft's inequality.

(2 $\Rightarrow$ 1) Let $T(x)$ be a time constructible function which is polynomial on $\mathbf{m}^t$-average. Then for some $\epsilon > 0$ we have

$$\sum_{x \in \Sigma^*} \frac{T(x)^\epsilon}{|x|} \mathbf{m}^t(x) < 1$$

Define $S_{i,j,n} = \{x \in \Sigma^n | 2^i \leq T(x) < 2^{i+1}$ and $K^t(x) = j\}$. Let 2^r be the approximate size of $S_{i,j,n}$. Then the Kolmogorov complexity of elements in $S_{i,j,n}$ is r up to an additive $\log n$ factor. The following claim (proof omitted) states this fact more formally.

Claim. For $i, j \leq n^2$, let $2^r \leq |S_{i,j,n}| < 2^{r+1}$. Then for any $x \in S_{i,j,n}$, $K(x) \leq r + O(\log n)$.

Consider the above sum restricted to elements in $S_{i,j,n}$. Then we have

$$\sum_{x \in S_{i,j,n}} \frac{T(x)^\epsilon}{|x|} \mathbf{m}^t(x) < 1$$

$T(x) \geq 2^i$, $\mathbf{m}^t(x) = 2^{-j}$ and there are at least 2^r elements in the above sum. Hence the above sum is lower-bounded by the expression $\frac{2^r.2^{i\epsilon}.2^{-j}}{|x|^c}$ for some constant c. This gives us

$$1 > \sum_{x \in S_{i,j,n}} \frac{T(x)^\epsilon}{|x|} \mathbf{m}^t(x)$$

$$\geq \frac{2^r.2^{i\epsilon}.2^{-j}}{|x|^c} = 2^{i\epsilon+r-j-c\log n}$$

That is $i\epsilon + r - j - c\log n < 1$. From Claim 3, it follows that there is a constant d, such that for all $x \in S_{i,j,n}$, $i\epsilon \leq depth_t(x) + d\log|x|$. Hence $T(x) \leq 2^{i+1} \leq 2^{\frac{d}{\epsilon}(depth_t(x)+\log|x|)}$.

Acknowledgment. We thank Paul Vitányi for useful discussions.

References

[AFvM01] Luis Antunes, Lance Fortnow, and Dieter van Melkebeek. Computational depth. In *Proceedings of the 16th IEEE Conference on Computational Complexity*, pages 266–273, 2001.

[BCGL92] S. Ben-David, B. Chor, O. Goldreich and M. Luby. On the theory of average case complexity. *J. Computer System Sci.*, 44(2):193–219, 1992.

[Ben88] Charles H. Bennett. Logical depth and physical complexity. In R. Herken, editor, *The Universal Turing Machine: A Half-Century Survey*, pages 227–257. Oxford University Press, 1988.

[HILL99] Johan Håstad, Russell Impagliazzo, Leonid A. Levin, and Michael Luby. A pseudorandom generator from any one-way function. *SIAM Journal on Computing*, 28(4):1364–1396, August 1999.

[Lev86] Leonid A. Levin. Average case complete problems. *SIAM Journal on Computing*, 15(1):285–286, 1986.

[Lev84] Leonid A. Levin. Randomness conservation inequalities: information and independence in mathematical theories. *Information and Control*, 61:15–37, 1984.

[LV92] Ming Li and Paul M. B. Vitanyi. Average case complexity under the universal distribution equals worst-case complexity. *Information Processing Letters*, 42(3):145–149, May 1992.

[LV97] Ming Li and Paul M. B. Vitányi. *An introduction to Kolmogorov complexity and its applications.* Springer, 2nd edition, 1997.

[Mil93] Peter Bro Miltersen. The complexity of malign measures. In *SIAM Journal on Computing*, 22(1):147–156, 1993.

[Sch99] Rainer Schuler. Universal distributions and time-bounded Kolmogorov complexity. In *Proc. 16th Annual Symposium on Theoretical Aspects of Computer Science*, pages 434–443, 1999.

[Wan97] Jie Wang. Average-case computational complexity theory. In *Alan L. Selman, Editor, Complexity Theory Retrospective*, volume 2. 1997.

Non-uniform Depth of Polynomial Time and Space Simulations

Richard J. Lipton[1] and Anastasios Viglas[2]

[1] College of Computing, Georgia Institute of Technology and
Telcordia Applied Research
rjl@cc.gatech.edu
[2] University of Toronto, Computer Science Department,
10 King's College Road, Toronto, ON M5S 3G4, Canada
aviglas@cs.toronto.edu

Abstract. We discuss some connections between polynomial time and non-uniform, small depth circuits. A connection is shown with simulating deterministic time in small space. The well known result of Hopcroft, Paul and Valiant [HPV77] showing that space is more powerful than time can be improved, by making an assumption about the connection of deterministic time computations and non-uniform, small depth circuits. To be more precise, we prove the following: If every linear time deterministic computation can be done by non-uniform circuits of polynomial size and sub-linear depth,then $\mathcal{DTIME}(t) \subseteq \mathcal{DSPACE}(t^{1-\epsilon})$ for some constant $\epsilon > 0$. We can also apply the same techniques to prove an unconditional result, a trade-off type of theorem for the size and depth of a non-uniform circuit that simulates a uniform computation.

Keywords: Space simulations, non-uniform depth, block respecting computation.

1   Introduction

We present an interesting connection between non-uniform characterizations of Polynomial time and time versus space results.

Hopcroft Paul and Valiant [HPV77] proved that space is more powerful than time: $\mathcal{DTIME}(t) \subseteq \mathcal{DSPACE}(t/\log t)$. The proof of this trade-off result is based on pebbling techniques and the notion of *block respecting* computation. Improving the space simulation of deterministic time has been a long standing open problem. Paul Tarjan and Celoni [PTC77] proved an $n/\log n$ lower bound for pebbling a certain family of graphs. This lower bound implies that the trade-off result $\mathcal{DTIME}(t) \subseteq \mathcal{DSPACE}(t/\log t)$ of [HPV77] cannot be improved using similar pebbling arguments.

In this work we present a connection between space simulations of deterministic time and the depth of non-uniform circuits simulating polynomial time computations. This connection gives a way to improve the space simulation result from [HPV77] mentioned above, by making a non-uniform assumption. If

A. Lingas and B.J. Nilsson (Eds.): FCT 2003, LNCS 2751, pp. 311–320, 2003.
© Springer-Verlag Berlin Heidelberg 2003

every problem in linear deterministic time can be solved by polynomial size non-uniform circuits of small (sub-linear) depth then every deterministic computation of time t can be simulated in space $t^{1-\epsilon}$ for some constant $\epsilon > 0$ (that depends only on our assumption about the non-uniform depth of linear time):

$$\mathcal{DTIME}(n) \subseteq \mathcal{SIZE\text{-}DEPTH}(poly(n), n^\delta)$$
$$\implies \mathcal{DTIME}(t) \subseteq \mathcal{DSPACE}(t^{1-\epsilon}) \tag{1}$$

where $\delta < 1$ and $\epsilon > 0$. Note that we allow the size of the non-uniform circuit to be *any* polynomial. Since $\mathcal{DTIME}(t) \subseteq \mathcal{SIZE}(t \cdot \log t)$ (proved in [PF79]), our assumption basically asks to reduce the depth of the non-uniform circuit by a small amount, allowing the size to increase by any polynomial factor.

It is interesting to note that in this result, a non-uniform assumption is used ($\mathcal{P}$ has small non-uniform depth) to prove a purely uniform result (deterministic time can be simulated in small space). This can also be considered as an interesting result for the power of non-uniformity: If non-uniformity is powerful enough to allow small depth circuits for linear time deterministic computations, then we can improve the space-bounded simulation of deterministic time given by Hopcroft Paul and Valiant.

A related result was shown by Sipser [Sip86,Sip88] from the point of view of reducing randomness required for randomized algorithms. His result considers the problem of constructing expanders with certain properties. Assuming that those expanders can be constructed efficiently, the main theorem proved is that either $\mathcal{P}$ is equal to $\mathcal{RP}$ *or* the space simulation of Hopcroft, Paul and Valiant [HPV77] can be improved: Under the hypothesis that certain expanders have explicit constructions, there exists an $\epsilon > 0$ such that

$$(\mathcal{P} = \mathcal{RP}) \text{ or } (\mathcal{DTIME}(t) \cap 1^*) \subseteq \mathcal{DSPACE}(t^{1-\epsilon}) \tag{2}$$

An explicit construction for the expanders mentioned above was given by Saks, Srinivasan and Zhou [SSZ98]. The theorem mentioned above reveals a deep connection between pseudo-randomness and efficient space simulations (for unary languages): either space bounded simulations for deterministic time can be improved, or we can construct (pseudorandom) sequences that can be used to improve the derandomization of certain algorithms. On the other hand, the result we are going to present in this work, gives a connection between the power of non-uniformity and the power of space bounded computations.

Other related results include Dymond and Tompa [DT85] where it is shown that $\mathcal{DTIME}(t) \subseteq \mathcal{ATIME}(t/\log t)$, improving the Hopcroft Paul Valiant theorem, and Paterson and Valiant [PV76] proving $\mathcal{SIZE}(t) \subseteq \mathcal{DEPTH}(t/\log t)$.

We also show how to apply the same techniques to prove an unconditional trade-off type of result for the size and depth of a non-uniform circuit that simulates a uniform computation. Any deterministic time t computation can be simulated by a non-uniform circuit of size roughly $2^{\sqrt{t}}$ and depth $\sqrt{t}$, which has "semi-unbounded" fan-in: all AND gates have polynomially bounded fan-in and OR gates are unbounded, or vice versa. Similar results were given in [DT85] showing that time t is in PRAM time $\sqrt{t}$.

2   Notation – Definitions

We use the standard notation for time and space complexity classes $\mathcal{DTIME}(t)$ and $\mathcal{DSPACE}(t)$. $\mathcal{SIZE{-}DEPTH}(s,d)$ will denote the class of non-uniform circuits with size (number of gates) $O(s)$ and depth $O(d)$. We also use $\mathcal{NC}/poly$ ($\mathcal{NC}$ with polynomial advice) to denote the class of non-uniform circuits of polynomial size and poly-logarithmic depth, $\mathcal{SIZE{-}DEPTH}(poly, polylog)$. At some points in the paper, we will also avoid writing poly-logarithmic factors in detail and use the notation $\tilde{O}(n)$ to denote $O(n \log^k n)$ for constant k. In this work we consider time complexity functions that are time constructible: A function $t(n)$ is called fully time constructible if there exists a deterministic Turing Machine that on input of length n halts after exactly $t(n)$ steps. In general a function $f(n)$ is t-time constructible, if there is a deterministic Turing Machine that on input x outputs $1^{f(|x|)}$ and runs in time $O(t)$. (t, s)-time-space constructible functions are defined similarly. We also use "TM" for "deterministic Turing Machine".

For the proof of the main result we use the notion of block respecting Turing machines introduced by Hopcroft Paul and Valiant in [HPV77].

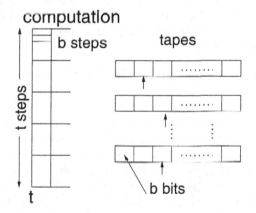

Fig. 1. Block respecting computation

Definition 1. *Let M be a machine running in time $t(n)$, where n is the length of its input x. Let the computation of M be partitioned in $a(n)$ segments, where each segment consists of $b(n)$ consecutive steps, $a(n) \cdot b(n) = t(n)$. Let also the tapes of M be partitioned into $a(n)$ blocks each consisting of $b(n)$ bits (cells) on each tape. We will call M block respecting if during each segment of its computation, each head visits only one block on each tape.*

Every Turing Machine can be converted to a block respecting machine with only a constant factor slow down in its running time. The construction is simple: Let M be a deterministic Turing Machine running in time t. Break the computation steps $(1 \ldots t)$ in segments of size B. Break the work tapes in blocks of the same size B. If at the start of a computation segment σ the work tape head is in block b_j, then during the computation steps (b steps) of that segment, the head could only visit the adjacent blocks, b_{j-1} or b_{j+1}. Keep a copy of those two blocks along with b_j and do all the computation of segment σ reading and updating from those copies (if needed). At the end of the computation of every segment, there is a clean-up step: update the blocks b_{j-1} and b_{j+1} and move the work tape head to the appropriate block to start the computation of the next segment. This construction can be done for different block sizes B. For our purposes B will be t^c for a small constant $c < 1$.

Block respecting Turing machines are also used in [PPST83] to prove that non-deterministic linear time is more powerful than deterministic linear time (see also [PR81] for a generalization of the results from [HPV77] for RAMs and other machine models).

3   Main Results

We show that if linear time has small non-uniform circuit depth (for polynomial size circuits) then $\mathcal{DTIME}(t) \subseteq \mathcal{DSPACE}(t^{1-\epsilon})$ for a constant $\epsilon > 0$.

To be more precise, the strongest form of the main result is the following: if (deterministic) linear time has polynomial size, non-uniform circuits of sublinear depth (for example depth n^δ for $0 < \delta < 1$), then $\mathcal{DTIME}(t) \subseteq \mathcal{DSPACE}(t^{1-\epsilon})$ for a small positive $\epsilon > 0$:

$$\mathcal{DTIME}(n) \subseteq \mathcal{SIZE\text{--}DEPTH}(\,poly, n^\delta) \implies \mathcal{DTIME}(t) \subseteq \mathcal{DSPACE}(t^{1-\epsilon}) \tag{3}$$

The main idea is the following: Start with a deterministic Turing machine M running in time t and convert it to a block respecting machine M_B with block size B. In each segment of the computation, M_B reads and/or writes in exactly one block on each tape. We will argue that we can check the computation in each such segment with the *same* sub-circuit and we can actually construct this sub-circuit with polynomial size and small (poly-logarithmic or sub-linear) depth. Combining all these sub-circuits together we can build a larger circuit that will check the entire computation of M_B in small depth. The final step is a technical lemma that shows how to evaluate this circuit in small space (equal to its depth).

We start by proving the main theorem using the assumption $\mathcal{P} \subseteq \mathcal{NC}/poly$. It is easy to see that an assumption of the form $\mathcal{DTIME}(n) \subseteq \mathcal{NC}/poly$ implies $\mathcal{P} \subseteq \mathcal{NC}/poly$ by padding arguments.

Theorem 1. *Let t be a polynomial time complexity function. If $\mathcal{P} \subseteq \mathcal{NC}/poly$ then $\mathcal{DTIME}(t) \subseteq \mathcal{DSPACE}(t^{1-\epsilon})$ for some constant $\epsilon > 0$.*

Proof. (Any "reasonable" time complexity function could be used in the statement of this theorem.) Consider any Turing Machine M running in deterministic time t. Here is how to simulate M in small space using the assumption that polynomial time has shallow (poly-logarithmic depth) polynomial size circuits:

1. Convert given TM in a block respecting machine with block size B.
2. Construct the graph that describes the computation. Each vertex corresponds to a computation segment of B steps.
3. The computation on each vertex can be checked by the *same* TM U that runs in polynomial time (linear time)
4. Since $\mathcal{P} \subseteq \mathcal{NC}/poly$, there is a circuit U_C that can replace U. U_C has polynomial size and polylogarithmic depth.
5. Construct U_C by trying all possible circuits.
6. Plug in the sub-circuit U_C to the entire graph. This graph is the description of a circuit of small depth, that corresponds to the computation of the given TM. Evaluate the circuit (in small space)

In more detail: Convert M to a block respecting machine M_B. Break the computation of M_B (on input x) in segments of size B each; the number of segments is t/B. Consider the directed graph G corresponding to the computation of the block respecting machine as described in [HPV77]: G has one vertex for every time segment (that is t/B vertices) and the edges are defined from the sequence of head positions. Let $v(\Delta)$ denotes the vertex corresponding to time segment Δ then and Δ_i is the last time segment before Δ during which the i-th head was scanning the same block as during segment Δ. Then the edges of G are $v(\Delta - 1) \to v(\Delta)$ and for all $1 \leq i \leq l$, $v(\Delta_i) \to v(\Delta)$. The number of edges can be at most $O(\frac{t}{B})$ and therefore the number of bits required to describe the graph is $O\left(\frac{t}{B} \log \frac{t}{B}\right)$. Figure 2 shows the idea behind the construction of the

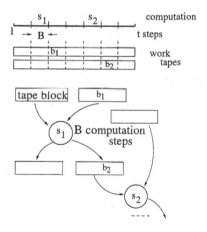

Fig. 2. Graph description of a block respecting computation.

graph for the block respecting computation. The computation is partitioned in segments of size B. Every segment corresponds to a vertex (denoted by a circle in figure 2). Each segment will access only one block on each tape. Figure 2 shows the tape blocks blocks which are read during a computation segment (input blocks for that vertex) and those that will be written during the same segment (shown as output blocks). If a block is written during a segment and the same block is read by another computation segment later in the computation, then the second segment depends directly from the previous one and there will be an edge connecting the corresponding vertices in our graph.

Each vertex of this graph corresponds to B computation steps of M_B. During this computation, M_B reads and writes only in one block from each tape. In order to check the computation that corresponds to a vertex of this graph, we would need to simulate M_B for B steps and check $O(B)$ bits from M_B's tapes. For each vertex we need to check/simulate a different segment of M_B's computation: this can be done by a Turing machine that will check the corresponding computation of M_B. We argue that the same Turing machine can be used on every vertex. The computation we need to do on each vertex of the graph is essentially the same: given the "input" and "output" contents of certain tape blocks, simulate the machine M_B for B steps and check if the output contents are correct. The only thing that changes is the actual segment of the computation of M_B that we are going to simulate (which B steps of M_B we should simulate). This means that the exact same "universal" Turing machine checks the computation for each segment/vertex, and this universal machine also takes as input the description (for example the index of the part of the computation of the initial machine M_B it will need to simulate or any reasonable encoding) of the computation that it needs to actually simulate on each vertex. Therefore we have the same machine U on all vertices of the graph which runs in deterministic polynomial time. If $\mathcal{P} \subseteq \mathcal{SIZE\text{-}DEPTH}(n^k, \log^l n)$ then U can be simulated by a circuit

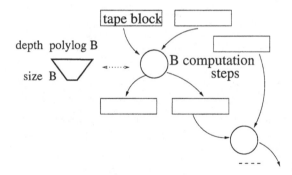

depth polylog B

size B

tape block

B computation steps

Fig. 3. Insert the (same) sub-circuit on all vertices

U_C of size $O(B^k)$ and small depth $O(\log^l B)$, for some k, l. The same circuit is used on all vertices of the graph. In order to construct this circuit, we can try all possible circuits and simulate them on all possible inputs. This requires exponential time, but only small amount of space: the size of the circuit is B^k and its depth polylogarithmic in B. We need $\tilde{O}(B^k)$ bits to write down the circuit and only polylog space to evaluate it (using lemma 1).

Once we have constructed U_C, we can build the entire circuit that will simulate M_B. This circuit derives directly from the (block-respecting) computation graph where each vertex is an instance of the sub-circuit U_C. The size of the entire circuit is too big to write down. We have up to t/B sub-circuits (U_C) that would require a size of $\tilde{O}(\frac{t}{B}B^k)$ for some constant k. But since it is the same sub-circuit U_C that appears throughout the graph, we can implicitly describe the entire circuit in much less space. For the evaluation of the circuit, we only need to be able to describe the exact position of a vertex in the graph, and determine the immediate neighbors of a given vertex (previous and next vertices). This can easily be done in space $\tilde{O}(t/B + B^k)$.

In order to complete the simulation we need to show how to evaluate a small-depth circuit in small space (see Borodin [Bor77]).

Lemma 1. *Consider a directed acyclic graph G with one source (root). Assume that the leaves are labeled from $\{0,1\}$, its inner nodes are either AND or OR nodes and the depth is at most d. Then we can evaluate the graph in space at most $O(d)$.*

Proof. (of lemma. See [Bor77] for more details).

Convert the graph to a tree (by making copies of the nodes). The tree will have much bigger size but the depth will remain the same. We can prove (by induction) that the value of the tree is the same as the value of the graph from which we started. Evaluating the tree corresponds to computing the value of its root. In order to find the value of any node v in the tree, proceed as follows: Let $u_1, \ldots, u_k$ denote the child-nodes of v.

If v is an AND node, then compute (recursively) the value of its first child u_1. If $value(u_1) = 0$ then the value of v is also 0. Otherwise continue with the next child. If the last child has value 1 then the value of v is 1. Notice that we do not need to remember the value of the child-nodes that we have evaluated. If v is an OR node, the same idea can be applied. We can use a stack for the evaluation of the tree. It is easy to see that the size of the stack will be at most $O(d)$, that is as big as the depth of the tree.   ∎

The total amount of space used is:

$$\tilde{O}\left(B^{2k} + \frac{t}{B} \log^l B\right) \tag{4}$$

To get the desired result, we need to choose the size B of the blocks appropriately to balance the two terms in (4). B will be $t^{1/c}$ for some constant c that is larger than k.   ∎

As mentioned above, the exact same proof would work even if we allow almost linear depth for the non-uniform circuits for just linear deterministic time instead of $\mathcal{P}$. The stronger theorem is the following:

Theorem 2. *If $\mathcal{DTIME}(n) \subseteq \mathcal{SIZE\text{–}DEPTH}(n^k, n^\delta)$ for some $k > 0$ and $\delta < 1$, then $\mathcal{DTIME}(t) \subseteq \mathcal{DSPACE}(t^{1-\epsilon})$ where $\epsilon = 1 - \frac{1-\delta}{2k+1}$.*

Proof. From the proof of theorem 1 we can calculate the space required for the simulation: In order to find the correct sub-circuit which has size B^k and depth

B^δ, we need $O(B^{2k} \log B)$ space to write it down and $O(B^\delta)$ to evaluate it. To evaluate the entire circuit which has depth $\frac{t}{B} \cdot B^\delta$) we are only using space

$$O(\frac{t}{B} \cdot B^\delta \log B + \frac{t}{B} \log t + B^{2k} \log B) \tag{5}$$

The first term in equation (5), is the space required to evaluate the entire circuit that has depth $\frac{t}{B} \cdot B^\delta$ and the second and third term is the space required to write down an implicit description of the entire circuit (description of the graph from the block respecting computation, and the description of the smaller sub-circuit)

Total space used (to find the correct sub-circuit and to evaluate the entire circuit) is

$$O(\frac{t}{B} \cdot B^\delta \log B + B^{2k} \log B) \tag{6}$$

If we set $B = t^{1/2k+1}$ then the space bound is

$$O(t^{1 - \frac{1-\delta}{2k+1}}) \tag{7}$$

In these calculations $2k + 1$ means just something greater than $2k$.   ∎

These proof ideas seem to fail if we try to simulate non-deterministic time in small space. In that case, evaluating the circuit would be more complicated: we would need to use more space in order to make sure that the non-deterministic guesses are consistent throughout the evaluation of the circuit.

4   Semi-unbounded Circuits

These simulation ideas using block respecting computation can also be used to prove an *unconditional* result relating uniform polynomial time and non-uniform small depth circuits. The simulation of the previous section implies unconditionally a trade-off type of result for the size and depth of non-uniform circuits that simulate uniform computations. The next theorem proves that any deterministic time t computation can be simulated by a non-uniform circuit of size $\sqrt{t} \cdot 2^{\sqrt{t}}$ or $2^{O(\sqrt{t})}$ and depth $\sqrt{t}$, which has "semi-unbounded" fan-in. Previous work by Dymond and Tompa [DT85] also present similar results showing that deterministic time t is in PRAM time $\sqrt{t}$.

Theorem 3. *Let t be a reasonable time complexity function. Then $\mathcal{DTIME}(t) \subseteq \mathcal{SIZE\text{–}DEPTH}(2^{O(\sqrt{t})}, \sqrt{t})$, and the simulating circuits require exponential fan-in for AND gates and polynomial for OR gates (or vice-versa)*

Proof. Given a Turing machine running in $\mathcal{DTIME}(t)$, construct the block respecting version, and repeat the exact same construction as the one presented in the proof of theorem 1: Construct the graph describing the block respecting computation, which has t/B nodes, and every node corresponds to a segment of B (we will chose the size B later in the proof) computation steps. Use this graph to construct the non-uniform circuit: For every node, build a circuit, say in DNF, that corresponds to the computation that takes place on that node. This circuit has size exponential in B in the worst case, $2^{O(B)}$, and depth 2. The entire graph describes a circuit of size $\frac{t}{B}2^{O(B)}$ and depth $O(B)$. Also, note that for every sub-circuit that corresponds to each node, the input gates (AND gates as described in the proof) have a fan-in of at most $O(B)$, while the second level might need exponential fan-in. This construction yields a circuit of "semi-unbounded" type fan-in. ∎

5 Discussion – Open Problems

In this work we have shown a connection between the power of non-uniformity and the power of space bounded computation. The proof of the main theorem is based on the notion of block respecting computation and various techniques for simulating Turing Machine computation. The main result states that if Polynomial time has small non-uniform depth then space can simulate deterministic time fast(-er). An interesting open question is to see if the same ideas can be used to prove a similar space simulation for non-deterministic time. It seems also possible that a result could be proved for probabilistic classes. A different approach would be to make a stronger assumption (about complexity classes) and reach a contradiction with some hierarchy theorem or other diagonalization result thus proving a complexity class separation.

Acknowledgments. We would like to thank Nicola Galesi, Toni Pitassi and Charlie Rackoff for many discussions on these ideas. Also many thanks to Dieter van Melkebeek and Lance Fortnow.

References

[Bor77]   A. Borodin. On relating time and space to size and depth. *SIAM Journal of Computing*, 6(4):733–744, December 1977.

[DT85]   Patrick W. Dymond and Martin Tompa. Speedups of deterministic machines by synchronous parallel machines. *Journal of Computer and System Sciences*, 30(2):149–161, April 1985.

[HPV77]   J. Hopcroft, W. Paul, and L. Valiant. On time versus space. *Journal of the ACM.*, 24(2):332–337, April 1977.

[PF79]   Nicholas Pippenger and Michael J. Fischer. Relations among complexity measures. *Journal of the ACM*, 26(2):361–381, April 1979.

[PPST83]   Wolfgang J. Paul, Nicholas Pippenger, Endre Szemerédi, and William T. Trotter. On determinism versus non-determinism and related problems (preliminary version). In *24th Annual Symposium on Foundations of Computer Science*, pages 429–438, Tucson, Arizona, 7–9 November 1983. IEEE.

[PR81]   W. Paul and R. Reischuk. On time versus space II. *Journal of Computer and System Sciences*, 22(3):312–327, June 1981.

[PTC77]   Wolfgang J. Paul, Robert Endre Tarjan, and James R. Celoni. Space bounds for a game on graphs. *Mathematical Systems Theory*, 10:239–251, 1977.

[PV76]   M. S. Paterson and L. G. Valiant. Circuit size is nonlinear in depth. *Theoretical Computer Science*, 2(3):397–400, September 1976.

[Sip86]   M. Sipser. Expanders, randomness, or time versus space. In Alan L. Selman, editor, *Proceedings of the Conference on Structure in Complexity Theory*, volume 223 of *LNCS*, pages 325–329, Berkeley, CA, June 1986. Springer.

[Sip88]   M. Sipser. Expanders, randomness, or time versus space. *Journal of Computer and System Sciences*, 36:379–383, 1988.

[SSZ98]   Michael Saks, Aravind Srinivasan, and Shiyu Zhou. Explicit OR-dispersers with polylogarithmic degree. *Journal of the ACM*, 45(1):123–154, January 1998.

Dimension- and Time-Hierarchies for Small Time Bounds

Martin Kutrib

Institute of Informatics, University of Giessen
Arndtstr. 2, D-35392 Giessen, Germany
kutrib@informatik.uni-giessen.de

Abstract. Recently, infinite time hierarchies of separated complexity classes in the range between real time and linear time have been shown. This result is generalized to arbitrary dimensions. Furthermore, for fixed time complexities of the form $id + r$, where $r \in o(id)$ is a sublinear function, proper dimension hierarchies are presented. The hierarchy results are established by counting arguments. For an equivalence relation and a family of witness languages the number of induced equivalence classes is compared to the number of equivalence classes distinguishable by the model in question. By contradiction the properness of the inclusions is proved.

1   Introduction

If one is particularly interested in computations with small time bounds, let us say in the range between real time and linear time, most of the relevant Turing machine results have been published in the early times of computational complexity. In the sequel we are concerned with time bounds of the form $id + r$, where id denotes the identity function on integers, and $r \in o(id)$ is a sublinear function. Most of the previous investigations in this area have been done in terms of one-dimensional Turing machines. Recently, infinite time hierarchies of separated complexity classes in the range in question have been shown [10].

In [2] it has been proved that the complexity class Q which is defined by nondeterministic multitape real-time computations is equal to the corresponding linear-time languages. Moreover, it has been shown that two working tapes and a one-way input tape are sufficient to accept the languages from Q in real time. On the other hand, in [13] an NP-complete language was exhibited which is accepted by a nondeterministic single-tape Turing machine in time $id + O(id^{\frac{1}{2}} \log)$ but not in real time. This interesting result stresses the power of nondeterminism impressively and motivates the exploration of the world below linear time once more.

For deterministic machines the situation is different. Though in [7] for one tape the identity $\mathsf{DTIME}_1(id) = \mathsf{DTIME}_1(\mathrm{LIN})$ has been proved, for a total of at least two tapes the real-time languages are strictly included in the linear-time languages.

A. Lingas and B.J. Nilsson (Eds.): FCT 2003, LNCS 2751, pp. 321–332, 2003.
© Springer-Verlag Berlin Heidelberg 2003

Another aspect that, at first glance, might attack the time range of interest is a possible speed-up. The well-known linear speed-up [6] from $t(n)$ to $id + \varepsilon \cdot t(n)$ for arbitrary $\varepsilon > 0$ yields complexity classes close to real time (i.e. DTIME(LIN) = DTIME($(1 + \varepsilon) \cdot id$)) for k-tape and multitape machines, but does not allow assertions on the range between real time and linear time. An application to the time bound $id + r$, $r \in o(id)$, would result in a slow-down to $id + \varepsilon \cdot (id + r) \geq id + \varepsilon \cdot id$.

Let us recall known time hierarchy results. For a number of $k \geq 2$ tapes in [5,14] the hierarchy $\mathsf{DTIME}_k(t') \subset \mathsf{DTIME}_k(t)$, if $t' \in o(t)$ and t constructible, has been shown. By the linear speed-up we obtain the necessity of the condition $t' \in o(t)$. The necessity of the constructibility property of t follows from the well-known Gap Theorem [9].

Since in case of multitape machines one needs to construct a Turing machine with a fixed number of tapes that simulates machines even with more tapes, the proof of a corresponding hierarchy involves a reduction of the number of tapes. This costs a factor log for the time complexity. The hierarchy $\mathsf{DTIME}(t') \subset \mathsf{DTIME}(t)$, if $t' \cdot \log(t') \in o(t)$ and t constructible, has been proved in [6]. Due to the necessary condition $t' \in o(t)$ resp. $t' \cdot \log(t') \in o(t)$, again, the range between real time and linear time is not affected by the known time hierarchy results. Moreover, it follows immediately from the condition $t' \in o(t)$ and the linear speed-up that there are no infinite hierarchies for time bounds of the form $t + r$, $r \in o(id)$, if $t \geq c \cdot id$, $c > 1$.

Related work concerning higher dimensional Turing machines can be found e.g. in [8], where for on-line computations the trade-off between time and dimensionality is investigated. Upper bounds for the reduction of the dimensions are dealt with e.g. in [12,15,16,19].

Here, on one hand, we are going to present infinite time hierarchies below linear time for any dimension. Such hierarchies are also known for one-dimensional iterative arrays [3]. On the other hand, dimension hierarchies are presented for each time bound in question. Thus, we obtain a double time-dimension hierarchy.

The basic notions and a preliminary result of technical flavor are the objects of the next section. Section 3 is devoted to the time hierarchies below linear time. They are established by counting arguments. For an equivalence relation and a family of witness languages the number of induced equivalence classes is compared to the number of equivalence classes distinguishable by the model in question. By contradiction the properness of the inclusions follows. In Section 4 for fixed time complexities of the form $id + r$, $r \in o(id)$ proper dimension hierarchies are proved.

2   Preliminaries

We denote the rational numbers by $\mathbb{Q}$, the integers by $\mathbb{Z}$, the positive integers $\{1, 2, ...\}$ by $\mathbb{N}$ and the set $\mathbb{N} \cup \{0\}$ by $\mathbb{N}_0$. The reversal of a word w is denoted by w^R. For the length of w we write $|w|$. We use $\subseteq$ for inclusions and $\subset$ if the inclusions are strict. Let $e_i = (0, \ldots, 0, 1, 0, \ldots, 0)$ (the 1 is at position i) denote the ith d-dimensional unit vector, then we define

$$E_d = \{e_i \mid 1 \le i \le d\} \cup \{-e_i \mid 1 \le i \le d\} \cup \{(0, \dots, 0)\}.$$

For a function $f : \mathbb{N}_0 \to \mathbb{N}$ we denote its i-fold composition by $f^{[i]}$, $i \in \mathbb{N}$. If f is increasing and unbounded, then its inverse is defined according to

$$f^{-1}(n) = \min\{m \in \mathbb{N} \mid f(m) \ge n\}.$$

The identity function $n \mapsto n$ is denoted by id. As usual we define the set of functions that grow strictly less than f by

$$o(f) = \{g : \mathbb{N}_0 \to \mathbb{N} \mid \lim_{n \to \infty} \frac{g(n)}{f(n)} = 0\}.$$

In terms of orders of magnitude, f is an upper bound of the set

$$O(f) = \{g : \mathbb{N}_0 \to \mathbb{N} \mid \exists\, n_0, c \in \mathbb{N} : \forall\, n \ge n_0 : g(n) \le c \cdot f(n)\}.$$

Conversely, f is a lower bound of the set $\Omega(f) = \{g : \mathbb{N}_0 \to \mathbb{N} \mid f \in O(g)\}$.

A d-dimensional Turing machine with $k \in \mathbb{N}$ tapes consists of a finite-state control, a read-only one-dimensional one-way input tape and k infinite d-dimensional working tapes. On the input tape a read-only head, and on each working tape a read-write head is positioned. At the outset of a computation, the Turing machine is in the designated initial state and the input is the inscription of the input tape, all the other tapes are blank. The head of the input tape scans the leftmost input symbol whereas all other heads are positioned on arbitrary tape cells. Dependent on the current state and the currently scanned symbols on the $k+1$ tapes, the Turing machine changes its state, rewrites the symbols at the head positions of the working tapes, and possibly moves the heads independently to a neighboring cell. The head of the input tape may only be moved to the right. With an eye towards language recognition, the machines have no extra output tape but the states are partitioned in accepting and rejecting states. More formally:

Definition 1. *A deterministic d-dimensional Turing machine with $k \in \mathbb{N}$ tapes* (DTM$_k^d$) *is a system $\langle S, T, A, \delta, s_0, F \rangle$, where*

1. *S is the finite set of internal states,*
2. *T is the finite set of tape symbols containing the blank symbol $\sqcup$,*
3. *$A \subseteq T \setminus \{\sqcup\}$ is the set of input symbols,*
4. *$s_0 \in S$ is the initial state,*
5. *$F \subseteq S$ is the set of accepting states,*
6. *$\delta : S \times (A \cup \{\sqcup\}) \times T^k \to S \times T^k \times \{0, 1\} \times E_d^k$ is the partial transition function.*

Since the input tape cannot be rewritten, we need no new symbol for its current tape cell. Due to the same fact, δ may only expect symbols from $A \cup \{\sqcup\}$ on it. The input tape is one dimensional and one way and, thus, its head moves according to $\{0, 1\}$. The set of rejecting states is implicitly given by the partitioning, i.e. $S \setminus F$. The unit vectors correspond to the possible moves of the read-write heads.

Let $\mathcal{M}$ be a DTM_k^d. A *configuration* of $\mathcal{M}$ at some time $t \geq 0$ is a description of its global state which is a $(2(k+1)+1)$-tuple $(s, f_0, f_1, \ldots, f_k, p_0, p_1, \ldots, p_k)$, where $s \in S$ is the current state, $f_0 : \mathbb{Z} \to A \cup \{\sqcup\}$ and $f_i : \mathbb{Z}^d \to T$ are functions that map the tape cells of the corresponding tape to their current contents, and $p_0 \in \mathbb{Z}$ and $p_i \in \mathbb{Z}^d$ are the current head positions, $1 \leq i \leq k$.

The initial configuration $(s_0, f_0, f_1, \ldots, f_k, 1, 0, \ldots, 0)$ at time 0 is defined by the input word $w = a_1 \cdots a_n \in A^*$, the initial state s_0, and blank working tapes:

$$f_0(m) = \begin{cases} a_m & \text{if } 1 \leq m \leq n \\ \sqcup & \text{otherwise} \end{cases}$$

$$f_i(m_1, \ldots, m_d) = \sqcup \quad \text{for } 1 \leq i \leq k$$

Successor configurations are computed according to the *global transition function* Δ: Let $(s, f_0, f_1, \ldots, f_k, p_0, p_1, \ldots, p_k)$ be a configuration. Then

$$(s', f_0', f_1', \ldots, f_k', p_0', p_1', \ldots, p_k') = \Delta(s, f_0, f_1, \ldots, f_k, p_0, p_1, \ldots, p_k)$$

if and only if

$$\delta(s, f_0(p_0), f_1(p_1), \ldots, f_k(p_k)) = (s', x_1, \ldots, x_k, j_0, j_1, \ldots, j_k)$$

such that

$$f_i'(m_1, \ldots, m_d) = \begin{cases} f_i(m_1, \ldots, m_d) & \text{if } (m_1, \ldots, m_d) \neq p_i \\ x_i & \text{if } (m_1, \ldots, m_d) = p_i \end{cases}$$

$$p_i' = p_i + j_i, \quad p_0' = p_0 + j_0$$

for $1 \leq i \leq k$. Thus, the global transition function Δ is induced by δ. Throughout the paper we are dealing with so-called multitape machines (DTM^d), where every machine has an arbitrary but fixed number of working tapes.

A Turing machine *halts* iff the transition function is undefined for the current configuration. An input word $w \in A^*$ is accepted by a Turing machine $\mathcal{M}$ if the machine halts at some time in an accepting state, otherwise it is rejected.

$L(\mathcal{M}) = \{w \in A^* \mid w \text{ is accepted by } \mathcal{M}\}$ is the *language accepted* by $\mathcal{M}$. If $t : \mathbb{N}_0 \to \mathbb{N}$, $t(n) \geq n$, is a function, then $\mathcal{M}$ is said to be t-*time-bounded* or of *time complexity* t iff it halts on all inputs w after at most $t(|w|)$ time steps.

If t equals the function id, acceptance is said to be in *real time*. The *linear-time* languages are defined according to time complexities $t = c \cdot id$, where $c \in \mathbb{Q}$ with $c \geq 1$. Since time complexities are mappings to positive integers and have to be greater than or equal to id, actually, $c \cdot id$ means $\max\{\lceil c \cdot id \rceil, id\}$. But for convenience we simplify the notation in the sequel.

The family of all languages which can be accepted by DTM^d with time complexity t is denoted by $\text{DTIME}^d(t)$.

In order to prove tight time hierarchies, in almost all cases well-behaved time bounding functions are required. Usually, the notion "well-behaved" is concretized in terms of computability or constructibility of the functions with respect to the device in question.

Definition 2. *Let $d \in \mathbb{N}$ be a constant. A function $f : \mathbb{N}_0 \to \mathbb{N}$ is said to be DTM^d constructible iff there exists a DTM^d which for every $n \in \mathbb{N}$ on input 1^n halts after exactly $f(n)$ time steps.*

Another common definition of constructibility demands the existence of an $O(f)$-time-bounded Turing machine that computes the binary representation of the value $f(n)$ on input 1^n. Both definitions have been proven to be equivalent for multitape machines [11].

The following definition summarizes the properties of well-behaved (in our sense) functions and names them.

Definition 3. *The set of all increasing, unbounded DTM^d-constructible functions f with the property $\forall c \in \mathbb{N} : \exists c' \in \mathbb{N} : c \cdot f(n) \leq f(c' \cdot n)$ is denoted by $\mathcal{T}(\mathrm{DTM}^d)$. The set of their inverses is $\mathcal{T}^{-1}(\mathrm{DTM}^d) = \{f^{-1} \mid f \in \mathcal{T}(\mathrm{DTM}^d)\}$.*

Since we are interested in time bounds of the form $id + r$, we need small functions r below the identity. The constructible functions are necessarily greater than the identity. Therefore, the inverses of constructible functions are used. The properties increasing and unbounded are straightforward. At first glance the property $\forall c \in \mathbb{N} : \exists c' \in \mathbb{N} : c \cdot f(n) \leq f(c' \cdot n)$ seems to be restrictive, but it is not. It is easily verified that almost all of the commonly considered bounding functions above the identity have this property (e.g, the identity itself, polynomials, exponential functions, etc.) As usual here we remark that even the family $T(\mathrm{DTM}^1)$ is very rich. More details can be found for example in [1,17,20].

In order to clarify later calculations, we observe the following: Let $r \in \mathcal{T}^{-1}(\mathrm{DTM}^d)$ be some function. Then there must exist a constructible function $\hat{r} \in \mathcal{T}(\mathrm{DTM}^d)$ such that $r = \hat{r}^{-1}$. Moreover, for all n we obtain $r(\hat{r}(n)) = n$ by definition: $r(\hat{r}(n)) = \min\{m \in \mathbb{N} \mid \hat{r}(m) \geq \hat{r}(n)\}$ implies $m = n$ and, thus, $r(\hat{r}(n)) = n$.

In general, we do not have equality for the converse $\hat{r}(r(n))$, but in the sequel we will need only the equality case.

The following equivalence relation is well known (cf. Myhill-Nerode Theorem on regular languages).

Definition 4. *Let $L \subseteq A^*$ be a language over an alphabet A and $l \in \mathbb{N}_0$ be a constant. Two words w and w' are l-equivalent with respect to L if and only if*

$$ww_l \in L \iff w'w_l \in L$$

for all $w_l \in A^l$. The number of l-equivalence classes of words of length $n - l$ with respect to L (i.e. $|ww_l| = n$) is denoted by $N(n, l, L)$.

The underlying idea is to bound the number of distinguishable equivalence classes. The following lemma gives a necessary condition for a language to be $(id + r)$-time acceptable by a DTM^d.

Lemma 5. *Let* $r : \mathbb{N}_0 \to \mathbb{N}$ *be a function and* $d \in \mathbb{N}$ *be a constant. If* $L \in$ $\mathrm{DTIME}^d(id + r)$, *then there exists a constant* $p > 1$ *such that:*

$$N(n, l, L) \le p^{(l+r(n))^d}$$

Proof. Let $\mathcal{M} = \langle S, T, A, \delta, s_0, F \rangle$ be a $(id + r)$-time DTM^d that accepts a language L.

In order to determine an upper bound for the number of l-equivalence classes, we consider the possible situations of $\mathcal{M}$ after reading all but l input symbols. The remaining computation depends on the current internal state and the contents of the at most $(2(l + r(n)) + 1)^d$ cells on each tape that are still reachable during the last at most $l + r(n)$ time steps.

Let $p_1 = \max\{|T|, |S|, 2\}$.

For the $(2(l + r(n)) + 1)^d$ cells per tape there are at most $p_1^{(2(l+r(n))+1)^d}$ different inscriptions. For some $k \in \mathbb{N}$ tapes we obtain altogether at most $p_1^{k(2(l+r(n))+1)^d + 1}$ different situations which bounds the number of l-equivalence classes. The lemma follows for $p = p_1^{(k+1) \cdot 3^d}$. $\qquad\square$

3   Time Hierarchies

In this section we will present the time hierarchies between real time and linear time for any dimension $d \in \mathbb{N}$.

Theorem 6. *Let* $r : \mathbb{N}_0 \to \mathbb{N}$ *and* $r' : \mathbb{N}_0 \to \mathbb{N}$ *be two increasing functions and* $d \in \mathbb{N}$ *be a constant. If* $r \in \mathcal{T}^{-1}(\mathrm{DTM}^d)$, $r \in O(id^{\frac{1}{d}})$, *and either* $r' \in o(r)$ *if* $d = 1$, *or* $r' \in o(r^{1-\varepsilon})$ *for an arbitrarily small* $\varepsilon > 0$ *if* $d > 1$, *then*

$$\mathrm{DTIME}^d(id + r') \subset \mathrm{DTIME}^d(id + r).$$

Before proving the theorem we give the following example which is naturally based on root functions. The dimension hierarchies to be proved in Theorem 8 are also depicted.

Example 7. Since $\mathcal{T}(\mathrm{DTM}^d)$ contains the polynomials id^c, $c \ge 1$, the functions $id^{\frac{1}{c}}$ are belonging to $\mathcal{T}^{-1}(\mathrm{DTM}^d)$. (Actually, the inverses of id^c are $\lceil id^{\frac{1}{c}} \rceil$ but as mentioned before we simplify the notation for convenience.)

For $d = 1$, trivially, $id^{\frac{1}{i+1}} \in o(id^{\frac{1}{i}})$.

For $d > 1$ we need to find an ε such that $id^{\frac{1}{i+1}} \in o(id^{\frac{1}{i}(1-\varepsilon)})$. The condition is fulfilled if and only if $\frac{1}{i+1} < \frac{1}{i}(1 - \varepsilon)$. Thus, if $\frac{i}{i+1} < 1 - \varepsilon$ and therefore, if $\varepsilon < 1 - \frac{i}{i+1}$. We conclude that the condition is fulfilled for all $\varepsilon < \frac{1}{i+1}$.

The hierarchy ist depicted in Figure 1. $\qquad\square$

Proof (of Theorem 6). At first we adjust a constant q dependent on ε. Choose q such that

$$\frac{d-1}{d^q + d} \le \varepsilon$$

for $d > 1$, and $q = 1$ for $d = 1$.

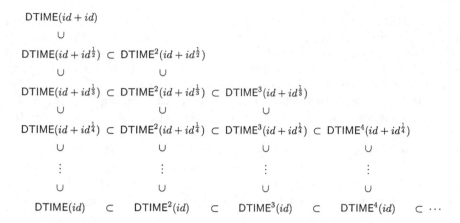

Fig. 1. Double hierarchy based on root functions.

Since $r \in \mathcal{T}^{-1}(\mathrm{DTM}^d)$, i.e. r is the inverse of a constructible function, there exists a constructible function $r^{-1} \in \mathcal{T}(\mathrm{DTM}^d)$ such that $r(r^{-1}(n)) = n$.

Now we are prepared to define a witness language L_1 for the assertion. The words of L_1 are of the form

$$\mathsf{a}^l \mathsf{b}^{r^{-1}(l^{1+d^{q-1}})} w_1 \$ w_1^R \mathsf{¢} w_2 \$ w_2^R \mathsf{¢} \cdots \mathsf{¢} w_s \$ w_s^R \mathsf{¢} d_1 \cdots d_m y,$$

where $l \in \mathbb{N}$ is a positive integer, $s = l^{d^q}$, $m = (d-1) \cdot l^{d^{q-1}}$, $y, w_i \in \{0,1\}^l$, $1 \le i \le s$, and $d_i \in E_{d-1}, 1 \le i \le m$.

The acceptance of such a word is best described by the behavior of an accepting DTM^d $\mathcal{M}$.

During a first phase, $\mathcal{M}$ reads a^l and stores it on a tape. Since d and q are constants, $f(l) = l^{1+d^{q-1}}$ is a polynomial and, thus, constructible. The function r^{-1} is constructible per assumption. The constructible functions are closed under composition. Therefore, during a second phase, $\mathcal{M}$ can simulate a constructor for $r^{-1}(f)$ on the stored input a^l and verify the number of b's.

Parallel to what follows, $\mathcal{M}$ verifies the lengths of the subwords w_i to be l (with the help of the stored a^l) and the numbers s and m ($s = l^{d^q}$ as well as $m = (d-1) \cdot l^{d^{q-1}}$ are constructible functions).

When w_1 appears in the input $\mathcal{M}$ begins to store the subwords w_i in a d-dimensional area of size $l^{d^{q-1}} \times \cdots \times l^{d^{q-1}} \times l^{1+d^{q-1}}$. Suppose the area to consist of l hypercubes with edge length $l^{d^{q-1}}$ that are stacked up. The subwords are stored along the last coordinate, such that $l^{d^{q-1}}$ subwords are stacked up, respectively.

If, for example, the head of the corresponding tape is located at coordinates $(m_1, \ldots, m_d)$, then the following subword w_i is stored into the cells

$$(m_1, \ldots, m_{d-1}, m_d), (m_1, \ldots, m_{d-1}, m_d + 1), \ldots, (m_1, \ldots, m_{d-1}, m_d + l - 1).$$

Temporarily, w_i is also stored on another tape. Now $\mathcal{M}$ has to decide where to store the next subword w_{i+1} (for this purpose it simulates appropriate constructors for $l^{d^{q-1}}$). In principle, there are two possibilities. The first one is that w_{i+1}

is stored as a neighbor of w_i. In this case the head has to move back to position $(m_1, \ldots, m_d)$ and to change the dth coordinate appropriately. The second one is that the subword w_{i+1} is stored below w_i. In this case the head has to keep its position $(m_1, \ldots, m_d + l)$. The head is possibly moved while reading w_i^R. In both cases w_i^R is verified with the temporarily stored w_i.

The last phase leads to acceptance or rejection. After storing all subwords w_i, we may assume that the last coordinate of the head position is $l^{1+d^{q-1}}$ (i.e., the head is on the bottom face of the area). While reading the d_i, $\mathcal{M}$ changes its head simply by adding d_i to the current position. Since $d_i \in E_{d-1}$ the dth coordinate is not affected. This phase leads to a head position $(m_1, \ldots, m_{d-1}, l^{1+d^{q-1}})$. Now the subword y is read and stored on two other tapes. Finally, $\mathcal{M}$ verifies whether or not y matches one of the subwords which have been stacked up in the cells

$$(m_1, \ldots, m_{d-1}, 0), \ldots, (m_1, \ldots, m_{d-1}, l^{1+d^{q-1}} - 1)$$

(if there are stored subwords in these cells at all). Continuous comparisons without delay are achieved by alternating moving one head from back to forth on one of the stored copies of y, while the other head moves from forth to back over the second copy. Machine $\mathcal{M}$ accepts if and only if it finds a matching subword.

Altogether, $\mathcal{M}$ needs n time steps for reading the whole input and at most another $l^{1+d^{q-1}}$ time steps for comparing the y with the stacked up subwords. The first part of the input contains $r^{-1}(l^{1+d^{q-1}})$ symbols b. Therefore, $n > r^{-1}(l^{1+d^{q-1}})$ and since r is increasing, $r(n) \geq r(r^{-1}(l^{1+d^{q-1}})) = l^{1+d^{q-1}}$. We conclude that $\mathcal{M}$ obeys the time complexity $id+r$ and, hence, $L_1 \in \mathsf{DTIME}^d(id + r)$.

Assume now L_1 is acceptable by some DTM^d $\mathcal{M}$ with time complexity $id+r'$. Two words

$$\mathsf{a}^l \mathsf{b}^{r^{-1}(l^{1+d^{q-1}})} w_1 \$ w_1^R \mathsf{¢} w_2 \$ w_2^R \mathsf{¢} \cdots \mathsf{¢} w_s \$ w_s^R \mathsf{¢}$$

and

$$\mathsf{a}^l \mathsf{b}^{r^{-1}(l^{1+d^{q-1}})} w_1' \$ w_1'^R \mathsf{¢} w_2' \$ w_2'^R \mathsf{¢} \cdots \mathsf{¢} w_s' \$ w_s'^R \mathsf{¢}$$

are not $(m + l)$-equivalent with respect to L_1 if the sets $\{w_1, \ldots, w_s\}$ and $\{w_1', \ldots, w_s'\}$ are different. There exist exactly $\binom{2^l}{l^{d^q}}$ different subsets of $\{0, 1\}^l$ with $s = l^{d^q}$ elements. For l large enough such that $\log(l^{d^q}) \leq \frac{1}{4}l$, it follows:

$$N(n, l + m, L_1) \geq \binom{2^l}{l^{d^q}} > \left(\frac{2^l - l^{d^q}}{l^{d^q}} \right)^{l^{d^q}}$$

$$\geq \left(\frac{2^{\frac{l}{2}}}{l^{d^q}} \right)^{l^{d^q}} = \left(2^{\frac{l}{2} - \log(l^{d^q})} \right)^{l^{d^q}}$$

$$\geq \left(2^{\frac{l}{4}} \right)^{l^{d^q}} = \left(2^{\frac{1}{4} l^{d^q}} \right) \in 2^{\Omega(l^{1+d^q})}$$

On the other hand, by Lemma 5 the number of equivalence classes distinguishable by $\mathcal{M}$, is bounded for a constant $p > 1$:

$$N(n, l + m, L_1) \leq p^{(l+m+r'(n))^d}$$

For n we have

$$n = l + r^{-1}(l^{1+d^{q-1}}) + (2l+2) \cdot l^{d^q} + (d-1) \cdot l^{d^{q-1}} + l$$
$$= O(l^{1+d^q}) + r^{-1}(l^{1+d^{q-1}}).$$

Since $r \in O(id^{\frac{1}{d}})$, it follows $r^{-1} \in \Omega(id^d)$. Therefore,

$$r^{-1}(l^{1+d^{q-1}}) \in \Omega(l^{d+d^q}).$$

We conclude

$$n \le c_1 \cdot r^{-1}(l^{1+d^{q-1}}) \text{ for some } c_1 \in \mathbb{N}.$$

Due to the property $\forall c \in \mathbb{N} : \exists c' \in \mathbb{N} : c \cdot r^{-1}(n) \le r^{-1}(c' \cdot n)$, we obtain

$$n \le r^{-1}(c_2 \cdot l^{1+d^{q-1}}) \text{ for some } c_2 \in \mathbb{N}.$$

From $1 - \varepsilon \le 1 - \frac{d-1}{d^q+d} = \frac{d^q+1}{d^q+d} = \frac{d^{q-1}+\frac{1}{d}}{d^{q-1}+1}$ and $r' \in o(r^{1-\varepsilon})$ it follows:

$$r'(n) < r'(r^{-1}(c_2 \cdot l^{1+d^{q-1}}))$$
$$\in o(r(r^{-1}(c_2 \cdot l^{1+d^{q-1}}))^{\frac{d^{q-1}+\frac{1}{d}}{d^{q-1}+1}})$$
$$= o(l^{\frac{1}{d}+d^{q-1}})$$

By $l + m = l + (d-1) \cdot l^{d^{q-1}} \in O(l^{d^{q-1}})$ it holds:

$$(l + m + r'(n))^d \in (O(l^{d^{q-1}}) + o(l^{\frac{1}{d}+d^{q-1}}))^d$$
$$= o(l^{\frac{1}{d}+d^{q-1}})^d = o(l^{1+d^q})$$

So the number of distinguishable equivalence classes is

$$N(n, l+m, L_1) \le p^{o(l^{1+d^q})} = 2^{o(l^{1+d^q})}.$$

Now we have the contradiction that previously $N(n, l+m, L_1)$ has been calculated to be at least $2^{\Omega(l^{1+d^q})}$ which proves $L_1 \notin \text{DTIME}^d(id + r')$.   □

For one-dimensional machines we have hierarchies from real time to linear time. Due to the possible speed-up from $id+r$ to $id+\varepsilon \cdot r$ the condition $r' \in o(r)$ cannot be relaxed.

4 Dimension Hierarchies

Now we are going to show that there exist infinite dimension hierarchies for all time complexities in question. So we obtain double hierarchies. It turns out that dimensions are more powerful than small time bounds.

Theorem 8. *Let* $r : \mathbb{N}_0 \to \mathbb{N}$ *be an increasing function and* $d \in \mathbb{N}$ *be a constant. If* $r \in o(id^{\frac{1}{d}})$, *then*

$$\mathrm{DTIME}^{d+1}(id) \setminus \mathrm{DTIME}^d(id + r) \neq \emptyset.$$

Again, before proving the theorem, we present an example based on natural functions. It shows another double hierarchy.

Example 9. Since $\mathcal{T}(\mathrm{DTM}^d)$ is closed under composition and contains 2^{id}, the functions $\log^{[i]}$, $i \geq 1$ are belonging to $\mathcal{T}^{-1}(\mathrm{DTM}^d)$.

For $d = 1$, trivially, $\log^{[i+1]} \in o(\log^{[i]})$.

For $d > 1$ we need to find an ε such that $\log^{[i+1]} \in o((\log^{[i]})^{1-\varepsilon})$. We have $\log(\log^{[i]})$ and $(\log^{[i]})^{1-\varepsilon}$ and, therefore, the condition is fulfilled for all $\varepsilon < 1$:

The hierarchy ist depicted in Figure 2.   □

DTIME$(id + id)$		DTIME$^2(id + id^{\frac{1}{2}})$		DTIME$^3(id + id^{\frac{1}{3}})$		DTIME$^4(id + id^{\frac{1}{4}})$	
∪		∪		∪		∪	
DTIME$(id + \log)$	⊂	DTIME$^2(id + \log)$	⊂	DTIME$^3(id + \log)$	⊂	DTIME$^4(id + \log)$	⊂ ⋯
∪		∪		∪		∪	
DTIME$(id + \log^{[2]})$	⊂	DTIME$^2(id + \log^{[2]})$	⊂	DTIME$^3(id + \log^{[2]})$	⊂	DTIME$^4(id + \log^{[2]})$	⊂ ⋯
∪		∪		∪		∪	
DTIME$(id + \log^{[3]})$	⊂	DTIME$^2(id + \log^{[3]})$	⊂	DTIME$^3(id + \log^{[3]})$	⊂	DTIME$^4(id + \log^{[3]})$	⊂ ⋯
⋮		⋮		⋮		⋮	
∪		∪		∪		∪	
DTIME(id)	⊂	DTIME$^2(id)$	⊂	DTIME$^3(id)$	⊂	DTIME$^4(id)$	⊂ ⋯

Fig. 2. Double hierarchy based on iterated logarithms.

Proof (of Theorem 8). The words of the witness language L_2 are of the form

$$w_1 \$ w_1^R \cent w_2 \$ w_2^R \cent \cdots \cent w_s \$ w_s^R \cent d_1 \cdots d_m y,$$

where $l \in \mathbb{N}$ is a positive integer, $s = l^d$, $m = d \cdot l$, $y, w_i \in \{0,1\}^l$, $1 \leq i \leq s$, and $d_i \in E_d$, $1 \leq i \leq m$.

An accepting $(d+1)$-dimensional real-time machine $\mathcal{M}$ works as follows. The subwords w_i are stored into a $(d+1)$-dimensional area of size $l \times l \times \cdots \times l$. The first symbols of the subwords w_i are stored at the l^d positions

$$(0, 0, \ldots, 0) \text{ to } (l-1, l-1, \ldots, l-1, 0).$$

The remaining symbols of each w_i are stored along the $(d+1)$st dimension, respectively.

After storing the subwords, $\mathcal{M}$ moves its corresponding head as requested by the d_i. Since the d_i are belonging to E_d, this movement is within the first d

dimensions only. Finally, when y appears in the input, $\mathcal{M}$ tries to compare it with the subword stored at the current position. $\mathcal{M}$ accepts if a subword has been stored at the current position at all and if the subword matches y. Thus, $L_2 \in \text{DTIME}^{d+1}(id)$.

In order to apply Lemma 5, we observe that, again, two words

$$w_1\$w_1^R\text{¢}w_2\$w_2^R\text{¢}\cdots\text{¢}w_s\$w_s^R\text{¢}$$

and

$$w_1'\$w_1'^R\text{¢}w_2\$w_2'^R\text{¢}\cdots\text{¢}w_s\$w_s'^R\text{¢}$$

are not $(m + l)$-equivalent with respect to L_2 if the sets $\{w_1, \ldots, w_s\}$ and $\{w_1', \ldots, w_s'\}$ are different. Therefore, L_2 induces at least

$$N(n, l+m, L_2) \geq \binom{2^l}{l^d} \geq 2^{\Omega(l^{d+1})}$$

equivalence classes for all sufficiently large l.

On the other hand, we obtain an upper bound of the number of distinguishable equivalence classes for an $(id + r)$-time DTM^d $\mathcal{M}$ as follows:

$$\begin{aligned}
N(n, l+m, L_2) &\leq p^{(l+m+r(n))^d} \\
&= p^{(l+d\cdot l+r((2l+2)\cdot l^d+l+d\cdot l))^d} \\
&\leq p^{(c_1\cdot l+r(c_2\cdot l^{d+1}))^d} \quad \text{for some } c_1, c_2 \in \mathbb{N} \\
&\in p^{(O(l)+o(c_2\cdot l^{d+1})^{\frac{1}{d}})^d} \quad \text{since } r \in o(id^{\frac{1}{d}}) \\
&= p^{(O(l)+o(l^{\frac{d+1}{d}}))^d} \\
&= p^{o(l^{\frac{d+1}{d}})^d} \\
&= p^{o(l^{d+1})} = 2^{o(l^{d+1})}
\end{aligned}$$

From the contradiction $L_2 \notin \text{DTIME}^d(id + r)$ follows. □

The inclusions $\text{DTIME}^{d+1}(id) \subseteq \text{DTIME}^{d+1}(id + r)$ and $\text{DTIME}^d(id + r) \subseteq \text{DTIME}^{d+1}(id+r)$ are trivial. An application of Theorem 8 yields the hierarchies:

Corollary 10. *Let $r : \mathbb{N}_0 \to \mathbb{N}$ be an increasing function and $d \in \mathbb{N}$ be a constant. If $r \in o(id^{\frac{1}{d}})$, then*

$$\text{DTIME}^d(id + r) \subset \text{DTIME}^{d+1}(id + r).$$

Note that despite the condition $r \in o(id^{\frac{1}{d}})$, the dimension hierarchies can touch $r = id^{\frac{1}{d}}$:

$id^{\frac{1}{d}} \in o(id^{\frac{1}{d-1}})$ and $\text{DTIME}^{d-1}(id + id^{\frac{1}{d}}) \subset \text{DTIME}^d(id + id^{\frac{1}{d}})$.

References

1. Balcázar, J. L., Díaz, J., and Gabarró, J. *Structural Complexity I*. Springer, Berlin, 1988.
2. Book, R. V. and Greibach, S. A. *Quasi-realtime languages*. Math. Systems Theory 4 (1970), 97–111.
3. Buchholz, T., Klein, A. and Kutrib, M. *Iterative arrays with small time bounds*, Mathematical Foundations of Computer Science (MFCS 2000), LNCS 1893, Springer 2000, pp. 243–252.
4. Cole, S. N. *Real-time computation by n-dimensional iterative arrays of finite-state machines*. IEEE Trans. Comput. C-18 (1969), 349–365.
5. Fürer, M. *The tight deterministic time hierarchy*. Proceedings of the Fourteenth Annual ACM Symposium on Theory of Computing (STOC '82), 1982, pp. 8–16.
6. Hartmanis, J. and Stearns, R. E. *On the computational complexity of algorithms*. Trans. Amer. Math. Soc. 117 (1965), 285–306.
7. Hennie, F. C. *One-tape, off-line turing machine computations*. Inform. Control 8 (1965), 553–578.
8. Hennie, F. C. *On-line turing machine computations*. IEEE Trans. Elect. Comput. EC-15 (1966), 35–44.
9. Hopcroft, J. E. and Ullman, J. D. *Introduction to Automata Theory, Language, and Computation*. Addison-Wesley, Reading, Massachusetts, 1979.
10. Klein A. and Kutrib, M. *Deterministic Turing machines in the range between real-time and linear-time*. Theoret. Comput. Sci. 289 (2002), 253–275.
11. Kobayashi, K. *On proving time constructibility of functions*. Theoret. Comput. Sci. 35 (1985), 215–225.
12. Loui, M. C. *Simulations among multidimensional turing machines*. Theoret. Comput. Sci. 21 (1982), 145–161.
13. Michel P. *An NP-complete language accepted in linear time by a one-tape Turing machine*. Theoret. Comput. Sci. 85 (1991), 205–212.
14. Paul, W. J. *On time hierarchies*. J. Comput. System Sci. 19 (1979), 197–202.
15. Paul, W., Seiferas, J. I., and Simon, J. *An information-theoretic approach to time bounds for on-line computation*. J. Comput. System Sci. 23 (1981), 108–126.
16. Pippenger, N. and Fischer, M. J. *Relations among complexity measures*. J. Assoc. Comput. Mach. 26 (1979), 361–381.
17. Reischuk, R. *Einführung in die Komplexitätstheorie*. Teubner, Stuttgart, 1990.
18. Rosenberg, A. L. *Real-time definable languages*. J. Assoc. Comput. Mach. 14 (1967), 645–662.
19. Stoß, H.-J. *Zwei-Band Simulation von Turingmaschinen*. Computing 7 (1971), 222–235.
20. Wagner, K. and Wechsung, G. *Computational Complexity*. Reidel, Dordrecht, 1986.

Baire's Categories on Small Complexity Classes

Philippe Moser

Computer Science Department, University of Geneva
moser@cui.unige.ch

Abstract. We generalize resource-bounded Baire's categories to small complexity classes such as P, QP and SUBEXP and to probabilistic classes such as BPP. We give an alternative characterization of small sets via resource-bounded Banach-Mazur games. As an application we show that for almost every language $A \in$ SUBEXP, in the sense of Baire's category, $P^A = BPP^A$.

1 Introduction

Resource-bounded measure and resource-bounded Baire's Category were introduced by Lutz in [1] and [2] for both complexity classes E and EXP. It provides a means of investigating the sizes of various subsets of E and EXP. In resource-bounded measure the small sets are those with measure zero, in resource-bounded Baire's Category the small sets are those of first category (meager sets). Both smallness notions satisfy the following three axioms. First every single language $L \in$ E is small, second the whole class E is large, and finally "easy infinite unions" of small sets are small. These axioms meet the essence of Lebegue's measure and Baire's category and ensure that it is impossible for a subset of E to be both large and small.

The first goal of Lutz's approach was to extend existence results, such as "there is a language in C satisfying property P", to abundance results such as "most languages in C satisfy property P", which is more informative since an abundance result reflects the typical behavior of languages in a class, whereas an existence result could as well correspond to an exception in the class. Both resource-bounded measure and resource-bounded Baire's Category have been successfully used to understand the structure of the exponential time classes E and EXP.

An important problem in resource-bounded measure theory was to generalize Lutz's measure theory to small complexity classes such as P, QP and SUBEXP and to probabilistic classes such as BPP and BPE. These issues have been solved in the following list of papers [3], [4], [5] and [6]. As noticed in [7], the same question in the Baire's category setting was still left unanswered.

In this paper we solve this problem by generalizing resource-bounded Baire's categories on small complexity classes such as P, QP and SUBEXP and to probabilistic classes such as BPP. We also give an alternative characterization of meager sets through Banach-Mazur games. As an application we improve the result of [3] where it was shown that for almost every language $A \in$ SUBEXP, in

A. Lingas and B.J. Nilsson (Eds.): FCT 2003, LNCS 2751, pp. 333–342, 2003.
© Springer-Verlag Berlin Heidelberg 2003

the sense of resource-bounded measure, $P^A = BPP$. The question whether the same result holds with $P^A = BPP^A$ was raised in [3]. We answer this question affirmatively in the resource-bounded Baire's category setting, by showing show that for almost every language $A \in SUBEXP$, in the sense of resource-bounded Baire's category, $P^A = BPP^A$.

The remainder of the paper is organized as follows. In section 3 we introduce resource-bounded Baire's category on P. In section 3.1 we give another characterization of small sets through resource-bounded Banach-Mazur games. In section 4 we introduce resource-bounded Baire's category on BPP with the corresponding resource-bounded Banach-Mazur games formulation. Finally in section 5 we prove the result on BPP mentioned above.

2   Preliminaries

We use standard notation for traditional complexity classes; see for instance [8] and [9], or [10]. For $\epsilon > 0$, denote by E_ϵ the class $E_\epsilon = \bigcup_{\delta < \epsilon} DTIME(2^{n^\delta})$. SUBEXP is the class $\cap_{\epsilon > 0} E_\epsilon$, and quasi polynomial time refers to the class $QP = \bigcup_{k \geq 1} DTIME(n^{\log^k n})$. Let us fix some notations for strings and languages. Let $s_0, s_1, \ldots$ be the standard enumeration of the strings in $\{0,1\}^*$ in lexicographical order, where $s_0 = \lambda$ denotes the empty string. A sequence is an element of $\{0,1\}^\infty$. If w is a string or a sequence and $1 \leq i \leq |w|$ then $w[i]$ and $w[s_i]$ denotes the ith bit of w. Similarly $w[i \ldots j]$ and $w[s_i \ldots s_j]$ denote the ith through jth bits, and $dom(w)$ the domain of w, where w is viewed as a partial function. We identify language L with its characteristic function χ_L, where χ_L is the sequence such that $\chi_L[i] = 1$ iff $s_i \in L$. For a string s_i define its position by $pos(s_i) = i$. If w_1 is a string and w_2 is a string or a sequence extending w_1, we write $w_1 \sqsubseteq w_2$. We write $w_1 \sqsubset w_2$ if $w_1 \sqsubseteq w_2$ and $w_1 \neq w_2$. For two strings $\tau, \sigma \in \{0,1\}^*$, we denote by $\tau^\wedge \sigma$ the concatenation of τ followed by σ. For $a, b \in \mathbb{N}$ let $a \dot{-} b$ denote $\max(a - b, 0)$. We identify $\mathbb{N}$ with $\{0,1\}^*$, thus we denote by $\mathbb{N}^\mathbb{N}$ the set of all functions mapping strings to strings.

2.1   Finite Extension Strategies

Whereas resource-bounded measure is defined via martingales, resource-bounded Baire's category is defined via finite extension strategies. Here is a definition.

Definition 1. *A function $h : \{0,1\}^* \to \{0,1\}^*$ is a finite extension strategy, or a constructor, if for every string $\tau \in \{0,1\}^*$, $\tau \sqsubseteq h(\tau)$.*

For simplicity we will use the word "strategy" for finite extension strategy. We will often consider indexed strategies. An indexed strategy is a function $h : \mathbb{N} \times \{0,1\}^* \to \{0,1\}^*$, such that $h_i := h(i, \cdot)$ is a strategy for every $i \in \mathbb{N}$. If h is a strategy and $\tau \in \{0,1\}^*$, define $ext\, h(\tau)$ as the unique string u such that $h(\tau) = \tau^\wedge u$. We say a strategy h avoids some language A (or language A avoids strategy h) if for every string $\tau \in \{0,1\}^*$ we have $h(\tau) \not\sqsubseteq \chi_A$. We say a strategy h meets some language A if h does not avoid A.

For the results in Section 5 we will need the following definition of the relativized hardness of a pseudorandom generator.

Definition 2. *Let A be any language. The hardness $H^A(G_{m,n})$ of a random generator $G_{m,n} : \{0,1\}^m \longrightarrow \{0,1\}^n$, is defined as the minimal s such that there exists an n-input circuit C with oracle gates to A, of size at most s, for which:*

$$| \Pr_{x \in \{0,1\}^m} [C(G_m(x)) = 1] - \Pr_{y \in \{0,1\}^n} [C(y) = 1]| \geq \frac{1}{s} . \tag{1}$$

Klivans and Melkebeek [11] noticed that Impagliazzo and Widgerson's [12] pseudorandom generator construction relativizes; i.e. for any language A, there is a deterministic polynomial time procedure that converts the truth table of a Boolean function that is hard to compute for circuits having oracle gates for A, into a pseudorandom generator that is pseudorandom for circuits with A oracle gates. More precisely,

Theorem 1 (Klivans-Melkebeek [11]).
Let A be any language. There is a polynomial-time computable function $F :$ $\{0,1\}^* \times \{0,1\}^* \to \{0,1\}^*$, with the following properties. For every $\epsilon > 0$, there exists $a, b \in \mathbb{N}$ such that

$$F : \{0,1\}^{n^a} \times \{0,1\}^{b \log n} \to \{0,1\}^n , \tag{2}$$

and if r is the truth table of a $(a \log n)$-variables Boolean function of A-oracle circuit complexity at least $n^{\epsilon a}$, then the function $G_r(s) = F(r, s)$ is a generator, mapping $\{0,1\}^{b \log n}$ into $\{0,1\}^n$, which has hardness $H^A(G_r) > n$.

3   Baire's Category on P

To define a resource bounded Baire's category on P, we will consider strategies computed by Turing machines which have random access to their inputs, i.e. on input τ, the machine can query any bit of τ to its oracle. For such a random Turing machine M running on input τ, we denote this convention by $M^\tau(\cdot)$. Note that random Turing machines can compute the lengths of their input τ in $O(\log |\tau|)$ steps, by using bisection. We will consider random Turing machines running in time polylog in the input's length $|\tau|$ or equivalently polynomial in $|s_{|\tau|}|$. Note that such machines cannot read their entire input, but only a sparse subset of it.

Definition 3. *An indexed strategy $h : \mathbb{N} \times \{0,1\}^* \to \{0,1\}^*$ is P-computable if there is a random access Turing machine M as above, such that for every $\tau \in \{0,1\}^*$ and every $i \in \mathbb{N}$,*

$$M^\tau(0^i) = ext\, h_i(\tau) \tag{3}$$

where M runs in time polynomial in $|s_{|\tau|}| + i$.

We say a class is small if there is a single indexed strategy that avoids every language in the class. More precisely,

Definition 4. *A class* C *of languages is* P-*meager if there exists a* P-*computable indexed strategy* h, *such that for every* $L \in$ C *there exists* $i \in \mathbb{N}$, *such that* h_i *avoids* L.

In order to formalize the third axiom we need to define "easy infinite unions" precisely.

Definition 5. $X = \bigcup_{i \in \mathbb{N}} X_i$ *is a* P-*union of* P-*meager sets, if there exists an indexed* P-*computable strategy* $h : \mathbb{N} \times \mathbb{N} \times \{0,1\}^* \to \{0,1\}^*$, *such that for every* $i \in \mathbb{N}$, $h_{i,\cdot}$ *witnesses* X_i's *meagerness.*

Let us prove the three basic axioms.

Theorem 2. *For any language* L *in* P, *the singleton* $\{L\}$ *is* P-*meager.*

Proof. Let $L \in$ P be any language. We describe a P-computable constructor h which avoids $\{L\}$. Consider the following Turing machine M computing h. On input string σ, M^σ simply outputs $1 - L(s_{|\sigma|+1})$. h is clearly P-computable, and h avoids $\{L\}$. □

The proof of the third axiom is straightforward.

Theorem 3.

1. *All subsets of a* P-*meager set are* P-*meager.*
2. *A* P-*union of* P-*meager sets is* P-*meager.*

Proof. Immediate by definition of P-meagerness. □

Let us prove the second axiom which says that the whole space P is not small.

Theorem 4. P *is not* P-*meager.*

Proof. Let h be an indexed P-computable constructor and let M be a Turing machine computing h. We construct a language $L \in$ P which meets h_i for every i. The idea is to construct a language L with the following characteristic function,

$$\chi_L = |\underbrace{0}_{B_0}|\underbrace{ext\, h_1(B_0)0\cdots0}_{B_1}|\underbrace{ext\, h_2(B_0{}^\wedge B_1)0\cdots0}_{B_2}|\cdots|\underbrace{ext\, h_i(B_0{}^\wedge B_1{}^\wedge\cdots{}^\wedge B_{i-1})0\cdots0}_{B_i}|$$

$$(4)$$

where block B_i corresponds to all strings of size i, and block B_i contains $ext\, h_i(B_0{}^\wedge B_1{}^\wedge \cdots {}^\wedge B_{i-1})$ followed by a padding with 0's. B_i is large enough to contain $ext\, h_i(B_0{}^\wedge B_1{}^\wedge \cdots {}^\wedge B_{i-1})$, because M's output's length is bounded by a polynomial in i.

Let us construct a polynomial time Turing machine N deciding L. On input x, where $|x| = n$,

1. Compute p where x is the pth word of length n.
2. For $i = 1$ to n simulate $M^{B_0 \wedge B_1 \wedge \cdots \wedge \bar{B}_{i-1}}(0^i)$. Answer M's queries with the previously stored binary sequences $\bar{B}_1, \bar{B}_2, \bar{B}_{i-1}$ in the following way. Suppose that during its simulation $M^{B_0 \wedge B_1 \wedge \cdots \wedge \bar{B}_{i-1}}(0^i)$ queries the kth bit of $B_0 \wedge B_1 \wedge \cdots \wedge B_{i-1}$ to its oracle. To answer this query, simply compute s_k and compute its lengths l_k and its position p_k among words of size l_k. Look up whether the stored binary sequence $\bar{B}_{l_k}$ contains a p_kth bit b_k. If this is the case answer M's query with b_k, else answer M's query with 0. Finally store the output of $M^{B_0 \wedge B_1 \wedge \cdots \wedge \bar{B}_{i-1}}(0^i)$ under $\bar{B}_i$.
3. If the stored binary sequence $\bar{B}_n$ contains a pth bit then output this bit, else output 0 (x is in the padded zone of B_n).

Let us check that L is in P. The first and third step are clearly computable in time polynomial in n. For the second step we have that for each of the n recursive steps there are at most a polynomial number of queries (because h is P-computable) and each simulation of M once the queries are answered takes time polynomial in n because M is polynomial. Note that all $\bar{B}_i$'s have size polynomial in n, therefore it's no problem to store them.   □

3.1   Resource-Bounded Banach-Mazur Games

We give an alternative characterization of small sets via resource-bounded Banach-Mazur games. Informally speaking, a Banach-Mazur game, is a game between two strategies f and g, where the game begins with the empty string λ. Then $g \circ f$ is applied successively on λ. Such a game yields a unique infinite string, or equivalently a language, called the result of the play between f and g. For a class C, we say that g is a winning strategy if it can force the result of the game with any strategy f to be a language not in C. We show that the existence of a winning strategy is equivalent to the meagerness of C. This equivalence result is useful in practice, since it is often easier to find a winning strategy, rather than a finite extension strategy.

Definition 6.

1. *A play of a Banach-Mazur game is a pair (f, g) of strategies such that for every string $\tau \in \{0, 1\}^*$, $\tau \sqsubseteq g(\tau)$.*
2. *The result $R(f, g)$ of the play (f, g) is the unique element of $\{0, 1\}^\infty$ that extends $(g \circ f)^i(\lambda)$ for every $i \in \mathbb{N}$.*

For a class of languages C and two function classes F_I and F_{II}, denote by $G[C, F_I, F_{II}]$ the Banach-Mazur game with distinguished set C, where player I must choose a strategy in F_I, and player II a strategy in F_{II}. We say player II wins the play (f, g) if $R(f, g) \notin C$, otherwise we say player I wins. We say player II has a winning strategy for the game $G[C, F_I, F_{II}]$, if there exists a strategy $g \in F_{II}$ such that for every strategy $f \in F_I$, player II wins (f, g)

The following result states that a class is meager iff there is a winning strategy for player II. This is very useful since in practice it is often easier to give a winning strategy for player II, than to exhibit a constructor avoiding every language in the class.

Theorem 5. *Let X be any class of languages. The following are equivalent.*

1. Player II has a winning strategy for $G[X, \mathbb{N}^{\mathbb{N}}, P]$.
2. X is P-meager.

Proof. Suppose the first statement holds and let g be a P-computable wining strategy for player II. Let M be a Turing machine computing g. We define an indexed P-computable constructor h. Let $k \in \mathbb{N}$ and $\sigma \in \{0, 1\}^*$,

$$h_k(\sigma) := g(\sigma') \quad \text{where } \sigma' = \sigma {}^\wedge 0^{k - |\sigma|} . \tag{5}$$

h is P-computable because computing $h_k(\sigma)$ simply requires to simulate $M^{\sigma'}$, answering M's queries in $dom(\sigma')\backslash dom(\sigma)$ by 0. We show that if language A meets h_k for every $k \in \mathbb{N}$, then $A \notin X$. This implies that X is P-meager as witnessed by h. To do this we show that for every $\alpha \sqsubset \chi_A$ there is a string β such that,

$$\alpha \sqsubseteq \beta \sqsubseteq g(\beta) \sqsubset \chi_A . \tag{6}$$

If this holds, then player I has a winning strategy yielding $R(f, g) = A$: for a given α player I extends it to obtain the corresponding β, thus forcing player II to extend to a prefix of χ_A. So let α be any prefix of χ_A, where $|\alpha| = k$. Since A meets h_k, there is a string $\sigma \sqsubset \chi_A$ such that

$$\sigma' \sqsubseteq g(\sigma') = h_k(\sigma) \sqsubset \chi_A \tag{7}$$

where $\sigma' = \sigma {}^\wedge 0^{k - |\sigma|}$. Since $|\alpha| \leq |\sigma'|$ and α, σ' are prefixes of χ_A, we have $\alpha \sqsubseteq \sigma'$. Define β to be σ'.

For the other direction, let X be P-meager as witnessed by h, i.e. for every $A \in X$ there exists $i \in \mathbb{N}$ such that h_i avoids A. Let N be a Turing machine computing h. We define a P-computable constructor g inducing a winning strategy for player II in the game $G[X, \mathbb{N}^{\mathbb{N}}, P]$. We show that for any strategy f, $R(f, g)$ meets h_i for every $i \in \mathbb{N}$, which implies $R(f, g) \notin X$. Here is a description of a Turing machine M computing g. For a string σ, M^σ does the following.

1. Compute $n_0 = \min_{t \leq n}[(\forall \tau \sqsubseteq \sigma \text{ such that } |\tau| \leq n) \ h_t(\tau) \not\sqsubseteq \sigma]$, where $n = |s_{|\sigma|}|$.
2. If no such n_0 exists output 0.
3. If n_0 exists (h_{n_0} is the next strategy to be met), simulate $N^{\sigma {}^\wedge 0}(0^{n_0})$ answering N's queries in $dom(\sigma {}^\wedge 0)\backslash dom(\sigma)$ with 0, denote N's answer by ω. Output $0 {}^\wedge \omega$.

g is clearly P-computable. We show that $R(f, g)$ meets every h_i for any strategy f. Suppose for a contradiction that this is not the case, i.e. there is a strategy f such that $R(f, g)$ does not meet h. Let n_0 be the smallest index such that $R(f, g)$ does not meet h_{n_0}. Since $R(f, g)$ meets h_{n_0-1} there is a string τ such that $h_{n_0-1}(\tau) \sqsubset R(f, g)$. Since g strictly extends strings at every round, after at most $2^{O(|\tau|)}$ rounds, f will output a string σ long enough to enable step 1 (of M's description) to find out that $h_{n_0-1}(\tau) \sqsubseteq \sigma \sqsubset R(f, g)$ thus incrementing $n_0 - 1$ to n_0. At this round we have $g(\sigma) = \sigma {}^\wedge 0 {}^\wedge ext \, h_{n_0}(\sigma {}^\wedge 0)$, i.e. $h_{n_0} \sqsubset R(f, g)$ which is a contradiction. □

It is easy to check that throughout Section 3, P can be replaced by QP or E_ϵ , thus yielding a Baire's category notion on both quasi-polynomial and subexponential time classes.

4 Baire's Category on BPP

To construct a notion of Baire's category on probabilistic classes, we will use the following probabilistic indexed strategies.

Definition 7. *An indexed strategy* $h : \mathbb{N} \times \{0,1\}^* \to \{0,1\}^*$ *is* BPP-*computable if there is a probabilistic oracle Turing machine M such that for every $\tau \in \{0,1\}^*$ and every $i, n \in \mathbb{N}$,*

$$\Pr[M^\tau(0^i, 0^n) = ext\, h_i(\tau)] \geq 1 - 2^{-n} \tag{8}$$

where the probability is taken over the internal coin tosses of M, and M runs in time polynomial in $|s_{|\tau|}| + i + n$.

By using standard Chernoff bound arguments it is easy to show that Definition 7 is robust, i.e. the error probability can range from $1/2 + 1/p(n)$ to $1 - 2^{-q(n)}$ for any polynomials p, q, without enlarging (resp. reducing) the class of strategies defined in Definition 7.

As in Section 3, a class is meager if there is a single probabilistic strategy that avoids every language in the class.

Definition 8. *A class of languages* C *is* BPP-*meager if there exists a* BPP-*computable indexed strategy h, such that for every $L \in$ C there exists $i \in \mathbb{N}$, such that h_i avoids L.*

As in section 3, we need to define "easy infinite unions" precisely in order to prove the third axiom.

Definition 9. $X = \bigcup_{i \in \mathbb{N}} X_i$ *is a* BPP-*union of* BPP-*meager sets, if there exists an indexed* BPP-*computable strategy $h : \mathbb{N} \times \mathbb{N} \times \{0,1\}^* \to \{0,1\}^*$, such that for every $i \in \mathbb{N}$, $h_{i,\cdot}$ witnesses X_i's meagerness.*

Let us prove that all three axioms hold for our Baire's category notion on BPP.

Theorem 6. *For any language L in* BPP, *$\{L\}$ is* BPP-*meager.*

Proof. The proof is similar to Theorem 2 except that the constructor h is computed with error probability smaller than 2^{-n}. $\qquad\square$

The third axiom holds by definition.

Theorem 7.

1. *All subsets of a* BPP-*meager set are* BPP-*meager.*
2. *A* BPP-*union of* BPP-*meager sets is* BPP-*meager.*

Proof. Immediate by definition of BPP-meagerness.   □

Let us prove the second axiom.

Theorem 8. BPP *is not* BPP-*meager.*

Proof. The proof is similar to Theorem 4 except for the second step of N's computation, where every simulation of M is performed with error probability smaller than 2^{-n}. Since there are n distinct simulation of M, the total error probability is smaller than $n2^{-n}$, which ensures that L is in BPP.   □

4.1   Resource-Bounded Banach-Mazur Games

Similarly to Section 3.1, we give an alternative characterization of meager sets through resource-bounded Banach-Mazur games.

Theorem 9. *Let X be any class of languages. The following are equivalent.*

1. *Player II has a winning strategy for $G[X, \mathbb{N}^{\mathbb{N}}, \text{BPP}]$.*
2. *X is* BPP-*meager.*

Proof. The 1. implies 2. direction is similar to Theorem 5, except that $h_k(\sigma)$ can be computed with error probability smaller than 2^{-n}.

For the other direction, the only difference with Theorem 5, is that the first and third step of M's computation can be performed with small error probability.   □

5   Application to the P = BPP Problem

It was shown in [3] that for every $\epsilon > 0$, almost every language $A \in \mathsf{E}_\epsilon$, in the sense of resource-bounded measure, satisfies $\mathsf{P}^A = \mathsf{BPP}$. We improve their result by showing that for every $\epsilon > 0$, almost every language $A \in \mathsf{E}_\epsilon$, in the sense of resource-bounded Baire's category, satisfies $\mathsf{P}^A = \mathsf{BPP}^A$.

Theorem 10. *For every $\epsilon > 0$, the set of languages A such that $\mathsf{P}^A \neq \mathsf{BPP}^A$ is E_ϵ -meager.*

Proof. Let $\epsilon > 0$. Let $0 < \delta < \max(\epsilon, 1/4)$ and $b > 2k\delta/\epsilon$, where k is some constant that will be determined later. Consider the following strategy h, computed by the following Turing machine M. On input σ, where $|s_{|\sigma|}| = n$, M does the following. At start $Z = \emptyset$, and $i = 1$. M computes z_i in the following way. Determine whether $pos(s_{|\sigma|+i}) = pos(0^{2^{b|u|}} u)$ for some string u of size $\log(n^{2/b})$. If not then $z_i = 0$, output z_i, and compute z_{i+1}; else denote by u_i the corresponding string u. Construct the set T_i of all truth tables of $|u_i|$-inputs Boolean circuits C with oracle gates for σ of size less than $2^{\delta|u_i|}$, such that $C(u_j) = z_j$ for every $(u_j, z_j) \in Z$. Compute $M_i = \text{Majority}_{C \in T_i}[C(u_i)]$, and let $z_i = 1 - M_i$. Add (u_i, z_i) to Z. Output z_i, and compute z_{i+1}, unless $u_i = 1^{\log(n^{2/b})}$ (i.e. u_i is the last string of size $\log(n^{2/b})$), in which case M stops.

Since there are $2^{n^{4\delta/b}}$ circuits to simulate, and simulating such a circuit takes time $O(n^{4\delta/b})$, by answering its queries to σ with the input σ, M runs in time $2^{n^{\epsilon'}}$, where $\epsilon' < \epsilon$. Finally computing the majority M_i takes time $2^{O(n^{4\delta/b})}$. Thus the total running time is less than $2^{n^{2c\delta/b}}$ for some constant c, which is less than $2^{n^{\epsilon'}}$ with $\epsilon' < \epsilon$ for an appropriate choice of k.

Let A be any language and consider $F(A) := \{u|0^{2^{b|u|}} u \in A\}$. It is clear that $F(A) \in \mathsf{E}^A$. Consider H_δ^A the set of languages with high circuit complexity, i.e. $H_\delta^A = \{L|$ every n-inputs circuits with oracle gates for A of size less than $2^{\delta n}$ fails to compute $L\}$. We have, $F(A) \cap H_\delta^A \neq \emptyset$ implies $\mathsf{P}^A = \mathsf{BPP}^A$, by Theorem 1.

We show that h avoids every language A such that $F(A) \cap H_\delta^A = \emptyset$. So let A be any such language, i.e. there is a n-inputs circuit family $\{C_n\}_{n>0}$, with oracle gates for A, of size less than $2^{\delta n}$ computing $F(A)$. We have

$$C(u_i) = 1 \text{ iff } 0^{2^{b|u_i|}} u_i \in A \text{ for every string } u_i \text{ such that } (u_i, z_i) \in Z. \quad (9)$$

(for simplicity we omit C's index). Consider the set D_n of all circuits with $\log(n^{2/b})$-inputs of size at most $n^{2\delta/b}$ with oracles gates for A satisfying Equation 9. We have $|D_n| \le 2^{4\delta/b}$. By construction, every z_i such that $(u_i, z_i) \in Z$ reduces the cardinal of D_n by a factor 2. Since there are $n^{2/b}$ z_i's such that $(u_i, z_i) \in Z$, we have $D_n \le 2^{4\delta/b} \cdot 2^{-n^{2/b}} < 1$, i.e. $D_n = \emptyset$. Therefore $h(\sigma) \not\sqsubseteq \chi_A$. $\qquad \square$

6   Conclusion

Theorem 4 shows that the class **SPARSE** of all languages with polynomial density is not P-meager. To remedy this situation we can improve the power of P-computable strategies by considering locally computable strategies, which can avoid **SPARSE** and even the class of language of subexponential density. This issue will be addressed in [13].

References

1. Lutz, J.: Category and measure in complexity classes. SIAM Journal on Computing **19**(1990) 1100–1131
2. Lutz, J.: Almost everywhere high nonuniform complexity. Journal of Computer and System Science **44**(1992) 220–258
3. Allender, E., Strauss, M.: Measure on small complexity classes, with application for BPP. Proceedings of the 35th Annual IEEE Symposium on Foundations of Computer Science (1994) 807–818
4. Strauss, M.: Measure on P-strength of the notion. Inform. and Comp. **136:1**(1997) 1–23
5. Regan, K., Sivakumar, D.: Probabilistic martingales and BPTIME classes. In Proc. 13th Annual IEEE Conference on Computational Complexity (1998) 186–200
6. Moser, P.: A generalization of Lutz's measure to probabilistic classes. submitted (2002)
7. Ambos-Spies, K.: Resource-bounded genericity. Proceedings of the Tenth Annual Structure in Complexity Theory Conference (1995) 162–181

8. Balcázar, J.L., Díaz, J., and Gabarró, J.: Structural Complexity I. EATCS Monographs on Theorical Computer Science Volume 11, Springer-Verlag (1995)
9. Balcázar, J.L., Díaz, J., and Gabarró, J.: Structural Complexity II. EATCS Monographs on Theorical Computer Science Volume 22, Springer-Verlag (1990)
10. Papadimitriou, C.: Computational complexity. Addisson-Wesley (1994)
11. Klivans, A., Melkebeek, D.: Graph nonisomorphism has subexponential size proofs unless the polynomial hierarchy collapses. Proceedings of the 31st Annual ACM Symposium on Theory of Computing (1999) 659–667
12. Impagliazzo, R., Widgerson, A.: P = BPP if E requires exponential circuits: derandomizing the XOR lemma. Proceedings of the 29th Annual ACM Symposium on Theory of Computing (1997) 220–229
13. Moser, P.: Locally computed Baire's categories on small complexity classes. submitted (2002)

Operations Preserving Recognizable Languages

Jean Berstel[1], Luc Boasson[2], Olivier Carton[2],
Bruno Petazzoni[3], and Jean-Éric Pin[2]

[1] Institut Gaspard Monge, Université de Marne-la-Vallée,
5, boulevard Descartes, Champs-sur-Marne, F-77454 Marne-la-Vallée Cedex 2,
berstel@univ-mlv.fr,
[2] LIAFA, Université Paris VII and CNRS, Case 7014,
2 Place Jussieu, F-75251 Paris Cedex 05, FRANCE[†]
{Olivier.Carton,Luc.Boasson,Jean-Eric.Pin}@liafa.jussieu.fr
[3] Lycée Marcelin Berthelot, Saint-Maur
bpetazzoni@ac-creteil.fr

Abstract. Given a subset S of $\mathbb{N}$, filtering a word $a_0a_1 \cdots a_n$ by S consists in deleting the letters a_i such that i is not in S. By a natural generalization, denote by $L[S]$, where L is a language, the set of all words of L filtered by S. The filtering problem is to characterize the filters S such that, for every recognizable language L, $L[S]$ is recognizable. In this paper, the filtering problem is solved, and a unified approach is provided to solve similar questions, including the removal problem considered by Seiferas and McNaughton. There are two main ingredients on our approach: the first one is the notion of residually ultimately periodic sequences, and the second one is the notion of representable transductions.

1 Introduction

The original motivation of this paper was to solve an automata-theoretic puzzle, proposed by the fourth author (see also [8]), that we shall refer to as the *filtering problem*. Given a subset S of $\mathbb{N}$, *filtering a word* $a_0a_1 \cdots a_n$ by S consists in deleting the letters a_i such that i is not in S. By a natural generalization, denote by $L[S]$, where L is a language, the set of all words of L filtered by S. The filtering problem is to characterize the filters S such that, for every recognizable language L, $L[S]$ is recognizable. The problem is non trivial since, for instance, it can be shown that the filter $\{n! \mid n \in \mathbb{N}\}$ preserves recognizable languages.

The quest for this problem led us to search for analogous questions in the literature. Similar puzzles were already investigated in the seminal paper of Stearns and Hartmanis [14], but the most relevant reference is the paper [12] of Seiferas and McNaughton, in which the so-called "removal problem" was solved: characterize the subsets S of $\mathbb{N}^2$ such that, for each recognizable language L, the language

$$P(S, L) = \{u \in A^* \mid \text{there exists } v \in A^* \text{ such that } (|u|, |v|) \in S \text{ and } uv \in L\}$$

is recognizable.

[†] Work supported by INTAS project 1224.

A. Lingas and B.J. Nilsson (Eds.): FCT 2003, LNCS 2751, pp. 343–354, 2003.
© Springer-Verlag Berlin Heidelberg 2003

The aim of this paper is to provide a unified approach to solve at the same time the filtering problem, the removal problem and similar questions. There are two main ingredients in our approach. The first one is the notion of residually ultimately periodic sequences, introduced in [12] as a generalization of a similar notion introduced by Siefkes [13]. The second one is the notion of representable transductions introduced in [9,10]. Complete proofs will be given in the extended version of this article.

Our paper is organized as follows. Section 2 introduces some basic definitions: rational and recognizable sets, etc. The precise formulation of the filtering problem is given in Section 3. Section 4 is dedicated to transductions. Residually ultimately periodic sequences are studied in Section 5 and the properties of differential sequences are analyzed in Section 6. Section 7 is devoted to residually representable transductions. Our main results are presented in Section 8. Further properties of residually ultimately periodic sequences are discussed in Section 9. The paper ends with a short conclusion.

2   Preliminaries and Background

2.1   Rational and Recognizable Sets

Given a multiplicative monoid M, the subsets of M form a semiring $\mathcal{P}(M)$ under union as addition and subset multiplication defined by $XY = \{xy \mid x \in X$ and $y \in Y\}$. Throughout this paper, we shall use the following convenient notation. If X is a subset of M, and K is a subset of $\mathbb{N}$, we set $X^K = \bigcup_{n \in K} X^n$.

Recall that the *rational* subsets of a monoid M form the smallest subset $\mathcal{R}$ of $\mathcal{P}(M)$ containing the finite subsets of M and closed under finite union, product, and star (where X^* is the submonoid generated by X). The set of rational subsets of M is denoted by $Rat(M)$. It is a subsemiring of $\mathcal{P}(M)$.

Recall that a subset P of a monoid M is *recognizable* if there exists a finite monoid F and a monoid morphism $\varphi : M \to F$ such that $P = \varphi^{-1}(\varphi(P))$. By Kleene's theorem, a subset of a finitely generated free monoid is recognizable if and only if it is rational. Various characterizations of the recognizable subsets of $\mathbb{N}$ are given in Proposition 1 below, but we need first to introduce some definitions.

A sequence $(s_n)_{n \geq 0}$ of elements of a set is *ultimately periodic* (u.p.) if there exist two integers $m \geq 0$ and $r > 0$ such that, for each $n \geq m$, $s_n = s_{n+r}$.

The *(first) differential sequence* of an integer sequence $(s_n)_{n \geq 0}$ is the sequence ∂s defined by $(\partial s)_n = s_{n+1} - s_n$. Note that the integration formula $s_n = s_0 + \sum_{0 \leq i \leq n-1} (\partial s)_i$ allows one to recover the original sequence from its differential and s_0. A sequence is *syndetic* if its differential sequence is bounded.

If S is an infinite subset of $\mathbb{N}$, the *enumerating sequence* of S is the unique strictly increasing sequence $(s_n)_{n \geq 0}$ such that $S = \{s_n \mid n \geq 0\}$. The differential sequence of this sequence is simply called the *differential sequence* of S. A set is *syndetic* if its enumerating sequence is syndetic.

The *characteristic sequence* of a subset S of $\mathbb{N}$ is the sequence c_n equal to 1 if $n \in S$ and to 0 otherwise. The following elementary result is folklore.

Proposition 1. *Let S be a set of non-negative integers. The following conditions are equivalent:*

(1) *S is recognizable,*

(2) *S is a finite union of arithmetic progressions,*

(3) *the characteristic sequence of S is ultimately periodic.*

If S is infinite, these conditions are also equivalent to the following conditions

(4) *the differential sequence of S is ultimately periodic.*

Example 1. Let $S = \{1, 3, 4, 9, 11\} \cup \{7 + 5n \mid n \geq 0\} \cup \{8 + 5n \mid n \geq 0\} = \{1, 3, 4, 7, 8, 9, 11, 12, 13, 17, 18, 22, 23, 27, 28, \dots\}$. Its characteristic sequence

$$0, 1, 0, 1, 1, 0, 0, 1, 1, 1, 0, 1, 1, 1, 0, 0, 0, 1, 1, 0, 0, 0, 1, 1, 0, 0, 0, 1, 1, \dots$$

and its differential sequence $2, 1, 3, 1, 1, 2, 1, 1, 4, 1, 4, 1, 4, \dots$ are ultimately periodic.

2.2 Relations

Given two sets E and F, a relation on E and F is a subset of $E \times F$. The *inverse* of a relation S on E and F is the relation S^{-1} on $F \times E$ defined by $(y, x) \in S^{-1}$ if and only if $(x, y) \in S$. A relation S on E and F can also be considered as a function from E into $\mathcal{P}(F)$, the set of subsets of F, by setting, for each $x \in E$, $S(x) = \{y \in F \mid (x, y) \in S\}$. It can also be viewed as a function from $\mathcal{P}(E)$ into $\mathcal{P}(F)$ by setting, for each subset X of E:

$$S(X) = \bigcup_{x \in X} S(x) = \{y \in F \mid \text{ there exists } x \in X \text{ such that } (x, y) \in S\}$$

Dually, S^{-1} can be viewed as a function from $\mathcal{P}(F)$ into $\mathcal{P}(E)$ defined, for each subset Y of F, by $S^{-1}(Y) = \{x \in E \mid S(x) \cap Y \neq \emptyset\}$. When this "dynamical" point of view is adopted, we say that S is a relation from E into F and we use the notation $S : E \rightarrow F$.

A relation $S : \mathbb{N} \rightarrow \mathbb{N}$ is *recognizability preserving* if, for each recognizable subset R of $\mathbb{N}$, the set $S^{-1}(R)$ is recognizable.

3 Filtering Languages

A *filter* is a finite or infinite increasing sequence s of non-negative integers. If $u = a_0 a_1 a_2 \cdots$ is an infinite word (the a_i are letters), we set $u[s] = a_{s_0} a_{s_1} \cdots$. Similarly, if $u = a_0 a_1 a_2 \cdots a_n$ is a finite word, we set $u[s] = a_{s_0} a_{s_1} \cdots a_{s_k}$, where k is the largest integer such that $s_k \leq n < s_{k+1}$. Thus, for instance, if s is the sequence of squares, $abracadabra[s] = abcr$.

By extension, if L is a language (resp. a set of infinite words), we set

$$L[s] = \{u[s] \mid u \in L\}$$

If s is the enumerative sequence of a subset S of $\mathbb{N}$, we also use the notation $L[S]$. If, for every recognizable language L, the set $L[s]$ is recognizable, we say that the filter S *preserves recognizability*. The *filtering problem* is to characterize the recognizability preserving filters.

4   Transductions

In this paper, we consider transductions that are relations from a free monoid A^* into a monoid M. Transductions were intensively studied in connection with context-free languages [1].

Some transductions can be realized by a non-deterministic automaton with output in $\mathcal{P}(M)$, called transducer. More precisely, a *transducer* is a 6-tuple $\mathcal{T} = (Q, A, M, I, F, E)$ where Q is a finite *set of states*, A is the *input alphabet*, M is the *output monoid*, $I = (I_q)_{q \in Q}$ and $F = (F_q)_{q \in Q}$ are arrays of elements of $\mathcal{P}(M)$, called respectively the *initial* and *final outputs*. The set of transitions E is a finite subset of $Q \times A \times \mathcal{P}(M) \times Q$. Intuitively, a transition (p, a, R, q) is interpreted as follows: if a is an input letter, the automaton moves from state p to state q and produces the output R.

A path is a sequence of consecutive transitions:

$$q_0 \xrightarrow{a_1|R_1} q_1 \xrightarrow{a_2|R_2} q_2 \cdots q_{n-1} \xrightarrow{a_n|R_n} q_n$$

The (input) *label* of the path is the word $a_1 a_2 \cdots a_n$. Its *output* is the set $I_{q_0} R_1 R_2 \cdots R_n F_{q_n}$. The transduction realized by $\mathcal{T}$ maps each word u of A^* onto the union of the outputs of all paths of input label u.

A transduction $\tau : A^* \to M$ is said to be *rational* if τ is a rational subset of the monoid $A^* \times M$. By the Kleene-Schützenberger theorem [1], a transduction $\tau : A^* \to M$ is rational if and only if it can be realized by a *rational transducer*, that is, a transducer with outputs in $Rat(M)$.

A transduction $\tau : A^* \to M$ is said to *preserve recognizability*, if, for each recognizable subset P of M, $\tau^{-1}(P)$ is a recognizable subset of A^*. It is well known that rational transductions preserve recognizability, but this property is also shared by the larger class of *representable transductions*, introduced in [9, 10].

Two types of transduction will play an important role in this paper, the *removal transductions* and the *filtering transductions*. Given a subset S of $\mathbb{N}^2$, considered as a relation on $\mathbb{N}$, the *removal transduction* of S is the transduction $\sigma_S : A^* \to A^*$ defined by $\sigma_S(u) = \bigcup_{(|u|,n) \in S} uA^n$. The *filtering transduction* of a filter s is the transduction $\tau_s : A^* \to A^*$ defined by $\tau_s(a_0 a_1 \cdots a_n) = A^{s_0} a_0 A^{s_1} a_1 \cdots A^{s_n} a_n A^{\{0,1,\ldots,s_{n+1}\}}$.

The main idea of [9,10] is to write an n-ary operator Ω on languages as the inverse of some transduction $\tau : A^* \to A^* \times \cdots \times A^*$, that is, $\Omega(L_1, \ldots, L_n) = \tau^{-1}(L_1 \times \cdots \times L_n)$. If the transduction τ turns out to be representable, the results of [9,10] give an explicit construction of a monoid recognizing $\Omega(L_1, \ldots, L_n)$, given monoids recognizing $L_1, \ldots, L_n$, respectively.

In our case, we claim that $P(S, L) = \sigma_S^{-1}(L)$ and $L[s] = \tau_{\partial s-1}^{-1}(L)$. Indeed, we have on the one hand

$$\sigma_S^{-1}(L) = \{u \in A^* \mid \left(\bigcup_{(|u|,n) \in S} uA^n \right) \cap L \neq \emptyset\}$$

$$= \{u \in A^* \mid \text{there exists } v \in A^* \text{ such that } (|u|, |v|) \in S \text{ and } uv \in L\}$$

$$= P(S, L)$$

and on the other hand

$$\tau_{\partial s-1}^{-1}(L) = \{a_0 a_1 \cdots a_n \in A^* \mid$$

$$A^{s_0-1} a_0 A^{s_1-s_0-1} a_1 \cdots A^{s_n-s_{n-1}-1} a_n A^{\{0,1,\ldots,s_{n+1}-s_n-1\}} \cap L \neq \emptyset\}$$

$$= L[s]$$

Unfortunately, the removal transductions and the filtering transductions are not in general representable. We shall see in Section 7 how to overcome this difficulty. But we first need to introduce our second major tool, the residually ultimately periodic sequences.

5   Residually Ultimately Periodic Sequences

Let M be a monoid. A sequence $(s_n)_{n \geq 0}$ of elements of M is *residually ultimately periodic* (r.u.p.) if, for each monoid morphism φ from M into a finite monoid F, the sequence $\varphi(s_n)$ is ultimately periodic.

We are mainly interested in the case where M is the additive monoid $\mathbb{N}$ of non-negative integers. The following connexion with recognizability preserving sequences was established in [5,7,12,16].

Proposition 2. *A sequence $(s_n)_{n \geq 0}$ of non-negative integers is residually ultimately periodic if and only if the function $n \to s_n$ preserves recognizability.*

For each non-negative integer t, define the *congruence threshold t* by setting:

$$x \equiv y \quad (\text{thr } t) \quad \text{if and only if } x = y < t \text{ or } x \geq t \text{ and } y \geq t.$$

Thus threshold counting can be viewed as a formalisation of children counting: zero, one, two, three, ... , many.

A function $s : \mathbb{N} \to \mathbb{N}$ is said to be *ultimately periodic modulo p* if, for each monoid morphism $\varphi : \mathbb{N} \to \mathbb{Z}/p\mathbb{Z}$, the sequence $u_n = \varphi(s(n))$ is ultimately periodic. It is equivalent to state that there exist two integers $m \geq 0$ and $r > 0$ such that, for each $n \geq m$, $u_n \equiv u_{n+r} \pmod{p}$. A sequence is said to be *cyclically ultimately periodic* (c.u.p.) if it is ultimately periodic modulo p for every $p > 0$. These functions are called "ultimately periodic reducible" in [12,13].

Similarly, function $s : \mathbb{N} \to \mathbb{N}$ is said to be *ultimately periodic threshold t* if, for each monoid morphism $\varphi : \mathbb{N} \to \mathbb{N}_{t,1}$, the sequence $u_n = \varphi(s(n))$ is ultimately periodic. It is equivalent to state that there exist two integers $m \geq 0$ and $r > 0$ such that, for each $n \geq m$, $u_n \equiv u_{n+r}$ (thr t).

Proposition 3. *A sequence of non-negative integers is residually ultimately periodic if and only if it is ultimately periodic modulo p for all p > 0 and ultimately periodic threshold t for all t ≥ 0.*

The next proposition gives a very simple criterion to generate sequences that are ultimately periodic threshold t for all t.

Proposition 4. *A sequence $(u_n)_{n \geq 0}$ of integers such that $\lim_{n \to \infty} u_n = +\infty$ is ultimately periodic threshold t for all t ≥ 0.*

Example 2. The sequence $n!$ is residually ultimately periodic. Indeed, let p be a positive integer. Then for each $n \geq p$, $n! \equiv 0 \bmod p$ and thus $n!$ is ultimately periodic modulo p. Furthermore, Proposition 4 shows that, for each $t \geq 0$, $n!$ is ultimately periodic threshold t.

The class of cyclically ultimately periodic functions has been thoroughly studied by Siefkes [13], who gave in particular a recursion scheme for producing such functions. Residually ultimately periodic sequences have been studied in [3, 5,7,12,15,16]. Their properties are summarized in the next proposition.

Theorem 1. *[16,3] Let $(u_n)_{n \geq 0}$ and $(v_n)_{n \geq 0}$ be r.u.p. sequences. Then the following sequences are also r.u.p.:*

(1) *(composition)* u_{v_n},

(2) *(sum)* $u_n + v_n$,

(3) *(product)* $u_n v_n$,

(4) *(difference)* $u_n - v_n$ *provided that* $u_n \geq v_n$ *and* $\lim_{n \to \infty} (u_n - v_n) = +\infty$,

(5) *(exponentation)* $u_n^{v_n}$,

(6) *(generalized sum)* $\sum_{0 \leq i \leq v_n} u_i$,

(7) *(generalized product)* $\prod_{0 \leq i \leq v_n} u_i$.

In particular, the sequences n^k and k^n (for a fixed k), are residually ultimately periodic. However, r.u.p. sequences are not closed under quotients. For instance, let u_n be the sequence equal to 1 if n is prime and to $n! + 1$ otherwise. Then $n u_n$ is r.u.p. but u_n is not r.u.p.. This answers a question left open in [15].

The sequence $2^{2^{2^{\cdots^2}}}$ (exponential stack of 2's of height n), considered in [12], is also a r.u.p. sequence, according to the following result.

Proposition 5. *Let k be a positive integer. Then the sequence u_n defined by $u_0 = 1$ and $u_{n+1} = k^{u_n}$ is r.u.p.*

The existence of non-recursive, r.u.p. sequences was established in [12]: if $\varphi : \mathbb{N} \to \mathbb{N}$ is a strictly increasing, non-recursive function, then the sequence $u_n = n! \varphi(n)$ is non-recursive but is residually ultimately periodic. The proof is similar to that of Example 2.

6   Differential Sequences

An integer sequence is called *differentially residually ultimately periodic* (d.r.u.p. in abbreviated form), if its differential sequence is residually ultimately periodic.

What are the connections between d.r.u.p. sequences and r.u.p. sequences? First, the following result holds:

Proposition 6. [3, Corollary 28] *Every d.r.u.p. sequence is r.u.p.*

However, the two notions are not equivalent: for instance, it was shown in [3] that if b_n is a non-ultimately periodic sequence of 0 and 1, the sequence $u_n = (\sum_{0 \le i < n} b_i)!$ is r.u.p. but is not d.r.u.p. It suffices to observe that $(\partial u)_n \equiv b_n$ threshold 1.

Note that, if only cyclic counting were used, it would make no difference:

Proposition 7. *Let p be a positive number. A sequence is ultimately periodic modulo p if and only if its differential sequence is ultimately periodic modulo p.*

There is a special case for which the notions of r.u.p. and d.r.u.p. sequences are equivalent. Indeed, if the differential sequence is bounded, Proposition 1 can be completed as follows.

Lemma 1. *If a syndetic sequence is residually ultimately periodic, then its differential sequence is ultimately periodic.*

Putting everything together, we obtain

Proposition 8. *Let s be a syndetic sequence of non-negative integers. The following conditions are equivalent:*

(1) *s is residually ultimately periodic,*

(2) *∂s is residually ultimately periodic,*

(3) *∂s is ultimately periodic.*

Proof. Proposition 6 shows that (2) implies (1). Furthermore (3) implies (2) is trivial. Finally, Lemma 1 shows that (1) implies (3).

Proposition 9. *Let S be an infinite syndetic subset of $\mathbb{N}$. The following conditions are equivalent:*

(1) *S is recognizable,*

(2) *the enumerating sequence of S is residually ultimately periodic,*

(3) *the differential sequence of S is residually ultimately periodic,*

(4) *the differential sequence of S is ultimately periodic.*

Proof. The last three conditions are equivalent by Proposition 8 and the equivalence of (1) and (4) follows from Proposition 1.

The class of d.r.u.p. sequences was thoroughly studied in [3].

Theorem 2. [3, Theorem 22] *Differential residually ultimately periodic sequences are closed under sum, product, exponentiation, generalized sum and generalized product. Furthermore, given two d.r.u.p. sequences $(u_n)_{n \ge 0}$ and $(v_n)_{n \ge 0}$ such that $u_n \ge v_n$ and $\lim_{n \to \infty} (\partial u)_n - (\partial v)_n = +\infty$, the sequence $u_n - v_n$ is d.r.u.p.*

7   Residually Representable Transductions

Let M be a monoid. A transduction $\tau : A^* \to M$ is *residually rational* (resp. *residually representable*) if, for every monoid morphism α from M into a finite monoid N, the transduction $\alpha \circ \tau : A^* \to N$ is rational (resp. representable).

Since a rational transduction is (linearly) representable, every residually rational transduction is residually representable. Furthermore, every representable transduction is residually representable. We now show that the removal transductions and the filtering transductions are residually rational. We first consider the removal transductions.

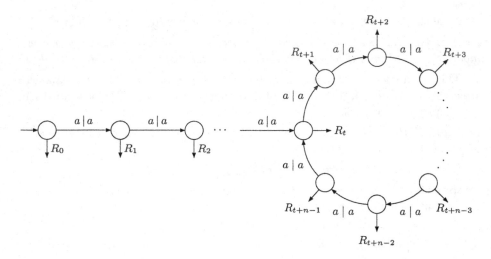

Fig. 1. A transducer realizing β.

Proposition 10. *Let S be a recognizability preserving relation on $\mathbb{N}$. The removal transduction of S is residually rational.*

Proof. Let α be a morphism from A^* into a finite monoid N. Let $\beta = \alpha \circ \tau_s$ and $R = \alpha(A)$. Since the monoid $\mathcal{P}(N)$ is finite, the sequence $(R^n)_{n \geq 0}$ is ultimately periodic. Therefore, there exist two integers $r \geq 0$ and $q > 0$ such that, for all $n \geq r$, $R^n = R^{n+q}$. Consider the following subsets of $\mathbb{N}$: $K_0 = \{0\}$, $K_1 = \{1\}$, $\dots$, $K_{r-1} = \{r - 1\}$, $K_r = \{r, r + q, r + 2q, \dots\}$, $K_{r+1} = \{r + 1, r + q + 1, r + 2q + 1, \dots\}$, $\dots$, $K_{r+q-1} = \{r + q - 1, r + 2q - 1, r + 3q - 1, \dots\}$. The sets K_i, for $i \in \{0, 1, \dots, r + q - 1\}$ are recognizable and since S is recognizability preserving, each set $S^{-1}(K_i)$ is also recognizable. By Proposition 1, there exist two integers $t_i \geq 0$ and $p_i > 0$ such that, for all $n \geq t_i$, $n \in S^{-1}(K_i)$ if and only if $n + p_i \in S^{-1}(K_i)$. Setting $t = \max_{0 \leq i \leq r+q-1} t_i$ and $p = \text{lcm}_{0 \leq i \leq r+q-1} p_i$, we conclude that, for all $n \geq t$ and for $0 \leq i \leq r + q - 1$, $n \in S^{-1}(K_i)$ if and only if $n + p \in S^{-1}(K_i)$, or equivalently

$$S(n) \cap K_i \neq \emptyset \Longleftrightarrow S(n + p) \cap K_i \neq \emptyset$$

It follows that the sequence R_n of $\mathcal{P}(N)$ defined by $R_n = R^{S(n)}$ is ultimately periodic of threshold t and period p, that is, $R_n = R_{n+p}$ for all $n \geq t$. Consequently, the transduction β can be realized by the transducer represented in Figure 1, in which a stands for a generic letter of A. Therefore β is rational and τ_s is residually rational.

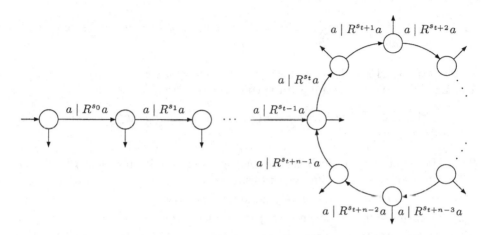

Fig. 2. A transducer realizing γ_s.

Proposition 11. *Let s be a residually ultimately periodic sequence. Then the filtering transduction τ_s is residually rational.*

Proof. Let α be a morphism from A^* into a finite monoid N. Let $\gamma_s = \alpha \circ \tau_s$ and $R = \alpha(A)$. Finally, let $\varphi : \mathbb{N} \to \mathcal{P}(N)$ be the morphism defined by $\varphi(n) = R^n$. Since $\mathcal{P}(N)$ is finite and s_n is residually ultimately periodic, the sequence $\varphi(s_n) = A^{s_n}$ is ultimately periodic. Therefore, there exist two integers $t \geq 0$ and $p > 0$ such that, for all $n \geq t$, $R^{s_{n+p}} = R^{s_n}$. It follows that the transduction γ_s can be realized by the transducer represented in Figure 2, in which a stands for a generic letter of A. Therefore γ_s is rational and thus τ_s is residually rational.

The fact that the two previous transducers preserve recognizability is now a direct consequence of the following general statement.

Theorem 3. *Let M be a monoid. Any residually rational transduction $\tau : A^* \to M$ preserves recognizability.*

Proof. Let P be a recognizable subset of M and let $\alpha : M \to N$ be a morphism recognizing P, where N is a finite monoid. By definition, $\alpha^{-1}(\alpha(P)) = P$. Since τ is residually rational, the transduction $\alpha \circ \tau : A^* \to N$ is rational. Since N is finite, every subset of N is recognizable. In particular, $\alpha(P)$ is recognizable and since τ preserves recognizability, $(\alpha \circ \tau)^{-1} \alpha(P)$ is recognizable. The theorem follows, since $(\alpha \circ \tau)^{-1} \alpha(P) = \tau^{-1}(\alpha^{-1}(\alpha(P))) = \tau^{-1}(P)$.

8 Main Results

The aim of this section is to provide a unified solution for the filtering problem and the removal problem.

8.1 The Filtering Problem

Theorem 4. *A filter preserves recognizability if and only if it is differentially residually ultimately periodic.*

Proposition 11 and Theorem 3 show that if a filter is d.r.u.p., then it preserves recognizability. We now establish the converse property.

Proposition 12. *Every recognizability preserving filter is differentially residually ultimately periodic.*

Proof. Let s be a recognizability preserving filter. By Proposition 3 and 7, it suffices to prove the following properties:

(1) for each $p > 0$, s is ultimately periodic modulo p,

(2) for each $t \geq 0$, ∂s is ultimately periodic threshold t.

(1) Let p be a positive integer and let $A = \{0, 1, ...(p-1)\}$. Let $u = a_0 a_1 \cdots$ be the infinite word whose i-th letter a_i is equal to s_i modulo p. At this stage, we shall need two elementary properties of ω-rational sets. The first one states that an infinite word u is ultimately periodic if and only if the ω-language $\{u\}$ is ω-rational. The second one states that, if L is a recognizable language of A^*, then $\overrightarrow{L}$ (the set of infinite words having infinitely many prefixes in L) is ω-rational.

We claim that u is ultimately periodic. Define L as the set of prefixes of the infinite word $(0123 \cdots (p-1))^\omega$. Then $L[s]$ is the set of prefixes of u. Since L is recognizable, $L[s]$ is recognizable, and thus the set $\overrightarrow{L[s]}$ is ω-rational. But this set reduces to $\{u\}$, which proves the claim. Therefore, the sequence $(s_n)_{n \geq 0}$ is ultimately periodic modulo p.

(2) The proof is quite similar to that of (1), but is sligthly more technical. Let t be a non-negative integer and let $B = \{0, 1, \dots, t\} \cup \{a\}$, where a is a special symbol. Let $d = d_0 d_1 \cdots$ be the infinite word whose i-th letter d_i is equal to $s_{i+1} - s_i - 1$ threshold t. Let us prove that d is ultimately periodic. Consider the recognizable prefix code $P = \{0, 1a, 2a^2, 3a^3, \dots, ta^t, a\}$. Then $P^*[s]$ is recognizable, and so is the language $R = P^*[s] \cap \{0, 1, \dots, t\}^*$. We claim that, for each $n > 0$, the word $p_n = d_0 d_1 \cdots d_{n-1}$ is the maximal word of R of length n in the lexicographic order induced by the natural order $0 < 1 < \dots < t$. First, $p_n = u[s]$, where $u = a^{s_0} d_0 a^{s_1 - s_0 - 1} d_1 \cdots d_{n-1} a^{s_n - s_{n-1} - 1}$ and thus $p_n \in R$. Next, let $p'_n = d'_0 d'_1 \cdots d'_{n-1}$ be another word of R of length n. Then $p'_n = u'[s]$ for some word $u' \in P^*$. Suppose that p'_n comes after p_n in the lexicographic order. We may assume that, for some index $i \leq n - 1$, $d_0 = d'_0$, $d_1 = d'_1$, $\dots$, $d_{i-1} = d'_{i-1}$ and $d_i < d'_i$. Since $u' \in P^*$, the letter d'_i, which occurs in position s_i in u', is followed by at least d'_i letters a. Now $d'_i > d_i$, whence $d_i < t$ and $d_i = s_{i+1} - s_i - 1$. It

follows in particular that in u', the letter in position s_{i+1} is an a, a contradiction, since $u'[s]$ contains no occurrence of a. This proves the claim.

Let now $\mathcal{A}$ be a finite deterministic trim automaton recognizing R. It follows from the claim that in order to read d in $\mathcal{A}$, starting from the initial state, it suffices to choose, in each state q, the unique transition with maximal label in the lexicographic order. It follows at once that d is ultimately periodic. Therefore, the sequence $(\partial s) - 1$ is ultimately periodic threshold t, and so is (∂s).

8.2   The Removal Problem

The solution of the removal problem was given in [12].

Theorem 5. *Let S be a subset of $\mathbb{N}^2$. The following conditions are equivalent:*

  (1) *for each recognizable language L, the language $P(S, L)$ is recognizable,*

  (2) *S is a recognizability preserving relation*

The most difficult part of the proof, (2) implies (1), follows immediately from Proposition 10 and Theorem 3.

9   Further Properties of d.r.u.p. Sequences

Coming back to the filtering problem, the question arises to characterize the filters S such that, for every recognizable language L, both $L[S]$ and $L[\mathbb{N} \setminus S]$ are recognizable. By Theorem 4, the sequences defined by S and its complement should be d.r.u.p. This implies that S is recognizable, according to the following slightly more general result.

Proposition 13. *Let S and S' be two infinite subsets of $\mathbb{N}$ such that $S \cup S'$ and $S \cap S'$ are recognizable. If the enumerating sequence of S is d.r.u.p. and if the enumerating sequence of S' is r.u.p., then S and S' are recognizable.*

One can show that the conclusion of Proposition 13 no longer holds if S' is only assumed to be residually ultimately periodic.

10   Conclusion

Our solution to the filtering problem was based on the fact that any residually rational transduction preserves recognizability. There are several advantages to our approach.

First, it gives a unified solution to apparently disconnected problems, like the filtering problem and the removal problem. Actually, most of – if not all – the automata-theoretic puzzles proposed in [4,5,6,7,9,10,11,12,14] and [15, Section 5.2], can be solved by using the strongest fact that any residually representable transduction preserves recognizability.

Next, refining the approach of [9,10], if $\tau : A^* \to A^* \times \cdots \times A^*$ is a residually representable transduction, one could give an explicit construction of a monoid recognizing $\tau^{-1}(L_1 \times \cdots \times L_n)$, given monoids recognizing $L_1, \ldots, L_n$, respectively (the details will be given in the full version of this paper). This information can be used, in turn, to see whether a given operation on languages preserves star-free languages, or other standard classes of rational languages.

Acknowledgements. Special thanks to Michèle Guerlain for her careful reading of a first version of this paper and to the anonymous referees for their suggestions.

References

1. J. Berstel, *Transductions and context-free languages*, Teubner, Stuttgart, (1979).
2. O. Carton and W. Thomas, The monadic theory of morphic infinite words and generalizations, in *MFCS 2000, Lecture Notes in Computer Science* **1893**, M. Nielsen and B. Rovan, eds, (2000), 275–284.
3. O. Carton and W. Thomas, The monadic theory of morphic infinite words and generalizations, Inform. Comput. **176**, (2002), 51–76.
4. S. R. Kosaraju, Finite state automata with markers, in *Proc. Fourth Annual Princeton Conference on Information Sciences and Systems*, Princeton, N. J. (1970), 380.
5. S. R. Kosaraju, Regularity preserving functions, *SIGACT News* **6 (2)**, (1974), 16-17. Correction to "Regularity preserving functions", *SIGACT News* **6 (3)**, (1974), 22.
6. S. R. Kosaraju, Context-free preserving functions, *Math. Systems Theory* **9**, (1975), 193–197.
7. D. Kozen, On regularity-preserving functions, Bull. Europ. Assoc. Theor. Comput. Sc. **58** (1996), 131–138. Erratum: On Regularity-Preserving Functions, Bull. Europ. Assoc. Theor. Comput. Sc. **59** (1996), 455.
8. A. B. Matos, Regularity-preserving letter selections, DCC-FCUP Internal Report.
9. J.-É. Pin and J. Sakarovitch, Operations and transductions that preserve rationality, in *6th GI Conference, Lecture Notes in Computer Science* **145**, Springer Verlag, Berlin, (1983), 617–628.
10. J.-É. Pin and J. Sakarovitch, Une application de la représentation matricielle des transductions, *Theoret. Comp. Sci.* **35** (1985), 271–293.
11. J. I. Seiferas, A note on prefixes of regular languages, *SIGACT News* **6**, (1974), 25–29.
12. J. I. Seiferas and R. McNaughton, Regularity-preserving relations, *Theoret. Comp. Sci.* **2**, (1976), 147–154.
13. D. Siefkes, Decidable extensions of monadic second order successor arithmetic, in: *Automatentheorie und formale Sprachen*, (Mannheim, 1970), J. Dörr and G. Hotz, Eds, B.I. Hochschultaschenbücher, 441–472.
14. R. E. Stearns and J. Hartmanis, Regularity preserving modifications of regular expressions, *Information and Control* **6**, (1963), 55–69.
15. Guo-Qiang Zhang, Automata, Boolean matrices, and ultimate periodicity, *Information and Computation*, **152**, (1999), 138–154.
16. Guo-Qiang Zhang, Periodic functions for finite semigroups, preprint.

Languages Defined by Generalized Equality Sets

Vesa Halava[1], Tero Harju[1], Hendrik Jan Hoogeboom[2], and Michel Latteux[3]

[1] Department of Mathematics and TUCS – Turku Centre for Computer Science,
University of Turku, FIN-20014, Turku, Finland
{vehalava,harju}@utu.fi
[2] Dept. of Comp. Science, Leiden University
P.O. Box 9512, 2300 RA Leiden, The Netherlands
hoogeboom@liacs.nl
[3] Université des Sciences et Technologies de Lille, Bâtiment M3,
59655 Villeneuve d'Ascq Cédex, France
latteux@lifl.fr

Abstract. We consider the generalized equality sets which are of the form $E_G(a, g_1, g_2) = \{w \mid g_1(w) = ag_2(w)\}$, determined by instances of the generalized Post Correspondence Problem, where the morphisms g_1 and g_2 are nonerasing and a is a letter. We are interested in the family consisting of the languages $h(E_G(J))$, where h is a coding and J is a shifted equality set of the above form. We prove several closure properties for this family.

1 Introduction

In formal language theory, languages are often determined by their generating grammars or accepting machines. It is also customary to say that languages generated by grammars of certain form or accepted by automata of specific type form a language family. Here we shall study a language family defined by simple generalized equality sets of the form $E_G(J)$, where $J = (a, g_1, g_2)$ is an instance of the *shifted Post Correspondence Problem* consisting of a letter a and two morphisms g_1 and g_2. Then the set $E_G(J)$ consists of the words w that satisfy $g_1(w) = ag_2(w)$.

Our motivation for these generalized equality sets comes partly from a result of [2], where it was proved that the family of regular valence languages is equal to the family of languages of the form $h(E_G(J))$, where h is a coding (i.e., a letter-to-letter morphism), and, moreover, in the instance $J = (a, g_1, g_2)$ the morphism g_2 is periodic. Here we shall consider general case where we do not assume g_2 to be periodic, but both morphisms to be nonerasing. We study the properties of this family $\mathcal{CE}$ of languages by studying its closure properties. In particular, we show that $\mathcal{CE}$ is closed under union, product, Kleene plus, intersection with regular sets. Also, more surprisingly, $\mathcal{CE}$ is closed under nonerasing morphisms inverse morphisms.

A. Lingas and B.J. Nilsson (Eds.): FCT 2003, LNCS 2751, pp. 355–363, 2003.
© Springer-Verlag Berlin Heidelberg 2003

2 Preliminaries

Let A be an alphabet, and denote by A^* the monoid of all finite words under the operation of catenation. Note that the *empty word*, denoted by ε, is in the monoid A^*. The semigroup $A^* \setminus \{\varepsilon\}$ generated by A is denoted by A^+.

For two words $u, v \in A^*$, u is a *prefix* of v if there exists a word $z \in A^*$ such that $v = uz$. This is denoted by $u \le v$. If $v = uz$, then we also write $u = vz^{-1}$ and $z = u^{-1}v$.

In the following, let A and B be alphabets and $g \colon A^* \to B^*$ a mapping. For a word $x \in B^*$, we denote by $g^{-1}(x) = \{w \in A^* \mid g(w) = x\}$ the inverse image of x under g. Then $g^{-1}(K) = \cup_{x \in K} g^{-1}(x)$ is the *inverse image* of $K \subseteq B^*$ under g, and $g(L) = \{g(w) \mid w \in L\}$ is the *image* of $L \subseteq A^*$ under g. Also, g is a *morphism* if $g(uv) = g(u)g(v)$ for all $u, v \in A^*$. A morphism g is a *coding*, if it maps letters to letters, that is, if $g(A) \subseteq B$. A morphism g is said to be *periodic*, if there exists a word $w \in B^*$ such that $g(A^*) \subseteq w^*$.

In the following section, for an alphabet A, the alphabet $\bar{A} = \{\bar{a} \mid a \in A\}$ is a *copy* of A, if $A \cap \bar{A} = \emptyset$.

In the *Post Correspondence Problem*, PCP for short, we are given two morphisms $g_1, g_2 \colon A^* \to B^*$ and it is asked whether or not there exists a nonempty word $w \in A^+$ such that $g_1(w) = g_2(w)$. Here the pair (g_1, g_2) is an *instance* of the PCP, and the word w is called a *solution*. As a general reference to the problems and results concerning the Post Correspondence Problem, we give [3].

For an instance $I = (g_1, g_2)$ of the PCP, let

$$E(I) = \{w \in A^* \mid g_1(w) = g_2(w)\}$$

be its *equality set*. It is easy to show that an equality set $E = E(g_1, g_2)$ is always a monoid, that is, $E = E^*$. In fact, it is a free monoid, and thus the algebraic structure of E is relatively simple, although the problem whether or not E is trivial is undecidable.

We shall now consider special instances of the *generalized* Post Correspondence Problem in order to have slightly more structured equality sets. In the *shifted Post Correspondence Problem*, or *shifted PCP* for short, we are given two morphisms $g_1, g_2 \colon A^* \to B^*$ and a letter $a \in B$, and it is asked whether there exists a word $w \in A^*$ such that

$$g_1(w) = ag_2(w). \tag{2.1}$$

The triple $J = (a, g_1, g_2)$ is called an *instance* of the shifted PCP and a word w satisfying equation (2.1) is called a *solution* of J. It is clear that a solution w is always nonempty. We let

$$E_G(J) = \{w \in A^+ \mid g_1(w) = ag_2(w)\}$$

be the *generalized equality set* of J.

We shall denote by $\mathcal{CE}$ the set of all languages $h(E_G(J))$, where h is a coding, and the morphisms in the instances J of the shifted PCP are both nonerasing.

In [2] $\mathcal{CE}_{\mathrm{per}}$ was defined as the family of languages $h(E_G(J))$, where h is a coding, and one of the morphisms in the instance J of the shifted PCP was assumed to be periodic. It was proved in [2] that $\mathcal{CE}_{\mathrm{per}}$ is equal to the family of languages defined by the regular valence grammars (see [6]). It is easy to see that the morphisms in the instances could have been assumed to be nonerasing in order to get the same result. Therefore, the family $\mathcal{CE}$ studied in this paper is a generalization of $\mathcal{CE}_{\mathrm{per}}$ or, actually, $\mathcal{CE}_{\mathrm{per}}$ is a subfamily of $\mathcal{CE}$.

3   Closure Properties of $\mathcal{CE}$

The closure properties of the family $\mathcal{CE}_{\mathrm{per}}$ follow from the known closure properties of regular valence languages. In this section, we study the closure properties of the more general family $\mathcal{CE}$ under various operations.

Before we start our journey through the closure results, we make first some assumptions of the instances of the shifted PCP defining the languages at hand.

First of all, we may always assume that in an instance $J = (a, g_1, g_2)$ of the shifted PCP the shift letter a is a special symbol that satisfies:

The shift letter a can appear only as the first letter in the images of g_1 and it does not occur at all in the images of g_2.

To see this, consider any language $L = h(E_G(a, g_1, g_2))$, where $g_1, g_2 \colon A^* \to B^*$ and $h \colon A^* \to C^*$. Let $\#$ be a new letter not in $A \cup B$. Construct a new instance $(\#, g_1', g_2')$, where $g_1', g_2' \colon (A \cup \bar{A})^* \to (B \cup \{\#\})^*$ and $\bar{A}$ is a copy of A, by setting for all $x \in A$ $g_2'(x) = g_2'(\bar{x}) = g_2(x)$, and $g_1'(x) = g_1(x)$ and

$$g_1'(\bar{x}) = \begin{cases} g_1(x), & \text{if } a \not\leq g_1(x), \\ \#w, & \text{if } g_1(x) = aw. \end{cases}$$

Define a new coding $h' \colon (A \cup \bar{A})^* \to C^*$ by $h'(x) = h'(\bar{x}) = h(x)$ for all $x \in A$. It is now obvious that $L = h'(E_G(\#, g_1', g_2'))$.

We shall call such an instance $(\#, g_1', g_2')$ *shift-fixed*, where the shift letter $\#$ is used only as the first letter.

The next lemma shows that we may also assume that the instance (g_1, g_2) does not have any nontrivial solutions, that is, $E(g_1, g_2) = \{\varepsilon\}$ for all instance $J = (a, g_1, g_2)$ defining the language $h(E_G(J))$.

For this result we introduce two mappings which are used for desynchronizing a pair of morphisms. Let d be a new letter. For a word $u = a_1 a_2 \cdots a_n$, where each a_i is a letter, define

$$\ell_d(u) = da_1 da_2 d \cdots da_n \quad \text{and} \quad r_d(u) = a_1 da_2 d \cdots da_n d.$$

In other words ℓ_d is a morphism that adds d in front of every letter and r_d is a morphism that adds d after every letter of a word.

Lemma 1 *For every instance J of the shifted PCP and coding h, there exists an instance $J' = (a, g_1', g_2')$ and a coding h' such that $h(E_G(J)) = h'(E_G(J'))$ and $E(g_1', g_2') = \{\varepsilon\}$.*

Proof. Let $J = (a, g_1, g_2)$ be a shift-fixed instance of the shifted PCP where $g_1, g_2 \colon A^* \to B^*$, and let $h \colon A^* \to C^*$ be a coding. We define new morphisms $g_1', g_2' \colon (A \cup \bar{A})^* \to (B \cup \{d\})^*$, where $d \notin B$ is a new letter and $\bar{A}$ is a copy of A, as follows. For all $x \in A$,

$$g_2'(x) = \ell_d(g_2(x)) \quad \text{and} \quad g_2'(\bar{x}) = \ell_d(g_2(x))d \tag{3.1}$$

$$g_1'(x) = g_1'(\bar{x}) = \begin{cases} \#d \cdot r_d(w), & \text{if } g_1(x) = \#w, \\ r_d(g_1(x)), & \text{if } a \not\leq g_1(x). \end{cases} \tag{3.2}$$

Note that the letters in $\bar{A}$ can be used only as the last letter of a solution of (a, g_1', g_2'). Since every image by g_2' begins with letter d and it is not a prefix of any image of g_1', we obtain that $E(g_1', g_2') = \{\varepsilon\}$. On the other hand, (a, g_1', g_2') has a solution $w\bar{x}$ if and only if wx is a solution of (a, g_1, g_2). Therefore, we define $h' \colon (A \cup \bar{A})^* \to C^*$ by $h'(x) = h'(\bar{x}) = h(x)$ for all $x \in A$. The claim of the lemma follows, since obviously $h(E_G(J)) = h'(E_G(J'))$.

We shall call an instance an instance (a, g_1, g_2) *reduced*, if it is shift-fixed and $E(g_1, g_2) = \{\varepsilon\}$.

3.1   Union and Product

Theorem 2 *The family $\mathcal{CE}$ is closed under union.*

Proof. Let $K, L \in \mathcal{CE}$ with $K = h_1(E_G(J_1))$ and $L = h_2(E_G(J_2))$, where $J_1 = (a_1, g_{11}, g_{12})$ and $J_2 = (a_2, g_{21}, g_{22})$ are reduced, and $g_{11}, g_{12} \colon \Sigma^* \to B_1^*$ and $g_{21}, g_{22} \colon \Omega^* \to B_2^*$. Without restriction we can suppose that $\Omega \cap \Sigma = \emptyset$. (Otherwise we take a primed copy of the alphabet Ω that is disjoint from Σ, and define a new instance J_2' by replacing the letter with primed copies.) Assume also that $B_1 \cap B_2 = \emptyset$.

Let $B = B_1 \cup B_2$, and let $\#$ be a new letter. First replace every appearance of the shift letters a_1 and a_2 in J_1 and J_2 with $\#$. Define morphisms $g_1, g_2 \colon (\Sigma \cup \Omega)^* \to B^*$ as follows: for all $x \in \Sigma \cup \Omega$,

$$g_1(x) = \begin{cases} g_{11}(x), & \text{if } x \in \Sigma \\ g_{21}(x), & \text{if } x \in \Omega \end{cases} \quad \text{and} \quad g_2(x) = \begin{cases} g_{12}(x), & \text{if } x \in \Sigma \\ g_{22}(x), & \text{if } x \in \Omega. \end{cases}$$

Define a coding $h \colon (\Sigma \cup \Omega)^* \to C^*$ similarly:

$$h(x) = \begin{cases} h_1(x), & \text{if } x \in \Sigma \\ h_2(x), & \text{if } x \in \Omega. \end{cases} \tag{3.3}$$

Since $\Sigma \cap \Omega = \emptyset$, and J_1 and J_2 are reduced (i.e., $E(g_{11}, g_{12}) = \{\varepsilon\} = E(g_{21}, g_{22})$), we see that the solutions in $E_G(J_1)$ and $E_G(J_2)$ cannot be combined or mixed. Thus, it is straightforward to show that $h(E_G(\#, g_1, g_2)) = K \cup L$.

Next we consider the product KL of languages.

Theorem 3 *The family $\mathcal{CE}$ is closed under product of languages.*

Proof. Let $K, L \in \mathcal{CE}$ with $K = h_1(E_G(J_1))$ and $L = h_2(E_G(J_2))$, where $J_1 = (a_1, g_{11}, g_{12})$ and $J_2 = (a_2, g_{21}, g_{22})$ are shift-fixed. Assume that $g_{11}, g_{12} \colon \Sigma^* \to B_1^*$, and $g_{21}, g_{22} \colon \Omega^* \to B_2^*$, where again we can assume that $\Sigma \cap \Omega = \emptyset$, and similarly that $B_1 \cap B_2 = \emptyset$. We also assume that the length of the images of the morphisms are at least 2 (actually, this is needed only for g_{11}). This can be assumed, for example, by the construction in Lemma 1.

We shall prove that $KL = \{uv \mid u \in K,\ v \in L\}$ is in $\mathcal{CE}$. For this, we define morphisms $g_1, g_2 \colon (\Sigma \cup \Omega)^* \to (B_1 \cup B_2)^*$ in the following way: for each $x \in \Sigma$,

$$
g_1(x) = \begin{cases} \ell_{a_2}(g_{11}(x)), & \text{if } a_1 \not\le g_{11}(x), \\ a_1 y \ell_{a_2}(w), & \text{if } g_{11}(x) = a_1 y w \quad (y \in B_1), \end{cases}
$$

and

$$
g_2(x) = r_{a_2}(g_{12}(x)),
$$

and for each $x \in \Omega$, $g_1(x) = g_{21}(x)$ and $g_2(x) = g_{22}(x)$. If we now define h by combining h_1 and h_2 as in (3.3), we obtain that $h(E_G(a_1, g_1, g_2)) = KL$.

We shall now extend the above result by proving that $\mathcal{CE}$ is closed under Kleene plus, i.e., if $K \in \mathcal{CE}$, then

$$
K^+ = \bigcup_{i \ge 1} K^i \in \mathcal{CE}.
$$

Clearly $\mathcal{CE}$ is not closed under Kleene star, since the empty word does not belong to any language in $\mathcal{CE}$.

Theorem 4 *The family $\mathcal{CE}$ is closed under Kleene plus.*

Proof. Let $K = h(E_G(\#, g_1, g_2))$, where $g_1, g_2 \colon A^* \to B^*$ are nonerasing morphisms, $h \colon A^* \to C^*$ is a coding and the instance $(\#, g_1, g_2)$ is shift-fixed. Also, let $\bar{A}$ be a copy of A, and define $\bar{g}_1, \bar{g}_2 \colon (A \cup \bar{A})^* \to B^*$ in the following way: for each $x \in A$,

$$
\bar{g}_1(x) = g_1(x) \quad \text{and} \quad \bar{g}_2(x) = g_2(x),
$$

$$
\bar{g}_1(\bar{x}) = \begin{cases} \ell_\#(g_1(x)), & \text{if } \# \not\le g_1(x), \\ \ell_\#(w), & \text{if } g_1(x) = \#w, \end{cases}
$$

$$
\bar{g}_2(\bar{x}) = r_\#(g_2(x)).
$$

Extend h also to $\bar{A}$ by setting $h(\bar{x}) = h(x)$ for all $x \in A$.

It is now clear that $h(E_G(\#, \bar{g}_1, \bar{g}_2)) = K^+$, since $\bar{g}_1(w) = \#\bar{g}_2(w)$ if and only if, $w = x_1 \cdots x_n x_{n+1}$, where $x_i \in \bar{A}^+$ for $1 \le i \le n$, $x_{n+1} \in A^+$, $\bar{g}_1(x_i)\# = \#\bar{g}_2(x_i)$ for $1 \le i \le n$ and $\bar{g}_1(x_{n+1}) = \#\bar{g}_2(x_{n+1})$. It is clear that after removing the bars form the letters x_i (by h), we obtain words in $E_G(\#, g_1, g_2)$.

3.2 Intersection with Regular Languages

We show now that $\mathcal{CE}$ is closed under intersections with regular languages. Note that for $\mathcal{CE}_{\mathrm{per}}$ this closure already follows from the closure of $\mathrm{Reg}(\mathbb{Z})$ languages.

Theorem 5 *The family $\mathcal{CE}$ is closed under intersections with regular languages.*

Proof. Let $J = (a, g_1, g_2)$ be an instance of the shifted PCP, $g_1, g_2 \colon \Sigma^* \to B^*$. Let $L = h(E_G(J))$, where $h \colon \Sigma^* \to C^*$ is coding.

We shall prove that $h(E_G(J)) \cap R$ is in $\mathcal{CE}$ for all regular $R \subseteq B^*$. We note first that $h(E_G(J)) \cap R = h(E_G(J) \cap h^{-1}(R))$, and therefore it is sufficient to show that, for all regular languages $R \subseteq \Sigma^*$, $h(E_G(J) \cap R)$ is in $\mathcal{CE}$. Therefore, we shall give a construction for instances J' of the shifted PCP such that $E_G(J') = E_G(J) \cap R$.

Assume $R \subseteq \Sigma^*$ is regular language, and let $G = (N, \Sigma, P, S)$ be a right linear grammar generating R (see [7]). Let $N = \{A_0, \ldots, A_{n-1}\}$, where $S = A_0$, and assume without restriction, that there are no productions having $S = A_0$ on the right hand side. We consider the set P of the productions as an alphabet.

Let $\#$ and d be new letters. We define new morphisms $g_1', g_2' \colon P^* \to (B \cup \{d, \#\})^*$ as follows. First assume that

$$g_1(a) = a_1 a_2 \ldots a_k \quad \text{and} \quad g_2(a) = b_1 b_2 \ldots b_m$$

for the (generic) letter a. We define

$$g_1'(\pi) = \begin{cases} \# d^n a_1 d^n a_2 d^n \ldots a_k d^j , & \text{if } \pi = (A_0 \to aA_j) \\ d^{n-i} a_1 d^n a_2 d^n \ldots a_k d^j , & \text{if } \pi = (A_i \to aA_j), \\ \# d^n a_1 d^n a_2 d^n \ldots a_k , & \text{if } \pi = (A_0 \to a), \\ d^{n-i} a_1 d^n a_2 d^n \ldots a_k , & \text{if } \pi = (A_i \to a). \end{cases}$$

and

$$g_2'(\pi) = d^n b_1 d^n b_2 \ldots d^n b_m, \quad \text{if } \pi = (A \to aX),$$

where $X \in N \cup \{\varepsilon\}$.

As in [4], $E_G(J') = E_G(J) \cap R$ for the new instance $J' = (\#, g_1', g_2')$. The claim follows from this.

3.3 Morphisms

Next we shall present a construction for the closure under nonerasing morphisms. This construction is a bit more complicated than the previous ones.

Theorem 6 *The family $\mathcal{CE}$ is closed under taking images of nonerasing morphisms.*

Proof. Let $J = (a, g_1, g_2)$ be an instance of the shifted PCP, where $g_1, g_2 \colon A^* \to B^*$. Let $L = h(E_G(J))$, where $h \colon A^* \to C^*$ is a coding. Assume that $f \colon C^* \to \Sigma^*$ is a nonerasing morphism. We shall construct h', g_1' and g_2' such that $f(L) = h'(E_G(J'))$ for the new instance $J' = (a, g_1', g_2')$.

First we show that we can restrict ourselves to cases where

$$\min\{|g_1(x)|, |g_2(x)|\} \geq |f(x)| \quad \text{for all} \ \ x \in A. \tag{3.4}$$

Indeed, suppose the instance J does not satisfy (3.4). We construct a new instance $\bar{J} = (\#, \bar{g}_1, \bar{g}_2)$ and a coding $\bar{h}$ such that $\bar{h}(E_G(\bar{J})) = h(E_G(J))$ and $\bar{g}_1$ and $\bar{g}_2$ do fulfill (3.4). Let $c \notin B$ be a new letter. Let $k = \max_{x \in A}\{|f(x)|\}$. We define $\bar{g}_1(x) = \ell_c^k(g_1(x))$ and $\bar{g}_2(x) = \ell_c^k(g_2(x))$ for all $x \in A$. We also need a new copy x' of each letter x for which a is a prefix of $g_1(x)$. If $g_1(x) = aw$, where $w \in B^*$, then define $\bar{g}_1(x') = \#\ell_c^k(w)$. It now follows that if $u \in E_G(\bar{J})$, then $u = x'v$ for some word $v \in A^*$ and $xv \in E_G(J)$. Therefore, by defining $\bar{h}$ as follows

$$\bar{h}(y) = \begin{cases} h(y), & \text{if } y \in A \\ h(x), & \text{if } y = x', \end{cases}$$

we have $\bar{h}(E_G(\bar{J})) = h(E_G(J))$ as required.

Now assume that (3.4) holds in $J = (a, g_1, g_2)$ and for f. Let us consider the nonerasing morphism $f \circ h \colon A^* \to \Sigma^*$. Note that also the morphism $f \circ h$ satisfies (3.4). In order to prove the claim, it is clearly sufficient to consider the case, where h is the identity mapping, that is, $f = f \circ h$.

First we define for every image $f(x)$, where $x \in A$, a new alphabet $A_x = \{b_x \mid b \in \Sigma\}$. We consider the words

$$(b_1 b_2 \ldots b_m)_x = (b_1)_x (b_2)_x \ldots (b_m)_x,$$

for $f(x) = b_1 \ldots b_m$.

Let c and d be new letters and let $n = \sum_{x \in A} |f(x)|$. Assume that $A = \{x_1, x_2, \ldots, x_q\}$.

Partition the integers $1, 2, \ldots, n$ into q sets such that for the letter x_i there corresponds a set, say $S_i = \{i_1, i_2, \ldots, i_{|f(x_i)|}\}$, of $|f(x_i)|$ integers.

Assume that $f(x_i) = b_1 \ldots b_m$, $g_1(x_i) = a_1 a_2 \ldots a_\ell$, and $g_2(x_i) = a_1' a_2' \ldots a_k'$. We define new morphisms g_1' and g_2' as follows:

$$g_1'((b_1)_{x_i}) = c^n d^n a_1 c^{i_1},$$
$$g_1'((b_j)_{x_i}) = c^{n-i_{j-1}} d^n a_j c^{i_j} \quad \text{for } j = 2, \ldots, m-1,$$
$$g_1'((b_m)_{x_i}) = c^{n-i_{m-1}} d^n a_m c^n d^n \ldots c^n d^n a_\ell,$$

and

$$g_2'((b_1)_{x_i}) = c^n d^n a_1 c^n d^{i_1},$$
$$g_2'((b_j)_{x_i}) = d^{n-i_{j-1}} a_j' c^n d^{i_j} \quad \text{for } j = 2, \ldots, m-1,$$
$$g_2'((b_m)_{x_i}) = c^n d^{n-i_{m-1}} a_m' c^n d^n \ldots c^n d^n a_k'.$$

Then

$$g_1'((b_1 \ldots b_m)_{x_i}) = c^n d^n a_1 c^n d^n a_2 \ldots c^n d^n a_\ell,$$
$$g_2'((b_1 \ldots b_m)_{x_i}) = c^n d^n a_1' c^n d^n a_2' \ldots c^n d^n a_k'.$$

The beginning has to be still fixed. For the cases, where $a_1 = a$, we need new letters $(b_1)_{x_i}'$, for which we define

$$g_1'((b_1)_{x_i}') = ac^{i_1} \text{ and } g_2'((b_1)_{x_i}') = c^n d^n a_j c^n d^{i_1}.$$

Now our constructions for the morphisms g_1' and g_2' are completed.

Next we define h', by setting $h'((b_i)_x) = b_i$ and $h'((b_1)_x') = b_1$ for all i and x. We obtain that $h'(E_G(J')) = f(h(E_G(J)))$, which proves the claim.

Next we shall prove that the family $\mathcal{CE}$ is closed under inverse of nonerasing morphisms.

Theorem 7 *The family $\mathcal{CE}$ is closed under nonerasing inverse morphisms.*

Proof. Consider an instance $h(E_G(J))$, where $J = (\#, g_1, g_2)$ with $g_i \colon A^* \to B^*$ and $h \colon A^* \to C^*$ is a coding. We may assume that $h(A) = C$.

Moreover, let $g \colon \Sigma^* \to C^*$ be a nonerasing morphism.

For each $a \in \Sigma$, let $h^{-1}g(a) = \{v_{a,1}, v_{a,2}, \ldots, v_{a,k_a}\}$ and let

$$\Sigma_a = \{a^{(1)}, \ldots, a^{(k_a)}\}$$

be a set of new letters for a. Denote $\Theta = \cup_{a \in \Sigma} \Sigma_a$, and define the morphisms $g_1', g_2' \colon \Theta^* \to B^*$ and the coding $t \colon \Theta^* \to \Sigma^*$ by

$$g_j'(a^{(i)}) = g_j(v_{a,i}) \text{ for } j = 1, 2, \text{ and } t(a^{(i)}) = a$$

for each $a^{(i)} \in \Theta$.

Consider the instance $J' = (\#, g_1', g_2')$.

Now, assume that $x = a_1 a_2 \ldots a_n \in g^{-1}h(E_G(J))$ (with $a_i \in \Sigma$). Then there exists a word $w = w_1 w_2 \ldots w_n$ such that $g_1(w) = \#g_2(w)$ and $a_i \in g^{-1}h(w_i)$, that is, $w_i = v_{a_i,r_i} \in h^{-1}g(a_i)$ for some r_i, and so $g_1'(w') = \#g_2'(w')$ for the word $w' = a_1^{(r_1)} a_2^{(r_2)} \ldots a_n^{(r_n)}$, for which $t(w') = x$. Therefore $x \in t(E_G(J'))$.

In converse inclusion, $t(E_G(J')) \subseteq g^{-1}h(E_G(J))$ is clear by the above constructions.

Let A and B be two alphabets. A mapping $\tau \colon A^* \to 2^{B^*}$, where 2^{B^*} denotes the set of all subsets of B^*, is a *substitution* if for all $u, v \in A^*$

$$\tau(uv) = \tau(u)\tau(v).$$

Note that τ is actually a morphisms from A^* to 2^{B^*}.

A substitution τ is called *finite* if $\tau(a)$ is finite for all $a \in A$, and *nonerasing* if $\emptyset \neq \tau(a) \neq \{\varepsilon\}$ for all $a \in A$.

Corollary 8 *The family* $C\mathcal{E}$ *is closed under nonerasing finite substitutions.*

Proof. Since $C\mathcal{E}$ is closed under nonerasing morphisms, inverse of nonerasing morphisms, that implies that it is closed under nonerasing finite substitutions that are compositions of inverse of a coding and a nonerasing morphism.

Note that $C\mathcal{E}$ is almost a *trio*, see [1], but it seems that it is not closed under *all* inverse morphisms. It is also almost a *bifaithful rational cone*, see [5], but since the languages do not contain ε, $C\mathcal{E}$ is not closed under the bifaithful finite transducers.

References

1. S. Ginsburg, *Algebraic and Automata-theoretic Properties of Formal Languages*, North-Holland, 1975.
2. V. Halava, T. Harju, H. J. Hoogeboom and M. Latteux, *Valence Languages Generated by Generalized Equality Sets*, Tech. Report **502**, Turku Centre for Computer Science, August 2002, submitted.
3. T. Harju and J. Karhumäki, *Morphisms*, Handbook of Formal Languages (G. Rozenberg and A. Salomaa, eds.), vol. 1, Springer-Verlag, 1997.
4. H. Latteux and J. Leguy, *On the composition of morphisms and inverse morphisms*, Lecture Notes in Comput. Sci. **154** (1983), 420–432.
5. H. Latteux and J. Leguy, *On Usefulness of Bifaithful Rational cones*, Math. Systems Theory **18** (1985), 19–32.
6. G. Păun, *A new generative device: valence grammars*, Revue Roumaine de Math. Pures et Appliquées **6** (1980), 911–924.
7. A. Salomaa, *Formal Languages*, Academic Press, New York, 1973.

Context-Sensitive Equivalences for Non-interference Based Protocol Analysis *

Michele Bugliesi, Ambra Ceccato, and Sabina Rossi

Dipartimento di Informatica, Università Ca' Foscari di Venezia
via Torino 155, 30172 Venezia, Italy
{bugliesi, ceccato, srossi}@dsi.unive.it

Abstract. We develop new proof techniques, based on non-interference, for the analysis of safety and liveness properties of cryptographic protocols expressed as terms of the process algebra CryptoSPA. Our approach draws on new notions of behavioral equivalence, built on top of a context-sensitive labelled transition system, that allow us to characterize the behavior of a process in the presence of any attacker with a given initial knowledge. We demonstrate the effectiveness of the approach with an example of a protocol of fair exchange.

1 Introduction

Non-Interference has been advocated by various authors [1, 9] as a powerful method for the analysis of cryptographic protocols. In [9], Focardi *et al.* propose a general schema for specifying security properties with a uniform and concise definition. The approach draws on earlier work by the same authors on characterizing information-flow security in terms of Non-Interference for the Security Process Algebra (SPA, for short). We briefly review the main ideas below.

SPA is a variant of CCS in which the set of actions is partitioned into two sets: L, for low, and H for high. A Non-Interference property $\mathcal{P}$ for a process E is expressed as follows:

$$E \in \mathcal{P} \text{ if } \forall \Pi \in \mathcal{E}_H : (E||\Pi) \setminus H \approx_{\mathcal{P}} E \setminus H \tag{1}$$

where $\mathcal{E}_H$ is the set of all high-level processes, $\approx_{\mathcal{P}}$ is an observation equivalence (parametric in $\mathcal{P}$), $||$ is parallel composition, and $\setminus$ is restriction. The processes $E \setminus H$ and $(E||\Pi) \setminus H$ represent the low-level views of E and of $E||\Pi$, respectively. The basic intuition is expressed by the slogan: "If no high-level process can change the low behavior, then no flow of information from high to low is possible".

In [9] this idea is refined to provide a general definition of security properties for cryptographic protocols described as terms of CryptoSPA, a process algebra that extends SPA with cryptographic primitives. Intuitively, the refinement amounts to viewing the participants to a protocol as low-level processes, while the high-level processes represent the external attackers. Then, Non-Interference implies that the attackers have no way to change the low (honest) behavior of the protocol.

---

* This work has been partially supported by the MIUR project "Modelli formali per la sicurezza (MEFISTO)" and the EU project IST-2001-32617 "Models and types for security in mobile distributed systems (MyThS)".

A. Lingas and B.J. Nilsson (Eds.): FCT 2003, LNCS 2751, pp. 364–375, 2003.
© Springer-Verlag Berlin Heidelberg 2003

There are two problems that need to be addressed to formalize this idea. First, the intruder should be assumed to have complete control over the public components of the network. Consequently, any step in a protocol involving a public channel should be classified as a high-level action. However, since a protocol specification is usually entirely determined by the exchange of messages over public channels, a characterization like (1) becomes trivial, as $(E||\Pi) \setminus H$ and $E \setminus H$ are simply the null processes. This is easily rectified by extending the protocol specification with low-level actions that are used to specify the desired security property.

A further problem arises from the formalization of the *perfect cryptography* assumption that is usually made in the analysis of the logical properties of cryptographic protocols. In [9] this assumption is expressed by making the definition of Non-Interference dependent on the initial knowledge of the attacker and on a deduction system by which the attacker may compute new information. The initial knowledge, noted ϕ, includes private data (e.g., the enemy's private keys) as well as any piece of publicly available information, such as names of entities and public keys. Property (1) is thus reformulated for a protocol P as follows:

$$P \in \mathcal{P} \text{ if } \forall \Pi \in \mathcal{E}_H^\phi : (P||\Pi) \setminus H \approx_\mathcal{P} P \setminus H. \tag{2}$$

where $\mathcal{E}_H^\phi$ is the set of the high-level processes Π which can perform only actions using the public channel names and whose messages (those syntactically appearing in Π) can be deduced from ϕ.

This framework is very general, and lends itself to the characterization of various security properties, obtained by instantiating the equivalence $\approx_\mathcal{P}$ in the schema above. Instead, it is less effective as a proof method, due to the universal quantification over the possible intruders Π in the class $\mathcal{E}_H^\phi$. In [9], the problem is circumvented by analyzing the protocol in presence of the "hardest attacker". However, In [9] this characterization is proved correct only for the class of relationships $\approx_\mathcal{P}$ that are behavioral preorders on processes. In particular, the proof method is not applicable for equivalences based on bisimulation, and consequently, for the analysis of certain, branching time, liveness properties, such as *fairness*.

We partially rectify the problem by developing a technique which does not require us to exhibit an explicit attacker (nor, in particular, it requires the existence of a hardest attacker). Our approach draws on ideas from [4] to represent the attacker indirectly, in terms of a context-sensitive labelled transition system. The labelled transitions take the form $\phi \triangleright P \xrightarrow{a} \phi' \triangleright P'$, where ϕ represents the context's knowledge prior to the transition, and ϕ' is the new knowledge resulting from P performing the action a. Building on this labelled transition system we provide quantification-free characterizations for different instantiations of (2), specifically when $\approx_\mathcal{P}$ is instantiated to trace equivalence, and to weak bisimulation equivalence. This allows us to apply our technique to the analysis of safety as well as liveness security properties. We demonstrate the latter with an example of a protocol of *fair exchange*.

The rest of the presentation proceeds as follows: Section 2 briefly reviews the process algebra CryptoSPA, Section 3 introduces context-sensitive labelled transition systems, Section 4 gives characterizations for various security properties, Section 5 illustrates the example, and Section 6 draws some conclusions.

All the results presented in this paper are described and proved in [7].

2   The CryptoSPA Language

The *Cryptographic Security Process Algebra* (CryptoSPA, for short) [9] is an extension
of SPA [8] with cryptographic primitives and constructs for value passing. The syntax
is based on the following elements: a set M of basic messages and a set K of encryption
keys with a function $\cdot^{-1} : K \longrightarrow K$ such that $(k^{-1})^{-1} = k$; a set $\mathcal{M}$, ranged over by
m, of all messages, defined as the least set containing $M \cup K$ and closed under the
deduction rules in Table 1 (more on this below); a set C of channels partitioned into
two sets H and L of high and low channels, respectively; a function Msg which maps
every channel c into the set of messages that can be sent and received on c and such
that $Msg(c) = Msg(\bar{c})$; a set $L = \{c(m) \mid m \in Msg(c)\} \cup \{\bar{c}m \mid m \in Msg(c)\}$ of visible
actions and the set $Act = L \cup \{\tau\}$ of all actions, ranged over by a, where τ is the internal
(invisible) action; a function $chan(a)$ which returns c if a is either $c(m)$ or $\bar{c}m$ and the
special channel $void$ when $a = \tau$; a set $Const$ of constants. By an abuse of notation, we
write $c(m), \bar{c}m \in H$ whenever $c, \bar{c} \in H$, and similarly for L.

The syntax of CryptoSPA *terms* (or *processes*) is defined as follows:

$$P ::= \mathbf{0} \mid c(x).P \mid \bar{c}m.P \mid \tau.P \mid P + P \mid P||P \mid P \setminus C \mid P[f] \mid$$
$$\mid A(m_1, ..., m_n) \mid [m = m']P; P \mid [\langle m_1 ... m_n \rangle \vdash_{rule} x]P; P$$

Both $c(x).P$ and $[\langle m_1 ... m_n \rangle \vdash_{rule} x]P; P'$ bind the variable x in P. Constants are defined
as: $A(x_1, ..., x_n) \overset{def}{=} P$, where P is a CryptoSPA process that may contain no free variables
except $x_1, ..., x_n$, which must be pairwise distinct.

Table 1. Inference system for message manipulation where $m, m' \in \mathcal{M}$ and $k, k^{-1} \in K$

$$\frac{m \quad m'}{(m, m')} \; (\vdash_{pair}) \qquad \frac{(m, m')}{m} \; (\vdash_{fst}) \qquad \frac{(m, m')}{m'} \; (\vdash_{snd})$$

$$\frac{m \quad k}{\{m\}_k} \; (\vdash_{enc}) \qquad \frac{\{m\}_k \quad k^{-1}}{m} \; (\vdash_{dec})$$

Intuitively, $\mathbf{0}$ is the empty process; $c(x).P$ waits for input m on channel c, and then
behaves as $P[m/x]$ (i.e., P with all the occurrences of x substituted by m); $\bar{c}(m).P$
outputs m on channel c and continues as P; $P_1 + P_2$ represents the nondeterministic
choice between P_1 and P_2; $P_1||P_2$ is parallel composition, where executions are inter-
leaved, possibly synchronized on complementary input/output actions, producing an
internal action τ; $P \setminus C$ is like P but prevented from sending and receiving messages

on channels in $C \subseteq \mathcal{C}$; in $P[f]$ every channel c is relabelled into $f(c)$; $A(m_1, ..., m_n)$ behaves like the respective definition where the variables $x_1, \cdots, x_n$ are substituted with messages $m_1, \cdots, m_n$; $[m = m']P_1; P_2$ behaves as P_1 if $m = m'$ and as P_2 otherwise; finally, $[\langle m_1 ... m_n \rangle \vdash_{rule} x]P_1; P_2$ tries to deduce an information z from the tuple $\langle m_1 ... m_n \rangle$ through rule $\vdash_{rule}$; if it succeeds then it behaves as $P_1[z/x]$, otherwise it behaves as P_2.

In formalizing the security properties of interest, we will find it convenient to rely on (an equivalent of) the *hiding* operator, of CSP, noted P/C with P process and $C \subseteq \mathcal{C}$, which turns all actions using channels in C into internal τ's. This operator can be defined in CryptoSPA as follows: given any set $C \subseteq \mathcal{C}$, $P/C \stackrel{\text{def}}{=} P[f_C]$ where $f_C(a) = a$ if $chan(a) \notin C$ and $f_C(a) = \tau$ if $chan(a) \in C$.

We denote by $\mathcal{E}$ the set of all CryptoSPA processes and by $\mathcal{E}_H$ the set of all high-level processes, i.e., those constructed only using actions in $H \cup \{\tau\}$.

The operational semantics of CryptoSPA is defined in terms of the labelled transition system (LTS) in Table 2. Most of the transitions are standard, and simply formalize the intuitive semantics of the process constructs discussed above. The two rules $(\vdash_i)$ connect the deduction system in in Table 1 with the transition system. The former system is used to model the ability of the attacker to deduce new information from its initial knowledge. Note, in particular, that secret keys, not initially known to the attacker, may not be deduced (hence we disregard cryptographic attacks, based on guessing secret keys). We say that m is *deducible* from a set of messages ϕ (and write $\phi \vdash m$) if m can be obtained from ϕ by applying the inference rules in Table 1. As in [9] we assume that $\vdash$ is decidable.

We complement the definition of the semantics with a corresponding notion of *observation equivalence*, which is used to establish equalities among processes and is based on the idea that two systems have the same semantics if and only if they cannot be distinguished by an external observer. The equivalences that are relevant to the present discussion are *trace equivalence*, noted $\approx_T$, and *weak bisimulation*, noted $\approx_B$ (see [13]).

In the next section, we introduce coarser versions of these equivalences, noted $\approx_T^\phi$ and $\approx_B^\phi$, which distinguish processes in contexts with initial knowledge ϕ. These context-sensitive notions of equivalence are built on a refined version of the labelled transition system, which we introduce next.

3   Context-Sensitive Equivalences

Following [4], we characterize the behavior of processes in terms of "context-sensitive labelled transitions" where each process transition depends on the knowledge of the context. To motivate, consider a process P that produces and sends a message $\{m\}_k$ reaching the state P', and assume that m and k are known to P but not to the context. Under these hypotheses, the context will never be able to reply the message m to P' (or any continuation thereof). Hence, if P' waits for further input, we can safely leave any input transition involving m out of the LTS, as the P' will never receive m from the context.

The states of the new labelled transition system are *configurations* of the form $\phi \triangleright P$, where P is a process and ϕ is the current knowledge of the context, represented through

Table 2. The operational rules for CryptoSPA

$(input)$ $\quad \dfrac{m \in Msg(c)}{c(x).P \xrightarrow{c(m)} P[m/x]}$ $\qquad (output)$ $\quad \dfrac{m \in Msg(c)}{\overline{c}m.P \xrightarrow{\overline{c}(m)} P}$

(tau) $\quad \dfrac{}{\tau.P \xrightarrow{\tau} P}$ $\qquad (+_1)$ $\quad \dfrac{P_1 \xrightarrow{a} P_1'}{P_1 + P_2 \xrightarrow{a} P_1'}$

$(\|_1)$ $\quad \dfrac{P_1 \xrightarrow{a} P_1'}{P_1 \| P_2 \xrightarrow{a} P_1' \| P_2}$ $\qquad (\|_2)$ $\quad \dfrac{P_1 \xrightarrow{c(m)} P_1' \quad P_2 \xrightarrow{\overline{c}(m)} P_2'}{P_1 \| P_2 \xrightarrow{\tau} P_1' \| P_2'}$

$(=_1)$ $\quad \dfrac{m \neq m' \quad P_2 \xrightarrow{a} P_2'}{[m = m']P_1 ; P_2 \xrightarrow{a} P_2'}$ $\qquad (=_2)$ $\quad \dfrac{m = m' \quad P_1 \xrightarrow{a} P_1'}{[m = m']P_1 ; P_2 \xrightarrow{a} P_1'}$

$([f])$ $\quad \dfrac{P \xrightarrow{a} P'}{P[f] \xrightarrow{f(a)} P'[f]}$ $\qquad (\backslash C)$ $\quad \dfrac{P \xrightarrow{a} P' \quad chan(a) \notin C}{P \backslash C \xrightarrow{a} P' \backslash C}$

$(constant)$ $\quad \dfrac{P[m_1/x_1, \ldots, m_n/x_n] \xrightarrow{a} P' \quad A(x_1, \ldots, x_n) \overset{def}{=} P}{A(m_1, \ldots, m_n) \xrightarrow{a} P'}$

$(\vdash_1)$ $\quad \dfrac{\langle m_1, \ldots, m_n \rangle \vdash_{rule} m \quad P_1[m/x] \xrightarrow{a} P_1'}{[\langle m_1, \ldots, m_n \rangle \vdash_{rule} x]P_1 ; P_2 \xrightarrow{a} P_1'}$

$(\vdash_2)$ $\quad \dfrac{\nexists m : \langle m_1, \ldots, m_n \rangle \vdash_{rule} m \quad P_2 \xrightarrow{a} P_2'}{[\langle m_1, \ldots, m_n \rangle \vdash_{rule} x]P_1 ; P_2 \xrightarrow{a} P_2'}$

Table 3. Inference rules for the ELTS

$$(output) \quad \frac{P \xrightarrow{\bar{c}m} P' \quad \bar{c}m \in H}{\phi \triangleright P \xrightarrow{\bar{c}(m)} \phi \cup \{m\} \triangleright P'} \qquad (input) \quad \frac{P \xrightarrow{c(m)} P' \quad \phi \vdash m \quad c(m) \in H}{\phi \triangleright P \xrightarrow{c(m)} \phi \triangleright P'}$$

$$(tau) \quad \frac{P \xrightarrow{\tau} P'}{\phi \triangleright P \xrightarrow{\tau} \phi \triangleright P'} \qquad (low) \quad \frac{P \xrightarrow{a} P' \quad a \in L}{\phi \triangleright P \xrightarrow{a} \phi \triangleright P'}$$

a set of messages. The transitions represent interactions between the process and the context and now take the form

$$\phi \triangleright P \xrightarrow{a} \phi' \triangleright P',$$

where a is the action executed by the process P and ϕ' is the new knowledge at disposal to the context for further interactions with P'.

The transitions between configurations, in Table 3, are defined rather directly starting from the corresponding transitions between processes. In rule (*output*), the context's knowledge is augmented with the information sent by the process. Dually, rule (*input*) assumes that the context performs an output action synchronizing with the input of the process. The message sent by the context must be completely deducible from the context's knowledge ϕ, otherwise the corresponding transition is impossible: this is how the new transitions provide an explicit account of the attacker's knowledge. The remaining rules, (*tau*) and (*low*) state that internal actions of the protocol, and low actions do not contribute to the knowledge of the context in any way.

In the rest of the presentation, we refer to the transition rules in Table 3 collectively as the *enriched LTS* (*ELTS*, for short). Also, we assume that the initial knowledge of the context includes only public information and the context's private names. This is a reasonable condition, since it simply corresponds to assuming that each protocol run starts with fresh keys and nonces, a condition that is readily guaranteed by relying on time-dependent elements (e.g., time-stamps) and assuming that session keys are distinct for every executions.

The notions of trace and weak bisimulation equivalences extend in the expected way from processes to ELTS configurations, as we discuss below.

We write $\phi \triangleright P \xRightarrow{a} \phi' \triangleright P'$ to denote the sequence of transitions $\phi \triangleright P (\xrightarrow{\tau})^* \phi \triangleright P_1 \xrightarrow{a} \phi' \triangleright P_2 (\xrightarrow{\tau})^* \phi' \triangleright P'$, where, as expected, $\phi = \phi'$ if $\xrightarrow{a}$ is an input, low or silent action. Furthermore, let $\gamma = a_1 \ldots a_n \in L^*$ be a sequence of (non silent) actions; then $\phi \triangleright P \xRightarrow{\gamma} \phi' \triangleright P'$ if there are $P_1, P_2, \ldots, P_{n-1} \in \mathcal{E}$ and $\phi_1, \phi_2, \ldots, \phi_{n-1}$ states such that $\phi \triangleright P \xRightarrow{a_1} \phi_1 \triangleright P_1 \xRightarrow{a_2} \ldots \xRightarrow{a_{n-1}} \phi_{n-1} \triangleright P_{n-1} \xRightarrow{a_n} \phi' \triangleright P'$. The notation $\phi \triangleright P \xRightarrow{\hat{a}} \phi' \triangleright P'$ stands for $\phi \triangleright P \xRightarrow{a} \phi' \triangleright P'$ if $a \in L$ and for $\phi \triangleright P (\xrightarrow{\tau})^* \phi \triangleright P'$ if $a = \tau$, as usual.

Definition 1 (Trace Equivalence over configurations).

- $T(\phi \triangleright P) = \{\gamma \in \mathcal{L}^* \mid \exists \phi', P' : \phi \triangleright P \overset{\gamma}{\Longrightarrow} \phi' \triangleright P'\}$ *is the set of* traces *associated with the configuration $\phi \triangleright P$.*
- *Two configurations $\phi_P \triangleright P$ and $\phi_Q \triangleright Q$ are* trace equivalent, *denoted by $\phi_P \triangleright P \approx_T^c$ $\phi_Q \triangleright Q$, if $T(\phi_P \triangleright P) = T(\phi_Q \triangleright Q)$.*

Based on trace equivalence over configurations we can then define a corresponding notion of process equivalence, for processes executing in an environment with initial knowledge ϕ. Formally, $P \approx_T^\phi Q$ whenever $\phi \triangleright P \approx_T^c \phi \triangleright Q$.

Definition 2 (Weak Bisimulation over configurations).

- *A binary relation $\mathcal{R}$ over configurations is a weak bisimulation if, assuming $(\phi_P \triangleright P, \phi_Q \triangleright Q) \in \mathcal{R}$, one has, for all $a \in Act$:*
  - *if $\phi_P \triangleright P \overset{a}{\longrightarrow} \phi_{P'} \triangleright P'$, then there exists a configuration $\phi_{Q'} \triangleright Q'$ such that $\phi_Q \triangleright Q \overset{\hat{a}}{\Longrightarrow} \phi_{Q'} \triangleright Q'$ and $(\phi_{P'} \triangleright P', \phi_{Q'} \triangleright Q') \in \mathcal{R}$;*
  - *if $\phi_Q \triangleright Q \overset{a}{\longrightarrow} \phi_{Q'} \triangleright Q'$, then there exists a configuration $\phi_{P'} \triangleright P'$ such that $\phi_P \triangleright P \overset{\hat{a}}{\Longrightarrow} \phi_{P'} \triangleright P'$ and $(\phi_{P'} \triangleright P', \phi_{Q'} \triangleright Q') \in \mathcal{R}$.*
- *Two configurations $\phi_P \triangleright P$ and $\phi_Q \triangleright Q$ are* weakly bisimilar, *denoted by $\phi_P \triangleright P \approx_B^c$ $\phi_Q \triangleright Q$, if there exists a weak bisimulation containing the pair $(\phi_P \triangleright P, \phi_Q \triangleright Q)$.*

It is not difficult to prove that relation $\approx_B^c$ is the largest weak bisimulation over configurations, and that it is an equivalence relation. As for trace equivalence, we can recover an equivalence relation on processes executing in a context with initial knowledge ϕ by defining $P \approx_B^\phi Q$ if and only if $\phi \triangleright P \approx_B^c \phi \triangleright Q$.

4 Non-interference Proof Techniques

We show that the new definitions of behavioral equivalence may be used to construct effective proof methods for various security properties within the general schema proposed in [9]. In particular, we show that making our equivalences dependent on the initial knowledge of the attacker provides us with security characterizations that are stated independently from the attacker itself.

The first property we study, known as NDC, results from instantiating $\approx_\mathcal{P}$ in (2) (see the introduction) to the trace equivalence relation $\approx_T$. As discussed in [9], NDC is a generalization of the classical idea of Non-Interference to non-deterministic systems and can be used for analyzing different security properties of cryptographic protocols such as secrecy, authentication and integrity. NDC can readily be extended to account for the context's knowledge as follows:

Definition 3 (NDC$^\phi$). $P \in NDC^\phi$ *if $P \setminus H \approx_T (P||\Pi) \setminus H, \, \forall \Pi \in \mathcal{E}_H^\phi$.*

A process P is NDC^ϕ if for every high-level process Π with initial knowledge ϕ a low level user cannot distinguish P from $(P||\Pi)$, i.e., if Π cannot interfere with the low-level execution of the process P.

Focardi *et al.* in [9] show that when ϕ is finite it is possible to find a most general intruder Top^ϕ so that verifying NDC^ϕ reduces to checking $P \setminus H \approx_T (P||Top^\phi) \setminus H$. Here we provide an alternative[1], quantification-free characterization of NDC^ϕ. Let P/H denote the process resulting from P, by replacing all high-level actions with the silent action τ (cf. Section 2).

Theorem 1 (NDC$^\phi$). $P \in NDC^\phi$ *if and only if* $P \setminus H \approx_T^\phi P/H$.

More interestingly, our approach allows us to find a sound proof method for the $BNDC^\phi$ property, which results from instantiating (2) in the introduction with the equivalence $\approx_B$ as follows:

Definition 4 (BNDC$^\phi$). $P \in BNDC^\phi$ *if* $P \setminus H \approx_B (P||\Pi) \setminus H$, $\forall \Pi \in \mathcal{E}_H^\phi$.

As for NDC^ϕ, the definition falls short of providing a proof method due to the universal quantification over Π. Here, however, the problem may not be circumvented by resorting to a hardest attacker, as the latter does not exist, being there no (known) preorder on processes corresponding to weak bisimilarity.

What we propose here is a partial solution that relies on providing a coinductive (and quantification free) characterization of a sound approximation of $BNDC^\phi$, based on the following *persistent* version of $BNDC^\phi$.

Definition 5 (P_BNDC$^\phi$). $P \in P\_BNDC^\phi$ *if* $P' \in BNDC^\phi$, $\forall P'$ *reachable from* P.

P_BNDC^ϕ is the context-sensitive version of the P_BNDC property studied in [10]. Following the technique in [10], one can show that P_BNDC^ϕ is a sound approximation of $BNDC^\phi$ which admits elegant quantification-free characterizations. Specifically, like P_BNDC, P_BNDC^ϕ can be characterized both in terms of a suitable weak bisimulation relation "up to high-level actions", noted $\approx_{\setminus H}^\phi$, and in terms of unwinding conditions, as discussed next. We first need the following definition:

Definition 6. *Let* $a \in Act$. *The transition relation* $\overset{\hat{a}}{\Longrightarrow}_{\setminus H}$ *is defined as follows:*

$$\overset{\hat{a}}{\Longrightarrow}_{\setminus H} = \begin{cases} \overset{\hat{a}}{\Longrightarrow} & \text{if } a \notin H \\ \overset{a}{\Longrightarrow} \text{ or } \overset{\hat{\tau}}{\Longrightarrow} & \text{if } a \in H \end{cases}$$

The transition relation $\overset{\hat{a}}{\Longrightarrow}_{\setminus H}$ is defined as $\overset{\hat{a}}{\Longrightarrow}$, except that it treats H-level actions as silent actions. Now, weak bisimulations up to H over configurations are defined as weak bisimulations over configurations except that they allow a high action to be matched by zero or more high actions. Formally:

Definition 7 (Weak Bisimulation up to H over configurations).

- *A binary relation* $\mathcal{R}$ *over configurations is a* weak bisimulation up to H *if* $(\phi_P \triangleright P, \phi_Q \triangleright Q) \in \mathcal{R}$ *implies that, for all* $a \in Act$,

---

[1] An analogous result has been recently presented by Gorrieri *et al.* in [11] for a *timed* extension of CryptoSPA. We discuss the relationships between our and their result in Section 6.

- if $\phi_P \rhd P \xrightarrow{a} \phi_{P'} \rhd P'$, then there exists a configuration $\phi_{Q'} \rhd Q'$ such that
  $\phi_Q \rhd Q \xRightarrow{\hat{a}}_{\backslash H} \phi_{Q'} \rhd Q'$ and $(\phi_{P'} \rhd P', \phi_{Q'} \rhd Q') \in \mathcal{R}$;
- if $\phi_Q \rhd Q \xrightarrow{a} \phi_{Q'} \rhd Q'$, then there exists a configuration $\phi_{P'} \rhd P'$ such that
  $\phi_P \rhd P \xRightarrow{\hat{a}}_{\backslash H} \phi_{P'} \rhd P'$ and $(\phi_{P'} \rhd P', \phi_{Q'} \rhd Q') \in \mathcal{R}$.

- Two configurations $\phi_P \rhd P$ and $\phi_Q \rhd Q$ are weakly bisimilar up to H, denoted by $\phi_P \rhd P \approx^c_{\backslash H} \phi_Q \rhd Q$, if there exists a weak bisimulation up to H containing the pair $(\phi_P \rhd P, \phi_Q \rhd Q)$.

Again, we can prove that the relation $\approx^c_{\backslash H}$ is the largest weak bisimulation up to H over configurations and that it is an equivalence relation. Also, as for previous relations over configurations, we can recover an associated relation over processes in a context with initial knowledge ϕ by defining

$$P \approx^\phi_{\backslash H} Q \text{ if and only if } \phi \rhd P \approx^c_{\backslash H} \phi \rhd Q.$$

We can finally state the two characterizations of P_BNDC^ϕ. The former characterization is expressed in terms of $\approx^\phi_{\backslash H}$ (with no quantification on the reachable states and on the high-level malicious processes).

Theorem 2 (P_BNDC$^\phi$ 1). $P \in P\_BNDC^\phi$ if and only if $P \backslash H \approx^\phi_{\backslash H} P$.

The second characterization of P_BNDC^ϕ is given in terms of *unwinding conditions* which demand properties of individual actions. Unwinding conditions aim at "distilling" the local effect of performing high-level actions and are useful to define both proof systems (see, e.g., [6]) and refinement operators that preserve security properties, as done in [12].

Theorem 3 (P_BNDC$^\phi$ 2). $P \in P\_BNDC^\phi$ if and only if for all $\phi_i \rhd P_i$ reachable from $\phi \rhd P$, if $\phi_i \rhd P_i \xrightarrow{h} \phi'_i \rhd P'_i$ for $h \in H$, then $\phi_i \rhd P_i \xRightarrow{\hat{\tau}} \phi''_i \rhd P''_i$ such that $\phi'_i \rhd P'_i \backslash H \approx^c_B \phi''_i \rhd P''_i \backslash H$.

Both the characterizations can be used for verifying cryptographic protocols. A concrete example of a fair exchange protocol is illustrated in the next section.

5   An Example: The ASW Fair Exchange Protocol

The ASW contract signing protocol [2] is used in electronic commerce transactions to enable two parties, named O (originator) and R (responder), to obtain each other's commitment on a previously agreed contractual text M. To deal with unfair situations, each party may appeal to a trusted third party T which can decide, on the basis of the data it has received, whether to issue a replacement contract or an abort token. If both O and R are honest, and they receive the messages sent to them, then they both obtain a valid contract upon the completion of the protocol.

We say that the protocol guarantees *fairness* to O (dually, to R) on message M, if whatever malicious R (O) is considered, if R (O) gets evidence that O (R) has originated M then also O (R) will eventually obtain the evidence that R (O) has received M.

Notice that this is a branching-time liveness property: we are requiring that something should happen if O (resp. R) gets his evidence —i.e., that also R (resp. O) should get his evidence— for all the execution traces in the protocol (cf. [9] for a thorough discussion on this point).

The protocol consists of three independent sub-protocols: *exchange*, *abort* and *resolve*. Here, we focus on the main *exchange* sub-protocol that is specified by the following four messages, where M is the contractual text on which we assume the two parties previously agreed, while SK_O and SK_R (PK_O and PK_R) are the private (public) keys of O and R, respectively.

$$O \to R : me_1 = \{M, h(N_O)\}_{SK_O}$$
$$R \to O : me_2 = \{\{M, h(N_O)\}_{SK_O}, h(N_R)\}_{SK_R}$$
$$O \to R : me_3 = N_O$$
$$R \to O : me_4 = N_R$$

In the first step, O commits to the contractual text by hashing a random number N_O, and signing a message that contains both $h(N_O)$ and M. While O does not actually reveal the value of its contract authenticator N_O to the recipient of message me_1, O is committed to it. As in a standard commitment protocol, we assume that it is not computationally feasible for O to find a different number N_O' such that $h(N_O') = h(N_O)$. In the second step, R replies with its own commitment. Finally, O and R exchange the actual contract authenticators.

We specify the sub-protocol in CryptoSPA (see the figure below), by introducing some low-level actions to verify the correctness of protocol's executions. We say that an execution is correct if we observe the sequence of low-level actions $\overline{received\_me_1}$, $\overline{received\_me_2}$, $\overline{received\_N_O}$, $\overline{received\_N_R}$ in this order.

$$O(M,N_O) \overset{def}{=} [\langle N_O, k_h \rangle \vdash_{enc} n] | [\langle (M,n), SK_O \rangle \vdash_{enc} p] \, \overline{c}p. \, c(v).$$
$$[\langle v, PK_R \rangle \vdash_{dec} i][i \vdash_{fst} p'][i \vdash_{snd} r'][p' = p] \, \overline{received\_v}.$$
$$\overline{c}N_O. \, c(j). \, [\langle j, k_h \rangle \vdash_{enc} r''][r'' = r'] \, \overline{received\_j}$$

$$R(M,N_R) \overset{def}{=} c(q). \, [\langle q, PK_O \rangle \vdash_{dec} s][s \vdash_{fst} m][s \vdash_{snd} n'][m = M] \, \overline{received\_q}.$$
$$[\langle N_R, k_h \rangle \vdash_{enc} r] | [\langle (q,r), SK_R \rangle \vdash_{enc} t] \, \overline{c}t. \, c(u).$$
$$[\langle u, k_h \rangle \vdash_{enc} n''][n'' = n'] \, \overline{received\_u}. \, \overline{c}N_R$$

$$P \overset{def}{=} O(M,N_O) \parallel R(M,N_R)$$

Fig. 1. The CryptoSPA specification of the *exchange* sub-protocol of ASW

We can demonstrate that the protocol does not satisfy property P_BNDC^ϕ when ϕ consists of public information and private data of possible attacker's. This can be easily checked by applying Theorem 3. Indeed, just observing the protocol ELTS, one can immediately notice that there exists a configuration transition $\phi \rhd P \xrightarrow{a} \phi' \rhd P'$, where $a = \overline{c}me_1$, but there isn't any ϕ'' and P'' such that $\phi \rhd P \overset{\hat{\tau}}{\Longrightarrow} \phi'' \rhd P''$ and $\phi' \rhd P' \setminus H \approx_B^c \phi_i'' \rhd P_i'' \setminus H$. In fact, it is easy to prove that $\phi' \rhd P' \setminus H \approx_B^c \mathbf{0}$ for all ϕ', while

$\phi'' \rhd P'' \setminus H \not\approx^c_B 0$ for all P'' and ϕ'' such that $\phi \rhd P \overset{\hat{\tau}}{\Longrightarrow} \phi'' \rhd P''$. However, the fact that, in this case, the ASW protocol does not satisfy P_BNDC^ϕ does not represent a real attack to the protocol since such a situation is resolved by inching the trusted party T.

More interestingly, we can analyze the protocol under the assumption that one of the participants is corrupt. This can be done by augmenting the knowledge ϕ with the corrupt party's private information such as its private key and its contract authenticator. We can show that the protocol does not satisfy P_BNDC^ϕ when O is corrupt, finding the attack already described in [14].

6   Conclusions and Related Work

We have studied context-sensitive equivalence relationships and relative proof techniques within the process algebra CryptoSPA to analyze protocols. Our approach builds on context-sensitive labelled transition systems, whose transitions are constrained by the knowledge of the environment. We showed that our technique can be used to analyze both safety and liveness properties of cryptographic protocols.

In a recent paper Gorrieri *et al.* [11] prove results related to ours, for a real-time extension of CryptoSPA. In particular, they prove an equivalent of Theorem 1: however, while the results are equivalent, the underlying proof techniques are not. More precisely, instead of using context-sensitive LTS's, [11] introduces a special hiding operator $/^\phi$ and prove that $P \in NDC^\phi$ if and only if $P \setminus H \approx_T P/^\phi H$. Process $P/^\phi H$ corresponds exactly to our configuration $\phi \rhd P/H$, in that the corresponding LTS's are isomorphic. However, the approach of [11] is still restricted to the class of observation equivalences that are behavioral preorders on processes and thus it does not extend to bisimulations.

As we pointed out since the outset, our approach is inspired by Boreale, De Nicola and Pugliese's work [4] on characterizing may test and barbed congruence in the spi calculus by means of trace and bisimulation equivalences built on top of context-sensitive LTS's. Based on the same technique, symbolic semantics and compositional proofs have been recently studied in [3,5], providing effective tools for the verification of cryptographic protocols. Symbolic description methods could be exploited to deal with the state-explosion problems which are intrinsic in the construction of context-sensitive labelled transition systems. Future plans include work in that direction.

References

1. M. Abadi. Security Protocols and Specifications. In W. Thomas, editor, *Proc. of the Second International Conference on Foundations of Software Science and Computation Structure (FoSSaCS'99)*, volume 1578 of *LNCS*, pages 1–13. Springer-Verlag, 1999.
2. N. Asokan, V. Shoup, and M. Waidener. Asynchronuous Protocols for Optimistic Fair Exchange. In *Proc. of the IEEE Symposium on Research in Security and Privacy*, pages 86–99. IEEE Computer Society Press, 1998.
3. M. Boreale and M. G. Buscemi. A Framework for the Analysis of Security Protocols. In *Proc. of the 13th International Conference on Concurrency Theory (CONCUR'02)*, volume 2421 of *LNCS*, pages 483–498. Springer-Verlag, 2002.

4. M. Boreale, R. De Nicola, and R. Pugliese. Proof Tecniques for Cryptographic Processes. In *Proc. of the 14th IEEE Symposium on Logic in Computer Science (LICS'99)*, pages 157–166. IEEE Computer Society Press, 1999.
5. M. Boreale and D. Gorla. On Compositional Reasoning in the spi-calculus. In *Proc. of the 5th International Conference on Foundations of Software Science and Computation Structures (FossaCS'02)*, volume 2303 of *LNCS*, pages 67–81. Springer-Verlag, 2002.
6. A. Bossi, R. Focardi, C. Piazza, and S. Rossi. A Proof System for Information Flow Security. In M. Leuschel, editor, *Proc. of Int. Workshop on Logic Based Program Development and Transformation*, LNCS. Springer-Verlag, 2002. To appear.
7. A. Ceccato. Analisi di protocolli crittografici in contesti ostili. Laurea thesis, Università Ca' Foscari di Venezia, 2001.
8. R. Focardi and R. Gorrieri. Classification of Security Properties (Part I: Information Flow). In R. Focardi and R. Gorrieri, editors, *Foundations of Security Analysis and Design*, volume 2171 of *LNCS*. Springer-Verlag, 2001.
9. R. Focardi, R. Gorrieri, and F. Martinelli. Non Interference for the Analysis of Cryptographic Protocols. In U. Montanari, J.D.P. Rolim, and E. Welzl, editors, *Proc. of Int. Colloquium on Automata, Languages and Programming (ICALP'00)*, volume 1853 of *LNCS*, pages 744–755. Springer-Verlag, 2000.
10. R. Focardi and S. Rossi. Information Flow Security in Dynamic Contexts. In *Proc. of the 15th IEEE Computer Security Foundations Workshop*, pages 307–319. IEEE Computer Society Press, 2002.
11. R. Gorrieri, E. Locatelli, and F. Martinelli. A Simple Language for Real-time Cryptographic Protocol Analysis. In *Proc. of 12th European Symposium on Programming Languages and Systems*, LNCS. Springer-Verlag, 2003. To appear.
12. H. Mantel. Unwinding Possibilistic Security Properties. In *Proc. of the European Symposium on Research in Computer Security*, volume 2895 of *LNCS*, pages 238–254. Springer-Verlag, 2000.
13. R. Milner. *Communication and Concurrency*. Prentice-Hall, 1989.
14. V. Shmatikov and J. C. Mitchell. Analysis of a Fair Exchange Protocol. In *Proc. of 7th Annual Symposium on Network and Distributed System Security (NDSS 2000)*, pages 119–128. Internet Society, 2000.

On the Exponentiation of Languages

Werner Kuich[1] and Klaus W. Wagner[2]

[1] Institut für Algebra und Computermathematik
Technische Universität Wien
Wiedner Hauptstraße 8, A 1040 Wien
kuich@tuwien.ac.at
[2] Institut für Informatik
Bayerische Julius-Maximilians-Universität Würzburg
Am Hubland, D-97074 Würzburg, Germany
wagner@informatik.uni-wuerzburg.de

Abstract. We characterize the exponentiation of languages by other language operations: In the presence of some "weak" operations, exponentiation is exactly as powerful as complement and ε-free morphism. This characterization implies, besides others, that a semi-AFL is closed under complement iff it is closed under exponentiation. As an application we characterize the exponentiation closure of the context-free languages. Furthermore, P is closed under exponentiation iff P = NP , and NP is closed under exponentiation iff NP = co-NP.

1 Introduction

Kuich, Sauer, Urbanek [4] defined addition + and multiplication × (different from concatenation) in such a way that equivalence classes of formal languages, defined by help of length preserving morphisms, form a lattice. They defined lattice families of formal languages and showed that, if $\mathfrak{F}$ is a lattice family of languages then $\mathcal{L}_{\mathfrak{F}}$ is a lattice with a least and a largest element. Here $\mathcal{L}_{\mathfrak{F}}$ is a set of equivalence classes defined by a family $\mathfrak{F}$ of languages.

Moreover, Kuich, Sauer, Urbanek [4] defined exponentiation of formal languages as a new operation. Then they defined stable families of languages (essentially, these are lattice families of languages closed under exponentiation) and showed that, if $\mathfrak{F}$ is a stable family of languages then $\mathcal{L}_{\mathfrak{F}}$ is a Heyting algebra with a largest element. Moreover, they proved that stable families $\mathfrak{F}$ of languages can be used to characterize the join and meet irreducibility of $\mathcal{L}_{\mathfrak{F}}$. (See Theorems 4.2 and 4.3 of Kuich, Sauer, Urbanek [4].)

From the point of view of lattice theory it is, by the results quoted above, very interesting to find families of languages that are lattice families or stable families.

The paper consists of this and four more sections. In Section 2, we introduce the language operations and language families (formal language classes as well as complexity classes) which are considered in this paper, and we cite from the literature the present knowledge on the closure properties of these classes.

A. Lingas and B.J. Nilsson (Eds.): FCT 2003, LNCS 2751, pp. 376–386, 2003.
© Springer-Verlag Berlin Heidelberg 2003

In Section 3 we examine which "classical" language operations are needed to generate the operations addition, multiplication and exponentiation. As corollaries we get lists of classes which are closed under these operations and which are lattice families or stable families, resp. It turns out that the regular languages, the context-sensitive languages, the rudimentary languages, the class PH of the polynomial time hierarchy, and the complexity classes PSPACE, DSPACE(s) for $s(n) \geq n$, and NSPACE(s) for space-constructible $s(n) \geq n$ are stable families and hence closed under exponentiation.

In Section 4 we prove that, for every family $\mathfrak{F}$ of languages that contains all regular languages, the closure of $\mathfrak{F}$ under union, inverse morphism and exponentiation coincides with the closure of $\mathfrak{F}$ under union, inverse morphism, ε-free morphism and complement. Since union and inverse morphism are weak operations which only smooth given language classes, this result can informally stated as follows: exponentiation is just as powerful as ε-free morphism and complement together. As one of the possible consequences we obtain: A semi-AFL is closed under exponentiation iff it is closed under complement.

In Section 5 we apply the results of Section 4 to various classes of languages which are not closed or not known to be closed under exponentiation. Kuich, Sauer, Urbanek [4] proved that the class CFL of context-free languages is not closed under exponentiation. We show that the closure of CFL under exponentiation and the weak operations of union and inverse morphism coincides with Smullyan's class RUD of rudimentary languages. Furthermore, we prove that the family of languages P (languages accepted by a deterministic Turing machine in polynomial time) is closed under exponentiation iff P = NP, and that the family of languages NP (languages accepted by a nondeterministic Turing machine in polynomial time) is closed under exponentiation iff NP = co-NP.

It is assumed that the reader has a basic knowledge of lattice theory (see Balbes, Dwinger [2]), formal language and automata theory (see Ginsburg [3]), and complexity theory (see Balcázar, Díaz, Gabarró [1] and Wagner, Wechsung [7]).

2   Families of Languages and Their Closure Properties

In this paper we consider several classical operations on languages. We use the symbol εh (lh, h^{-1}, lh^{-1}, $\cap$REG, and $^-$, resp.) for the operation of ε-free morphism (length preserving morphism, inverse morphism, inverse length preserving morphism, intersection with regular languages, and complement, resp.).

Given operations $\mathcal{O}_1, \mathcal{O}_2, \ldots, \mathcal{O}_r$ on languages, we introduce the closure operator $\Gamma_{\mathcal{O}_1, \mathcal{O}_2, \ldots, \mathcal{O}_r}$ on families of languages as follows: For a family $\mathfrak{F}$ of languages, $\Gamma_{\mathcal{O}_1, \mathcal{O}_2, \ldots, \mathcal{O}_r}(\mathfrak{F})$ is the closure of $\mathfrak{F}$ under the operations $\mathcal{O}_1, \mathcal{O}_2, \ldots, \mathcal{O}_r$, i.e., the least family of languages containing $\mathfrak{F}$ and being closed under the operations $\mathcal{O}_1, \mathcal{O}_2, \ldots, \mathcal{O}_r$.

Let REG, CFL, and CSL be the classes of regular, context-free, and context-sensitive languages, resp. The class LOGCFL consists of the languages which are logarithmic-space many-one reducible to context-free languages. The class RUD of *rudimentary* languages is the smallest class of languages that contains CFL and is closed under ε-free morphism and complement, i.e., RUD $= \Gamma_{\varepsilon\text{h},-}(\text{CFL})$.

The classes P and NP consist of all languages which can be accepted in polynomial time by deterministic and nondeterministic, resp., Turing machines. Let co-NP be the class of all languages whose complement is in NP. With Q we denote the classes of languages which can be accepted in linear time by nondeterministic Turing machines.

The classes L and NL consist of all languages which can be accepted in logarithmic space by deterministic and nondeterministic, resp., Turing machines. The class PSPACE consists of all languages which can be accepted in polynomial space by deterministic Turing machines.

Let Σ_k^{P} and Π_k^{P}, $k \geq 1$, be the classes of the polynomial-time hierarchy, i.e., $\Sigma_1^{\mathrm{P}} = NP$, $\Sigma_{k+1}^{\mathrm{P}}$ is the class of all languages which are nondeterministically polynomial-time Turing-reducible to languages from Σ_k^{P}, and Π_k^{P} is the class of all languages whose complement is in Σ_k^{P} ($k \geq 1$). Finally, PH is the union of all these classes Σ_k^{P} and Π_k^{P}. Notice that PH $\subseteq$ PSPACE.

For a function $t : \mathbb{N} \to \mathbb{N}$, the classes DTIME($t$) and NTIME($t$) consist of all languages which can be accepted in time t by deterministic and nondeterministic, resp., Turing machines. For a function $s : \mathbb{N} \to \mathbb{N}$, the classes DSPACE($s$) and NSPACE($s$) consist of all languages which can be accepted in space s by deterministic and nondeterministic, resp., Turing machines.

For exact definitions and more information about these classes see e.g. [1] and [7]. The following table shows the known closure properties of these classes (cf. [7]).

Theorem 21 *An entry + (-, ?, resp.) in the following table means that the class in this row is closed (not closed, not known to be closed, resp.) under the operation in this column.*

3   Lattice Families and Stable Families of Languages

In this section we introduce the operations of addition, multiplication and exponentiation of languages, and we see how they can be generated by "classical" operations on languages.

Throughout this paper the symbol Σ (possibly provided with indices) denotes a finite subalphabet of some infinite alphabet Σ_∞ of symbols.

Let $L_1 \subseteq \Sigma_1^*$ and $L_2 \subseteq \Sigma_2^*$. Define $L_1 \leq L_2$ if $h(L_1) \subseteq L_2$ for some length preserving morphism $h : \Sigma_1^* \to \Sigma_2^*$ and $L_1 \sim L_2$ if $L_1 \leq L_2$ and $L_2 \leq L_1$. Then $\sim$ is an equivalence relation. If $L_1 \sim L_1'$ and $L_2 \sim L_2'$ then $L_1 \leq L_2$ iff $L_1' \leq L_2'$. It follows that $\leq$ is a partial order relation on the $\sim$-equivalence classes. Let $[L]$ be the $\sim$-equivalence class including the language L.

Let $L_1 \subseteq \Sigma_1^*$ and $L_2 \subseteq \Sigma_2^*$. Define $L_1 \times L_2 = \{(a_1, b_1) \dots (a_n, b_n) \mid a_1 \dots a_n \in L_1,\ b_1 \dots b_n \in L_2\} \subseteq (\Sigma_1 \times \Sigma_2)^*$, and let $L_1 + L_2$ be the disjoint union of L_1 and L_2. That is the language defined as $L_1 \cup L_2$ given that $\Sigma_1 \cap \Sigma_2 = \emptyset$. If $\Sigma_1 \cap \Sigma_2 \neq \emptyset$

language classes	$\cup$	$\cap$REG	$\cap$	$^-$	εh	lh^{-1}	h^{-1}
			operations				
REG	+	+	+	+	+	+	+
CFL	+	+	−	−	+	+	+
CSL	+	+	+	+	+	+	+
LOGCFL	+	+	+	+	?	+	+
RUD	+	+	+	+	+	+	+
L	+	+	+	+	?	+	+
NL	+	+	+	+	?	+	+
P	+	+	+	+	?	+	+
Q	+	+	+	?	+	+	+
NP	+	+	+	?	+	+	+
co-NP	+	+	+	?	?	+	+
Σ_k^P ($k \geq 1$)	+	+	+	?	+	+	+
Π_k^P ($k \geq 1$)	+	+	+	?	?	+	+
PH	+	+	+	+	+	+	+
PSPACE	+	+	+	+	+	+	+
DTIME(t) ($t(n) \geq n$)	+	+	+	+	?	+	$+^1$
NTIME(t) ($t(n) \geq n$)	+	+	+	?	+	+	$+^1$
DSPACE(s) ($s(n) \geq n$)	+	+	+	+	+	+	$+^2$
NSPACE(s) ($s(n) \geq n$)	+	+	−	$-^3$	+	+	$+^?$

The functions t and s are assumed to be increasing.

$+^1$ - Replace t with $t(\mathcal{O}(n))$

$+^2$ - Replace s with $s(\mathcal{O}(n))$

$+^3$ - Assume that s is space-constructible, i.e., the computa-
   tion $x \to s(|x|)$ can be carried out in space $s(|x|)$.

then create the new alphabet $\bar{\Sigma} = \{\bar{a} \mid a \subset \Sigma_2\}$ such that $\Sigma_1 \cap \bar{\Sigma} = \emptyset$ and a
copy $\bar{L} \subseteq \bar{\Sigma}^*$ of L_2 and take $L_1 + L_2 = L_1 \cup \bar{L}$.

It is easy to see that if $L_1 \sim L_3$ and $L_2 \sim L_4$ then $L_1 + L_2 \sim L_3 + L_4$ and
$L_1 \times L_2 \sim L_3 \times L_4$. It follows that the operations $+$ and $\times$ lift consistently to $\sim$-
equivalence classes of languages. It is clear that multiplication $\times$ and addition $+$
on $\sim$-equivalence classes are commutative and associative operations. We denote
the set of $\sim$-equivalence classes of languages by $\mathcal{L}$. If $\mathfrak{F}$ is a family of languages
then we denote $\mathcal{L}_{\mathfrak{F}} = \{[L] \cap \mathfrak{F} \mid L \in \mathfrak{F}\}$. By $\overset{\circ}{1} \in \mathcal{L}$ we denote the $\sim$-equivalence
class containing the language $\{a\}^*$ for some $a \in \Sigma_\infty$ and by $\emptyset \in \mathcal{L}$ we denote
the $\sim$-equivalence class containing the language $\emptyset$.

A *lattice* $\langle P; \leq, +, \times \rangle$ is a partially ordered set in which for every two elements
$a, b \in P$ there exists a least upper bound, denoted by $a + b$, and a greatest lower
bound, denoted by $a \times b$.

A family $\mathfrak{F}$ of languages is called *lattice family* if $\mathfrak{F}$ is closed under isomorphism, plus $+$ and times $\times$, and contains $\emptyset$ and Σ^* for all finite $\Sigma \subset \Sigma_\infty$.

Theorem 31 (Kuich, Sauer, Urbanek [4]) $\langle \mathcal{L}; \leq, +, \times \rangle$ *is a lattice with least element* $\emptyset$ *and largest element* $\overset{\circ}{1}$. *If* $\mathfrak{F}$ *is a lattice family of languages then* $\langle \mathcal{L}_{\mathfrak{F}}; \leq, +, \times \rangle$ *is a lattice with least element* $\emptyset$ *and largest element* $\overset{\circ}{1}$.

Lemma 32 *For all* $L_1 \subseteq \Sigma_1^*$ *and* $L_2 \subseteq \Sigma_2^*$ *there exist length preserving morphisms* H, H_1, H_2 *such that*

$$L_1 + L_2 = L_1 \cup H^{-1}(L_2) \quad and \quad L_1 \times L_2 = H_1^{-1}(L_1) \cap H_2^{-1}(L_2).$$

Proof. (i) If $\Sigma_1 \cap \Sigma_2 = \emptyset$ then $H : \Sigma_2^* \to \Sigma_2^*$ is the identity. If $\Sigma_1 \cap \Sigma_2 \neq \emptyset$ then create the new alphabet $\bar{\Sigma} = \{\bar{a} \mid a \in \Sigma_2\}$ and define $H : \bar{\Sigma}^* \to \Sigma_2^*$ by $H(\bar{a}) = a$, $a \in \Sigma_2$.

(ii) Define $H_i : (\Sigma_1 \times \Sigma_2)^* \to \Sigma_i^*$, $i = 1, 2$, by $H_1([a, b]) = a$ and $H_2([a, b]) = b$, $a \in \Sigma_1$, $b \in \Sigma_2$. Then $L_1 \times L_2 = H_1^{-1}(L_1) \cap H_2^{-1}(L_2)$. $\qquad\square$

From this and the previous theorem we conclude the following theorem.

Theorem 33 *1. If* $\mathfrak{F}$ *is a family of languages closed under union, intersection and inverse length preserving morphism then* $\mathfrak{F}$ *is also closed under addition and multiplication.*

  2. If $\mathfrak{F}$ *is a family of languages that contains* $\emptyset$ *and* Σ^* *for all finite* $\Sigma \subseteq \Sigma_\infty$ *and that is closed under union, intersection, and inverse length preserving morphism then* $\mathfrak{F}$ *is a lattice family.*

Corollary 34 *The following families of languages are lattice families:*
(i)   REG, CSL, LOGCFL, and RUD.
(ii) L, NL, P, Q, NP, and PSPACE.
(iii) Σ_k^{P}, Π_k^{P} for $k \geq 1$, and PH.
(iv) DTIME(t) and NTIME(t) for $t(n) \geq n$.
(v)   DSPACE(s) and NSPACE(s) for $s(n) \geq n$.

Proof. This is an immediate consequence of Theorem 21 $\qquad\qquad\square$

Let $\Sigma = \{h \mid h : \Sigma_1 \to \Sigma_2\}$ be the set of all functions $h : \Sigma_1 \to \Sigma_2$ considered as an alphabet. This alphabet is denoted by $\Sigma_2^{\Sigma_1}$. For $f = h_1 \ldots h_n \in \Sigma^n$ and $w = a_1 \ldots a_m \in \Sigma_1^m$ define

$$f(w) = \begin{cases} h_1(a_1) \ldots h_n(a_n) & \text{if } n = m \\ \text{undefined} & \text{if } n \neq m. \end{cases}$$

(and $\varepsilon(\varepsilon) = \varepsilon$ if $n = 0$). For $L_1 \subseteq \Sigma_1^*$, $L_2 \subseteq \Sigma_2^*$ define

$$L_2^{L_1} = \{f \in \Sigma^* \mid f(w) \in L_2 \text{ for all } w \in L_1 \text{ for which } f(w) \text{ is defined}\}.$$

Observe that $L_2^{L_1}$ depends on the sets Σ_1 and Σ_2.

The notion of exponentiation lifts to $\sim$-equivalence classes of languages. Hence, for $\sim$-equivalence classes of languages L_1 and L_2 the class $L_2^{L_1}$ is independent of the alphabets.

A lattice $\langle P; \leq, +, \times \rangle$ is called *Heyting algebra* if (i) for all $a, b \in P$ there exists a greatest $c \in P$ such that $a \times c \leq b$. This element c is denoted by b^a. It is called the *exponentiation* of b by a. (ii) There exists a least element 0 in P.

A family $\mathfrak{F}$ of languages is *stable* if it is a lattice family and closed under exponentiation and intersection with regular languages.

Theorem 35 (Kuich, Sauer, Urbanek [4]) *Let $\mathfrak{F}$ be a stable family of languages. Then $\langle \mathcal{L}_{\mathfrak{F}}; \leq, +, \times \rangle$ is a Heyting algebra, where the class $\emptyset$ is the 0-element and $\overset{\circ}{1}$ is the largest element.*

Hence, for the equivalence classes of $\mathcal{L}_{\mathfrak{F}}$, where $\mathfrak{F}$ is a stable family of languages, the computation rules given in Kuich, Sauer, Urbanek [4], Corollary 2.3, are valid, e. g., $L^{L_1 + L_2} = L^{L_1} \times L^{L_2}$, $(L^{L_1})^{L_2} = L^{L_1 \times L_2}$, $(L_1 \times L_2)^L = L_1^L \times L_2^L$ for all $L, L_1, L_2 \in \mathcal{L}_{\mathfrak{F}}$.

For $L \subseteq \Sigma^*$ we define the complement of L by $\mathrm{compl}_{\Sigma}(L) = \Sigma^* - L$.

Lemma 36 *For all $L_1 \subseteq \Sigma_1^*$ and $L_2 \subseteq \Sigma_2^*$ there exist length preserving morphisms H_1, H_2, H_3 such that*

$$L_2^{L_1} = \mathrm{compl}_{\Sigma}(H_3(H_1^{-1}(L_1) \cap H_2^{-1}(\mathrm{compl}_{\Sigma_2}(L_2)))),$$

where $\Sigma = \Sigma_2^{\Sigma_1}$.

Proof. Define the morphisms $H_1 : (\Sigma \times \Sigma_1)^* \to \Sigma_1^*$, $H_2 : (\Sigma \times \Sigma_1)^* \to \Sigma_2^*$ and $H_3 : (\Sigma \times \Sigma_1)^* \to \Sigma^*$ by $H_1([h, a]) = a$, $H_2([h, a]) = h(a)$ and $H_3([h, a]) = h$ for all $h \in \Sigma$ and $a \in \Sigma_1$. Then, for all $h_1, \ldots, h_n \in \Sigma$, $n \geq 0$,

$$
\begin{aligned}
h_1 \ldots h_n \in \mathrm{compl}_{\Sigma}(L_2^{L_1}) &\Leftrightarrow \exists a_1, \ldots, a_n (a_1 \ldots a_n \in L_1 \wedge h_1(a_1) \ldots h_n(a_n) \in \mathrm{compl}_{\Sigma_2}(L_2)) \\
&\Leftrightarrow \exists a_1, \ldots, a_n (H_1([h_1, a_1]) \ldots H_1([h_n, a_n]) \in L_1 \\
&\qquad\qquad\qquad \wedge H_2([h_1, a_1]) \ldots H_2([h_n, a_n]) \in \mathrm{compl}_{\Sigma_2}(L_2)) \\
&\Leftrightarrow \exists a_1, \ldots, a_n ([h_1, a_1] \ldots [h_n, a_n] \in H_1^{-1}(L_1) \cap H_2^{-1}(\mathrm{compl}_{\Sigma_2}(L_2))) \\
&\Leftrightarrow h_1 \ldots h_n \in H_3(H_1^{-1}(L_1) \cap H_2^{-1}(\mathrm{compl}_{\Sigma_2}(L_2))). \qquad \square
\end{aligned}
$$

From this and the previous theorem we conclude the following theorem.

Theorem 37 *1. If $\mathfrak{F}$ is a family of languages closed under union, complement, inverse length preserving morphism and length preserving morphism then $\mathfrak{F}$ is also closed under exponentiation.*

2. If $\mathfrak{F}$ is a family of languages that contains $\emptyset$ and Σ^* for all finite $\Sigma \subseteq \Sigma_{\infty}$ and that is closed under union, complement, inverse length preserving morphism, length preserving morphism and intersection with regular languages then $\mathfrak{F}$ is stable.

From this and Theorem 21 we obtain

Corollary 38 *The following families of languages are stable (and hence closed under exponentiation):*

(i)   REG, CSL, *and* RUD.
(ii) PH *and* PSPACE.
(iii) DSPACE(*s*) *for* $s(n) \geq n$.
(iv) NSPACE(*s*) *for space-constructible* $s(n) \geq n$.

4   On the Power of Exponentiation

In this section we will compare the power of exponentiation with the power of complement and ε-free morphism. In this comparision some other operations play a role, namely union, intersection with regular languages, and inverse morphism. However, these operations are weak in the sense that they do not really add power to language classes, they only smooth them. Practically all formal language classes and complexity classes are closed under these operations. On the other side, the operations of length preserving morphism and complement are more powerful: ε-free morphisms introduce nondeterminism, and the class of context free languages, for example, is not closed under complement.

In this section we prove that, in the presence of the above mentioned weak operations, ε-free morphism and complement on the one side and exponentiation on the other side are equally powerful.

We start with two lemmas showing how length preserving morphism and complementation can be generated by exponentiation.

For $\Sigma \subset \Sigma_\infty$ we define $E_\Sigma \subseteq (\Sigma \times \Sigma)^*$ by $E_\Sigma = \{[x,x] \mid x \in \Sigma\}^+$. Observe that $\text{compl}_{\Sigma \times \Sigma}(E_\Sigma)$ is a regular language.

Lemma 41 *For $L \subseteq \Sigma^*$ there exists a length preserving morphism $H : \Sigma^* \to ((\Sigma \times \Sigma)^\Sigma)^*$ such that $\text{compl}_\Sigma(L) = H^{-1}((\text{compl}_{\Sigma \times \Sigma}(E_\Sigma))^L)$.*

Proof. We define $h_b : \Sigma \to \Sigma \times \Sigma$ by $h_b(a) = [a,b]$ and the morphism H by $H(b) = h_b$ for all $a,b \in \Sigma$. Then, for $b_1,\dots,b_n \in \Sigma$, the equivalence

$$b_1 \dots b_n \in L \Leftrightarrow \exists a_1,\dots,a_n(a_1 \dots a_n \in L \wedge a_1 \dots a_n = b_1 \dots b_n)$$

implies the equivalences

$$
\begin{aligned}
b_1 \dots b_n \in \text{compl}_\Sigma(L) &\Leftrightarrow \forall a_1,\dots,a_n(a_1 \dots a_n \in L \Rightarrow a_1 \dots a_n \neq b_1 \dots b_n) \\
&\Leftrightarrow \forall a_1,\dots,a_n(a_1 \dots a_n \in L \Rightarrow [a_1,b_1]\dots[a_n,b_n] \in \text{compl}_{\Sigma \times \Sigma}(E_\Sigma)) \\
&\Leftrightarrow \forall a_1,\dots,a_n(a_1 \dots a_n \in L \Rightarrow h_{b_1}(a_1)\dots h_{b_n}(a_n) \in \text{compl}_{\Sigma \times \Sigma}(E_\Sigma)) \\
&\Leftrightarrow h_{b_1} \dots h_{b_n} \in \text{compl}_{\Sigma \times \Sigma}(E_\Sigma)^L \\
&\Leftrightarrow H(b_1 \dots b_n) \in \text{compl}_{\Sigma \times \Sigma}(E_\Sigma)^L \\
&\Leftrightarrow b_1 \dots b_n \in H^{-1}(\text{compl}_{\Sigma \times \Sigma}(E_\Sigma)^L) \qquad\qquad \square
\end{aligned}
$$

For a length preserving morphism $h : \Sigma_1^* \to \Sigma_2^*$ we define $E_h = \{[x,h(x)] \mid x \in \Sigma_1\}^+$. Observe that E_h is a regular language.

Lemma 42 *For $L \subseteq \Sigma_1^*$ and a length preserving morphism $h : \Sigma_1^* \to \Sigma_2^*$ there exist length preserving morphisms $H_1 : \Sigma_2^* \to ((\Sigma_1 \times \Sigma_2)^{\Sigma_1})^*$ and $H_2 : (\Sigma_1 \times \Sigma_2)^* \to \Sigma_1^*$ such that*

$$h(L) = \text{compl}_{\Sigma_2}(H_1^{-1}(\text{compl}_{\Sigma_1 \times \Sigma_2}(E_h \cap H_2^{-1}(L))^{\Sigma_1^*})).$$

Proof. We define $h_b : \Sigma_1 \to \Sigma_1 \times \Sigma_2$ by $h_b(a) = [a, b]$, H_1 by $H_1(b) = h_b$, and H_2 by $H_2([a, b]) = a$ for all $a \in \Sigma_1$, $b \in \Sigma_2$. Then, for $b_1, \ldots, b_n \in \Sigma_2$, the equivalence

$$b_1 \ldots b_n \in h(L) \Leftrightarrow \exists a_1, \ldots, a_n (h(a_1 \ldots a_n) = b_1 \ldots b_n \wedge a_1 \ldots a_n \in L)$$

implies the equivalences
$b_1 \ldots b_n \in \mathrm{compl}_{\Sigma_2}(h(L)) \Leftrightarrow$

$$\Leftrightarrow \forall a_1, \ldots, a_n (a_1 \ldots a_n \in \Sigma_1^* \Rightarrow \neg(h(a_1 \ldots a_n) = b_1 \ldots b_n \wedge a_1 \ldots a_n \in L))$$
$$\Leftrightarrow \forall a_1, \ldots, a_n (a_1 \ldots a_n \in \Sigma_1^* \Rightarrow [a_1, b_1] \ldots [a_n, b_n] \notin E_h \cap H_2^{-1}(L))$$
$$\Leftrightarrow \forall a_1, \ldots, a_n (a_1 \ldots a_n \in \Sigma_1^*$$
$$\Rightarrow h_{b_1}(a_1) \ldots h_{b_n}(a_n) \in \mathrm{compl}_{\Sigma_1 \times \Sigma_2}(E_h \cap H_2^{-1}(L)))$$
$$\Leftrightarrow h_{b_1} \ldots h_{b_n} \in \mathrm{compl}_{\Sigma_1 \times \Sigma_2}(E_h \cap H_2^{-1}(L))^{\Sigma_1^*}$$
$$\Leftrightarrow H_1(b_1 \ldots b_n) \in \mathrm{compl}_{\Sigma_1 \times \Sigma_2}(E_h \cap H_2^{-1}(L))^{\Sigma_1^*}$$
$$\Leftrightarrow b_1 \ldots b_n \in H_1^{-1}(\mathrm{compl}_{\Sigma_1 \times \Sigma_2}(E_h \cap H_2^{-1}(L))^{\Sigma_1^*}). \qquad \square$$

The next lemma shows how ε-free morphisms can be generated by length preserving morphisms (cf. [3]).

Lemma 43 *Consider $L \subseteq \Sigma_1^*$ and an ε-free morphism $h : \Sigma_1^* \to \Sigma_2^*$. Then there exists a length preserving morphism $h' : \Sigma^* \to \Sigma_2^*$, a morphism $H : \Sigma^* \to \Sigma_1^*$, and a regular set $R \subseteq \Sigma^*$ such that*

$$h(L) = h'(H^{-1}(L) \cap R).$$

Proof. Let $\Sigma_1 = \{a_1, \ldots, a_k\}$, and let $h(a_i) = b_{i1} b_{i2} \ldots b_{ir_i}$ for $i = 1, \ldots, k$. Define the alphabet Σ by $\Sigma = \{a_{ij} \mid i = 1, \ldots, k \text{ and } j = 1, \ldots, r_i\}$, the length preserving morphism h' by $h'(a_{ij}) = b_{ij}$ for $i = 1, \ldots, k$ and $j = 1, \ldots, r_i$, the morphism H by $H(a_{i1}) = a_i$, $H(a_{ij}) = \varepsilon$ for $i = 1, \ldots, k$ and $j = 2, \ldots, r_i$, and the regular set R by $R = \{a_{i1} a_{i2} \ldots a_{ir_i} \mid i = 1, \ldots, k\}^*$. Then we obtain

$$
\begin{aligned}
h(L) &= \{h(a_{i_1} a_{i_2} \ldots a_{i_n}) \mid a_{i_1} a_{i_2} \ldots a_{i_n} \in L\} \\
&= \{b_{i_1 1} \ldots b_{i_1 r_{i_1}} b_{i_2 1} \ldots b_{i_2 r_{i_2}} \ldots b_{i_n 1} \ldots b_{i_n r_{i_n}} \mid a_{i_1} a_{i_2} \ldots a_{i_n} \in L\} \\
&= \{h'(a_{i_1 1} \ldots a_{i_1 r_{i_1}} a_{i_2 1} \ldots a_{i_2 r_{i_2}} \ldots a_{i_n 1} \ldots a_{i_n r_{i_n}}) \mid a_{i_1} a_{i_2} \ldots a_{i_n} \in L\} \\
&= h'(\{a_{i_1 1} \ldots a_{i_1 r_{i_1}} a_{i_2 1} \ldots a_{i_2 r_{i_2}} \ldots a_{i_n 1} \ldots a_{i_n r_{i_n}} \mid a_{i_1} a_{i_2} \ldots a_{i_n} \in L\}) \\
&= h'(H^{-1}(L) \cap R). \qquad \square
\end{aligned}
$$

Using this notation we immediately obtain the following consequences from Lemma 36, Lemma 41, Lemma 42, and Lemma 43.

Corollary 44 *For any family $\mathfrak{F}$ of languages there holds:*

1. $\Gamma_{\exp}(\mathfrak{F}) \subseteq \Gamma_{\cup, \mathrm{lh}^{-1}, \mathrm{lh}, -}(\mathfrak{F})$
2. $\Gamma_{-}(\mathfrak{F}) \subseteq \Gamma_{\mathrm{lh}^{-1}, \exp}(\mathfrak{F} \cup \mathrm{REG})$
3. $\Gamma_{\mathrm{lh}}(\mathfrak{F}) \subseteq \Gamma_{\cap \mathrm{REG}, \mathrm{lh}^{-1}, -, \exp}(\mathfrak{F})$
4. $\Gamma_{\varepsilon \mathrm{h}}(\mathfrak{F}) \subseteq \Gamma_{\cap \mathrm{REG}, \mathrm{h}^{-1}, \mathrm{lh}}(\mathfrak{F})$

Now we can prove the main theorem of this section. Informally it says that, in the presence of the weak operations $\cup$ and h^{-1}, the operation exp is as powerful as the operations $\varepsilon \mathrm{h}$ and $^{-}$ (lh and $^{-}$, resp.).

Theorem 45 *For a family $\mathfrak{F}$ of languages that contains* REG, *there holds*

1. $\Gamma_{\cup,\mathrm{lh}^{-1},\mathrm{lh},-}(\mathfrak{F}) = \Gamma_{\cup,\mathrm{lh}^{-1},\exp}(\mathfrak{F})$.
2. $\Gamma_{\cup,\mathrm{h}^{-1},\varepsilon\mathrm{h},-}(\mathfrak{F}) = \Gamma_{\cup,\mathrm{h}^{-1},\mathrm{lh},-}(\mathfrak{F}) = \Gamma_{\cup,\mathrm{h}^{-1},\exp}(\mathfrak{F})$.

Proof. We conclude

$$
\begin{aligned}
\Gamma_{\cup,\mathrm{lh}^{-1},\mathrm{lh},-}(\mathfrak{F}) &\subseteq \Gamma_{\cup,\mathrm{lh}^{-1},\cap\mathrm{REG},-,\exp}(\mathfrak{F}) = \Gamma_{\cup,\mathrm{lh}^{-1},-,\exp}(\mathfrak{F}) && \text{(Lemma 44.3)} \\
&\subseteq \Gamma_{\cup,\mathrm{lh}^{-1},\exp}(\mathfrak{F}) && \text{(Lemma 44.2)} \\
&\subseteq \Gamma_{\cup,\mathrm{lh}^{-1},\mathrm{lh},-}(\mathfrak{F}) && \text{(Lemma 44.1)}
\end{aligned}
$$

and

$$
\begin{aligned}
\Gamma_{\cup,\mathrm{h}^{-1},\varepsilon\mathrm{h},-}(\mathfrak{F}) &\subseteq \Gamma_{\cup,\mathrm{h}^{-1},\cap\mathrm{REG},\mathrm{lh},-}(\mathfrak{F}) = \Gamma_{\cup,\mathrm{h}^{-1},\mathrm{lh},-}(\mathfrak{F}) && \text{(Lemma 44.4)} \\
&\subseteq \Gamma_{\cup,\mathrm{h}^{-1},\cap\mathrm{REG},-,\exp}(\mathfrak{F}) = \Gamma_{\cup,\mathrm{h}^{-1},-,\exp}(\mathfrak{F}) && \text{(Lemma 44.3)} \\
&\subseteq \Gamma_{\cup,\mathrm{h}^{-1},\exp}(\mathfrak{F}) && \text{(Lemma 44.2)} \\
&\subseteq \Gamma_{\cup,\mathrm{h}^{-1},\mathrm{lh},-}(\mathfrak{F}) && \text{(Lemma 44.1)} \\
&\subseteq \Gamma_{\cup,\mathrm{h}^{-1},\varepsilon\mathrm{h},-}(\mathfrak{F}) && \square
\end{aligned}
$$

Corollary 46   1. *Let $\mathfrak{F}$ be a family of languages that contains* REG *and is closed under union and inverse length preserving morphism. Then $\mathfrak{F}$ is closed under exponentiation iff it is closed under length preserving morphism and complement.*

2. *Let $\mathfrak{F}$ be a family of languages that contains* REG *and is closed under union and inverse morphism. Then $\mathfrak{F}$ is closed under exponentiation iff it is closed under ε-free morphism and complement.*

From this corollary we get directly the following three corollaries.

Corollary 47 *Let $\mathfrak{F}$ be a family of languages that contains* REG *and is closed under union, inverse length preserving morphism, and length preserving morphism. Then $\mathfrak{F}$ is closed under complement iff it is closed under exponentiation.*

A family of languages is called a *semi-AFL* if it is closed under union, inverse morphism, ε-free morphism, and intersection with regular languages and if it contains $\emptyset$ and Σ^* for all $\Sigma \subseteq \Sigma_\infty$. (see [3]).

Corollary 48 *A semi-AFL is closed under complement iff it is closed under exponentiation.*

Corollary 49   1. *Let $\mathfrak{F}$ be a family of languages that contains* REG *and is closed under union, complement and inverse length preserving morphism. Then $\mathfrak{F}$ is closed under length preserving morphism iff it is closed under exponentiation.*

2. *Let $\mathfrak{F}$ be a family of languages that contains* REG *and is closed under union, complement and inverse morphism. Then $\mathfrak{F}$ is closed under ε-free morphism iff it is closed under exponentiation.*

5   Application to Language Classes

In this section we apply the results of the previous section to the language classes mentioned in Section 2. In the case that a class is not closed under exponentiation we will characterize the closure of this class under exponentiation. In the case that it is not known whether the class is closed under exponentiation we will give equivalent conditions for the class being closed under exponentiation.

Let us start with the class CFL of context-free languages. By Lemma 2.1 of Kuich, Sauer, Urbanek [4], the context-free languages are not closed under exponentiation. We are now able to determine the closure of CFL under exponentiation (together with some "weak" operations).

The class RUD of rudimentary languages, introduced by Smullyan in [5], can be considered as the linear time analogon of the class PH of the polynomial time hierarchy. From Theorem 45.2 and Theorem 21 we obtain the following theorem.

Theorem 51 *The class* RUD *coincides with the closure of* CFL *under union, inverse morphism and exponentiation.*

Now we turn to classes which are not known to be closed under exponentiation. We start with some classes between L and P.

Theorem 52 *Let $\mathfrak{F}$ be a family of languages that is closed under union, complement, and logarithmic space many-one reducibility and that fulfills $L \subseteq \mathfrak{F} \subseteq NP$. Then $\mathfrak{F}$ is closed under exponentiation iff $\mathfrak{F} = NP$.*

Proof. Obviously, closure under logarithmic space many-one reducibility implies closure under inverse morphism. By Corollary 49.2 we obtain that $\mathfrak{F}$ is closed under exponentiation iff it is closed under ε-free morphism.

If $\mathfrak{F}$ is closed under ε-free morphism, then we obtain $\mathfrak{F} = \Gamma_{\varepsilon h}(\mathfrak{F}) \supseteq \Gamma_{\varepsilon h}(L)$. A result by Springsteel [6] says that $\Gamma_{\varepsilon h}(L) \supseteq Q$. Hence $\mathfrak{F} \supseteq Q$. The class Q contains sets which are logarithmic space many-one complete for NP. Since $\mathfrak{F}$ is closed under logarithmic space many-one reducibility we get $\mathfrak{F} \supseteq NP$ and hence $\mathfrak{F} = NP$.

On the other side, if $\mathfrak{F} = NP$ then, by Theorem 21, $\mathfrak{F}$ is closed under ε-free morphism.   □

Since the classes L, NL, LOGCFL, P, and NP∩coNP are closed under union, complement, and logarithmic space many-one reducibility, we obtain the following corollary.

Corollary 53   *1. L is closed under exponentiation iff $L = NP$.*

2. NL is closed under exponentiation iff $NL = NP$.

3. LOGCFL is closed under exponentiation iff $LOGCFL = NP$.

4. P is closed under exponentiation iff $P = NP$.

5. NP ∩ coNP is closed under exponentiation iff $NP = coNP$.

The classes in the previous corollary are closed under complement but not known to be closed under ε-free morphism. For the nondeterministic time classes Q, NP, NTIME(t) and Σ_k^{P} the opposite is true. Here we can apply Corollary 47.

Theorem 54 *1. Q is closed under exponentiation iff* Q = co-Q.

2. NP is closed under exponentiation iff NP = co-NP.

3. For every increasing $t : \mathbb{N} \to \mathbb{N}$ such that $t(n) \geq n$,
NTIME(t) is closed under exponentiation iff NTIME(t) = co-NTIME(t).

4. Σ_k^{P} is closed under exponentiation iff $\Sigma_k^{\mathrm{P}} = \Pi_k^{\mathrm{P}}$.

Note that Q = co-Q implies NP = co-NP, and NP = co-NP implies $\Sigma_k^{\mathrm{P}} = \Pi_k^{\mathrm{P}}$ for $k \geq 2$ (cf. [7]).

Finally we consider the classes Π_k^{P} of the polynomial-time hierarchy.

Theorem 55 *For $k \geq 1$, the class Π_k^{P} is closed under exponentiation iff $\Pi_k^{\mathrm{P}} = \Sigma_k^{\mathrm{P}}$.*

Proof. If Π_k^{P} is closed under exponentiation then, by Corollary 44.2 and Theorem 21, Π_k^{P} is closed under complementation, i.e., $\Pi_k^{\mathrm{P}} = \Sigma_k^{\mathrm{P}}$.

On the other side, if $\Pi_k^{\mathrm{P}} = \Sigma_k^{\mathrm{P}}$ then $\Pi_k^{\mathrm{P}} = \mathrm{PH}$. By Corollary 38 we obtain that Π_k^{P} is closed under exponentiation. $\square$

References

[1] Balcázar J.L., Díaz   J., Gabarró   J.: Structural Complexity I. Second edition. Springer-Verlag Berlin, 1995.

[2] Balbes R., Dwinger P.: Distributive Lattices. University of Missouri Press, 1974.

[3] Ginsburg S.: Algebraic and Automata-Theoretic Properties of Formal Languages. North-Holland, 1975.

[4] Kuich W., Sauer N., Urbanek F.: Heyting algebras and formal languages. J.UCS 8(2002), 722–736.

[5] Smullyan R.: Theory of Formal Systems. Annals of Mathematical Studies vol. 47. Princeton University Press, 1961.

[6] Springsteel F.N.: On the pre-AFL of log$n$ space and related families of languages. Theoretical Computer Science 2(1976), 295–303.

[7] Wagner K., Wechsung G.: Computational Complexity. Deutscher Verlag der Wissenschaften, 1986.

Kleene's Theorem for Weighted Tree-Automata

Christian Pech*

Technische Universität Dresden
Fakultät für Mathematik und Naturwissenschaften
D-01062 Dresden, Germany
pech@math.tu-dresden.de

Abstract. We sketch the proof of a Kleene-type theorem for formal tree-series over commutative semirings. That is, for a suitable set of rational operations we show that the proper rational formal tree-series coincide with the recognizable ones. A complete proof is part of the PhD-thesis of the author, which is available at [9].

Keywords: tree, automata, weight, language, Kleene's theorem, Schützenberger's theorem, rational expression.

A formal tree-series is a function from the set T_Σ of trees over a given ranked alphabet Σ into a semiring K. The classical notion of formal tree-languages is obtained if K is chosen to be the Boolean semiring.

Rational operations on formal tree-languages like sum, topcatenation, a-multiplication etc. have been used by Thatcher and Wright [11] to characterize the recognizable formal tree-languages by rational expressions. Thus they generalized the classical Kleene-theorem [6] stating that rational and recognizable formal languages coincide.

The rational operations on tree-languages can be generalized to formal tree-series. We would like to know the generating power of these operations. There are several results on this problem—each for some restricted class of semirings—saying that for formal tree-series the rational series coincide with the recognizable series, too. In particular it was shown by Kuich [7] for complete, commutative semirings, by Bozapalidis [3] for ω-additive, commutative semirings, by Bloom and Ésik [2] for commutative Conway-semirings and by Droste and Vogler [5] for idempotent, commutative semirings. The necessary restrictions on the semiring are in contrast with the generality of Schützenbergers theorem for formal power series (i.e. functions from Σ^* into a semiring) [10] that is completely independent of the semiring.

Here we develop a technique how to restrict the list of requirements to a minimum. The main idea is that instead of working directly with formal tree-series, we introduce the notion of weighted tree-languages. They form a category which algebraically is closer related to formal tree-languages than to formal tree-series. The environment that we obtain allows us to translate the known constructions of the rational operations directly to weighted tree-languages.

---

* This work was supported by the German Research Council (DFG, GRK 433/2).

A. Lingas and B.J. Nilsson (Eds.): FCT 2003, LNCS 2751, pp. 387–399, 2003.
© Springer-Verlag Berlin Heidelberg 2003

On the level of weighted tree-languages we can prove a Kleene-type theorem . Its proof is rather conventional and often uses classical automata-theoretic constructions tailored to the new categorical setting of weighted tree-languages.

Upto this point the results do not depend on the semiring at all. Only when translating our results to formal tree-series the unavoidable restriction to the semiring becomes apparent. Luckily we only need to require the coefficient-semiring to be commutative—a very mild restriction given that almost all semirings, that are actually used in applications like image compression (cf. [4]) or natural language processing (cf. [8]) are commutative.

1   Preliminaries

A *ranked alphabet* (or *ranked set*) is a pair (Σ, rk) where Σ is a set of letters (an alphabet) and rk : $\Sigma \to \mathbb{N}$ assigns to each letter its rank. With $\Sigma^{(n)}$ we denote the set of letters from Σ with rank n. For any set X disjoint from Σ we define $\Sigma(X) := (\Sigma \cup X, \text{rk}')$ where $\text{rk}'|_\Sigma := \text{rk}$ and $\text{rk}'(x) := 0$ for all $x \in X$. If X consists just of one element x then we also write $\Sigma(x)$ instead of $\Sigma(\{x\})$.

The set T_Σ of *trees* is the smallest set of words such that $\Sigma^{(0)} \subseteq T_\Sigma$ and if $f \in \Sigma^{(n)}$, $t_1, \ldots, t_n \in T_\Sigma$, then $f\langle t_1, \ldots, t_n \rangle \in T_\Sigma$.

A *semiring* is a quintuple $(K, \oplus, \odot, 0, 1)$ such that $(K, \oplus, 0)$ is a commutative monoid, $(K, \odot, 1)$ is a monoid and the following identities hold: $(x \oplus y) \odot z = (x \odot z) \oplus (y \odot z)$, $x \odot (y \oplus z) = (x \odot y) \oplus (x \odot z)$ and $x \odot 0 = 0 \odot x = 0$.

The set WT_Σ of *weighted trees* is the smallest set of words such that $[a|c] \in \text{WT}_\Sigma$ for all $a \in \Sigma^{(0)}$, $c \in K$ and if $f \in \Sigma^{(n)}$, $t_1, \ldots, t_n \in \text{WT}_\Sigma$, $c \in K$, then $[f|c]\langle t_1, \ldots, t_n \rangle \in \text{WT}_\Sigma$. Each weighted tree t has an *underlying tree* $\text{ut}(t)$. This tree is obtained from t be deleting all weights from the nodes. Let $a \in \Sigma^{(0)}$. To each tree $s \in T_\Sigma$ we associate its *a-rank* $\text{rk}_a(s) \in \mathbb{N}$. This is just the number of occurrences of the letter a in s. The a-rank can be lifted to weighted trees according to $\text{rk}_a(t) := \text{rk}_a(\text{ut}(t))$ (for $t \in \text{WTL}_\Sigma$).

The semiring K acts naturally on WT_Σ from the left. In particular, for every $c, d \in K$: $d \cdot [a|c] := [a|d \odot c]$, $d \cdot [f|c]\langle t_1, \ldots, t_n \rangle := [f|d \odot c]\langle t_1, \ldots, t_n \rangle$. Obviously $(c \odot d) \cdot t = c \cdot (d \cdot t)$ for $c, d \in K$ and $t \in \text{WT}_\Sigma$.

For $a \in \Sigma^{(0)}$ we define the operation of *a-substitution* on WT_Σ. In particular, for $t \in \text{WT}_\Sigma$, $t_1, \ldots, t_{\text{rk}_a(t)} \in \text{WT}_\Sigma$ we define $t \circ_a \langle t_1, \ldots, t_{\text{rk}_a(t)} \rangle$ by induction on the structure of t: $[a|c] \circ_a \langle t_1 \rangle := c \cdot t_1$, $[b|c] \circ_a \langle \rangle := [b|c]$ (where $b \neq a$) and $[f|c]\langle t_1, \ldots, t_n \rangle \circ_a \langle s_{1,1}, \ldots, s_{n,m_n} \rangle := [f|c]\langle t_1 \circ_a \langle s_{1,1}, \ldots, s_{1,m_1} \rangle, \ldots, t_n \circ_a \langle s_{n,1}, \ldots, s_{n,m_n} \rangle \rangle$.

Next we equip WT_Σ with the structure of a ranked monoid[1]. Before we can do that, we need to introduce further notions:

A *ranked semigroup* is a triple $(S, \text{rk}, \circ)$ where (S, rk) is a ranked set and where $\circ = (\circ_i)_{i \in \mathbb{N}}$ is a family of composition operations $\circ_i : S^{(i)} \times S^i \to S$ where $\circ_i : (f, (g_1, \ldots, g_i)) \mapsto f \circ \langle g_1, \ldots, g_i \rangle$ such that $\text{rk}(f \circ \langle g_1, \ldots, g_i \rangle) = \text{rk}(g_1) + \cdots + \text{rk}(g_i)$, and

---

[1] These structures were already used by Berstel and Reutenauer [1] under the name "magma"; however, this leads to a name clash with another type of algebraic structures.

$$(f \circ \langle g_1, \ldots, g_n \rangle) \circ \langle h_{1,1}, \ldots, h_{1,m_1}, \ldots, h_{n,1}, \ldots, h_{n,m_n} \rangle$$
$$= f \circ \langle g_1 \circ \langle h_{1,1}, \ldots, h_{1,m_1} \rangle, \ldots, g_n \circ \langle h_{n,1}, \ldots, h_{n,m_n} \rangle \rangle.$$

The latter is called *superassociativity law*.

A *ranked monoid* is a tuple $(S, \mathrm{rk}, \circ, 1)$ where $(S, \mathrm{rk}, \circ)$ is a ranked semigroup and $1 \in S^{(1)}$ is a left- and right-unit of $\circ$. That is $x \circ \langle 1, \ldots, 1 \rangle = x$ and $1 \circ \langle y \rangle = y$ for all $x, y \in S$. Examples of ranked monoids are $(T_\Sigma, \mathrm{rk}_a, \circ_a, a)$ for $a \in \Sigma^{(0)}$ and $(\mathrm{WT}_\Sigma, \mathrm{rk}_a, \circ_a, [a|1])$.

Homomorphisms between ranked semigroups (monoids) are defined in the evident way—as rank-preserving functions between the carriers that additionally preserve the composition-operation (and the unit 1). Ranked semigroups and ranked monoids may be considered as a special kind of many-sorted algebras where the sorts are the natural numbers. Hence there exist free structures. The *free ranked monoid freely generated by a ranked alphabet* Σ will be denoted by $(\Sigma, \mathrm{rk})^*$. With $\Sigma' := \Sigma(\varepsilon)$ (where ε is a letter that is not in Σ) we have that

$$(\Sigma, \mathrm{rk})^* = (T_{\Sigma'}, \mathrm{rk}_\varepsilon, \circ_\varepsilon, \varepsilon). \tag{1}$$

2 Weighted Tree-Languages

Let K be a semiring and $\Sigma = (\Sigma, \mathrm{rk})$ be a ranked alphabet. A *weighted tree-language* is a pair $\mathcal{L} = (L, |.|)$ where L is a set, $|.| : L \to \mathrm{WT}_\Sigma : s \mapsto |s|$. Let $\mathcal{L}_1 = (L_1, |.|_1)$, $\mathcal{L}_2 = (L_2, |.|_2)$ be weighted tree-languages. A function $h : L_1 \to L_2$ is called *homomorphism* from $\mathcal{L}_1$ to $\mathcal{L}_2$ if for all $t \in L_1$ holds $|t|_1 = |h(t)|_2$. Thus the weighted tree-languages form a category which will be denoted by WTL_Σ. This category is complete and cocomplete. The forgetful functor $U : \mathrm{WTL}_\Sigma \to \mathrm{Set}$ creates colimits. Moreover WTL_Σ has an initial object $(\emptyset, \emptyset)$ and a terminal object $(\mathrm{WT}_\Sigma, 1_{\mathrm{WT}_\Sigma})$.

The *action of* K on WT_Σ may be extended to a functor on WTL_Σ. In particular, for $c \in K$ we define the functor $[c \cdot -] : \mathcal{L} \mapsto c \cdot \mathcal{L}, h \mapsto h$ where $c \cdot (L, |.|) := (L, |.|')$ such that $|.|' : t \mapsto c \cdot |t|$.

Next we define the *topcatenation*. Let $f \in \Sigma^{(n)}, c \in K$. Then we define the functor $[f|c]\langle -_1, \ldots, -_n \rangle : (\mathcal{L}_1, \ldots, \mathcal{L}_n) \mapsto [f|c]\langle \mathcal{L}_1, \ldots, \mathcal{L}_n \rangle, (h_1, \ldots, h_n) \mapsto h_1 \times \cdots \times h_n$ where for $\mathcal{L}_i = (L_i, |.|_i)$ $(i = 1, \ldots, n)$ $[f|c]\langle \mathcal{L}_1, \ldots, \mathcal{L}_n \rangle := (L_1 \times \cdots \times L_n, |.|)$ such that $|(t_1, \ldots, t_n)| := [f|c]\langle |t_1|_1, \ldots, |t_n|_n \rangle$.

Let $a \in \Sigma^{(0)}$. We will lift now the *a-substitution* from weighted trees to weighted tree-languages. We do this in two steps. First we define $t \cdot_a \mathcal{L}$ for $t \in \mathrm{WT}_\Sigma, \mathcal{L} \in \mathrm{WTL}_\Sigma$. Later we will define $\mathcal{L}_1 \cdot_a \mathcal{L}_2$ for $\mathcal{L}_1, \mathcal{L}_2 \in \mathrm{WTL}_\Sigma$. As usual we proceed by induction: $[[a|c] \cdot_a -] := [c \cdot -]$, $[[b|c] \cdot_a -] := C_{\{[b|c]\}}$ where $C_{\{[b|c]\}}$ is the constant functor that maps each language to $\{[b|c]\}$ and each homomorphism to the unit-homomorphism of $\{[b|c]\}$. $[[f|c]\langle t_1, \ldots, t_n \rangle \cdot_a -] := [f|c]\langle t_1 \cdot_a -, \ldots, t_n \cdot_a - \rangle$. The connection of this operation with the a-substitution on weighted trees is as follows. Let $t \in \mathrm{WT}_\Sigma$ with $\mathrm{rk}_a(t) = n$. Let $\mathcal{L} = (L, |.|) \in \mathrm{WTL}_\Sigma$. Then $t \cdot_a \mathcal{L} \cong (L^n, |.|_{t,a})$ where $|(t_1, \ldots, t_n)|_{t,a} := t \circ_a \langle |t_1|, \ldots, |t_n| \rangle$.

The *a-product* of two weighted tree-languages is now obtained by $[- \cdot_a \mathcal{L}_2] : \mathcal{L}_1 \mapsto \coprod_{t \in \mathcal{L}_1} |t|_1 \cdot_a \mathcal{L}_2$. The definition of this functor on homomorphisms is done pointwise in the evident way. Of course we can give a more transparent construction of this operation: Let L the set of words defined according to $L := \{t \circ_a \langle s_1, \ldots, s_{\mathrm{rk}_a(t)} \rangle \mid t \in L_1, s_1, \ldots, s_{\mathrm{rk}_a(t)} \in L_2\}$ and define a structure map $|.|$ on L according to $|t \circ_a \langle s_1, \ldots, s_{\mathrm{rk}_a(t)} \rangle| := |t|_1 \circ_a \langle |s_1|_2, \ldots, |s_{\mathrm{rk}_a(t)}|_2 \rangle$. Then $\mathcal{L}_1 \cdot_a \mathcal{L}_2 \cong (L, |.|)$.

As special case of the a-product we define $[-_{\neg a}] : \mathcal{L} \mapsto \mathcal{L}_{\neg a}$ where $\mathcal{L}_{\neg a} := \mathcal{L} \cdot_a \emptyset$. This operation is called a-annihilation.

Proposition 1. $[c \cdot -], [f|c]\langle -_1, \ldots, -_n \rangle, [-\cdot_a \mathcal{L}]$ and $[-_{\neg a}]$ preserve arbitrary colimits and monos. $[t \cdot_a -]$ and $[\mathcal{L} \cdot_a -]$ preserve directed colimits and monos.   □

Apart from the already defined operations on WTL_Σ we also have the coproduct-functor $[-_1 + -_2]$. We note that this functor also preserves directed colimits and monos. Recall that the composition of functors preserving directed colimits will again preserve directed colimits (the same holds for monos-preserving functors).

Our next step is to introduce some iteration operations on WTL_Σ. This can be done as for usual tree-languages, only using the appropriate categorical notions. Let us start with the a-iteration—a generalization of the Kleene-star for formal languages to weighted tree-languages. Define $\mathrm{S}_a : \mathrm{WTL}_\Sigma^2 \to \mathrm{WTL}_\Sigma : (X, \mathcal{L}) \mapsto (\mathcal{L} \cdot_a X) + \{[a|1]\}$. Then this functor preserves directed colimits. Since WTL_Σ has an initial object (the empty language), there exists an initial $\mathrm{S}_a(-, \mathcal{L})$-algebra $\mu X.\mathrm{S}_a(X, \mathcal{L})$. Its carrier may be chosen to be the colimit of the initial sequence

$$\emptyset \to \mathrm{S}_a(\emptyset, \mathcal{L}) \to \mathrm{S}_a^2(\emptyset, \mathcal{L}) \to \cdots$$

It is called the a-iteration of $\mathcal{L}$ and is denoted by $\mathcal{L}_a^*$. Next we will reveal a very nice connection between a-iteration and ranked monoids. It is this connection that makes the a-iteration a generalization of the Kleene-star.

Proposition 2. Given $\mathcal{L} = (L, |.|) \in \mathrm{WTL}_\Sigma$. For $t \in L$ set $\mathrm{rk}_a(t) := \mathrm{rk}_a(|t|)$. Let $(L, \mathrm{rk}_a)^*$ be the free ranked monoid generated by (L, rk_a). Let L_a^* be its carrier and let $|.|_a^*$ be the initial homomorphism from $(L, \mathrm{rk}_a)^*$ to $(\mathrm{WT}_\Sigma, \mathrm{rk}_a, \circ_a, [a|1])$ induced by $|.|$. Then $(L_a^*, |.|_a^*) \cong \mathcal{L}_a^*$.

Another important iteration operation is obtained from $\mathrm{R}_a : \mathrm{WTL}_\Sigma^2 \to \mathrm{WTL}_\Sigma : (X, \mathcal{L}) \mapsto \mathcal{L} \cdot_a X$. We call its initial algebra carrier $\mu X.\mathrm{R}_a(X, -)$ a-recursion. The a-recursion of a weighted tree-language $\mathcal{L}$ will be denoted by $\mathcal{L}_a^\mu$. A close relation of a-recursion to a-iteration is given by the fact that $\mathcal{L}_a^\mu \cong (\mathcal{L}_a^*)_{\neg a}$ for any $\mathcal{L} \in \mathrm{WTL}_\Sigma$.

Let us introduce a last iteration operation. Set $\mathrm{P}_a : \mathrm{WTL}_\Sigma^2 \to \mathrm{WTL}_\Sigma : (X, \mathcal{L}) \mapsto \mathcal{L} \cdot_a (X + \{[a|1]\})$. Then the initial algebra carrier $\mu X.\mathrm{P}_a(X, -) : \mathcal{L} \mapsto \mathcal{L}_a^+$ will be called a-semiiteration. The relation of this operation to a-iteration is given by $\mathcal{L}_a^* \cong \mathcal{L}_a^+ + \{[a|1]\}$. An immediate consequence is that $\mathcal{L}_a^\mu \cong (\mathcal{L}_a^+)_{\neg a}$.

The following two properties of weighted tree-languages will be important later, when we associate formal tree-series to weighted tree-languages. A weighted tree-language $\mathcal{L} = (L, |.|)$ is called finitary if for all $t \in \mathrm{T}_\Sigma$ the set $\{s \in L \mid \mathrm{ut}(|s|) = t\}$ is finite. It is called a-quasiregular (for some $a \in \Sigma^{(0)}$) if it does not contain any element s with $\mathrm{ut}(|s|) = a$. The full subcategory of WTL_Σ of all finitary weighted tree-languages will be denoted by WTL_Σ^f.

Proposition 3. *Let* $\mathcal{L}_1, \ldots, \mathcal{L}_n \in \text{WTL}_\Sigma^f$, $c \in K$, $f \in \Sigma^{(n)}$. *Then* $\mathcal{L}_1 + \mathcal{L}_2$, $c \cdot \mathcal{L}_1$, $[f|c]\langle \mathcal{L}_1, \ldots, \mathcal{L}_n\rangle$, $\mathcal{L}_1 \cdot_a \mathcal{L}_2$, $(\mathcal{L}_1)_{\neg a}$ *are all finitary again.* $\qquad\qquad\square$

Proposition 4. *Let* $\mathcal{L} \in \text{WTL}_\Sigma^f$. *Then* $\mathcal{L}_a^*$ *is finitary if and only if* $\mathcal{L}$ *is a-quasiregular.* $\qquad\qquad\square$

3   Weighted Tree-Automata

Given a ranked alphabet Σ and a semiring K, a finite *weak weighted tree-automaton* (wWTA) is a 7-tuple $(Q, I, \iota, T, \lambda, S, \sigma)$ where Q is a finite set of states, $I \subseteq Q$ is a set of initial states, $\iota : I \to K$ describes the initial weights, T is a finite ranked set of transition-symbols and λ is a function assigning to each transition-symbol $\tau \in T$ a transition where for $\tau \in T^{(n)}$ a transition is a tuple $(q, f, q_1, \ldots, q_n, c)$ such that $q, q_1, \ldots, q_n \in Q$, $f \in \Sigma^{(n)}$ and $c \in K$. Moreover, S is a finite set of silent transition-symbols and σ assigns to each silent transition-symbol a silent transition where a silent transition is a triple (q_1, q_2, c) for $q_1, q_2 \in Q, c \in K$. Let $\mathcal{A}$ be a wWTA. For convenience reasons for $\tau \in T$ with $\lambda(\tau) = (q, f, q_1, \ldots, q_n, c)$ we define $\text{lab}(\tau) := f, \text{wt}(\tau) := c$, $\text{dom}(\tau) := q$, $\text{cdom}_i(\tau) := q_i$ and $\text{cdom}(\tau) := \{q_1, \ldots, q_n\}$ and for $s \in S$ with $\sigma(s) = (q_1, q_2, c)$ we define $\text{dom}(s) := q_1$, $\text{cdom}(s) := q_2$ and $\text{wt}(s) := c$.

Let $\mathcal{A}$ be a wWTA. Runs through $\mathcal{A}$ are defined inductively: If $\tau \in T$, $\lambda(\tau) = (q, a, c)$, then τ is a run of $\mathcal{A}$ with root q along a. If $s \in S$, $\sigma(s) = (q, q', c)$ and p is a run of $\mathcal{A}$ with root q' along t. Then $s \cdot p$ is a run of $\mathcal{A}$ with root q along t. If finally $\tau \in T$, $\lambda(\tau) = (q, f, q_1, \ldots, q_n, c)$ and $p_1, \ldots, p_n$ are runs of $\mathcal{A}$ with root $q_1, \ldots, q_n$ along trees $t_1, \ldots, t_n$, respectively, then $\tau\langle p_1, \ldots, p_n\rangle$ is a run of $\mathcal{A}$ with root q along $f\langle t_1, \ldots, t_n\rangle$. The root of a run p will be denoted by $\text{root}(p)$. A run is called *initial* if its root is in I. With $\text{run}_t(\mathcal{A})$ we denote the set of all initial runs in $\mathcal{A}$ along t and with $\text{run}(\mathcal{A})$ we denote the set of all initial runs of $\mathcal{A}$. A (silent) transition symbol is called *reachable* if it is involved in some initial run of $\mathcal{A}$. A state of $\mathcal{A}$ is called *reachable* if it is the domain of some reachable (silent) transition-symbol.

To each run p of $\mathcal{A}$ we may associate a weighted tree $|p|$. This is done by induction on the structure of p. If $p = \tau$, $\lambda(\tau) = (q, a, c)$, then $|\tau| := [a|c]$. If $p = s \cdot p'$ with $\sigma(s) = (q_1, q_2, c)$, then $|p| := c \cdot |p'|$ and if $p = \tau\langle p_1, \ldots, p_n\rangle$, $\lambda(\tau) = (q, f, q_1, \ldots, q_n, c)$, then $|p| := [f|c]\langle |p_1|, \ldots, |p_n|\rangle$. The weighted tree-language *recognized by* $\mathcal{A}$ is defined as $\mathcal{L}_\mathcal{A} := (\text{run}(\mathcal{A}), |.|_\mathcal{A})$ where $|p|_\mathcal{A} := \iota(\text{root}(p)) \cdot |p|$. A weighted tree-language $\mathcal{L}$ is called *weakly recognizable* if there is a finite wWTA $\mathcal{A}$ with $\mathcal{L} \cong \mathcal{L}_\mathcal{A}$. Two wWTAs $\mathcal{A}_1, \mathcal{A}_2$ are called *equivalent* (denoted by $\mathcal{A}_1 \equiv \mathcal{A}_2$) if $\mathcal{L}_{\mathcal{A}_1} \cong \mathcal{L}_{\mathcal{A}_2}$. A wWTA $\mathcal{A}$ is called *reduced* if each of its states and (silent) transition-symbols is reachable. It is called *normalized* if it has precisely one initial state and the initial weight of this state is equal to 1. It is easy to see that for every wWTA $\mathcal{A}$ there is a reduced, normalized wWTA $\mathcal{A}'$ such that $\mathcal{A} \equiv \mathcal{A}'$. Therefore, from now on we will only consider normalized wWTAs.

Since the description of wWTAs by a tuple of certain set and mappings is tedious we sometimes prefer a graphical representation. In such a representation each transition-symbol τ with $\lambda(\tau) = (q, f, q_1, \ldots, q_n, c)$ will be depicted by

The output-arms are always ordered counterclockwise starting directly after the input-arm. The initial weights are depicted by arrows to the initial states carrying weights. Silent transition symbols are represented by arrows between states that are equipped with a weight. In normalized wWTAs we usually omit the arrow with the initial weight. Let us give a small example of a wWTA:

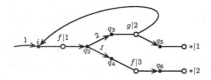

A *weighted tree-automaton* (WTA) is a wWTA with empty set of silent transition-symbols. A weighted tree-language $\mathcal{L}$ is called *recognizable* if there is a WTA $\mathcal{A}$ such that $\mathcal{L} \cong \mathcal{L}_{\mathcal{A}}$.

Proposition 5. *Let $\mathcal{L}_1, \ldots, \mathcal{L}_n$ be recognizable weighted tree-languages, $c \in K$, $f \in \Sigma^{(n)}$, $a \in \Sigma^{(0)}$. Then the $c \cdot \mathcal{L}_1$, $\mathcal{L}_1 + \mathcal{L}_2$, $[f|c]\langle \mathcal{L}_1, \ldots, \mathcal{L}_n \rangle$ and $\mathcal{L}_1 \cdot_a \mathcal{L}_2$ are also recognizable.* □

Note that recognizable weighted tree-languages are always finitary. In particular we can see already that the recognizable weighted tree-languages will not be closed with respect to a-iteration (e.g. $\{[a|c]\}_a^*$ is not recognizable).

It is clear that recognizability implies weak recognizability. However, the converse does not hold. In the next few paragraphs we will give necessary and sufficient conditions for a wWTA to recognize a recognizable weighted tree-language.

A word $\mathbf{s} = s_1 \cdots s_k \in S^*$ of silent transitions of $\mathcal{A}$ is called *silent path* if $\text{cdom}(s_i) = \text{dom}(s_{i+1})$ $(1 \le i < k)$. By convention, the empty word ε counts also as a silent path. We may extend dom and cdom to non-empty silent paths according to $\text{dom}(\mathbf{s}) := \text{dom}(s_1)$, $\text{cdom}(\mathbf{s}) := \text{cdom}(s_k)$. A silent path $\mathbf{s}$ with $\text{dom}(\mathbf{s}) = \text{cdom}(\mathbf{s})$ is called *silent cycle*. If any silent transition of a silent cycle is *reachable* then the cycle is called reachable. The set of all silent paths of $\mathcal{A}$ is denoted by $\text{sP}_{\mathcal{A}}$.

To each silent path $\mathbf{s} \in \text{sP}_{\mathcal{A}}$ we assign a weight $\text{wt}(\mathbf{s}) \in K$ according to $\text{wt}(\varepsilon) := 1$, $\text{wt}(s \cdot \mathbf{s}) := \text{wt}(s) \odot \text{wt}(\mathbf{s})$.

Silent cycles play a crucial role in the characterization of the finitary weakly recognizable weighted tree-languages.

Proposition 6. *Let $\mathcal{A}$ be a wWTA. Then $\mathcal{L}_{\mathcal{A}}$ is finitary if and only if $\mathcal{A}$ does not contain a reachable silent cycle.* □

Proposition 7. *Let $\mathcal{A}$ be a wWTA without reachable silent cycles. Then there is a WTA $\mathcal{A}'$ such that $\mathcal{L}_{\mathcal{A}} \cong \mathcal{L}_{\mathcal{A}'}$.*

Proof. Since the normalization and reduction of wWTAs do not introduce new silent cycles, we can assume that $\mathcal{A} = (Q, \{i\}, \iota, T, \lambda, S, \sigma)$ is normalized and reduced. Let $\mathcal{A}$ have no silent cycles. Then we claim that $\mathrm{sP}_\mathcal{A}$ is finite, for assume it is not, then it contains words of arbitrary length (because S is finite). Hence it would also contain a word of length $> |Q|$ but such a word contains necessarily a cycle—contradiction.

Let us construct the WTA $\mathcal{A}'$ now. Its state set is Q and the set of transitions T' of $\mathcal{A}'$ is defined as follows: $T' := \{(\mathbf{s}, t) \mid \mathbf{s} \in \mathrm{sP}_\mathcal{A}, t \in T, \mathbf{s} = \varepsilon \text{ or } \mathrm{cdom}(\mathbf{s}) = \mathrm{dom}(t)\}$ and $\lambda'(\mathbf{s}, t) := (q', f, q_1, \dots, q_n, c')$ where $\lambda(t) = (q, f, q_1, \dots, q_n, c)$ and $c' := \mathrm{wt}(\mathbf{s}) \odot c$ and where $q' = q$ if $\mathbf{s} = \varepsilon$ and $q' = \mathrm{dom}(\mathbf{s})$ else. Altogether $\mathcal{A}' = (Q, \{i\}, \iota, T', \lambda', \emptyset, \emptyset)$. We skip the proof that $\mathcal{A}'$ is indeed equivalent to $\mathcal{A}$. $\square$

As immediate consequence we get that a weakly recognizable weighted tree-language is recognizable if and only if it is finitary. Another important question is how to decide whether a given wWTA recognizes an a-quasiregular weighted tree-language. A short thought reveals that a wWTA $\mathcal{A}$ fails to be a-quasiregular if and only if either there is some $t \in T$ with $\mathrm{dom}(t) \in I$, $\mathrm{lab}(t) = a$ or there exists a silent path $\mathbf{s}$ starting in I and ending in a state that is the domain of a transition $t \in T$ with $\mathrm{lab}(t) = a$.

Proposition 8. *Let $\mathcal{L}_1, \dots, \mathcal{L}_n$ be weakly recognizable weighted tree-languages, $c \in K$, $f \in \Sigma^{(n)}$, $a \in \Sigma^{(0)}$. Then the $c \cdot \mathcal{L}_1$, $\mathcal{L}_1 + \mathcal{L}_2$, $[f|c]\langle \mathcal{L}_1, \dots, \mathcal{L}_n \rangle$, $\mathcal{L}_1 \cdot_a \mathcal{L}_2$, $(\mathcal{L}_1)_a^*$, $(\mathcal{L}_1)_a^+$ and $(\mathcal{L}_1)_a^\mu$ are also weakly recognizable.*

Proof. Each operation is defined as construction on wWTAs. Then we argue that the assignment $\mathcal{A} \mapsto \mathcal{L}_\mathcal{A}$ preserves the operations up to isomorphism.

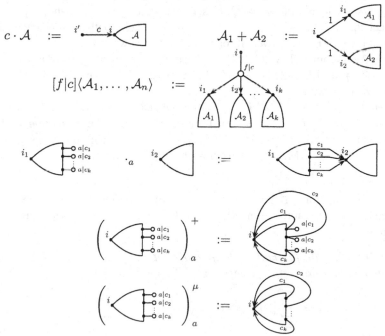

The a-iteration of a wWTA $\mathcal{A}$ can now be defined according to $\mathcal{A}_a^* := \mathcal{A}_a^+ + \mathcal{A}'$ where $\mathcal{A}'$ is a wWTA that recognizes $\{[a|1]\}$. $\square$

4 A Kleene-Type Result

Let X be a set of variable symbols disjoint from Σ and let K be a semiring. The set $\mathrm{Rat}(\Sigma, K, X)$ of *rational expressions* over Σ, X and K is the set E of words given by the following grammar:

$$E ::= a \mid x \mid c \cdot E \mid E + E \mid f\langle E, \dots, E \rangle \mid \mu x.(E) \quad a \in \Sigma^{(0)}, x \in X, f \in \Sigma$$

where in $f\langle E, \dots, E \rangle$ the number of E's is equal to the rank of f.

The semantics of rational expressions is given in terms of weighted tree-languages over the ranked alphabet $\Sigma(X)$. It is defined inductively: $\llbracket a \rrbracket := \{[a|1]\}$, $\llbracket x \rrbracket := \{[x|1]\}$, $\llbracket f\langle e_1, \dots, e_n \rangle \rrbracket := [f|1]\langle \llbracket e_1 \rrbracket, \dots, \llbracket e_n \rrbracket \rangle$, $\llbracket c \cdot e \rrbracket := c \cdot \llbracket e \rrbracket$, $\llbracket e_1 + e_2 \rrbracket := \llbracket e_1 \rrbracket + \llbracket e_2 \rrbracket$ and $\llbracket \mu x.(e) \rrbracket := \llbracket e \rrbracket_x^\mu$.

We have already seen that the semantics of each rational expression is weakly recognizable. Showing the opposite direction—namely that each weakly recognizable weighted tree-language is isomorphic to the semantics of a rational expression—will be our goal in the next few paragraphs. As first step into this direction we introduce the accessibility graph of wWTAs.

Let $\mathcal{A} = (Q, \{i\}, \iota, T, \lambda, S, \sigma)$ be a normalized wWTA. Let $E_1 := \bigcup_{j \in \mathbb{N} \setminus \{0\}} T^{(j)} \times$ $\{1, 2, \dots, j\}$, $E := E_1 \dot\cup S$, and define $s : E \to Q$ according to $s(e) = \mathrm{dom}(t)$ if $e = (t, j)$, $t \in T$ and $s(e) = \mathrm{dom}(e)$ if $e \in S$. Moreover define $d : E \to Q$ according to $d(e) := \mathrm{cdom}_j(t)$ if $e = (t, j)$, $t \in T$ and $d(e) := \mathrm{cdom}(e)$ if $e \in S$. Then the multigraph $\Gamma_\mathcal{A} = (Q, E, s, d)$ is called *accessibility-graph* of $\mathcal{A}$.[2]

A *path* of length n in $\Gamma_\mathcal{A} = (Q, E, s, d)$ is a word $e_1 e_2 \cdots e_n$ where $e_1, \dots, e_n \in E$ and such that $d(e_j) = s(e_{j+1})\,(j = 1, \dots, n-1)$. Such a path is called *cyclic* if $d(e_n) = s(e_1)$. It is called *minimal cycle* if for all $1 \le j, k \le n$ we have $s(e_j) = s(e_k) \Rightarrow j = k$. The number of minimal cycles of $\Gamma_\mathcal{A}$ is called the *cyclicity* of $\mathcal{A}$. It is denoted by $\mathrm{cyc}(\mathcal{A})$.

A state q of $\mathcal{A}$ is called *source* if it is a source of $\Gamma_\mathcal{A}$. That is, there does not exist any arc e of $\Gamma_\mathcal{A}$ with $d(e) = q$.

Let $\mathcal{A} = (Q, \{i\}, \iota, T, \lambda, S, \sigma)$ be a normalized wWTA. Let $\tau \in T$ with domain i. Assume $\lambda(\tau) = (i, f, q_1, \dots, q_n, c)$. Then for $1 \le k \le n$ the *derivation of $\mathcal{A}$ by (q, k)* is the reduction of the automaton $(Q, \{q_k\}, \iota', T, \lambda, S, \sigma)$ where ι' maps q_k to 1. It will be denoted by $\frac{\partial \mathcal{A}}{\partial(\tau, k)}$. Moreover we define the *complete derivation* of $\mathcal{A}$ by τ as the tuple

$$\frac{\partial \mathcal{A}}{\partial \tau} := \left(\frac{\partial \mathcal{A}}{\partial(\tau, 1)}, \dots, \frac{\partial \mathcal{A}}{\partial(\tau, n)} \right).$$

Analogously, for $s \in S$ with $\sigma(s) = (i, q, c)$ we define the *derivation of $\mathcal{A}$ by s* as the reduction of the automaton $(Q, \{q\}, \iota', T, \lambda, S, \sigma)$ where ι' maps q to 1. It will be denoted by $\frac{\partial \mathcal{A}}{\partial s}$.

Proposition 9. *With the notions from above let $T_i \subseteq T$, $S_i \subseteq S$ be the sets of all transition-symbols and silent transition-symbols with domain i, respectively. Then*

$$\mathcal{A} \equiv \sum_{\tau \in T_i} [\mathrm{lab}(\tau)|\mathrm{wt}(\tau)] \left\langle \frac{\partial \mathcal{A}}{\partial \tau} \right\rangle + \sum_{s \in S_i} \mathrm{wt}(s) \frac{\partial \mathcal{A}}{\partial s} \qquad \square$$

---

[2] The function names s and d are abbreviations for "source" and "destination" of arcs, respectively.

Proposition 10. *Let* $\mathcal{A} = (Q, \{i\}, \iota, T, \lambda, S, \sigma)$ *be a reduced and normalized* wWTA *whose initial state* i *is not a source. Let* x *be a variable symbol that does not occur in* $\mathcal{A}$. *Define* $Q' := Q + \{q'\}$ *and* $T' := T + \{\tau'\}$ *and let* $\varphi : Q \to Q'$ *such that* $\varphi(q) = q$ *if* $q \neq i$ *and* $\varphi(i) = q'$. *For* τ *in* T *with* $\lambda(\tau) = (q, f, q_1, \ldots, q_n, c)$ *define* $\lambda'(\tau) := (q, f, \varphi(q_1), \ldots, \varphi(q_n), c)$ *and for* s *in* S *with* $\sigma(s) = (q_1, q_2, c)$ *define* $\sigma'(s) := (q_1, \varphi(q_2), c)$. *Finally define* $\lambda'(\tau') := (q', x, 1)$. *Then the* wWTA $\mathcal{A}' = (Q', \{i\}, \iota, T', \lambda', S, \sigma')$ *is still normalized and reduced with* i *being a source. Moreover* $(\mathcal{A}')_x^\mu \equiv \mathcal{A}$.

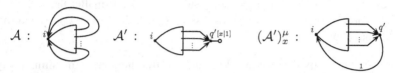

Theorem 11. *Every weakly recognizable weighted tree-language is definable by a rational expression.*

Proof. We prove inductively that each wWTA recognizes a rationally definable weighted tree-language.

To each normalized automaton $\mathcal{A} = (Q, \{i\}, \iota, T, \lambda, S, \sigma)$ we associate the pair of integers $(\mathrm{cyc}(\mathcal{A}), |Q|)$. On these integer-pairs we consider the lexicographical order: $(x, y) \leq (u, v) \iff x < u \vee (x = u \wedge y \leq v)$ and take this as an induction-index.

Since any wWTA has an initial state, the smallest possible index is $(0, 1)$. Such an automaton has $Q = \{i\}$ and $S = \emptyset$. Moreover if $T = \{t_1, \ldots, t_n\}$ then there are $a_1, \ldots, a_n \in \Sigma^{(0)} \cup X$ and $c_1, \ldots, c_n \in K$ such that $\lambda(t_k) = (i, a_k, c_k)$ $(k = 1, \ldots, n)$. The weighted tree-language that is recognized by such an automaton is $\{[a_1|c_1], \ldots, [a_n|c_n]\}$ this is definable by the following rational expression:

$$\sum_{k=1}^{n} c_k \cdot a_k.$$

Suppose now the claim holds for all wWTAs with index less than (n, m). Let $\mathcal{A} = (Q, \{i\}, \iota, T, \lambda, S, \sigma)$ be a normalized wWTA with $\mathrm{cyc}(\mathcal{A}) = n$ and $|Q| = m$.

If i is a source, then we use Proposition 9 and obtain

$$\mathcal{A} \equiv \sum_{\tau \in T_i} [\mathrm{lab}(\tau)|\mathrm{wt}(\tau)] \left\langle \frac{\partial \mathcal{A}}{\partial \tau} \right\rangle + \sum_{s \in S_i} \mathrm{wt}(s) \frac{\partial \mathcal{A}}{\partial s}$$

For $\tau \in T_i$ of arity k let $\mathcal{A}_{\tau, k} := \frac{\partial \mathcal{A}}{\partial (\tau, k)}$ and for $s \in S_i$ let $\mathcal{A}_s := \frac{\partial \mathcal{A}}{\partial s}$.

Since the number of states of $\mathcal{A}_{\tau, k}$ is strictly smaller than that of $\mathcal{A}$ and the cyclicity of $\mathcal{A}_{\tau, k}$ is not greater that that of $\mathcal{A}$, we conclude that the index of $\mathcal{A}_{\tau, k}$ is strictly smaller than that of $\mathcal{A}$. Hence the weighted tree-language that is recognized by $\mathcal{A}_{\tau, k}$ is rationally definable. The same holds for the derivations by silent transitions.

For $j \in \mathbb{N}, \tau \in T_i^{(j)}, 1 \leq k \leq j$ let $e_{\tau, k}$ be a rational expression defining a weighted tree-language isomorphic to the one recognized by $\mathcal{A}_{\tau, k}$. Moreover, for $s \in S_i$ let e_s

be a rational expression defining a weighted tree-language that is isomorphic to the one recognized by $\mathcal{A}_s$. Then

$$\sum_{j \in \mathbb{N}} \sum_{\tau \in T_i^{(j)}} [\mathrm{lab}(\tau)|\mathrm{wt}(\tau)]\langle e_{t,1}, \dots, e_{t,j}\rangle \quad + \quad \sum_{s \in S_i} \mathrm{wt}(s) \cdot e_s$$

is a rational expression defining a weighted tree-language isomorphic to $\mathcal{L}_{\mathcal{A}}$.

If i is not a source then we use Proposition 10 and obtain a wWTA $\mathcal{A}'$ such that $(\mathcal{A}')_x^\mu \equiv \mathcal{A}$. Clearly, $\mathcal{A}'$ has a smaller cyclicity and hence also a smaller index than $\mathcal{A}$. By induction hypothesis there is a rational expression e such that $[\![e]\!] \equiv \mathcal{L}_{\mathcal{A}'}$. Therefore $\mu x.(e)$ is a rational expression for $\mathcal{L}_{\mathcal{A}}$. $\square$

If we want to characterize the recognizable weighted tree-languages in a similar way, then we must take care about the problem that only the a-recursion of a-quasiregular recognizable weighted tree-languages is guaranteed to be recognizable again. Therefore we restrict the set of rational expressions: The set $\mathrm{pRat}(\Sigma, X, K)$ of proper rational expressions shall consist of all words of the language E defined by the following grammar:

$$E ::= a \mid x \mid c \cdot E \mid E + E \mid f\langle E, \dots, E\rangle \mid \mu x.(E_x)$$
$$E_x ::= a \mid y \mid c \cdot E_x \mid E_x + E_x \mid f\langle E, \dots, E\rangle \mid \mu x.(E_x).$$

where $a \in \Sigma^{(0)}$, $x, y \in X$, $x \neq y$, $c \in K$ and $f \in \Sigma$. The semantics of proper rational expressions is the same as for rational expressions. The essential difference between Rat and pRat is that an expression $\mu x.(e)$ is in pRat only if $[\![e]\!]$ is x-quasiregular. Therefore it is clear that the semantics of proper rational expressions are always going to be recognizable.

Theorem 12. *For every recognizable weighted tree-language $\mathcal{L}$ there is a proper rational expression e such that $\mathcal{L} \cong [\![e]\!]$.*

Proof. $\mathcal{L}$ is recognized by a wWTA without silent cycles. The decomposition steps to obtain a rational expression for $\mathcal{L}$ never introduce new silent cycles (in fact they never introduce any cycles). Therefore the construction from the proof of Theorem 11 produces a proper rational expression. $\square$

5   Formal Tree-Series

Given a ranked alphabet (Σ, rk) and a semiring $(K, \oplus, \odot, 0, 1)$ let T_Σ be the set of all trees over Σ. A function $S : T_\Sigma \to K$ is called *formal tree-series*. We will adopt the usual notation and write (S, t) for the image of t under S. With $K\langle\!\langle \Sigma \rangle\!\rangle$ we will denote the set of all formal tree-series over Σ.

Let WT_Σ be the set of all weighted trees over Σ with weights from K. To each weighted tree t we associate its weight $\mathrm{wt}(t) \in K$ and its underlying tree $\mathrm{ut}(t) \in T_\Sigma$. The function ut we already defined above. The function $\mathrm{wt} : \mathrm{WT}_\Sigma \to K$ is defined inductively: $\mathrm{wt}([a|c]) := c$ and $\mathrm{wt}([f|c]\langle t_1, \dots, t_n\rangle) := c \odot \bigodot_{i=1}^{n} \mathrm{wt}(t_i)$.

An easy property of wt is that $\mathrm{wt}(c \cdot t) = c \odot \mathrm{wt}(t)$ for all $t \in \mathrm{WT}_\Sigma$. Another very crucial property only holds if K is commutative, namely for $t \in \mathrm{WT}_\Sigma$ with $\mathrm{rk}_a(t) = n$ and for $s_1, \ldots, s_n \in \mathrm{WT}_\Sigma$:

$$\mathrm{wt}(t \circ_a \langle s_1, \ldots, s_n \rangle) = \mathrm{wt}(t) \odot \bigodot_{i=1}^{n} \mathrm{wt}(s_i).$$

From now on we assume that K is commutative.

Given now a finitary $\mathcal{L} = (L, |.|) \in \mathrm{WTL}_\Sigma$ we associate a formal tree-series $S_\mathcal{L}$ with $\mathcal{L}$ according to:

$$(S_\mathcal{L}, t) := \bigoplus_{\substack{s \in L \\ \mathrm{ut}(|s|) = t}} \mathrm{wt}(|s|) \qquad (t \in T_\Sigma).$$

Since $\mathcal{L}$ is finitary, $S_\mathcal{L}$ is welldefined.

We call $S \in K\langle\!\langle \Sigma \rangle\!\rangle$ a-quasiregular if $(S, a) = 0$ and we call S recognizable if there is a recognizable $\mathcal{L} \in \mathrm{WTL}_\Sigma$ with $S_\mathcal{L} = S$. It is easy to see that if a finitary weighted tree-language $\mathcal{L}$ is a-quasiregular, then $S_\mathcal{L}$ is a-quasiregular.

The operations of *sum* and *product with scalars* can be introduced for formal tree-series pointwise. That is $(S_1 + S_2, t) := (S_1, t) \oplus (S_2, t)$ and $(c \cdot S_1, t) := c \odot (S_1, t)$ for any $S_1, S_2 \in K\langle\!\langle \Sigma \rangle\!\rangle$, $c \in K$. It is not surprising that for any $\mathcal{L}_1, \mathcal{L}_2 \in \mathrm{WTL}_\Sigma^\mathrm{f}$ we have $S_{\mathcal{L}_1 + \mathcal{L}_2} = S_{\mathcal{L}_1} + S_{\mathcal{L}_2}$ and $S_{c \cdot \mathcal{L}_1} = c \cdot S_{\mathcal{L}_1}$.

Next we define the a-product of formal tree-series S_1, S_2 for $a \in \Sigma^{(0)}$ according to

$$(S_1 \cdot_a S_2, t) := \bigoplus_{\substack{s \in T_\Sigma \\ s_1, \ldots, s_{\mathrm{rk}_a(s)} \in T_\Sigma \\ t = s \circ_a \langle s_1, \ldots, s_{\mathrm{rk}_a(s)} \rangle}} (S_1, s) \odot \bigodot_{i=1}^{\mathrm{rk}_a(s)} (S_2, s_i)$$

Whenever K is commutative, then for $\mathcal{L}_1, \mathcal{L}_2 \in \mathrm{WTL}_\Sigma^\mathrm{f}$ we have $S_{\mathcal{L}_1 \cdot_a \mathcal{L}_2} = S_{\mathcal{L}_1} \cdot_a S_{\mathcal{L}_2}$.

Let $f \in \Sigma^{(n)}$, $c \in K$, $S_1, \ldots, S_n \in K\langle\!\langle \Sigma \rangle\!\rangle$. Then we define the *topcatenation* $[f|c]\langle S_1, \ldots, S_n \rangle$ according to

$$([f|c]\langle S_1, \ldots, S_n \rangle, t) = \begin{cases} c \odot \prod_{i=1}^{n}(S_i, t_i) & \text{if } t = f\langle t_1, \ldots, t_n \rangle \\ 0 & \text{else.} \end{cases}$$

Again, some thought reveals that we have $S_{[f|c]\langle \mathcal{L}_1, \ldots, \mathcal{L}_n \rangle} = [f|c]\langle S_{\mathcal{L}_1}, \ldots, S_{\mathcal{L}_n} \rangle$ for all $\mathcal{L}_1, \ldots, \mathcal{L}_n \in \mathrm{WTL}_\Sigma^\mathrm{f}$. Note that here the semiring does not need to be commutative.

The most delicate operation to define for formal tree-series is the a-iteration. Luckily we showed its close relation to free ranked monoids. This relationship we use to define the a-iteration on formal tree-series. Let $S \in K\langle\!\langle \Sigma \rangle\!\rangle$, $a \in \Sigma^{(0)}$ such that $(S, a) = 0$. Let $(T_\Sigma, \mathrm{rk}_a)^*$ be the free ranked monoid generated by $(T_\Sigma, \mathrm{rk}_a)$ (cf. (1) in Section 1). Let T_Σ^* be its carrier and ε be its neutral element. Let $\varphi : (T_\Sigma, \mathrm{rk}_a)^* \to (T_\Sigma, \mathrm{rk}_a, \circ_a, a)$ be the unique homomorphism induced by the identity map of T_Σ. On T_Σ^* we define a weight-function wt_S^* inductively:

$$
\mathrm{wt}_S^*(s) := \begin{cases} 1 & s = \varepsilon \\ (S, s) & s \in T_\Sigma \\ (S, t) \odot \displaystyle\bigodot_{i=1}^{\mathrm{rk}_a(t)} \mathrm{wt}_S^*(t_i) & s = t\langle t_1, \dots, t_{\mathrm{rk}_a(t)}\rangle,\ t \in T_\Sigma, \\ & t_1, \dots, t_{\mathrm{rk}_a(t)} \in T_\Sigma^*. \end{cases}
$$

Then we define $S_a^* \in K\langle\!\langle \Sigma \rangle\!\rangle$ according to

$$
(S_a^*, t) := \bigoplus_{\substack{s \in T_\Sigma^* \\ \varphi(s) = t}} \mathrm{wt}_S^*(s).
$$

Assume K is commutative and $\mathcal{L} \in \mathrm{WTL}_\Sigma^f$ is a-quasiregular. Then $S_{\mathcal{L}_a^*} = (S_\mathcal{L})_a^*$. Summing up we obtain:

Proposition 13. *If K is commutative, then the assignment $\mathcal{L} \mapsto S_\mathcal{L}$ preserves sum, product with scalars, a-product, topcatenation and a-iteration.* □

For $S \in K\langle\!\langle \Sigma \rangle\!\rangle$ we define the a-annihilation by $S_{\neg a} := S \cdot_a 0$ where 0 denotes the series that maps each tree to 0. Clearly we have $S_{\mathcal{L}_{\neg a}} = (S_\mathcal{L})_{\neg a}$ for any $\mathcal{L} \in \mathrm{WTL}_\Sigma^f$.

The a-recursion of formal tree-series can also be introduced easily now. Let $S \in K\langle\!\langle \Sigma \rangle\!\rangle$ be a-quasiregular. Then we define $S_a^\mu := (S_a^*)_{\neg a}$. Using the characterization of the a-recursion through the a-iteration of weighted tree-languages, it is evident that $S_{\mathcal{L}_a^\mu} = (S_\mathcal{L})_a^\mu$.

It is clear that for any $e \in \mathrm{pRat}(\Sigma, X, K)$ we get that $S_{[\![e]\!]}$ is a recognizable element of $K\langle\!\langle \Sigma(X) \rangle\!\rangle$. From Theorem 12 and from Proposition 13 we obtain immediately the following result

Theorem 14. *Let K be commutative and let $S \in K\langle\!\langle \Sigma(X) \rangle\!\rangle$ be recognizable. Then there is a proper rational expression e with $S = S_{[\![e]\!]}$.* □

Using that a-product preserves recognizability and that the a-recursion may be simulated by a-iteration and a-product, we can also formulate a more conventional Kleene-type result:

Corollary 15. *Let K be commutative. Then the set of all recognizable formal tree-series over $\Sigma(X)$ is the smallest subset of $K\langle\!\langle \Sigma(X) \rangle\!\rangle$ that contains all polynomials and that is closed with respect to sum, product with scalars, x-product ($x \in X$) and x-iteration ($x \in X$).* □

References

1. Berstel, J., Reutenauer, C.: Recognizable formal power series on trees. Theoretical Computer Science **18** (1982) 115–148
2. Bloom, S.L., Ésik, Z.: An extension theorem with an application to formal tree series. BRICS Report Series RS-02-19, University of Aarhus (2002)

3. Bozapalidis, S.: Equational elements in additive algebras. Theory Comput. Systems **32** (1999) 1–33

4. Culik, K., Kari, J.: Image compression using weighted finite automata. Computer and Graphics **17** (1993) 305–313

5. Droste, M., Vogler, H.: A Kleene theorem for weighted tree automata. technical report TUD-FI02-04, Technische Universität Dresden (2002)

6. Kleene, S.E.: Representation of events in nerve nets and finite automata. In Shannon, C.E., McCarthy, J., eds.: Automata Studies. Princeton University Press, Princeton, N.J. (1956) 3–42

7. Kuich, W.: Formal power series over trees. In: Proc. of the 3rd International Conference Developments in Language Theory, Aristotle University of Thesaloniki (1997) 60–101

8. Mohri, M.: Finite-state transducers in language and speech processing. Computational Linguistics **23** (1997) 269–311

9. Pech, C.: Kleene-type results for weighted tree-automata. Dissertation, TU-Dresden (2003) http://www.math.tu-dresden.de/~pech/diss.ps.

10. Schützenberger, M.P.: On the definition of a family of automata. Information and Control **4** (1961) 245–270

11. Thatcher, J.W., Wright, J.B.: Generalized finite automata theory with application to a decision problem of second-order logic. Math. Systems Theory **2** (1968) 57–81

Weak Cardinality Theorems for First-Order Logic

(Extended Abstract)

Till Tantau

Fakultät IV – Elektrotechnik und Informatik
Technische Universität Berlin
Franklinstraße 28/29, D-10587 Berlin, Germany
tantau@cs.tu-berlin.de

Abstract. Kummer's cardinality theorem states that a language A is recursive if a Turing machine can exclude for any n words $w_1, \ldots, w_n$ one of the $n+1$ possibilities for the cardinality of $\{w_1, \ldots, w_n\} \cap A$. It is known that this theorem does not hold for polynomial-time computations, but there is evidence that it holds for finite automata: at least weak cardinality theorems hold for them. This paper shows that some of the weak recursion-theoretic and automata-theoretic cardinality theorems are instantiations of purely logical theorems. Apart from unifying previous results in a single framework, the logical approach allows us to prove new theorems for other computational models. For example, weak cardinality theorems hold for Presburger arithmetic.

1 Introduction

Given a language A and n input words, we often wish to know which of these words are in the language. For languages like the satisfiability problem this problem is presumably difficult to solve, for languages like the halting problem it is impossible to solve. To tackle such problems, Gasarch [7] has proposed to study a simpler problem instead: we just *count how many* of the input words are elements of A. To make things even easier, we do not require this number to be computed exactly, but only approximately. Indeed, let us just try to *exclude* one possibility for the number of input words in A.

In recursion theory, Kummer's cardinality theorem [16] states that, using a Turing machine, excluding one possibility for the number of input words in A is just as hard as deciding A. It is not known whether this statement carries over to automata theory, that is, it is not known whether a language A must be regular if a finite automaton can always exclude one possibility for the number of input words in A. However, several *weak* forms of this theorem are known to hold for automata theory. For example, the finite automata cardinality theorem is known [25] to hold for $n = 2$.

These parallels between recursion and automata theory are surprising insofar as computational models 'in between' exhibit a different behaviour: there are

A. Lingas and B.J. Nilsson (Eds.): FCT 2003, LNCS 2751, pp. 400–411, 2003.
© Springer-Verlag Berlin Heidelberg 2003

languages A outside the class P of problems decidable in polynomial time for which we *can* always exclude, in polynomial time, for any $n \geq 2$ words one possibility for their number in A.

The present paper explains (at least partly) *why* the parallels between recursion and automata theory exist and why they are not shared by the models in between. Basically, the weak cardinality theorems for Turing machines and finite automata are just different instantiations of the same logical theorems. These logical theorems cannot be instantiated for polynomial time, because polynomial time lacks a logical characterisation in terms of elementary definitions.

Using logic for the formulation and proof of the weak cardinality theorems has another advantage, apart from unifying previous results. Theorems formulated for arbitrary logical structures can be instantiated in novel ways: the weak cardinality theorems all hold for Presburger arithmetic and the nonspeedup theorem also holds for ordinal number arithmetic.

In the logical setting 'computational models' are replaced by 'logical structures' and 'computations' are replaced by 'elementary definitions'. For example, the cardinality theorem for $n = 2$ now becomes the following statement: Let S be a logical structure with universe U satisfying certain requirements and let $A \subseteq U$. If there exists a function $f \colon U \times U \to \{0, 1, 2\}$ with $f(x, y) \neq |\{x, y\} \cap A|$ for all $x, y \in U$ that is elementarily definable in S, then A is elementarily definable in S.

Cardinality computations have applications in the study of separability. As argued in [26], 'cardinality theorems are separability results in disguise'. In recursion theory and in automata theory one can rephrase the weak cardinality theorems as separability results. Such a rephrasing is also possible for the logical versions and we can formulate purely logical separability theorems that are interesting in their own right. An example of such a theorem is the following statement: Let S be a logical structure with universe U satisfying certain requirements and let $A \subseteq U$. If there exist elementarily definable supersets of $A \times A$, $A \times \bar{A}$, and $\bar{A} \times \bar{A}$ whose intersection is empty, then A is elementarily definable in S.

This paper is organised as follows. In section 2 the history of the cardinality theorem is retraced and the weak cardinality theorems are formulated rigorously. Section 3 prepares the logical formulation of the weak cardinality theorems. It is shown how the class of regular languages and the class of recursively enumerable languages can be characterised in terms of appropriate elementary definitions. In section 4 the weak cardinality theorems for first-order logic are formulated. In section 5 applications of the theorems to separability are discussed.

This extended abstract does not include any proofs due to lack of space. They can be found in the full technical report version of the paper [27].

2   History of the Cardinality Theorem

2.1   The Cardinality Theorem for Recursion Theory

For a set A, the *cardinality function* $\#_A^n$ takes n words as input and yields the number of words in A as output, that is, $\#_A^n(w_1, \ldots, w_n) = |\{w_1, \ldots, w_n\} \cap A|$.

The cardinality function and the idea of 'counting input words', which is due to Gasarch [7] in its general form, play an important role in a variety of proofs both in complexity theory [9,12,14,18,23] and recursion theory [4,16,17]. For example, the core idea of the Immerman–Szelepcsényi theorem is to *count* the number of reachable vertices in a graph in order to *decide* a reachability problem.

One way of quantifying the complexity of $\#_A^n$ is to consider its *enumeration complexity*, which is the smallest number m such that $\#_A^n$ is m-enumerable. Enumerability, which was first defined by Cai and Hemaspaandra [6] in the context of polynomial-time computations and which was later transferred to recursive computations, can be regarded as 'generalised approximability'. It is defined as follows: a function f, taking n tuples of words as input, is m-*Turing-enumerable* if there exists a Turing machine that on input $w_1, \ldots, w_n$ starts a possibly infinite computation during which it prints words onto an output tape. At most m different words may be printed and one of them must be $f(w_1, \ldots, w_n)$.

Intuitively, the larger m, the easier it should be to m-Turing-enumerate $\#_A^n$. *This intuition is wrong.* Kummer's cardinality theorem, see below, states that even n-Turing-enumerating $\#_A^n$ is just as hard as deciding A. In other words, excluding just one possibility for $\#_A^n(w_1, \ldots, w_n)$ is just as hard as deciding A. Intriguingly, the intuition *is* correct for polynomial-time computations since the work of Gasarch, Hoene, and Nickelsen [7,11,20] shows that a polynomial-time version of the cardinality theorem does not hold for $n \geq 2$.

Theorem 2.1 (Cardinality theorem [16]). *If $\#_A^n$ is n-Turing-enumerable, then A is recursive.*

The cardinality theorem has applications for instance in the study of semirecursive sets [13], which play a key role in the solution of Post's problem [22]. The proof of the cardinality theorem is difficult. Several less general results had already been proved when Kummer wrote his paper 'A proof of Beigel's cardinality conjecture' [16]. The title of Kummer's paper refers to the fact that Richard Beigel was the first to conjecture the cardinality theorem as a generalisation of his so-called 'nonspeedup theorem' [3]. In the following formulation of the nonspeedup theorem χ_A^n denotes the n-fold characteristic function of A, which maps any n words $w_1, \ldots, w_n$ to a bitstring whose ith bit is 1 iff $w_i \in A$. The nonspeedup theorem is a simple consequence of the cardinality theorem.

Theorem 2.2 (Nonspeedup theorem [3]). *If χ_A^n is n-Turing-enumerable, then A is recursive.*

Owings [21] succeeded in proving the cardinality theorem for $n = 2$. For larger n he could only show that if $\#_A^n$ is n-Turing-enumerable, then A is recursive in the halting problem. Harizanov et al. [8] have formulated a restricted cardinality theorem, whose proof is somewhat simpler than the proof of the full cardinality theorem.

Theorem 2.3 (Restricted cardinality theorem [8]). *If $\#_A^n$ is n-Turing-enumerable via a Turing machine that never enumerates both 0 and n simultaneously, then A is recursive.*

2.2   Weak Cardinality Theorems for Automata Theory

If we restrict the computational power of Turing machines, the cardinality theorem no longer holds [7,11,20]: there are languages $A \notin P$ for which we can always exclude one possibility for $\#_A^n(w_1, \ldots, w_n)$ in polynomial time for $n \geq 2$. However, if we restrict the computational power even further, namely if we consider finite automata, there is strong evidence that the cardinality theorem holds once more, see the following conjecture:

Conjecture 2.4 ([25]). If $\#_A^n$ is n-fa-enumerable, then A is regular.

The conjecture refers to the notion of *m-enumerability by finite automata.* This notion was introduced in [24] and is defined as follows: A function f is *m-fa-enumerable* if there exists a finite automaton for which for every input tuple $(w_1, \ldots, w_n)$ the output attached to the last state reached is a set of size at most m that contains $f(w_1, \ldots, w_n)$. The different components of the tuple are put onto n different tapes, shorter words padded with blanks, and the automaton scans the tapes synchronously, which means that all heads advance exactly one symbol in each step. The same method of feeding multiple words to a finite automaton has been used in [1,2,15].

In a line of research [1,2,15,24,25,26], the following three theorems were established. They support the above conjecture by showing that all of the historically earlier, weak forms of the recursion-theoretic cardinality theorem hold for finite automata.

Theorem 2.5 ([24]). *If χ_A^n is n-fa-enumerable, then A is regular.*

Theorem 2.6 ([25]). *If $\#_A^2$ is 2-fa-enumerable, then A is regular.*

Theorem 2.7 ([25,2]). *If $\#_A^n$ is n-fa-enumerable via a finite automaton that never enumerates both 0 and n simultaneously, then A is regular.*

3   Computational Models as Logical Structures

The aim of formulating purely logical versions of the weak cardinality theorems is to abstract from concrete computational models. The present section explains which logical abstraction is used.

3.1   Presburger Arithmetic

Let us start with an easy example: Presburger arithmetic. This notion is easily transferred to a logical setting since it is defined in terms of first-order logic in the first place. A set A of natural numbers is called *definable in Presburger arithmetic* if there exists a first-order formula $\phi(x)$ over the signature $\{+^2\}$ with the following property: A contains exactly those numbers a that make $\phi(x)$

true if we interpret x as a and the symbol $+$ as the normal addition of natural numbers. For example, the set of even natural numbers is definable in Presburger arithmetic using the formula $\phi(x) = \exists y\,(y + y = x)$.

In the abstract logical setting used in the next sections the 'computational model Presburger arithmetic' is represented by the logical structure $(\mathbb{N}, +)$. The class of sets that are 'computable in Presburger arithmetic' is given by the class of sets that are elementarily definable in $(\mathbb{N}, +)$. Recall that a relation R is called *elementarily definable in a logical structure* S if there exists a first-order formula $\phi(x_1, \ldots, x_n)$ such that $(a_1, \ldots, a_n) \in R$ iff $\phi(x_1, \ldots, x_n)$ holds in S if we interpret each x_i as a_i.

3.2 Finite Automata

In order to make finite automata and regular languages accessible to a logical setting, for a given alphabet Σ we need to find a logical structure $S_{\mathrm{REG},\Sigma}$ with the following property: a language $A \subseteq \Sigma^*$ is regular iff it is elementarily definable in $S_{\mathrm{REG},\Sigma}$.

It is known that such a structure $S_{\mathrm{REG},\Sigma}$ exists: Büchi has proposed one [5], though a small correction is necessary as pointed out by McNaughton [19]. However, the elements of Büchi's structure are natural numbers, not words, and thus a reencoding is necessary. A more directly applicable structure is discussed in [26], where it is shown that for non-unary alphabets the structure $(\Sigma^*, I_{\sigma_1}, \ldots, I_{\sigma_{|\Sigma|}})$ has the desired properties. The relations I_{σ_i}, one for each symbol $\sigma_i \in \Sigma$, are binary relations that hold for a pair (u, v) of words if the $|v|$-th letter of u is σ_i. For unary alphabets, an appropriate structure $S_{\mathrm{REG},\Sigma}$ can also be constructed.

3.3 Polynomially Time-Bounded Turing Machines

There is no logical structure S such that the class of languages that are elementarily definable in S is exactly the class P of languages decidable in polynomial time. To see this, consider the relation $R = \{(M, t) \mid M$ halts on input M after t steps$\}$. This relation is in P, but the language defined by the first-order formula $\phi(M) = \exists t\, R(M, t)$ is exactly the halting problem. Thus in any logical structure in which we can elementarily define R we can also elementarily define the halting problem.

3.4 Resource-Unbounded Turing Machines

On the one hand, the class of recursive languages cannot be defined elementarily: the argument for polynomial-time machines also applies here. On the other hand, the arithmetical hierarchy contains exactly the sets that are elementarily definable in $(\mathbb{N}, +, \cdot)$.

The most interesting case, the class of recursively enumerable languages, is more subtle. Since the class is not closed under complement, it cannot be characterised by elementary definitions. However, it can be characterised by *positive* elementary definitions, which are elementary definitions that do not contain negations: For every alphabet Σ there is a structure $\mathcal{S}_{\mathrm{RE},\Sigma}$ such that a language $A \subseteq \Sigma^*$ is recursively enumerable iff it is positively elementarily definable in $\mathcal{S}_{\mathrm{RE},\Sigma}$. An example of such a structure $\mathcal{S}_{\mathrm{RE},\Sigma}$ is the following: its universe is Σ^* and it contains all recursively enumerable relations over the alphabet Σ^*.

4   Logical Versions of the Weak Cardinality Theorems

In this section the weak cardinality theorems for first-order logic are presented. The theorems are first formulated for elementary definitions, which allows us to apply them to all computational models that can be characterised in terms of elementary definitions. As argued in the previous section, this includes Presburger arithmetic, finite automata, and the arithmetical hierarchy, but misses the recursively enumerable languages. This is remedied later in this section, where positive elementary definitions are discussed. It is shown that at least the non-speedup theorem can be formulated in a 'positive' way. At the end of the section higher-order logics are briefly touched.

We are still missing one crucial definition for the formulation of the weak cardinality theorems: What does it mean that a function is 'm-enumerable in a logical structure'?

Definition 4.1. *Let $\mathcal{S}$ be a logical structure with universe U and m a positive integer. A function $f \colon U \to U$ is* (positively) *elementarily m-enumerable in $\mathcal{S}$ if there exists a relation $R \subseteq U \times U$ with the following properties:*

1. *R is (positively) elementarily definable in $\mathcal{S}$,*
2. *the graph of f is contained in R,*
3. *R is m-bounded, that is, for every $x \in U$ there exist at most m different y with $(x, y) \in R$.*

The definition is easily adapted to functions f that take more than one input or yield more than one output. This definition does, indeed, reflect the notion of enumerability: A function with finite range is m-fa-enumerable iff it is elementarily m-enumerable in $\mathcal{S}_{\mathrm{REG},\Sigma}$; a function is m-Turing-enumerable iff it is positively elementarily m-enumerable in $\mathcal{S}_{\mathrm{RE},\Sigma}$.

4.1   The Non-positive First-Order Case

We are now ready to formulate the weak cardinality theorems for first-order logic. In the following theorems, a logical structure is called *well-orderable* if a well-ordering of its universe can be defined elementarily. For example $(\mathbb{N}, +)$ is well-orderable using the formula $\phi_\leq(x, y) = \exists z\, (x + z = y)$. The *cross product* of two function f and g is defined in the usual way by $(f \times g)(u, v) = \big(f(u), g(v)\big)$.

The first of the weak cardinality theorems, the nonspeedup theorem, is actually just a corollary of a more general theorem that is formulated first: the cross product theorem.

Theorem 4.2 (Cross product theorem). *Let S be a well-orderable logical structure with universe U. Let $f, g\colon U \to U$ be functions. If $f \times g$ is elementarily $(n + m)$-enumerable in S, then f is elementarily n-enumerable in S or g is elementarily m-enumerable in S.*

Theorem 4.3 (Nonspeedup theorem). *Let S be a well-orderable logical structure with universe U. Let $A \subseteq U$. If χ_A^n is elementarily n-enumerable in S, then A is elementarily definable in S.*

Theorem 4.4 (Cardinality theorem for two words). *Let S be a well-orderable logical structure with universe U. Let every finite relation on U be elementarily definable in S. Let $A \subseteq U$. If $\#_A^2$ is elementarily 2-enumerable in S, then A is elementarily definable in S.*

Theorem 4.5 (Restricted cardinality theorem). *Let S be a well-orderable logical structure with universe U. Let every finite relation on U be elementarily definable in S. Let $A \subseteq U$. If $\#_A^n$ is elementarily n-enumerable in S via a relation R that never 'enumerates' 0 and n simultaneously, then A is elementarily definable in S.*

The premises of the first two and the last two of the above theorems differ in the following way: for the last two theorems we require that every finite relation on S is elementarily definable in S. An example of a logical structure where this is not the case is $(\omega_1, +, \cdot)$, where ω_1 is the first uncountable ordinal number and $+$ and $\cdot$ denote ordinal number addition and multiplication. Since this structure is uncountable, there exist a singleton set $A = \{\alpha\}$ with $\alpha \in \omega_1$ that is not elementarily definable in $(\omega_1, +, \cdot)$. For this structure theorems 4.4 and 4.5 do not hold: $\#_A^2$ is elementarily 2-enumerable in $(\omega_1, +, \cdot)$ since $\#_A^2(x, y) \in \{0, 1\}$ for all $x, y \in \omega_1$, but A is not elementarily definable in $(\omega_1, +, \cdot)$.

4.2   The Positive First-Order Case

The above theorems cannot be applied to Turing enumerability since they refer to elementary definitions, not to *positive* elementary definitions. Unfortunately, the proofs of the theorems cannot simply be reformulated in a 'positive' way. They use negations to define the smallest element in a set B with respect to a well-ordering $<$. The defining formula is given by $\phi(x) = B(x) \wedge \neg \exists x'\, (x' < x \wedge B(x'))$.

This is a fundamental problem: the set $\{(M, x) \mid x$ is the smallest word accepted by $M\}$ is not recursively enumerable. Thus if we insist on finding the smallest element in every recursively enumerable set, we will not be able to apply the theorems to Turing machines. Fortunately, a closer examination of the proofs shows that we do not actually need the *smallest* element in B, but just *any* element of B as long as the same element is always chosen.

This is not as easy as it may sound—as is well-recognised in set theory, where the axiom of choice is needed for this choosing operation. Suppose you and a friend wish to agree on a certain element of B, but neither you nor your friend know the set B beforehand. Rather, you must decide on a generic method of picking an element such that, when the set B becomes known to you and your friend, you will both pick the same element. Agreements like 'pick some element from B' will not guarantee that you both pick the same element, except if the set happens to be a singleton.

We need a (partial) recursive *choice function* that assigns a word that is accepted by M to every Turing machine M, provided such a word exists. Such a choice function does, indeed, exist: it maps M to the first word that is accepted by M during a dovetailed simulation of M on all words.

In the following, first-order logic is augmented by choice operators. Choice operators have been used for example in [10], but the following definitions are adapted to the purposes of this paper and differ from the formalism used in [10]. On the sematic side we augment logical structures by a choice function; on the syntactic side we augment first-order logic by a choice operator ε:

Definition 4.6. *A* choice function *on a set U is a function $\zeta \colon \mathcal{P}(U) \to U$ such that $\zeta(B) \in B$ for all nonempty $B \subseteq U$.*

Definition 4.7. *A* choice structure *is a pair (S, ζ) consisting of a logical structure S and a choice function ζ on the universe of S.*

Definition 4.8 (Syntax of the choice operator). First-order formulas with choice *are defined inductively in the usual way with one addition: if x is a variable and ϕ is a first-order formula with choice, so is $\varepsilon(x, \phi)$.*

In the next definition $\phi^{(S,\zeta)}(x) = \{ u \in U \mid (S, \zeta) \models \phi[x = a] \}$ denotes the set of all u that make ϕ hold in (S, ζ) when plugged in for the variable x.

Definition 4.9 (Semantics of the choice operator). *The semantics of first-order logic with choice operator is defined in the usual way with the following addition: a formula of the form $\varepsilon(x, \phi)$ holds in a choice structure (S, ζ) for an assignment α if $\phi^{(S,\zeta)}(x)$ is nonempty and $\alpha(x) = \zeta(\phi^{(S,\zeta)}(x))$.*

As an example, consider the logical structure $S = (\mathbb{N}, +, \cdot, <, 0)$ and let ζ map every nonempty set of natural numbers to its smallest element. Let $\phi(x, y, z) = \varepsilon(z, 0 < z \wedge \exists a \, (x \cdot a = z) \wedge \exists b \, (y \cdot b = z))$. Then $\phi^{(S,\zeta)}(x, y, z)$ is the set of all triples (n, m, k) such that k is the least common multiple of n and m: the formula $0 < z \wedge \exists a \, (x \cdot a = z) \wedge \exists b \, (y \cdot b = z)$ is true for all positive z that are multiples of both x and y; thus the choice operator picks the smallest one of these.

The following theorem shows that the class of recursively enumerable sets can be characterised in terms of first-order logic with choice.

Theorem 4.10. *For every alphabet Σ there exists a choice structure $(S_{\mathrm{RE},\Sigma}, \zeta)$ such that a language $A \subseteq \Sigma^*$ is recursively enumerable iff it is positively elementarily definable with choice in $(S_{\mathrm{RE},\Sigma}, \zeta)$.*

We can now formulate the cross product theorem and the nonspeedup theorem in such a way that they can be applied both to finite automata and to Turing machines.

Theorem 4.11 (Cross product theorem, positive version). *Let $(\mathcal{S}, \zeta)$ be a choice structure with universe U. Let the inequality relation on U be positively elementarily definable in $(\mathcal{S}, \zeta)$. Let every finite relation on U that is elementarily definable with choice in $(\mathcal{S}, \zeta)$ be positively elementarily definable with choice in $(\mathcal{S}, \zeta)$. Let $f, g: U \to U$ be functions. If $f \times g$ is positively $(n + m)$-enumerable with choice in $(\mathcal{S}, \zeta)$, then f is positively n-enumerable with choice in $(\mathcal{S}, \zeta)$ or g is positively m-enumerable with choice in $(\mathcal{S}, \zeta)$.*

Theorem 4.12 (Nonspeedup theorem, positive version). *Let $(\mathcal{S}, \zeta)$ be a choice structure with universe U. Let the inequality relation on U be positively elementarily definable in $(\mathcal{S}, \zeta)$. Let every finite relation on U that is elementarily definable with choice in $(\mathcal{S}, \zeta)$ be positively elementarily definable with choice in $(\mathcal{S}, \zeta)$. Let $A \subseteq U$. If χ_A^n is positively n-enumerable with choice in $(\mathcal{S}, \zeta)$, then A is positively elementarily definable with choice in $(\mathcal{S}, \zeta)$.*

The cross product theorem, theorem 4.2, is a consequence of its positive version, theorem 4.11. (And not the other way round, as one might perhaps expect.) The same is true for the nonspeedup theorem. To see this, consider a well-orderable structure $\mathcal{S}$ whose existence is postulated in theorem 4.2. Define a choice structure $(\mathcal{S}', \zeta)$ as follows: $\mathcal{S}'$ has the same universe as $\mathcal{S}$ and contains *all relations that are elementarily definable in $\mathcal{S}$.* The function ζ maps each set A to its smallest element with respect the well-ordering of $\mathcal{S}$'s universe. With these definitions, a relation is positively elementarily definable with choice in $(\mathcal{S}', \zeta)$ iff it is elementarily definable in $\mathcal{S}$.

4.3 The Higher-Order Case

We just saw that the cross product theorem for a certain logic, namely first-order logic, is a consequence of the cross product theorem for a less powerful logic, namely positive first-order logic. We may ask whether we can similarly apply the theorems for first-order logic to higher-order logics.

This is indeed possible and we can use the same kind of argument as above: Consider any logical structure $\mathcal{S}$. Define a new structure $\mathcal{S}'$ as follows: it has the same universe as $\mathcal{S}$ and it contains every relation that is higher-order definable in $\mathcal{S}$. Then a relation is elementarily definable in $\mathcal{S}'$ iff it is higher-order definable in $\mathcal{S}$. This allows us to transfer the cross product theorem and all of the weak cardinality theorems *to all logics that are at least as powerful as first-order logic.* Just one example of such a transfer is the following:

Theorem 4.13 (Cross product theorem for higher-order logic). *Let $\mathcal{S}$ be a well-orderable logical structure with universe U. Let $f, g: U \to U$ be functions. If $f \times g$ is higher-order $(n + m)$-enumerable in $\mathcal{S}$, then f is higher-order n-enumerable in $\mathcal{S}$ or g is higher-order m-enumerable in $\mathcal{S}$.*

5   Separability Theorems for First-Order Logic

Kummer's cardinality theorem can be reformulated in terms of separability. In [26] it is shown that it is equivalent to the following statement, where $A^{\binom{n}{k}}$ denotes the set of all n-tuples of distinct words such that exactly k of them are in A.

Theorem 5.1 (Separability version of Kummer's cardinality theorem).
Let A be a language. Suppose there exist recursively enumerable supersets of $A^{\binom{n}{0}}$, $A^{\binom{n}{1}}$, ... , $A^{\binom{n}{n}}$ whose intersection is empty. Then A is recursive.

In [26] it is also shown that the above statement is still true if we replace 'recursive enumerable' by 'co-recursively enumerable'.

The weak cardinality theorems for first-order logic can be reformulated in a similar way. Let us start with the cardinality theorem for two words. It can be stated equivalently as follows, where $\bar{A} = U \setminus A$ denotes the complement of A.

Theorem 5.2. *Let S be a well-orderable logical structure with universe U. Let every finite relation on U be elementarily definable in S. Let $A \subseteq U$. Suppose there exist elementarily definable supersets of $A \times A$, $A \times \bar{A}$, and $\bar{A} \times \bar{A}$ whose intersection is empty. Then A is elementarily definable in S.*

The restricted cardinality theorem can be reformulated in terms of elementary separability. Let us call two sets A and B *elementarily separable* in a structure S if there exists a set C with $A \subseteq C \subseteq \bar{B}$ that is elementarily definable in S.

Theorem 5.3. *Let S be a well-orderable structure with universe U. Let every finite relation on U be elementarily definable in S. Let $A \subseteq U$. If $A^{\binom{n}{0}}$ and $A^{\binom{n}{n}}$ are elementarily separable in S, then A is elementarily definable in S.*

6   Conclusion

This paper proposed a new, logic-based approach to the proof of (weak) cardinality theorems. The approach has two advantages:

1. It unifies previous results in a single framework.
2. The results can easily be applied to other computational models.

Regarding the first advantage, only the cross product theorem and the non-speedup theorem are completely 'unified' by the theorems presented in this paper: the Turing machine versions and the finite automata versions of these theorems are just different instantiations of theorems 4.11 and 4.12.

For the cardinality theorem for two words and for the restricted cardinality theorem the situation is (currently) more complex. These theorem hold for Turing machines and for finite automata, but different proofs are used. In particular, the logical theorems cannot be instantiated for Turing enumerability. Nevertheless, the logical approach is fruitful here: the logical theorem *can* be instantiated for new models like Presburger arithmetics.

Organised by computational model, the results of this paper can be summarised as follows: the cross product theorem and the nonspeedup theorem

- hold for Presburger arithmetic,
- hold for finite automata,
- do not hold for polynomial-time machines,
- hold for Turing machines,
- hold for natural number arithmetic,
- hold for ordinal number arithmetic.

The cardinality theorem for two inputs and the restricted cardinality theorem

- hold for Presburger arithmetic,
- hold for finite automata,
- do not hold for polynomial-time machines,
- hold for Turing machines,
- hold for natural number arithmetic,
- do not hold for ordinal number arithmetic.

The behaviour of ordinal number arithmetic is interesting: the cardinality theorem for two inputs and the restricted cardinality theorem fail since there exist ordinal numbers that are not elementarily definable in ordinal number arithmetic—but this is not a 'problem' for the cross product theorem and the nonspeedup theorem.

The results of this paper raise the question of whether the cardinality theorem holds for first-order logic. I conjecture that this is the case, that is, I conjecture that for well-orderable structures S in which all finite relations can be elementarily defined, if $\#_A^n$ is elementarily n-enumerable then A is elementarily definable. Proving this conjecture would also settle the open problem of whether the cardinality theorem holds for finite automata.

References

1. H. Austinat, V. Diekert, and U. Hertrampf. A structural property of regular frequency computations. *Theoretical Comput. Sci.*, 292(1):33–43, 2003.
2. H. Austinat, V. Diekert, U. Hertrampf, and H. Petersen. Regular frequency computations. In *Proc. RIMS Symposium on Algebraic Systems, Formal Languages and Computation*, volume 1166 of *RIMS Kokyuroku*, pages 35–42. Research Inst. for Mathematical Sci., Kyoto Univ., Japan, 2000.
3. R. Beigel. *Query-Limited Reducibilities*. PhD thesis, Stanford Univ., USA, 1987.
4. R. Beigel, W. I. Gasarch, M. Kummer, G. Martin, T. McNicholl, and F. Stephan. The complexity of ODD_n^A. *J. Symbolic Logic*, 65(1):1–18, 2000.
5. J. R. Büchi. On a decision method in restricted second-order arithmetic. In *Proc. 1960 International Congress on Logic, Methodology and Philosophy of Sci.*, pages 1–11. Stanford Univ. Press, 1962.

6. J.-Y. Cai and L. A. Hemachandra. Enumerative counting is hard. *Inf. Computation*, 82(1):34–44, 1989.
7. W. I. Gasarch. Bounded queries in recursion theory: A survey. In *Proceedings of the Sixth Annual Structure in Complexity Theory Conference*, pages 62–78. IEEE Computer Soc. Press, 1991.
8. V. Harizanov, M. Kummer, and J. Owings. Frequency computations and the cardinality theorem. *J. Symbolic Logic*, 52(2):682–687, 1992.
9. L. A. Hemachandra. The strong exponential hierarchy collapses. *J. Comput. Syst. Sci.*, 39(3):299–322, 1989.
10. D. Hilbert and P. Bernay. *Grundlagen der Mathematik II*, volume 50 of *Die Grundlehren der mathematischen Wissenschaft in Einzeldarstellungen*. Springer-Verlag, second edition, 1970.
11. A. Hoene and A. Nickelsen. Counting, selecting, and sorting by query-bounded machines. In *Proc. 10th International Symposium on Theoretical Aspects of Comp. Sci.*, volume 665 of *Lecture Notes on Comp. Sci.*, pages 196–205. Springer-Verlag, 1993.
12. N. Immerman. Nondeterministic space is closed under complementation. *SIAM J. Comput.*, 17(5):935–938, 1988.
13. C. G. Jockusch, Jr. *Reducibilities in Recursive Function Theory*. PhD thesis, Massachusetts Inst. of Technology, USA, 1966.
14. J. Kadin. $P^{NP[O(\log n)]}$ and sparse Turing-complete sets for NP. *J. Comput. Syst. Sci.*, 39(3):282–298, 1989.
15. E. B. Kinber. Frequency computations in finite automata. *Cybernetics*, 2:179–187, 1976.
16. M. Kummer. A proof of Beigel's cardinality conjecture. *J. Symbolic Logic*, 57(2):677–681, 1992.
17. M. Kummer and F. Stephan. Effecitive search problems. *Mathematical Logic Quarterly*, 40(2):224–236, 1994.
18. S. R. Mahaney. Sparse complete sets for NP: Solution of a conjecture of Berman and Hartmanis. *J. Comput. Syst. Sci.*, 25(2):130–143, 1982.
19. R. McNaughton. Review of [5]. *J. Symbolic Logic*, 28(1):100–102, 1963.
20. A. Nickelsen. On polynomially D-verbose sets. In *Proceedings of the 14th International Symposium on Theoretical Aspects of Computer Science*, volume 1200 of *Lecture Notes on Comp. Sci.*, pages 307–318. Springer-Verlag, 1997.
21. J. C. Owings, Jr. A cardinality version of Beigel's nonspeedup theorem. *J. Symbolic Logic*, 54(3):761–767, 1989.
22. E. L. Post. Recursively enumerable sets of positive integers and their decision problems. *Bulletin of the American Mathematical Society*, 50:284–316, 1944.
23. R. Szelepcsényi. The method of forced enumeration for nondeterministic automata. *Acta Informatica*, 23(3):279–284, 1988.
24. T. Tantau. Comparing verboseness for finite automata and Turing machines. In *Proc. 19th International Symposium on Theoretical Aspects of Comp. Sci.*, volume 2285 of *Lecture Notes on Comp. Sci.*, pages 465–476. Springer-Verlag, 2002.
25. T. Tantau. Towards a cardinality theorem for finite automata. In *Proc. 27th International Symposium on Mathematical Foundations of Comp. Sci.*, volume 2420 of *Lecture Notes on Comp. Sci.*, pages 625–636. Springer-Verlag, 2002.
26. T. Tantau. *On Structural Similarities of Finite Automata and Turing Machine Enumerability Classes*. PhD thesis, Technical Univ. Berlin, Germany, 2003.
27. T. Tantau. Weak cardinality theorems for first-order logic. Technical Report TR03-024, Electronic Colloquium on Computational Complexity, www.eccc.uni-trier.de/eccc, 2003.

Compositionality of Hennessy-Milner Logic through Structural Operational Semantics

Wan Fokkink[1,2], Rob van Glabbeek[1], and Paulien de Wind[2]

[1] CWI, Department of Software Engineering
PO Box 94079, 1090 GB Amsterdam, The Netherlands
[2] Vrije Universiteit Amsterdam, Department of Theoretical Computer Science
De Boelelaan 1081a, 1081 HV Amsterdam, The Netherlands
wan@cwi.nl, http://www.cwi.nl/~wan/
rvg@cs.stanford.edu, http://theory.stanford.edu/~rvg/
pdwind@cs.vu.nl, http://www.cs.vu.nl/~pdwind/

Abstract. This paper presents a method for the decomposition of HML formulae. It can be used to decide whether a process algebra term satisfies a HML formula, by checking whether subterms satisfy certain formulae, obtained by decomposing the original formula. The method uses the structural operational semantics of the process algebra. The main contribution of this paper is that an earlier decomposition method from LARSEN [14] for the De Simone format is extended to the more general ntyft/ntyxt format without lookahead.

1 Introduction

In the past two decades, compositional methods have been developed for checking the validity of assertions in modal logics, used to describe the behaviour of processes. This means that the truth of an assertion for a composition of processes can be deduced from the truth of certain assertions for the components of the composition. Most research papers in this area focus on a particular process algebra.

BARRINGER, KUIPER & PNUELI [3] present (a preliminary version of) a compositional proof system for concurrent programs, which is based on a rich temporal logic, including operators from process logic [10] and LTL [20]. For modelling concurrent programs they define a language including assignment, conditional and while statements. Interaction between parallel components is done via shared variables.

In STIRLING [22] modal proof systems are developed for subsets of CCS [16] (with and without silent actions) including only sequential and alternative composition, to decide the validity of formulae from Hennessy-Milner Logic (HML) [11]. In STIRLING [23,24] the results from [22] are extended, creating proof systems for subsets of CCS and SCCS [18] including asynchronous and synchronous parallelism and infinite behaviour, using ideas from [3]. In STIRLING [25] the proposals in [23,24] are generalised to be able to cope with the restriction operator.

A. Lingas and B.J. Nilsson (Eds.): FCT 2003, LNCS 2751, pp. 412–422, 2003.
© Springer-Verlag Berlin Heidelberg 2003

In WINSKEL [26] a method is given to decompose formulae with respect to each operation in SCCS. The language of assertions is HML with infinite conjunction and disjunction. This decomposition provides the foundations of Winskel's proof system for SCCS with modal assertions. In [27], [2] and [1] processes are described by specification languages inspired by CCS and CSP [6]. The articles describe compositional methods for deciding whether processes satisfy assertions from a modal μ-calculus [13].

LARSEN [14] developed a more general compositional method for deciding whether a process satisfies a certain property. Unlike the aforementioned methods, this method is not oriented towards a particular process algebra, but it is based on structural operational semantics [19], which provides process algebras and specification languages with an interpretation. A transition system specification, consisting of an algebraic signature and a set of transition rules of the form $\frac{\text{premises}}{\text{conclusion}}$, generates a transition relation between the closed terms over the signature. An example of a transition rule, for alternative composition, is

$$\frac{x_1 \xrightarrow{a} y}{x_1 + x_2 \xrightarrow{a} y}$$

meaning for states t_1, t_2 and u that if state t_1 can evolve into state u by the execution of action a, then so can state $t_1 + t_2$. Larsen showed how to decompose HML formulae with respect to a transition system specification in the De Simone format [21]. This format was originally put forward to guarantee that the bisimulation equivalence associated with a transition system specification is a congruence, meaning that bisimulation equivalence is preserved by all functions in the signature. LARSEN AND XINXIN [15] extended this decomposition method to HML with recursion (which is equivalent to the modal μ-calculus).

Since modal proof systems for specific process algebras are tailor-made, they may be more concise than the ones generated by the general decomposition method of Larsen (e.g., [23,24,25]). However, in some cases the general decomposition method does produce modal proof systems that are similar in spirit to those in the literature (e.g., [22,26]).

In BLOOM, FOKKINK & VAN GLABBEEK [4] a method is given for decomposing formulae from a fragment of HML with infinite conjunctions, with respect to terms from any process algebra that has a structural operational semantics in ntyft/ntyxt format [9] without lookahead. This format is a generalisation of the De Simone format, and still guarantees that bisimulation equivalence is a congruence. The decomposition method is not presented in its own right, but is used in the derivation of congruence formats for a range of behavioural equivalences from VAN GLABBEEK [8].

In this paper the decomposition method from [4] is extended to full HML with infinite conjunction, again with respect to terms from any process algebra that has a structural operational semantics in ntyft/ntyxt format without lookahead.

2   Preliminaries

In this section we give the basic notions of structural operational semantics and Hennessy-Milner Logic (HML) that are needed to define our decomposition method.

2.1   Structural Operational Semantics

Structural operational semantics [19] provides a framework to give an operational semantics to programming and specification languages. In particular, because of its intuitive appeal and flexibility, structural operational semantics has found considerable application in the study of the semantics of concurrent processes.

Let V be an infinite set of variables. A syntactic object is called *closed* if it does not contain any variables from V.

Definition 1 (signature). *A* signature *is a collection Σ of function symbols $f \notin V$, equipped with a function $ar : \Sigma \to \mathbb{N}$. The set $\mathbb{T}(\Sigma)$ of terms over a signature Σ is defined recursively by:*

- $V \subseteq \mathbb{T}(\Sigma)$,
- *if $f \in \Sigma$ and $t_1, \ldots, t_{ar(f)} \in \mathbb{T}(\Sigma)$, then $f(t_1, \ldots, t_{ar(f)}) \in \mathbb{T}(\Sigma)$.*

A term $c()$ is abbreviated as c. For $t \in \mathbb{T}(\Sigma)$, $var(t)$ denotes the set of variables that occur in t. $T(\Sigma)$ is the set of closed terms over Σ, i.e. the terms $t \in \mathbb{T}(\Sigma)$ with $var(t) = \emptyset$. A Σ-substitution σ is a partial function from V to $\mathbb{T}(\Sigma)$. If σ is a Σ-substitution and S is any syntactic object, then $\sigma(S)$ denotes the object obtained from S by replacing, for x in the domain of σ, every occurrence of x in S by $\sigma(x)$. In that case $\sigma(S)$ is called a substitution instance *of S. A Σ-substitution is* closed *if it is a total function from V to $T(\Sigma)$.*

In the remainder, let Σ denote a signature and A a set of actions, satisfying $|\Sigma| \leq |V|$ and $|A| \leq |V|$.

Definition 2 (literal). *A positive Σ-literal is an expression $t \xrightarrow{a} t'$ and a negative Σ-literal an expression $t \xnrightarrow{a}$ with $t, t' \in \mathbb{T}(\Sigma)$ and $a \in A$. For $t, t' \in \mathbb{T}(\Sigma)$ and $a \in A$, the literals $t \xrightarrow{a} t'$ and $t \xnrightarrow{a}$ are said to* deny *each other.*

Definition 3 (transition rule). *A transition rule over Σ is an expression of the form $\frac{H}{\alpha}$ with H a set of Σ-literals (the premises of the the rule) and α a positive Σ-literal (the conclusion). The left- and right-hand side of α are called the* source *and the* target *of the rule, respectively. A rule $\frac{H}{\alpha}$ with $H = \emptyset$ is also written α.*

Definition 4 (transition system specification). *A transition system specification (TSS) is a pair (Σ, R) with R a collection of transition rules over Σ.*

Definition 5 (proof). *Let $P = (\Sigma, R)$ be a TSS. A proof of a transition rule $\frac{H}{\alpha}$ from P is a well-founded, upwardly branching tree of which the nodes are labelled by Σ-literals, and some of the leaves are marked "hypothesis", such that:*

- *the root is labelled by α,*
- *H contains the labels of the hypotheses, and*
- *if β is the label of a node q which is not an hypothesis and K is the set of labels of the nodes directly above q, then $\frac{K}{\beta}$ is a substitution instance of a transition rule in R.*

If a proof of $\frac{K}{\alpha}$ from P exists, then $\frac{K}{\alpha}$ is provable from P, notation $P \vdash \frac{K}{\alpha}$.

Definition 6 (transition relation). *A transition relation over Σ is a relation $\rightarrow\ \subseteq T(\Sigma) \times A \times T(\Sigma)$. We write $p \xrightarrow{a} q$ for $(p, a, q) \in\ \rightarrow$ and $p \not\xrightarrow{a}$ for $\neg \exists q \in T(\Sigma) : p \xrightarrow{a} q$.*

Thus a transition relation over Σ can be regarded as a set of closed positive Σ-literals (*transitions*). A TSS with only positive premises specifies a transition relation in a straightforward way as the set of all provable transitions. But it is much less trivial to associate a transition relation to a TSS with negative premises. Several solutions are proposed in GROOTE [9], BOL & GROOTE [5] and VAN GLABBEEK [7]. From the latter we adopt the notion of a well-supported proof and a complete TSS.

Definition 7 (well-supported proof). *Let $P = (\Sigma, R)$ be a TSS. A well-supported proof of a closed literal α from P is a well-founded, upwardly branching tree of which the nodes are labelled by closed Σ-literals, such that:*

- *the root is labelled by α, and*
- *if β is the label of a node q and K is the set of labels of the nodes directly above q, then*
    1. *either $\frac{K}{\beta}$ is a closed substitution instance of a transition rule in R*
    2. *or β is negative and for every set N of negative closed literals such that $P \vdash \frac{N}{\gamma}$ for γ a closed literal denying β, a literal in K denies one in N.*

We say α is ws-provable from P, notation $P \vdash_{ws} \alpha$, if a well-supported proof of α from P exists.

In [7] it was noted that $\vdash_{ws}$ is *consistent*, in the sense that no standard TSS admits well-supported proofs of two literals that deny each other.

Definition 8 (completeness). *A TSS P is complete if for any closed literal $p \not\xrightarrow{a}$ either $P \vdash_{ws} p \xrightarrow{a} p'$ for some closed term p' or $P \vdash_{ws} p \not\xrightarrow{a}$.*

Now a TSS specifies a transition relation if and only if it is complete. The specified transition relation is then the set of all *ws*-provable transitions.

2.2   Hennessy-Milner Logic

A variety of modal logics have been developed to express properties of transition relations. Modal logic aims to formulate properties of process terms, and to identify terms that satisfy the same properties. HENNESSY & MILNER [11] have defined a modal language, often called Hennessy-Milner Logic (HML), which characterises the bisimulation equivalence relation on process terms, assuming that each term has only finitely many outgoing transitions. This assumption can be discarded if infinite conjunctions are allowed [17,12].

Definition 9 (Hennessy-Milner Logic). *Assume an action set A. The set $\mathbb{O}$ of* potential observations *or* modal formulae *is recursively defined by*

$$\varphi ::= \bigwedge_{i \in I} \varphi_i \mid \langle a \rangle \varphi \mid \neg \varphi$$

with $a \in A$ and I some index set.

Definition 10 (satisfaction relation). *Let $P = (\Sigma, R)$ be a TSS. The satis-faction relation $\models_P \subseteq T(\Sigma) \times \mathbb{O}$ is defined as follows, with $p \in T(\Sigma)$:*

$$p \models_P \bigwedge_{i \in I} \varphi_i \text{ iff } p \models_P \varphi_i \text{ for all } i \in I$$
$$p \models_P \langle a \rangle \varphi \quad \text{iff there is a } q \in T(\Sigma) \text{ such that } P \vdash_{ws} p \xrightarrow{a} q \text{ and } q \models_P \varphi$$
$$p \models_P \neg \varphi \quad \text{iff } p \not\models_P \varphi$$

We will use the binary conjunction $\varphi_1 \wedge \varphi_2$ as an abbreviation of $\bigwedge_{i \in \{1,2\}} \varphi_i$, whereas $\top$ is an abbreviation for the empty conjunction. We identify formulae that are logically equivalent using the laws $\top \wedge \varphi \cong \varphi$, $\bigwedge_{i \in I}(\bigwedge_{j \in J_i} \varphi_j) \cong \bigwedge_{i \in I, j \in J_i} \varphi_j$ and $\neg\neg\varphi \cong \varphi$. This is justified because $\varphi \cong \psi$ implies $p \models_P \varphi \Leftrightarrow p \models_P \psi$.

3   Decomposing HML Formulae

In this section we will see how one can decompose HML formulae with respect to process terms. The TSS defining the transition relation on these terms should be in *ready simulation format* [4], allowing only ntyft/ntyxt rules [9] without lookahead.

Definition 11 (ntyxt,ntyft,nxytt). *An ntytt rule is a transition rule in which the right-hand sides of positive premises are variables that are all distinct, and that do not occur in the source. An ntytt rule is an* ntyxt *rule if its source is a variable, and an* ntyft *rule if its source contains exactly one function symbol and no multiple occurrences of variables. An ntytt rule is an* nxytt *rule if the left-hand sides of its premises are variables.*

Definition 12 (lookahead). *A transition rule has* no lookahead *if the variables occurring in the right-hand sides of its positive premises do not occur in the left-hand sides of its premises.*

Definition 13 (ready simulation format). *A TSS is in* ready simulation format *if its transition rules are ntyft or ntyxt rules that have no lookahead.*

Definition 14 (free). *A variable occurring in a transition rule is* free *if it does not occur in the source nor in the right-hand sides of the positive premises of this rule.*

Definition 15 (decent). *A transition rule is* decent *if it has no lookahead and does not contain free variables.*

In BLOOM, FOKKINK & VAN GLABBEEK [4] for any TSS P in ready simulation format the collection of *P-ruloids* is defined. These are decent nxytt rules for which the following holds:

Theorem 1. *[4] Let P be a TSS in ready simulation format. Then $P \vdash_{ws}$ $\sigma(t) \xrightarrow{a} p$ for t a term, p a closed term and σ a closed substitution, iff there are a P-ruloid $\dfrac{H}{t \xrightarrow{a} u}$ and a closed substitution σ' with $P \vdash_{ws} \sigma'(\alpha)$ for $\alpha \in H$, $\sigma'(t) = \sigma(t)$ and $\sigma'(u) = p$.*

Given a TSS $P = (\Sigma, R)$ in ready simulation format, the following definition assigns to each term $t \in \mathbb{T}(\Sigma)$ and each observation $\varphi \in \mathbb{O}$ a collection $t_P^{-1}(\varphi)$ of *decomposition mappings* $\psi : V \to \mathbb{O}$. Each of these mappings $\psi \in t_P^{-1}(\varphi)$ guarantees, given a closed substitution σ, that $\sigma(t)$ satisfies φ if $\sigma(x)$ satisfies the formula $\psi(x)$ for all $x \in var(t)$. Moreover, whenever for some closed substitution σ the term $\sigma(t)$ satisfies φ, there must be a decomposition mapping $\psi \in t_P^{-1}(\varphi)$ with $\sigma(x)$ satisfying $\psi(x)$ for all $x \in var(t)$. This is formalised in Theorem 2 and proven thereafter.

Definition 16. *Let $P = (\Sigma, R)$ be a TSS in ready simulation format. Then $\cdot_P^{-1} : \mathbb{T}(\Sigma) \to (\mathbb{O} \to \mathcal{P}(V \to \mathbb{O}))$ is defined by:*

- $\psi \in t_P^{-1}(\langle a \rangle \varphi)$ *iff there is a P-ruloid $\dfrac{H}{t \xrightarrow{a} u}$ and a $\chi \in u_P^{-1}(\varphi)$ and $\psi : V \to \mathbb{O}$ is given by*

$$\psi(x) = \begin{cases} \chi(x) \wedge \bigwedge_{(x \xrightarrow{b} y) \in H} \langle b \rangle \chi(y) \wedge \bigwedge_{(x \xrightarrow{c} \not\rightarrow) \in H} \neg \langle c \rangle \top & \text{if } x \in var(t) \\ \top & \text{if } x \notin var(t) \end{cases}$$

- $\psi \in t_P^{-1}(\bigwedge_{i \in I} \varphi_i)$ *iff*

$$\psi(x) = \bigwedge_{i \in I} \psi_i(x)$$

where $\psi_i \in t_P^{-1}(\varphi_i)$ for $i \in I$.
- $\psi \in t_P^{-1}(\neg \varphi)$ *iff there is a function $h : t_P^{-1}(\varphi) \to var(t)$ and $\psi : V \to \mathbb{O}$ is given by*

$$\psi(x) = \bigwedge_{\chi \in h^{-1}(x)} \neg \chi(x)$$

When clear from the context, the subscript P will be omitted.

It is not hard to see that if $\psi \in t_P^{-1}(\varphi)$ then $\psi(x) = \top$ for all $x \notin var(t)$.

Theorem 2. *Let $P = (\Sigma, R)$ be a complete TSS in ready simulation format. Let $\varphi \in \mathbb{O}$. For any term $t \in \mathbb{T}(\Sigma)$ and closed substitution $\sigma : V \to T(\Sigma)$ one has*

$$\sigma(t) \models \varphi \quad \Leftrightarrow \quad \exists \psi \in t^{-1}(\varphi) \forall x \in var(t)\big(\sigma(x) \models \psi(x)\big)$$

Proof. With induction on the structure of φ.

- $\varphi = \langle a \rangle \varphi'$

  $\boxed{\Rightarrow}$ Suppose $\sigma(t) \models \langle a \rangle \varphi'$. Then by Definition 10 there is a $p \in T(\Sigma)$ with $P \vdash_{ws} \sigma(t) \xrightarrow{a} p$ and $p \models \varphi'$. Thus, by Theorem 1 there must be a P-ruloid $\frac{H}{t \xrightarrow{a} u}$ and a closed substitution σ' with $P \vdash_{ws} \sigma'(\alpha)$ for $\alpha \in H$, $\sigma'(t) = \sigma(t)$, i.e. $\sigma'(x) = \sigma(x)$ for $x \in var(t)$, and $\sigma'(u) = p$. Since $\sigma'(u) \models \varphi'$, the induction hypothesis can be applied, and there must be a $\chi \in u^{-1}(\varphi')$ such that $\sigma'(z) \models \chi(z)$ for all $z \in var(u)$. Furthermore $\sigma'(z) \models \chi(z) = \top$ for all $z \notin var(u)$. Now define ψ as indicated in Definition 16. By definition, $\psi \in t^{-1}(\langle a \rangle \varphi')$. Let $x \in var(t)$. For $(x \xrightarrow{b} y) \in H$ one has $P \vdash_{ws} \sigma'(x) \xrightarrow{b} \sigma'(y)$ and $\sigma'(y) \models \chi(y)$, so $\sigma'(x) \models \langle b \rangle \chi(y)$. Moreover, for $(x \xrightarrow{c} \!\!\!\!\!/) \in H$ one has $P \vdash_{ws} \sigma'(x) \xrightarrow{c} \!\!\!\!\!/$, so the consistency of $\vdash_{ws}$ yields $P \nvdash_{ws} \sigma'(x) \xrightarrow{c} q$ for all $q \in T(\Sigma)$, and thus $\sigma'(x) \models \neg\langle c \rangle\top$. It follows that $\sigma(x) = \sigma'(x) \models \psi(x)$.

  $\boxed{\Leftarrow}$ Now suppose that there is a $\psi \in t^{-1}(\langle a \rangle \varphi')$ such that $\sigma(x) \models \psi(x)$ for all $x \in var(t)$. This means that there is a P-ruloid

$$\frac{\{x \xrightarrow{a_i} y_i \mid i \in I_x, \ x \in var(t)\} \cup \{x \xrightarrow{b_j} \!\!\!\!\!/ \mid j \in J_x, \ x \in var(t)\}}{t \xrightarrow{a} u}$$

and a decomposition mapping $\chi \in u^{-1}(\varphi')$ such that, for all $x \in var(t)$,

$$\sigma(x) \models \chi(x) \wedge \bigwedge_{i \in I_x} \langle a_i \rangle \chi(y_i) \wedge \bigwedge_{j \in J_x} \neg\langle b_j \rangle\top$$

By Definition 10 it follows that, for $x \in var(t)$ and $i \in I_x$, $P \vdash_{ws} \sigma(x) \xrightarrow{a_i} p_i$ for some $p_i \in T(\Sigma)$ with $p_i \models \chi(y_i)$. Moreover, for $x \in var(t)$ and $j \in J_x$, $P \nvdash_{ws} \sigma(x) \xrightarrow{b_j} q$ for all $q \in T(\Sigma)$, so by the completeness of P, $P \vdash_{ws} \sigma(x) \xrightarrow{b_j} \!\!\!\!\!/$. Let σ' be a closed substitution with $\sigma'(x) = \sigma(x)$ for $x \in var(t)$ and $\sigma'(y_i) = p_i$ for $i \in I_x$ and $x \in var(t)$. Here we use that the variables x and y_i are all different. Now $\sigma'(z) \models \chi(z)$ for $z \in var(u)$, using that u contains only variables that occur in t or in the premises of the ruloid. Thus the induction hypothesis can be applied, and $\sigma'(u) \models \varphi'$. Moreover, $P \vdash_{ws} \sigma'(x) \xrightarrow{a_i} \sigma'(y_i)$ for $x \in var(t)$ and $i \in I_x$, and $P \vdash_{ws} \sigma'(x) \xrightarrow{b_j} \!\!\!\!\!/$ for $x \in var(t)$ and $j \in J_x$. So, by Theorem 1, $P \vdash_{ws} \sigma'(t) \xrightarrow{a} \sigma'(u)$, which implies $\sigma(t) = \sigma'(t) \models \langle a \rangle \varphi'$.

- $\varphi = \bigwedge_{i \in I} \varphi_i$

$$\sigma(t) \models \bigwedge_{i \in I} \varphi_i \Leftrightarrow \forall i \in I : \sigma(t) \models \varphi_i$$
$$\Leftrightarrow \forall i \in I \ \exists \psi_i \in t^{-1}(\varphi_i) \ \forall x \in var(t) : \sigma(x) \models \psi_i(x)$$
$$\Leftrightarrow \exists \psi \in t^{-1}(\bigwedge_{i \in I} \varphi_i) \ \forall x \in var(t) : \sigma(x) \models \psi(x).$$

- $\varphi = \neg \varphi'$

$\boxed{\Rightarrow}$ Suppose $\sigma(t) \models \neg \varphi'$. Then by Definition 10 we have $\sigma(t) \not\models \varphi'$. Using the induction hypothesis, there is no $\chi \in t^{-1}(\varphi')$ such that $\sigma(x) \models \chi(x)$ for all $x \in var(t)$. So for all $\chi \in t^{-1}(\varphi')$ there is an $x \in var(t)$ such that $\sigma(x) \models \neg\chi(x)$. Let us denote this x as $h(\chi)$, so that we obtain a function $h : t^{-1}(\varphi') \to var(t)$ such that $\sigma(h(\chi)) \models \neg\chi(h(\chi))$ for all $\chi \in t^{-1}(\varphi')$. Define $\psi \in t^{-1}(\neg\varphi')$ as indicated in Definition 16, using h. Let $x \in var(t)$. If $x = h(\chi)$ for some $\chi \in t^{-1}(\varphi')$ then $\sigma(x) \models \neg\chi(x)$. Hence, $\sigma(x) \models \bigwedge_{\chi \in h^{-1}(x)} \neg\chi(x) = \psi(x)$.

$\boxed{\Leftarrow}$ Suppose that there is a $\psi \in t^{-1}(\neg\varphi')$ such that $\sigma(x) \models \psi(x)$ for all $x \in var(t)$. By Definition 16 there is a function $h : t^{-1}(\varphi') \to var(t)$ such that $\psi(x) = \bigwedge_{\chi \in h^{-1}(x)} \neg\chi(x)$ for all $x \in var(t)$. So for all $x \in var(t)$ and for all $\chi \in h^{-1}(x)$ we have that $\sigma(x) \models \neg\chi(x)$. In other words, for all $\chi \in t^{-1}(\varphi')$, we have $\sigma(h(\chi)) \models \neg\chi(h(\chi))$. So $\neg\exists\chi \in t^{-1}(\varphi') \forall x \in var(t)(\sigma(x) \models \chi(x))$. Then using the induction hypothesis, we have $\sigma(t) \not\models \varphi'$, so $\sigma(t) \models \neg\varphi'$.

We give a few examples of the application of Definition 16.

Example 1. Let $A = \{a, b\}$ and let $P = (\Sigma, R)$ with Σ consisting of the constant c and the binary function symbol f and R is:

$$c \xrightarrow{a} c \qquad \frac{x_1 \xrightarrow{a} y}{f(x_1, x_2) \xrightarrow{b} y} \qquad \frac{x_2 \xrightarrow{a} y \quad x_1 \xcancel{\xrightarrow{b}}}{f(x_1, x_2) \xrightarrow{b} y}$$

This TSS is complete and in ready simulation format. We proceed to compute $f(x_1, x_2)^{-1}(\langle b \rangle \top)$. There are two P-ruloids with a conclusion of the form $f(x_1, x_2) \xrightarrow{b} \_$, namely $\frac{x_1 \xrightarrow{a} y}{f(x_1,x_2) \xrightarrow{b} y}$ and $\frac{x_2 \xrightarrow{a} y \quad x_1 \xcancel{\xrightarrow{b}}}{f(x_1,x_2) \xrightarrow{b} y}$. According to Definition 16, we have $f(x_1, x_2)^{-1}(\langle b \rangle \top) = \{\psi_1, \psi_2\}$ with ψ_1 and ψ_2 as defined below, using $\chi \in y^{-1}(\top)$ (so $\chi(x) = \top$ for all variables $x \in V$):

$$\psi_1(x_1) = \chi(x_1) \wedge \langle a \rangle \chi(y) = \top \wedge \langle a \rangle \top = \langle a \rangle \top$$
$$\psi_1(x_2) = \chi(x_2) = \top$$
$$\psi_1(x) = \top \text{ for } x \notin var(f(x_1, x_2))$$

$$\psi_2(x_1) = \chi(x_1) \wedge \neg\langle b \rangle \top = \top \wedge \neg\langle b \rangle \top = \neg\langle b \rangle \top$$
$$\psi_2(x_2) = \chi(x_2) \wedge \langle a \rangle \chi(y) = \top \wedge \langle a \rangle \top = \langle a \rangle \top$$
$$\psi_2(x) = \top \text{ for } x \notin var(f(x_1, x_2))$$

By Theorem 2 a closed term $f(u_1, u_2)$ can execute a b if and only if the closed term u_1 can execute an a, or the closed term u_1 can not execute a b and the closed term u_2 can execute an a. Looking at the premises, this is what we would expect.

Example 2. Using the TSS and the mappings $\psi_1, \psi_2 \in f(x_1, x_2)^{-1}(\langle b \rangle \top)$ from Example 1, we can compute $f(x_1, x_2)^{-1}(\neg \langle b \rangle \top)$. There are four possible functions $h : f(x_1, x_2)^{-1}(\langle b \rangle \top) \to var(f(x_1, x_2))$, yielding four possible definitions of $\psi \in f(x_1, x_2)^{-1}(\neg \langle b \rangle \top)$.

1. If $h(\psi_1) = h(\psi_2) = x_1$ then

$$\psi(x_1) = \neg \psi_1(x_1) \wedge \neg \psi_2(x_1) = \neg \langle a \rangle \top \wedge \neg \neg \langle b \rangle \top = \neg \langle a \rangle \top \wedge \langle b \rangle \top$$
$$\psi(x_2) = \top$$

2. If $h(\psi_1) = h(\psi_2) = x_2$ then

$$\psi(x_1) = \top$$
$$\psi(x_2) = \neg \psi_1(x_2) \wedge \neg \psi_2(x_2) = \neg \top \wedge \neg \langle a \rangle \top$$

3. If $h(\psi_1) = x_1$ and $h(\psi_2) = x_2$ then

$$\psi(x_1) = \neg \psi_1(x_1) = \neg \langle a \rangle \top$$
$$\psi(x_2) = \neg \psi_2(x_2) = \neg \langle a \rangle \top$$

4. If $h(\psi_1) = x_2$ and $h(\psi_2) = x_1$ then

$$\psi(x_1) = \neg \psi_2(x_1) = \neg \neg \langle b \rangle \top = \langle b \rangle \top$$
$$\psi(x_2) = \neg \psi_1(x_2) = \neg \top$$

By Theorem 2 a closed term $f(u_1, u_2)$ can *not* execute a b if and only if (1) the closed term u_1 can execute a b but not an a, or (3) the closed term u_1 can not execute an a and the closed term u_2 can not execute an a. Looking at the premises, this is again what we would expect. The other two possibilities (2) and (4) do not qualify, since no term can ever satisfy $\neg \top$.

A little less obvious example is the following:

Example 3. Let $A = \{a, b\}$ and let $P = (\Sigma, R)$ with Σ consisting of the constant c and the unary function symbol f and R is:

$$c \xrightarrow{a} c \qquad \frac{x \xrightarrow{a} y}{f(x) \xrightarrow{b} y} \qquad \frac{x \xrightarrow{b} y}{f(x) \xrightarrow{a} f(y)}$$

This TSS is complete and in ready simulation format. We proceed to compute $f(f(x))^{-1}(\langle b \rangle \langle a \rangle \top)$. The only P-ruloid that has a conclusion $f(f(x)) \xrightarrow{b}$ _ is $\frac{x \xrightarrow{b} y}{f(f(x)) \xrightarrow{b} f(y)}$. So for each $\psi \in f(f(x))^{-1}(\langle b \rangle \langle a \rangle \top)$, $\psi(x) = \chi(x) \wedge \langle b \rangle \chi(y)$ with $\chi \in f(y)^{-1}(\langle a \rangle \top)$. The only P-ruloid that has a conclusion $f(y) \xrightarrow{a}$ _ is $\frac{y \xrightarrow{b} z}{f(y) \xrightarrow{a} f(z)}$. So $\chi(y) = \chi'(y) \wedge \langle b \rangle \chi'(z)$ with $\chi' \in f(z)^{-1}(\top)$. Since $\chi'(y) = \chi'(z) = \top$ we have $\chi(y) = \langle b \rangle \top$. Moreover $x \notin var(f(y))$ implies $\chi(x) = \top$. Hence $\psi(x) = \langle b \rangle \langle b \rangle \top$.

By Theorem 2 a closed term $f(f(u))$ can execute a b followed by an a if and only if the closed term u can execute two consecutive b's.

The following example shows that in Theorem 2 it is essential that the TSS is complete. That is, the theorem would fail if we would take the transition relation induced by a TSS to consist of those transitions for which a well-supported proof exists.

Example 4. Let $A = \{a, b\}$ and let $P = (\Sigma, R)$ with Σ consisting of the constant c and the unary function symbol f and R is:

$$\frac{x \not\xrightarrow{a}}{f(x) \xrightarrow{b} c} \qquad \frac{c \not\xrightarrow{a}}{c \xrightarrow{a} c}$$

This TSS, which is in ready simulation format, is incomplete. For example, neither $P \vdash_{ws} c \xrightarrow{a} t$ for a closed term t nor $P \vdash_{ws} c \not\xrightarrow{a}$.

Let us assume that the transition relation induced by this TSS consists of those transitions for which a well-supported proof exists. Then there is no a-transition for c and no b-transition for $f(c)$, so $c \not\models \langle a \rangle \top$ and $f(c) \not\models \langle b \rangle \top$.

The only P-ruloid is $\dfrac{x \not\xrightarrow{a}}{f(x) \xrightarrow{b} c}$. Hence Theorem 2 would yield $f(c) \models \langle b \rangle \top \Leftrightarrow c \models \neg \langle a \rangle \top \Leftrightarrow c \not\models \langle a \rangle \top$. Since this is false, Theorem 2 would fail with respect to P.

References

1. H. R. ANDERSEN, C. STIRLING & G. WINSKEL (1994): *A compositional proof system for the modal μ-calculus.* In Proceedings, Ninth Annual IEEE Symposium on Logic in Computer Science, IEEE Computer Society Press, Paris, France, pp. 144–153.
2. H. R. ANDERSEN & G. WINSKEL (1992): *Compositional checking of satisfaction.* Formal Methods in System Design 1(4), pp. 323–354.
3. H. BARRINGER, R. KUIPER & A. PNUELI (1984): *Now you may compose temporal logic specifications.* In ACM Symposium on Theory of Computing (STOC '84), ACM Press, Baltimore, USA, pp. 51–63.
4. B. BLOOM, W. J. FOKKINK & R. J. VAN GLABBEEK (2003): *Precongruence formats for decorated trace semantics.* ACM Transactions on Computational Logic. To appear.
5. R. BOL & J. F. GROOTE (1996): *The meaning of negative premises in transition system specifications.* Journal of the ACM 43(5), pp. 863–914.
6. S. D. BROOKES, C. A. R. HOARE & A. W. ROSCOE (1984): *A theory of communicating sequential processes.* Journal of the ACM 31(3), pp. 560–599.
7. R. J. VAN GLABBEEK (1996): *The meaning of negative premises in transition system specifications II.* In F. Meyer auf der Heide & B. Monien, editors: Automata, Languages and Programming, 23rd Colloquium (ICALP '96), Lecture Notes in Computer Science 1099, Springer-Verlag, Paderborn, Germany, pp. 502–513.
8. R. J. VAN GLABBEEK (2001): *The linear time – branching time spectrum I: The semantics of concrete, sequential processes.* In J. A. Bergstra, A. Ponse & S. A. Smolka, editors: Handbook of Process Algebra, chapter 1, Elsevier, pp. 3–99.
9. J. F. GROOTE (1993): *Transition system specifications with negative premises.* Theoretical Computer Science 118(2), pp. 263–299.

10. D. HAREL, D. KOZEN & R. PARIKH (1982): *Process logic: Expressiveness, decidability, completeness.* Journal of Computer and System Sciences 25(2), pp. 144–170.

11. M. C. B. HENNESSY & R. MILNER (1985): *Algebraic laws for non-determinism and concurrency.* Journal of the ACM 32(1), pp. 137–161.

12. M. C. B. HENNESSY & C. STIRLING (1985): *The power of the future perfect in program logics.* Information and Control 67(1–3), pp. 23–52.

13. D. KOZEN (1983): *Results on the propositional μ-calculus.* Theoretical Computer Science 27(3), pp. 333–354.

14. K. G. LARSEN (1986): *Context-Dependent Bisimulation between Processes.* PhD thesis, University of Edinburgh, Edinburgh.

15. K. G. LARSEN & L. XINXIN (1991): *Compositionality through an operational semantics of contexts.* Journal of Logic and Computation 1(6), pp. 761–795.

16. R. MILNER (1980): *A Calculus of Communicating Systems.* Springer-Verlag. Volume 92 of *Lecture Notes in Computer Science.*

17. R. MILNER (1981): *A modal characterization of observable machine-behaviour.* In E. Astesiano & C. Böhm, editors: *CAAP '81: Trees in Algebra and Programming, 6th Colloquium, Lecture Notes in Computer Science* 112, Springer-Verlag, Genoa, pp. 25–34.

18. R. MILNER (1983): *Calculi for synchrony and asynchrony.* Theoretical Computer Science 25(3), pp. 267–310.

19. G. D. PLOTKIN (1981): *A structural approach to operational semantics.* Technical Report DAIMI FN-19, Computer Science Department, Aarhus University, Aarhus, Denmark.

20. A. PNUELI (1981): *The temporal logic of concurrent programs.* Theoretical Computer Science 13, pp. 45–60.

21. R. DE SIMONE (1985): *Higher-level synchronising devices in* MEIJE–SCCS. Theoretical Computer Science 37(3), pp. 245–267.

22. C. STIRLING (1985): *A proof-theoretic characterization of observational equivalence.* Theoretical Computer Science 39(1), pp. 27–45.

23. C. STIRLING (1985): *A complete compositional modal proof system for a subset of CCS.* In W. Brauer, editor: *Automata, Languages and Programming, 12th Colloquium (ICALP '85), Lecture Notes in Computer Science* 194, Springer-Verlag, pp. 475–486.

24. C. STIRLING (1985): *A complete modal proof system for a subset of SCCS.* In H. Ehrig, C. Floyd, M. Nivat & J. W. Thatcher, editors: *Mathematical Foundations of Software Development: Proceedings of the Joint Conference on Theory and Practice of Software Development (TAPSOFT), Volume 1: Colloquium on Trees in Algebra and Programming (CAAP '85), Lecture Notes in Computer Science* 185, Springer-Verlag, pp. 253–266.

25. C. STIRLING (1987): *Modal logics for communicating systems.* Theoretical Computer Science 49(2-3), pp. 311–347.

26. G. WINSKEL (1986): *A complete proof system for SCCS with modal assertions.* Fundamenta Informaticae IX, pp. 401–420.

27. G. WINSKEL (1990): *On the compositional checking of validity (extended abstract).* In J. C. M. Baeten & J. W. Klop, editors: *CONCUR '90: Theories of Concurrency: Unification and Extension, Lecture Notes in Computer Science* 458, Springer-Verlag, Amsterdam, The Netherlands, pp. 481–501.

On a Logical Approach to Estimating Computational Complexity of Potentially Intractable Problems[*]

Andrzej Szałas

The College of Economics and Computer Science, Olsztyn, Poland
and
Department of Computer Science, University of Linköping, Sweden
andsz@ida.liu.se

Abstract. In the paper we present a purely logical approach to estimating computational complexity of potentially intractable problems. The approach is based on descriptive complexity and second-order quantifier elimination techniques.

We illustrate the approach on the case of the transversal hypergraph problem, TRANSHYP, which has attracted a great deal of attention. The complexity of the problem remains unsolved for over twenty years. Given two hypergraphs, $\mathcal{G}$ and $\mathcal{H}$, TRANSHYP depends on checking whether $\mathcal{G} = \mathcal{H}^d$, where $\mathcal{H}^d$ is the transversal hypergraph of $\mathcal{H}$.

In the paper we provide a logical characterization of minimal transversals of a given hypergraph and prove that checking whether $\mathcal{G} \subseteq \mathcal{H}^d$ is tractable. For the opposite inclusion the problem still remains open. However, we interpret the resulting quantifier sequences in terms of determinism and bounded nondeterminism. The results give better upper bounds than those known from the literature, e.g., in the case when hypergraph $\mathcal{H}$ has a sub-logarithmic number of hyperedges and (for the deterministic case) all hyperedges have the cardinality bounded by a function sub-linear wrt maximum of sizes of $\mathcal{G}$ and $\mathcal{H}$.

Keywords: second-order logic, second-order quantifier elimination, descriptive complexity, transversal hypergraph problem

1 Introduction

In the current paper we propose a rather general methodology for estimating the complexity of potentially intractable problems. The methodology consists of the following steps[1]:

1. Specify the problem in the second-order logic.
   The complexity of checking validity of second-order formulas in a finite model is PSPACE-complete wrt the size of the model. Thus, for all problems in PSPACE such a description exists. The existential fragment of the second-order logic[2] is NPTIME-

---

[*] Supported in part by the KBN grant 8 T11C 00919.

[1] Below and throughout the paper we apply well-known results of descriptive complexity theory. For the relevant details see, e.g., [5,12].

[2] I.e., the fragment consisting of formulas in which all second-order quantifiers are existential and appear only in prefixes of formulas.

A. Lingas and B.J. Nilsson (Eds.): FCT 2003, LNCS 2751, pp. 423–431, 2003.
© Springer-Verlag Berlin Heidelberg 2003

complete over finite models. Dually, the universal fragment of second-order logic is CO-NPTIME-complete over finite models.

2. Try to eliminate second-order quantifiers.

An application of known methods, if successful, might result in[3]:

- a formula of the first-order logic, validity of which (over finite models) is in PTIME and LOGSPACE. Here one can apply, e.g., the Ackermann lemma (see Lemma 2.4) or the SCAN algorithm of [10];
- a formula of the fixpoint logic, validity of which (over finite models) is in PTIME.[4] Here one can apply the elimination theorem of [15].

3. If the second-order quantifier elimination is not successful, which is likely to happen for NPTIME, CO-NPTIME or PSPACE-complete problems, one can try to identify subclasses of the problem, for which elimination of second-order quantifiers is guaranteed. In such cases tractable (or quasi-polynomial) subproblems of the main problem can be identified.

Below we apply the methodology to the transversal hypergraph problem and show that inclusion in one direction is in PTIME. We also identify some tractable and almost tractable cases for verifying the opposite inclusion, and relate the results to a bounded nondeterminism. Let us, however, emphasize that our main goal is to show how logic can help in analyzing the complexity of problems which can be naturally expressed by means of the second-order logic. The hypergraph problem is chosen mainly as a case study.

Hypergraph theory [2] has may applications in computer science and artificial intelligence (see, e.g., [3,6,7,11,13]). In particular, the transversal hypergraph problem, TRANSHYP, has attracted a great deal of attention. Many important problems of databases, knowledge representation, Boolean circuits, duality theory, diagnosis, machine learning, data mining, explanation finding, etc. can be reduced to TRANSHYP (see, e.g., [7]). However, the precise complexity of this problem remains open for over twenty years. The best known algorithm, provided in [9], runs in quasi-polynomial time wrt the size of the input hypergraphs. More precisely, if n is the size of the input hypergraphs, then the algorithm of [9] requires $n^{o(\log n)}$ steps. The paper [8] provides a result that relates TRANSHYP to a limited nondeterminism by showing that the complement of the problem can be solved in polynomial time with $O(\chi(n) * \log n)$ guessed bits, where $\chi(n)^{\chi(n)} = n$. As observed in [9], $\chi(n) \approx \log n / \log \log n = o(\log n)$.

2   Preliminaries

Let us first define notions related to the TRANSHYP problem. We provide definitions slightly adapted for further logical characterization. However, the definitions are equivalent to those considered in the literature.

Definition 2.1. *By a* hypergraph *we mean a triple* $\mathcal{H} = \langle V, E, M \rangle$, *where*

---

[3] For an overview of known second-order quantifier elimination techniques see, e.g., [14].

[4] Recall that fixpoint logic captures all problems solvable in deterministic polynomial time, provided that the underlying domain is linearly ordered.

    – *V and E are finite disjoint sets of* elements *and* hyperedges, *respectively*
    – $M \subseteq E \times V$ *is an* edge membership relation.

A transversal *of* $\mathcal{H}$ *is any set* $T \subseteq V$ *such that for any hyperedge* $e \in E$ *there is* $v \in V$ *such that* $(T(v) \wedge M(e,v))$ *holds. A* transversal *is* minimal *iff it is minimal wrt set inclusion.* ∎

    In the sequel we sometimes identify hyperedges with sets of their members, i.e., any hyperedge $e \in E$ of hypergraph $\mathcal{H} = \langle V, E, M \rangle$ is identified with set

$$\{v \in V : M(e,v) \text{ holds}\}.$$

Definition 2.2. *By the* transversal hypergraph[5] *of a hypergraph* $\mathcal{H}$ *we mean hypergraph* $\mathcal{H}^d$ *whose hyperedges are all minimal transversals of* $\mathcal{H}$. ∎

Definition 2.3. *By the* transversal hypergraph problem, *denoted by* TRANSHYP, *we mean a problem of checking, for given hypergraphs* $\mathcal{G}$ *and* $\mathcal{H}$, *whether* $\mathcal{G} = \mathcal{H}^d$. ∎

    We say that a formula Φ is *positive* w.r.t. a predicate P iff any occurrence of P in Φ appears within the scope of an even number of negations only[6]. Dually, we say that Φ is *negative* w.r.t. P iff any occurrence of P in Φ appears within the scope of an odd number of negations only.

    By $\Psi\left[P(\bar{t}) := [\Phi]_{\bar{t}}^{\bar{x}}\right]$ we understand formula obtained from Ψ by replacing every occurrence of P in by Φ where in each replacement the actual argument of P, say $\bar{t}$, replaces the variables of $\bar{x}$ in Φ (with renaming bound variables, whenever necessary).

    The following lemma is substantial for the technique we propose.

Lemma 2.4. *Let P be a predicate variable and let Φ and $\Psi(P)$ be first–order formulas such that $\Psi(P)$ is positive w.r.t. P and Φ contains no occurrences of P. Then*

$$\exists P \, \forall \bar{x} \, (P(\bar{x}) \rightarrow \Phi(\bar{x})) \wedge \Psi(P) \quad \equiv \quad \Psi\left[P(\bar{t}) := [\Phi]_{\bar{t}}^{\bar{x}}\right]$$

and similarly if the sign of P is switched to $\neg$ and Ψ is negative w.r.t. P. ∎

Lemma 2.4 was proved by Ackermann in [1]. It can also be found in [16] and, in the context of circumscription[7], in [4]. A substantially stronger elimination theorem extending this lemma is given in [15].

    We shall also need the following simple proposition.

Proposition 2.5. *Let P be a predicate variable and let Φ, Ψ be first–order formulas. Assume that P does not occur in Φ. Then*

$$\exists P \, \forall \bar{x} \, (P(\bar{x}) \equiv \Phi(\bar{x})) \wedge \Psi(P) \quad \equiv \quad \Psi\left[P(\bar{t}) := [\Phi]_{\bar{t}}^{\bar{x}}\right]$$ ∎

---

[5] Called also a *dual hypergraph*.

[6] Under the standard convention stating that implication $(\Psi_1 \rightarrow \Psi_2)$ is treated as the disjunction $(\neg \Psi_1 \vee \Psi_2)$, and equivalence $(\Psi_1 \equiv \Psi_2)$ is treated as formula $[(\Psi_1 \wedge \Psi_2) \vee (\neg \Psi_1 \wedge \neg \Psi_2)]$.

[7] Observe that the conjunction (1)∧(2), substantial for our considerations, is simply the circumscribed formula (1), where T is minimized.

3   Characterization of Minimal Transversals of Hypergraphs

Obviously, T is a transversal of hypergraph $\mathcal{H} = \langle V, E, M \rangle$ iff

$$\forall e \in E \exists v \in V (T(v) \wedge M(e, v)).$$

It is a minimal transversal iff

$$\forall e \in E \exists v \in V (T(v) \wedge M(e, v)) \wedge \tag{1}$$

$$\forall T' \{ [\forall e \in E \exists v \in V (T'(v) \wedge M(e, v)) \wedge \forall w \in V (T'(w) \rightarrow T(w))] \rightarrow \tag{2}$$
$$\forall u \in V (T(u) \rightarrow T'(u)) \}$$

Formula (2) is a universal second-order formula. Application of this formula to the verification whether a given transversal is minimal, is thus in CO-NPTIME. On the other hand, one can eliminate the second-order quantification by applying Lemma 2.4. To do this, we first negate (2):

$$\exists T' \{ [\forall e \in E \exists v \in V (T'(v) \wedge M(e, v)) \wedge \forall w \in V (T'(w) \rightarrow T(w))] \wedge \tag{3}$$
$$\exists u \in V (T(u) \wedge \neg T'(u)) \}$$

Formula (3) is equivalent to

$$\exists u \in V \exists T' [\forall w \in V (T'(w) \rightarrow T(w)) \wedge \tag{4}$$
$$\forall e \in E \exists v \in V (T'(v) \wedge M(e, v)) \wedge T(u) \wedge \neg T'(u)],$$

i.e., to

$$\exists u \in V \exists T' [\forall w \in V (T'(w) \rightarrow T(w)) \wedge$$
$$\forall e \in E \exists v \in V (T'(v) \wedge M(e, v)) \wedge T(u) \wedge$$
$$\forall w \in V (T'(w) \rightarrow w \neq u)],$$

and finally, to

$$\exists u \in V \exists T' [\forall w \in V (T'(w) \rightarrow (T(w) \wedge w \neq u)) \wedge \tag{5}$$
$$\forall e \in E \exists v \in V (T'(v) \wedge M(e, v)) \wedge T(u)].$$

After the application of Lemma 2.4 we obtain the following formula equivalent to (5):

$$\exists u \in V [\forall e \in E \exists v \in V (T(v) \wedge v \neq u \wedge M(e, v)) \wedge T(u)]. \tag{6}$$

After negating formula (6) and rearranging the result, we obtain the following first-order formula equivalent to (2):

$$\forall u \in V [T(u) \rightarrow \exists e \in E \forall v \in V ((T(v) \wedge M(e, v)) \rightarrow v = u)]. \tag{7}$$

Let $\mathcal{H} = \langle V, E, M \rangle$ be a hypergraph. In the sequel we use notation $Min_{\mathcal{H}}(T)$, defined by

$$Min_{\mathcal{H}}(T) \stackrel{\text{def}}{\equiv} \tag{8}$$
$$\forall e \in E \exists v \in V (T(v) \wedge M(e, v)) \wedge$$
$$\forall u \in V [T(u) \rightarrow \exists e \in E \forall v \in V ((T(v) \wedge M(e, v)) \rightarrow v = u)].$$

We now have the following lemma.

Lemma 3.1. *For any hypergraph* $\mathcal{H} = \langle V, E, M \rangle$, *T is a minimal transversal of* $\mathcal{H}$ *iff it satisfies formula* $Min_{\mathcal{H}}(T)$. *In consequence[8], checking whether a given T is a minimal transversal of a hypergraph is in* PTIME *and* LOGSPACE *wrt the size of the hypergraph.* ∎

4 Specification of the TRANSHYP Problem In Logic

4.1 Specification of the TRANSHYP Problem in the Second-Order Logic

Let $\mathcal{G} = \langle V, E_{\mathcal{G}}, M_{\mathcal{G}} \rangle$ and $\mathcal{H} = \langle V, E_{\mathcal{H}}, M_{\mathcal{H}} \rangle$ be hypergraphs. In order to check whether $\mathcal{G} = \mathcal{H}^d$, we verify inclusions $\mathcal{G} \subseteq \mathcal{H}^d$ and $\mathcal{H}^d \subseteq \mathcal{G}$. The inclusions can be characterized in the second-order logic as follows:

$$\forall e \in E_{\mathcal{G}} \, \exists e' \in E_{\mathcal{H}}^d \forall v \in V (M_{\mathcal{G}}(e, v) \equiv M_{\mathcal{H}}^d(e', v)) \tag{9}$$

$$\forall e' \in E_{\mathcal{H}}^d \, \exists e \in E_{\mathcal{G}} \forall v \in V (M_{\mathcal{G}}(e, v) \equiv M_{\mathcal{H}}^d(e', v)). \tag{10}$$

According to Lemma 3.1, formulas (9) and (10) can be expressed as

$$\forall e \in E_{\mathcal{G}} \, \exists T [Min_{\mathcal{H}}(T) \wedge \forall v \in V (M_{\mathcal{G}}(e, v) \equiv T(v))] \tag{11}$$

$$\forall T [Min_{\mathcal{H}}(T) \rightarrow \exists e \in E_{\mathcal{G}} \forall v \in V (M_{\mathcal{G}}(e, v) \equiv T(v))]. \tag{12}$$

The above specification leads to intractable algorithms (unless PTIME = NPTIME). In the following sections we attempt to reduce the complexity by eliminating second-order quantifiers from formulas (11) and (12).

4.2 The Case of Inclusion $\mathcal{G} \subseteq \mathcal{H}^d$

Consider the second-order part of formula (11), i.e.,

$$\exists T [Min_{\mathcal{H}}(T) \wedge \forall v \in V (M_{\mathcal{G}}(e, v) \equiv T(v))]. \tag{13}$$

Due to equivalence (8), Lemma 3.1 and Proposition 2.5,[9] formula (13) is equivalent to

$$\forall e' \in E_{\mathcal{H}} \exists v \in V (M_{\mathcal{G}}(e, v) \wedge M_{\mathcal{H}}(e', v)) \wedge \tag{14}$$
$$\forall u \in V [M_{\mathcal{G}}(e, u) \rightarrow$$
$$\exists e' \in E_{\mathcal{H}} \forall v \in V ((M_{\mathcal{G}}(e, v) \wedge M_{\mathcal{H}}(e', v)) \rightarrow v = u)].$$

In consequence, formula (11) is equivalent to

$$\forall e \in E_{\mathcal{G}} \forall e' \in E_{\mathcal{H}} \exists v \in V (M_{\mathcal{G}}(e, v) \wedge M_{\mathcal{H}}(e', v)) \wedge \tag{15}$$
$$\forall e \in E_{\mathcal{G}} \forall u \in V [M_{\mathcal{G}}(e, u) \rightarrow$$
$$\exists e' \in E_{\mathcal{H}} \forall v \in V ((M_{\mathcal{G}}(e, v) \wedge M_{\mathcal{H}}(e', v)) \rightarrow v = u)].$$

Thus the inclusion $\mathcal{G} \subseteq \mathcal{H}^d$ is first-order definable by formula (15). We then have the following corollary.

Corollary 4.1. *For any hypergraphs* $\mathcal{G} = \langle V, E_{\mathcal{G}}, M_{\mathcal{G}} \rangle$ *and* $\mathcal{H} = \langle V, E_{\mathcal{H}}, M_{\mathcal{H}} \rangle$, *checking whether* $\mathcal{G} \subseteq \mathcal{H}^d$, *is in* PTIME *and* LOGSPACE *wrt the maximum of sizes of hypergraphs* $\mathcal{G}$ *and* $\mathcal{H}$. ∎

---

[8] This easily follows from the equivalence (8) by which $Min_{\mathcal{H}}(T)$ is characterized by a first-order formula.

[9] Note that in order to apply Proposition 2.5, bound variable e is renamed into e'

4.3 The Case of Inclusion $\mathcal{H}^d \subseteq \mathcal{G}$

Unfortunately, no known second-order quantifier elimination method is successful for the inclusion (12). We thus equivalently transform formula (12) to a form where Lemma 2.4 is applicable. The verification of the resulting formula in finite models is, in general, of exponential complexity. However, when some restrictions are assumed, the complexity reduces to the deterministic polynomial or quasi-polynomial time, as shown below.

By (8), formula (12) is equivalent to

$$\forall T\big\{[\forall e \in E_{\mathcal{H}} \exists v \in V(T(v) \wedge M_{\mathcal{H}}(e,v)) \wedge \tag{16}$$
$$\forall u \in V[T(u) \rightarrow \exists e \in E_{\mathcal{H}} \forall v \in V((T(v) \wedge M_{\mathcal{H}}(e,v)) \rightarrow v = u)]] \rightarrow$$
$$\exists e \in E_{\mathcal{G}} \forall v \in V(M_{\mathcal{G}}(e,v) \equiv T(v))\big\}$$

Let us assume that the inclusion $\mathcal{G} \subseteq \mathcal{H}^d$ holds. If not, then the answer to TRANSHYP for this particular instance is negative. Under this assumption, formula (16) is equivalent to[10]

$$\forall T\big\{[\forall e \in E_{\mathcal{H}} \exists v \in V(T(v) \wedge M_{\mathcal{H}}(e,v)) \wedge \tag{17}$$
$$\forall u \in V[T(u) \rightarrow \exists e \in E_{\mathcal{H}} \forall v \in V((T(v) \wedge M_{\mathcal{H}}(e,v)) \rightarrow v = u)]] \rightarrow$$
$$\exists e \in E_{\mathcal{G}} \forall v \in V(M_{\mathcal{G}}(e,v) \rightarrow T(v))\big\}.$$

In order to apply Lemma 2.4 we first negate (17):

$$\exists T\big\{\forall e \in E_{\mathcal{H}} \exists v \in V(T(v) \wedge M_{\mathcal{H}}(e,v)) \wedge \tag{18}$$
$$\forall u \in V[T(u) \rightarrow \exists e \in E_{\mathcal{H}} \forall v \in V((T(v) \wedge M_{\mathcal{H}}(e,v)) \rightarrow v = u)] \wedge$$
$$\forall e \in E_{\mathcal{G}} \exists v \in V(M_{\mathcal{G}}(e,v) \wedge \neg T(v))\big\}.$$

In order to simplify calculations, by $\Gamma(T)$ we denote the conjunction of formulas given in the last two lines of (18). Formula (18) is then expressed by

$$\exists T\big\{\forall e \in E_{\mathcal{H}} \exists v \in V(T(v) \wedge M_{\mathcal{H}}(e,v)) \wedge \Gamma(T)\big\}. \tag{19}$$

Observe that $\Gamma(T)$ is negative wrt T. Thus the main obstacle for applying Lemma 2.4 is created by the existential quantifier $\exists v \in V$ appearing within the scope of $\forall e \in E_{\mathcal{H}}$. Assume $E_{\mathcal{H}} = \{e_1, \ldots, e_k\}$. Denote by $V_e \stackrel{\text{def}}{=} \{x : M_{\mathcal{H}}(e,x)$ holds$\}$. Formula (19) can then be expressed by

$$\exists T\big\{\exists v_1 \in V_{e_1} T(v_1) \wedge \ldots \wedge \exists v_k \in V_{e_k} T(v_k) \wedge \Gamma(T)\big\},$$

i.e., by

$$\exists v_1 \in V_{e_1} \ldots \exists v_k \in V_{e_k} \exists T\big\{T(v_1) \wedge \ldots \wedge T(v_k) \wedge \Gamma(T)\big\},$$

which is equivalent to

$$\exists v_1 \in V_{e_1} \ldots \exists v_k \in V_{e_k} \exists T\big\{\forall v \in V[(v = v_1 \vee \ldots \vee v = v_k) \rightarrow T(v)] \wedge \Gamma(T)\big\}.$$

---

[10] By minimality of $\mathcal{H}^d$, and the assumption $\mathcal{G} \subseteq \mathcal{H}^d$, inclusion expressed by $\forall v \in V(M_{\mathcal{G}}(e,v) \rightarrow T(v))$ is equivalent to the set equality, expressed by $\forall v \in V(M_{\mathcal{G}}(e,v) \equiv T(v))$.

The application of Lemma 2.4 results in the following first-order formula:

$$\exists v_1 \in V_{e_1} \ldots \exists v_k \in V_{e_k} \{ \Gamma \, [T(t) := [(v = v_1 \vee \ldots \vee v = v_k)]_t^v \,] \}.$$

In consequence, formula (17) is equivalent to

$$\forall v_1 \in V_{e_1} \ldots \forall v_k \in V_{e_k} \{ \neg \Gamma \, [T(t) := [(v = v_1 \vee \ldots \vee v = v_k)]_t^v \,] \},$$

i.e., to

$$\forall v_1 \in V_{e_1} \ldots \forall v_k \in V_{e_k} \forall u \in V \Big\{ [(u = v_1 \vee \ldots \vee u = v_k) \rightarrow \qquad (20)$$
$$\exists e \in E_{\mathcal{H}} \forall v \in V [((v = v_1 \vee \ldots \vee v = v_k) \wedge M_{\mathcal{H}}(e,v)) \rightarrow v = u]] \rightarrow$$
$$\exists e \in E_{\mathcal{G}} \forall v \in V (M_{\mathcal{G}}(e,v) \rightarrow (v = v_1 \vee \ldots \vee v = v_k)) \Big\}.$$

The major complexity of checking whether given hypergraphs satisfy formula (20) is caused by the sequence of quantifiers $\forall v_1 \in V_{e_1} \ldots \forall v_k \in V_{e_k} \forall u$. We then have the following theorem.

Theorem 4.2. *For given hypergraphs $\mathcal{G}$ and $\mathcal{H}$, such that $\mathcal{G} \subseteq \mathcal{H}^d$, the problem of checking whether $\mathcal{H}^d \subseteq \mathcal{G}$ is solvable in time $O(|V_1| * \ldots * |V_k| * p(n))$, where $p(n)$ is a polynomial[11], n is the maximum of sizes of $\mathcal{G}$ and $\mathcal{H}$, k is the number of edges in $\mathcal{H}$, and for $e = 1, \ldots, k$, $|V_e|$ denotes the cardinality of set $\{x : M_{\mathcal{H}}(e, x) \text{ holds}\}$.* ∎

Accordingly we have the following corollary.

Corollary 4.3. *Under assumptions of Theorem 4.2, if cardinalities $|V_1|, \ldots, |V_k|$ are bounded by a function $f(n)$ then the problem of checking whether $\mathcal{H}^d \subseteq \mathcal{G}$ is solvable in time $O(f(n)^k * p(n))$.* ∎

In the view of the result given in [9], Corollary 4.3 can be useful if k is bounded by a (sub-) logarithmic function, and $f(n)$ is (sub-)linear wrt n. For instance, if both k and $f(n)$ are bounded by $\log n$ then the corollary gives us an upper bound $O((\log n)^{\log n} * p(n))$ which is better than that offered by algorithm of [9]. Let us emphasize that in many cases $|V|$ and consequently $f(n)$ is bounded by $\log n$, since the dual hypergraph might be of size exponential wrt $|V|$.

The characterization provided by formula (20) is also related to the bounded non-determinism. Namely, consider the complement of TRANSHYP problem. The sequence of quantifiers $\forall v_1 \in V_{e_1} \ldots \forall v_k \in V_{e_k}$ appearing in formula (20) is transformed into $\exists v_1 \in V_{e_1} \ldots \exists v_k \in V_{e_k}$. In order to verify the negated formula it is then sufficient to guess k sequences of bits of size not greater than $\log \max_{e=1,\ldots,k} \{|V_e|\}$. Thus, in the worst case, it suffices to guess $k * \log |V|$ bits. By the result of [8], mentioned in Section 1, $O(\log^2 n)$ guessed bits suffice to further solve the TRANSHYP problem in deterministic polynomial time. Thus the observation we just made is useful, e.g., when one considers the input graph $\mathcal{H}$ with the number of edges (sub-)logarithmic wrt n. Observe, however, that often n is exponentially larger than $|V|$.[12]

---

[11] Reflecting the complexity introduced by quantifiers inside formula (20).

[12] This frequently happens in the duality theory, where the number of prime implicants and implicates is exponential wrt the size of the input formula.

5 Conclusions

In the paper we presented a purely logical approach to estimating computational complexity of potentially intractable problems. We illustrated the approach on the case of the complexity of the TRANSHYP problem. We provided a logical characterization of minimal transversals of a given hypergraph and proved that checking the inclusion $G \subseteq \mathcal{H}^d$ is tractable. For the opposite inclusion the problem still remains open. However, we interpreted the resulting quantifier sequences in terms of determinism and bounded nondeterminism. The results give better upper bounds than those known from the literature in the case when hypergraph $\mathcal{H}$ has a sub-logarithmic number of hyperedges and (for the deterministic case) all hyperedges have the cardinality bounded by a function sub-linear wrt the maximum of sizes of the input hypergraphs.

Let us also emphasize that the simplest second-order quantifier elimination techniques were applied. In some cases it might be useful to apply theorem of [15] which results in a fixpoint formula, i.e., much stronger but still tractable formalism.

References

1. W. Ackermann. Untersuchungen über das eliminationsproblem der mathematischen logik. *Mathematische Annalen*, 110:390–413, 1935.
2. C. Berge. *Hypergraphs*, volume 45 of *North-Holland Mathematical Library*. Elsevier, 1989.
3. E. Boros, V. Gurvich, L. Khachiyan, and K. Makino. Generating partial and multiple transversals of a hypergraph. In *Automata, Languages and Programming*, volume 1853 of *Lecture Notes in Computer Science*, pages 588–599. Springer, 2000.
4. P. Doherty, W. Lukaszewicz, and A. Szałas. Computing circumscription revisited. *Journal of Automated Reasoning*, 18(3):297–336, 1997.
5. H-D. Ebbinghaus and J. Flum. *Finite Model Theory*. Springer-Verlag, Heidelberg, 1995.
6. T. Eiter and G. Gottlob. Identifying the minimal transversals of a hypergraph and related problems. *SIAM Journal on Computing*, 24(6):1278–1304, 1995.
7. T. Eiter and G. Gottlob. Hypergraph transversal computation and related problems in logic and AI. In M. Flesca, S. Greco, N. Leone, and G. Ianni, editors, *Proceedings of the 8th Conference JELIA 2002*, LNAI 2424, pages 549–564. Springer-Verlag, 2002.
8. T. Eiter, G. Gottlob, and K. Makino. New results on monotone dualization and generating hypergraph transversals. In *ACM STOC 2002*, pages 14–22, 2002.
9. M.L. Fredman and L. Khachiyan. On the complexity of dualization of monotone disjunctive normal forms. *Journal of Algorithms*, 21:618–628, 1996.
10. D. M. Gabbay and H. J. Ohlbach. Quantifier elimination in second-order predicate logic. In B. Nebel, C. Rich, and W. Swartout, editors, *Principles of Knowledge representation and reasoning, KR 92*, pages 425–435. Morgan Kauffman, 1992.
11. G. Gogic, C.H. Papadimitriou, and M. Sideri. Incremental recompilation of knowledge. *Journal of Artificial Intelligence Research*, 8:23–37, 1998.
12. N. Immerman. *Descriptive Complexity*. Springer-Verlag, New York, Berlin, 1998.
13. D.J. Kavvadias and E.C. Stavropoulos. Evaluation of an algorithm for the transversal hypergraph problem. In J. Scott Vitter and C. D. Zaroliagis, editors, *Algorithm Engineering, 3rd International Workshop, WAE '99*, volume 1668 of *Lecture Notes in Computer Science*, pages 72–84. Springer, 1999.
14. A. Nonnengart, H.J. Ohlbach, and A. Szałas. Elimination of predicate quantifiers. In H.J. Ohlbach and U. Reyle, editors, *Logic, Language and Reasoning. Essays in Honor of Dov Gabbay, Part I*, pages 159–181. Kluwer, 1999.

15. A. Nonnengart and A. Szałas. A fixpoint approach to second-order quantifier elimination with applications to correspondence theory. In E. Orłowska, editor, *Logic at Work: Essays Dedicated to the Memory of Helena Rasiowa*, volume 24 of *Studies in Fuzziness and Soft Computing*, pages 307–328. Springer Physica-Verlag, 1998.
16. A. Szałas. On the correspondence between modal and classical logic: An automated approach. *Journal of Logic and Computation*, 3:605–620, 1993.

Author Index

Lecture Notes in Computer Science

For information about Vols. 1–2659
please contact your bookseller or Springer-Verlag

Vol. 2660: P.M.A. Sloot, D. Abramson, A.V. Bogdanov, J.J. Dongarra, A.Y. Zomaya, Y.E. Gorbachev (Eds.), Computational Science – ICCS 2003. Proceedings, Part IV. 2003. LVI, 1161 pages. 2003.

Vol. 2663: E. Menasalvas, J. Segovia, P.S. Szczepaniak (Eds.), Advances in Web Intelligence. Proceedings, 2003. XII, 350 pages. 2003. (Subseries LNAI).

Vol. 2664: M. Leuschel (Ed.), Logic Based Program Synthesis and Transformation. Proceedings, 2002. X, 281 pages. 2003.

Vol. 2665: H. Chen, R. Miranda, D.D. Zeng, C. Demchak, J. Schroeder, T. Madhusudan (Eds.), Intelligence and Security Informatics. Proceedings, 2003. XIV, 392 pages. 2003.

Vol. 2666: C. Guerra, S. Istrail (Eds.), Mathematical Methods for Protein Structure Analysis and Design. Proceedings, 2000. XI, 157 pages. 2003. (Subseries LNBI).

Vol. 2667: V. Kumar, M.L. Gavrilova, C.J.K. Tan, P. L'Ecuyer (Eds.), Computational Science and Its Applications – ICCSA 2003. Proceedings, Part I. 2003. XXXIV, 1060 pages. 2003.

Vol. 2668: V. Kumar, M.L. Gavrilova, C.J.K. Tan, P. L'Ecuyer (Eds.), Computational Science and Its Applications – ICCSA 2003. Proceedings, Part II. 2003. XXXIV, 942 pages. 2003.

Vol. 2669: V. Kumar, M.L. Gavrilova, C.J.K. Tan, P. L'Ecuyer (Eds.), Computational Science and Its Applications – ICCSA 2003. Proceedings, Part III. 2003. XXXIV, 948 pages. 2003.

Vol. 2670: R. Peña, T. Arts (Eds.), Implementation of Functional Languages. Proceedings, 2002. X, 249 pages. 2003.

Vol. 2671: Y. Xiang, B. Chaib-draa (Eds.), Advances in Artificial Intelligence. Proceedings, 2003. XIV, 642 pages. 2003. (Subseries LNAI).

Vol. 2672: M. Endler, D. Schmidt (Eds.), Middleware 2003. Proceedings, 2003. XIII, 513 pages. 2003.

Vol. 2673: N. Ayache, H. Delingette (Eds.), Surgery Simulation and Soft Tissue Modeling. Proceedings, 2003. XII, 386 pages. 2003.

Vol. 2674: I.E. Magnin, J. Montagnat, P. Clarysse, J. Nenonen, T. Katila (Eds.), Functional Imaging and Modeling of the Heart. Proceedings, 2003. XI, 308 pages. 2003.

Vol. 2675: M. Marchesi, G. Succi (Eds.), Extreme Programming and Agile Processes in Software Engineering. Proceedings, 2003. XV, 464 pages. 2003.

Vol. 2676: R. Baeza-Yates, E. Chávez, M. Crochemore (Eds.), Combinatorial Pattern Matching. Proceedings, 2003. XI, 403 pages. 2003.

Vol. 2678: W. van der Aalst, A. ter Hofstede, M. Weske (Eds.), Business Process Management. Proceedings, 2003. XI, 391 pages. 2003.

Vol. 2679: W. van der Aalst, E. Best (Eds.), Applications and Theory of Petri Nets 2003. Proceedings, 2003. XI, 508 pages. 2003.

Vol. 2680: P. Blackburn, C. Ghidini, R.M. Turner, F. Giunchiglia (Eds.), Modeling and Using Context. Proceedings, 2003. XII, 525 pages. 2003. (Subseries LNAI).

Vol. 2681: J. Eder, M. Missikoff (Eds.), Advanced Information Systems Engineering. Proceedings, 2003. XV, 740 pages. 2003.

Vol. 2683: A. Rangarajan, M. Figueiredo, J. Zerubia (Eds.), Energy Minimization Methods in Computer Vision and Pattern Recognition. Proceedings, 2003. XI, 534 pages. 2003.

Vol. 2684: M.V. Butz, O. Sigaud, P. Gérard (Eds.), Anticipatory Behavior in Adaptive Learning Systems. X, 303 pages. 2003. (Subseries LNAI).

Vol. 2685: C. Freksa, W. Brauer, C. Habel, K.F. Wender (Eds.), Spatial Cognition III. X, 415 pages. 2003. (Subseries LNAI).

Vol. 2686: J. Mira, J.R. Álvarez (Eds.), Computational Methods in Neural Modeling. Proceedings, Part I. 2003. XXVII, 764 pages. 2003.

Vol. 2687: J. Mira, J.R. Álvarez (Eds.), Artificial Neural Nets Problem Solving Methods. Proceedings, Part II. 2003. XXVII, 820 pages. 2003.

Vol. 2688: J. Kittler, M.S. Nixon (Eds.), Audio- and Video-Based Biometric Person Authentication. Proceedings, 2003. XVII, 978 pages. 2003.

Vol. 2689: K.D. Ashley, D.G. Bridge (Eds.), Case-Based Reasoning Research and Development. Proceedings, 2003. XV, 734 pages. 2003. (Subseries LNAI).

Vol. 2691: V. Mařík, J. Müller, M. Pěchouček (Eds.), Multi-Agent Systems and Applications III. Proceedings, 2003. XIV, 660 pages. 2003. (Subseries LNAI).

Vol. 2692: P. Nixon, S. Terzis (Eds.), Trust Management. Proceedings, 2003. X, 349 pages. 2003.

Vol. 2693: A. Cechich, M. Piattini, A. Vallecillo (Eds.), Component-Based Software Quality. X, 403 pages. 2003.

Vol. 2694: R. Cousot (Ed.), Static Analysis. Proceedings, 2003. XIV, 505 pages. 2003.

Vol. 2695: L.D. Griffin, M. Lillholm (Eds.), Scale Space Methods in Computer Vision. Proceedings, 2003. XII, 816 pages. 2003.

Vol. 2696: J. Feigenbaum (Ed.), Digital Rights Management. Proceedings, 2002. X, 221 pages. 2003.

Vol. 2697: T. Warnow, B. Zhu (Eds.), Computing and Combinatorics. Proceedings, 2003. XIII, 560 pages. 2003.

Vol. 2698: W. Burakowski, B. Koch, A. Bęben (Eds.), Architectures for Quality of Service in the Internet. Proceedings, 2003. XI, 305 pages. 2003.

Vol. 2701: M. Hofmann (Ed.), Typed Lambda Calculi and Applications. Proceedings, 2003. VIII, 317 pages. 2003.

Vol. 2702: P. Brusilovsky, A. Corbett, F. de Rosis (Eds.), User Modeling 2003. Proceedings, 2003. XIV, 436 pages. 2003. (Subseries LNAI).

Vol. 2704: S.-T. Huang, T. Herman (Eds.), Self-Stabilizing Systems. Proceedings, 2003. X, 215 pages. 2003.

Vol. 2706: R. Nieuwenhuis (Ed.), Rewriting Techniques and Applications. Proceedings, 2003. XI, 515 pages. 2003.

Vol. 2707: K. Jeffay, I. Stoica, K. Wehrle (Eds.), Quality of Service – IWQoS 2003. Proceedings, 2003. XI, 517 pages. 2003.

Vol. 2708: R. Reed, J. Reed (Eds.), SDL 2003: System Design. Proceedings, 2003. XI, 405 pages. 2003.

Vol. 2709: T. Windeatt, F. Roli (Eds.), Multiple Classifier Systems. Proceedings, 2003. X, 406 pages. 2003.

Vol. 2710: Z. Ésik, Z, Fülöp (Eds.), Developments in Language Theory. Proceedings, 2003. XI, 437 pages. 2003.

Vol. 2711: T.D. Nielsen, N.L. Zhang (Eds.), Symbolic and Quantitative Approaches to Reasoning with Uncertainty. Proceedings, 2003. XII, 608 pages. 2003. (Subseries LNAI).

Vol. 2712: A. James, B. Lings, M. Younas (Eds.), New Horizons in Information Management. Proceedings, 2003. XII, 281 pages. 2003.

Vol. 2713: C.-W. Chung, C.-K. Kim, W. Kim, T.-W. Ling, K.-H. Song (Eds.), Web and Communication Technologies and Internet-Related Social Issues – HSI 2003. Proceedings, 2003. XXII, 773 pages. 2003.

Vol. 2714: O. Kaynak, E. Alpaydin, E. Oja, L. Xu (Eds.), Artificial Neural Networks and Neural Information Processing – ICANN/ICONIP 2003. Proceedings, 2003. XXII, 1188 pages. 2003.

Vol. 2715: T. Bilgiç, B. De Baets, O. Kaynak (Eds.), Fuzzy Sets and Systems – IFSA 2003. Proceedings, 2003. XV, 735 pages. 2003. (Subseries LNAI).

Vol. 2716: M.J. Voss (Ed.), OpenMP Shared Memory Parallel Programming. Proceedings, 2003. VIII, 271 pages. 2003.

Vol. 2718: P. W. H. Chung, C. Hinde, M. Ali (Eds.), Developments in Applied Artificial Intelligence. Proceedings, 2003. XIV, 817 pages. 2003. (Subseries LNAI).

Vol. 2719: J.C.M. Baeten, J.K. Lenstra, J. Parrow, G.J. Woeginger (Eds.), Automata, Languages and Programming. Proceedings, 2003. XVIII, 1199 pages. 2003.

Vol. 2720: M. Marques Freire, P. Lorenz, M.M.-O. Lee (Eds.), High-Speed Networks and Multimedia Communications. Proceedings, 2003. XIII, 582 pages. 2003.

Vol. 2721: N.J. Mamede, J. Baptista, I. Trancoso, M. das Graças Volpe Nunes (Eds.), Computational Processing of the Portuguese Language. Proceedings, 2003. XIV, 268 pages. 2003. (Subseries LNAI).

Vol. 2722: J.M. Cueva Lovelle, B.M. González Rodríguez, L. Joyanes Aguilar, J.E. Labra Gayo, M. del Puerto Paule Ruiz (Eds.), Web Engineering. Proceedings, 2003. XIX, 554 pages. 2003.

Vol. 2723: E. Cantú-Paz, J.A. Foster, K. Deb, L.D. Davis, R. Roy, U.-M. O'Reilly, H.-G. Beyer, R. Standish, G. Kendall, S. Wilson, M. Harman, J. Wegener, D. Dasgupta, M.A. Potter, A.C. Schultz, K.A. Dowsland, N. Jonoska, J. Miller (Eds.), Genetic and Evolutionary Computation – GECCO 2003. Proceedings, Part I. 2003. XLVII, 1252 pages. 2003.

Vol. 2724: E. Cantú-Paz, J.A. Foster, K. Deb, L.D. Davis, R. Roy, U.-M. O'Reilly, H.-G. Beyer, R. Standish, G. Kendall, S. Wilson, M. Harman, J. Wegener, D. Dasgupta, M.A. Potter, A.C. Schultz, K.A. Dowsland, N. Jonoska, J. Miller (Eds.), Genetic and Evolutionary Computation – GECCO 2003. Proceedings, Part II. 2003. XLVII, 1274 pages. 2003.

Vol. 2725: W.A. Hunt, Jr., F. Somenzi (Eds.), Computer Aided Verification. Proceedings, 2003. XII, 462 pages. 2003.

Vol. 2726: E. Hancock, M. Vento (Eds.), Graph Based Representations in Pattern Recognition. Proceedings, 2003. VIII, 271 pages. 2003.

Vol. 2727: R. Safavi-Naini, J. Seberry (Eds.), Information Security and Privacy. Proceedings, 2003. XII, 534 pages. 2003.

Vol. 2728: E.M. Bakker, T.S. Huang, M.S. Lew, N. Sebe, X.S. Zhou (Eds.), Image and Video Retrieval. Proceedings, 2003. XIII, 512 pages. 2003.

Vol. 2731: C.S. Calude, M.J. Dinneen, V. Vajnovszki (Eds.), Discrete Mathematics and Theoretical Computer Science. Proceedings, 2003. VIII, 301 pages. 2003.

Vol. 2732: C. Taylor, J.A. Noble (Eds.), Information Processing in Medical Imaging. Proceedings, 2003. XVI, 698 pages. 2003.

Vol. 2733: A. Butz, A. Krüger, P. Olivier (Eds.), Smart Graphics. Proceedings, 2003. XI, 261 pages. 2003.

Vol. 2734: P. Perner, A. Rosenfeld (Eds.), Machine Learning and Data Mining in Pattern Recognition. Proceedings, 2003. XII, 440 pages. 2003. (Subseries LNAI).

Vol. 2741: F. Baader (Ed.), Automated Deduction – CADE-19. Proceedings, 2003. XII, 503 pages. 2003. (Subseries LNAI).

Vol. 2743: L. Cardelli (Ed.), ECOOP 2003 – Object-Oriented Programming. Proceedings, 2003. X, 501 pages. 2003.

Vol. 2745: M. Guo, L.T. Yang (Eds.), Parallel and Distributed Processing and Applications. Proceedings, 2003. XII, 450 pages. 2003.

Vol. 2746: A. de Moor, W. Lex, B. Ganter (Eds.), Conceptual Structures for Knowledge Creation and Communication. Proceedings, 2003. XI, 405 pages. 2003. (Subseries LNAI).

Vol. 2748: F. Dehne, J.-R. Sack, M. Smid (Eds.), Algorithms and Data Structures. Proceedings, 2003. XII, 522 pages. 2003.

Vol. 2749: J. Bigun, T. Gustavsson (Eds.), Image Analysis. Proceedings, 2003. XXII, 1174 pages. 2003.

Vol. 2750: T. Hadzilacos, Y. Manolopoulos, J.F. Roddick, Y. Theodoridis (Eds.), Advances in Spatial and Temporal Databases. Proceedings, 2003. XIII, 525 pages. 2003.

Vol. 2751: A. Lingas, B.J. Nilsson (Eds.), Fundamentals of Computation Theory. Proceedings, 2003. XII, 433 pages. 2003.

Vol. 2753: F. Maurer, D. Wells (Eds.), Extreme Programming and Agile Methods – XP/Agile Universe 2003. Proceedings, 2003. XI, 215 pages. 2003.

Vol. 2759: O.H. Ibarra, Z. Dang (Eds.), Implementation and Application of Automata. Proceedings, 2003. XI, 312 pages. 2003.